FORMATION DAMAGE DURING IMPROVED OIL RECOVERY

FORMATION DAMAGE DURING IMPROVED OIL RECOVERY
Fundamentals and Applications

Edited by

BIN YUAN
The University of Oklahoma, Norman, OK, United States

DAVID A. WOOD
DWA Energy Limited, London, United Kingdom

LIBRARY OF
CONGRESS
SURPLUS
DUPLICATE

ELSEVIER

Gulf Professional Publishing
An imprint of Elsevier

Gulf Professional Publishing is an imprint of Elsevier
50 Hampshire Street, 5th Floor, Cambridge, MA 02139, United States
The Boulevard, Langford Lane, Kidlington, Oxford, OX5 1GB, United Kingdom

Copyright © 2018 Elsevier Inc. All rights reserved.

No part of this publication may be reproduced or transmitted in any form or by any means,
electronic or mechanical, including photocopying, recording, or any information storage and retrieval
system, without permission in writing from the publisher. Details on how to seek permission, further
information about the Publisher's permissions policies and our arrangements with organizations such
as the Copyright Clearance Center and the Copyright Licensing Agency, can be found at our
website: www.elsevier.com/permissions.

This book and the individual contributions contained in it are protected under copyright by the
Publisher (other than as may be noted herein).

Notices
Knowledge and best practice in this field are constantly changing. As new research and experience
broaden our understanding, changes in research methods, professional practices, or medical treatment
may become necessary.

Practitioners and researchers must always rely on their own experience and knowledge in evaluating
and using any information, methods, compounds, or experiments described herein. In using such
information or methods they should be mindful of their own safety and the safety of others, including
parties for whom they have a professional responsibility.

To the fullest extent of the law, neither the Publisher nor the authors, contributors, or editors, assume
any liability for any injury and/or damage to persons or property as a matter of products liability,
negligence or otherwise, or from any use or operation of any methods, products, instructions, or ideas
contained in the material herein.

British Library Cataloguing-in-Publication Data
A catalogue record for this book is available from the British Library

Library of Congress Cataloging-in-Publication Data
A catalog record for this book is available from the Library of Congress

ISBN: 978-0-12-813782-6

For Information on all Gulf Professional Publishing publications
visit our website at https://www.elsevier.com/books-and-journals

Working together
to grow libraries in
developing countries

www.elsevier.com • www.bookaid.org

Publishing Director: Joe Hayton
Senior Acquisition Editor: Katie Hammon
Editorial Project Manager: Andrea Gallego Ortiz
Production Project Manager: Kamesh Ramajogi
Cover Designer: Miles Hitchen

Typeset by MPS Limited, Chennai, India

DEDICATION

To all those who spend their careers trying to work out what on earth is going on down there! And, occasionally, coming up with valuable insight that improves our understanding and operating performance. Our highest respects go to all those devoted to the subsurface, geothermal, and coal industries.

CONTENTS

13. Formation Damage in Coalbed Methane Recovery   **499**

Jack C. Pashin, Sarada P. Pradhan and Vikram Vishal

14. Special Focus on Produced Water in Oil and Gas Fields: Origin, Management, and Reinjection Practice   **515**

Yu Liang, Yang Ning, Lulu Liao and Bin Yuan

15. Integrated Risks Assessment and Management of IOR/EOR Projects: A Formation Damage View   **587**

David A. Wood and Bin Yuan

LIST OF CONTRIBUTORS

Emmanuel Akita
The University of Oklahoma, Norman, OK, United States

Alexander Badalyan
The University of Adelaide, Adelaide, SA, Australia

Pavel Bedrikovetsky
The University of Adelaide, Adelaide, SA, Australia

Sara Borazjani
The University of Adelaide, Adelaide, SA, Australia

Steven Carpenter
University of Wyoming, Casper, WY, United States

Zhangxin Chen
University of Calgary, Calgary, AB, Canada

Larissa Chequer
The University of Adelaide, Adelaide, SA, Australia

Caili Dai
China University of Petroleum (East China), Qindao, China

Davud Davudov
The University of Oklahoma, Norman, OK, United States

Andrew Finley
Goolsby, Finley & Associates, LLC, Casper, WY, United States

Xuebing Fu
University of Wyoming, Laramie, WY, United States

Yu Liang
The University of Texas at Austin, Austin, TX, United States

Lulu Liao
Sinopec Research Institute of Petroleum Engineering, Beijing, China

Rouzbeh G. Moghanloo
The University of Oklahoma, Norman, OK, United States

Yang Ning
University of Houston, Houston, TX, United States

Jack C. Pashin
Oklahoma State University, Stillwater, OK, United States

Sarada P. Pradhan
Indian Institute of Technology Roorkee, Roorkee, Uttarakhand, India

Thomas Russell
The University of Adelaide, Adelaide, SA, Australia

Vikram Vishal
Indian Institute of Technology Bombay, Mumbai, India

David A. Wood
DWA Energy Limited, Lincoln, United Kingdom

Xingru Wu
The University of Oklahoma, Norman, OK, United States

Yulong Yang
The University of Adelaide, Adelaide, SA, Australia

Zhenjiang You
The University of Adelaide, Adelaide, SA, Australia

Bin Yuan
The University of Oklahoma, Norman, OK, United States; Coven Energy Technology Research Institute, Qingdao, Shandong, China

Abbas Zeinijahromi
The University of Adelaide, Adelaide, SA, Australia

Jiaming Zhang
CNPC Economics & Technology Research Institute, Beijing, China

PREFACE

The concept for this book was originated in 2016. We realized, from the valuable body of published work exists focused on improved/enhanced oil recovery (IOR/EOR) and formation damage, that most of that work did not link or integrate the knowledge available for those two topics. Formation damage, of some form, is frequently associated with most EOR projects, so it makes sense to consider them with a more integrated approach. That is the aim of this book.

During 2017 we assembled a team of respected experts in the various EOR techniques to produce concise and focused chapters on how formation damage impacts various EOR applications. These include causes, effects, and potential mitigation and exploitation strategies. This book could certainly not have been written and compiled without the valuable contributions of each chapter author, and their dedication to the demanding schedule we imposed upon them. We thank them for their contributions.

Formation damage in subsurface reservoirs, typically caused by changes in the chemical—physical environment during IOR/EOR, such as low-salinity water flooding, inorganic/organic scaling, chemical flooding, gas flooding, thermal recovery of heavy oil, and produced water reinjection, can involve major negative consequences for the reservoirs, e.g., water blockage, fines migration, wettability alteration, scaling, wax, asphaltene deposition, etc. These impacts need to be mitigated. However, in some cases, formation damage does lead to beneficial outcomes that enhance oil recovery, e.g., isolation of high-permeability water flow, improved reservoir sweep, reduction of fines migration, etc. Such benefits need to be exploited to improve reservoir performance and profitability.

Several of the chapters include detailed quantitative analysis supported by laboratory analysis, field case studies, and simulation and reservoir modeling. However, our approach here is to bridge the gap between theoretical knowledge and field practices while highlighting the risks and opportunities associated with formation damage relating to IOR/EOR associated with both conventional and unconventional oil and gas reservoirs, geothermal reservoirs, deepwater oilfields, and coal-bed methane. Indeed, we describe an integrated risk and opportunity assessment and management framework, designed to improve outcomes and awareness of the diverse possible outcomes associated with IOR/EOR projects.

Finally, we would like to thank Elsevier for providing us with the opportunity to realize our concept in the form of this book. In particular, Katie Hammon for helping us to refine the book content, and Andrea Gallego Ortiz, who project managed the process with such attention to detail and efficiency.

We hope you not only enjoy the book, but, having read it, will ultimately share our enthusiasm for combining these two hitherto separately considered topics.

Bin Yuan and David A. Wood

CHAPTER ONE

Overview of Formation Damage During Improved and Enhanced Oil Recovery

Bin Yuan[1] and David A. Wood[2]

[1]The University of Oklahoma, Norman, OK, United States
[2]DWA Energy Limited, Lincoln, United Kingdom

Contents

1.1 INTRODUCTION

To meet the growing demands and constraints on energy resources, oil operators have to exploit more complex reservoirs; but, to do so requires more advanced technologies. Several recently improved techniques and technologies are now commercially applied to increase production and ultimate recovery from both conventional and unconventional resources (Lake et al., 2014). In low–oil–price markets and competitive environments, it is particularly desirable to improve the efficiency of improved oil recovery (IOR) techniques by minimizing any potential risks and costs.

Formation Damage during Improved Oil Recovery.
DOI: https://doi.org/10.1016/B978-0-12-813782-6.00001-4

© 2018 Elsevier Inc.
All rights reserved.

Formation damage during enhanced oil recovery (EOR) refers to the impairment of physical, chemical or mechanical properties of petroleum-bearing formations, primarily due to reservoir permeability degradation impeding oil and gas flow and recovery. Civan (2015) summarized the relevant causes of formation damage and its consequences, and various approaches and techniques for formation assessment, control and remediation. The changes of chemical-physical-thermodynamic conditions associated with EOR techniques can result in various types of formation damage, i.e., water and bubble blockage, fines migration, fluids-rock interactions, organic and inorganic precipitation and deposition, scale formation, alterations of pore surface properties, pore structures and mechanic characteristics etc. In some cases, formation damage may itself lead to some benefits which enhance oil recovery, e.g., improve sweep efficiency thorough selective blockage of high-permeability regions cause by fines migration Yuan et al. (2017a,b,c); however, more usually, it reduces the efficiency of secondary and tertiary recovery processes from the reservoir and impairs well injectivity and/or productivity dramatically. Porter (1989) and Civan (2015) stated that it is better to avoid formation than to make tremendous efforts to remediate its effects. Types of formation damage are realized during drilling, completion, workover, stimulation, and fluid injection and production operations. The studies of formation susceptibility to damage have limited practical value if conducted without linking them to the associated engineering activities, which may or may not lead to potential damage. Hence, it is an essential topic for oil operators to understand formation damage specific to various EOR approaches, because it enables them to maximize oil recovery both technically and economically by applying EOR techniques to specific reservoirs with due consideration given to the relevant formation damage issues. To do that, the comprehensive integration of modelling, simulation, laboratory experiments and field testing are required to predict, characterize and control any risks of formation damage. Such an approach also aids the development of new advanced technologies and methodologies capable of addressing formation damage during secondary and tertiary recovery processes in both conventional and unconventional reservoirs. The intention of this work is to provide a better understanding of the pros and cons of diverse EOR techniques, and to provide guidelines on how to control or take advantages of formation damage issues by optimizing the design of EOR projects.

1.2 SUMMARY OF FORMATION DAMAGE DURING EOR

In the most general sense, formation damage can be defined as the various damage mechanisms affecting the properties of reservoir formations (matrix, pore space and fractures) through which the transport efficiency of multi-phase fluids (oil, gas, water, particles, droplet, foam and emulsion) is determined. It is usually diagnosed as the changes of well performance in terms of well injectivity/productivity and oil recovery factor. The major damage mechanisms can be categorized as four types, i.e., mechanically, chemically, thermally and biologically induced formation damage. Among these categories, the types of chemical damage mechanisms can be further classified as (Civan, 2015): (1) fluid-fluid incompatibility (such as, inorganic scale deposit, organic asphaltene deposit, foam/emulsion blockage, hydrate formation etc.); (2) rock-fluid incompatibility (such as, clay swelling/deflocculation, wettability alteration, and ionic/surfactant/polymer adsorption). The mechanically induced damage mechanisms mainly include fines/sands or any other types of particle migration, phase trapping caused by high capillary force in multiphase flow, and rock compaction or dilatation caused by changes of pressure. Changes of temperature (thermally-induced mechanisms) also lead to the dissolution and or deposition of minerals, transformation of minerals, and temperature-dependent wettability alternation. In addition, the biological activities of bacteria in reservoirs can cause the souring of crude oil, erosion of minerals, and blockage of pore-throats. For the different approaches of IOR to be successful, various changes to the physical, chemical, thermal-electrical, mechanical and/or biological environment of the reservoir must be initiated. The following sections group the mechanisms of formation damage associated with diverse techniques of IOR, and also provide a special focus on the potential formation damage in the development of deepwater, geothermal, and unconventional reservoirs, including shales and coal-bed methane.

1.3 LOW-SALINITY WATER FLOODING (LSWF)

LSWF has been justified as an effective EOR method by numerous experimental studies and field trials for both tertiary (residual oil) and secondary (initial water condition) modes of water flooding. Several field examples have been reported using LSWF in North Slope of Alsaka (Seccombe et al., 2010), Powder River Basin in Wyoming (Robertson, 2010), Norwegian continental shelf fields (Skrettingland et al., 2010) and El-Morgan field, Gulf of Suez, Egypt (Darihim et al., 2013). Despite the success of LSWF in core experiments and field pilots, the multiple mechanisms impacting reservoirs and oil recovery during this technique still remain debatable and lack of understanding.

Several major mechanism of formation damage and EOR have been identified during LSWF, including: (1) wettability alteration toward more water-wet conditions by releasing original mixed-wet particles (Alagic and Skauge, 2010; Skauge, 2008); (2) wettability alteration due to mineral dissolution and ion-exchange reactions (Lager et al. 2006); (3) emulsion and clay swelling by multicomponent ionic exchanges among crude oil, brine, and clay particles (Sorbie and Collins, 2010); (4) local pH increases at water-clay interfaces that desorb organic materials from pore surfaces (Austad et al., 2008); (5) fines migration (Aksulu et al., 2012) carrying small amounts of residual oil through the detachment of oil-coated particles from rock grains. In addition, selected blockage of high-permeability layers by fines migration provides a simple mobility-control method to enhance sweep efficiency (Zeinijahromi, 2013; Yuan and Moghanloo, 2017b, 2018a).

To offset the potential damage of fines migration, different types of acid systems have been developed to remove the formation fines plugged in the near-wellbore region. In addition, gravel packs and sand-control screens have been applied in the well bores under various downhole conditions (Khilar and Fogler, 1998). Amorim et al. (2007) studied the performance of various salts to inhibit clay swelling on various clays, and confirmed that $CaCl_2$ concentrations were the most effective in inhibiting clay swelling. They also defined the critical salt concentration (CSC) required to inhibit clay swelling in the samples studied using different salts, as: 0.5 M for NaCl, 0.4 M for KCl but only 0.2 M for $CaCl_2$. Nanoparticles can effectively mitigate formation damage caused by fine particles and clogging pore-throats through enhancing attractive forces among fine particles and grains (Arab and Pourafshary, 2013). Yuan et al. (2018b,c) justified the positive effects of nanoparticle treatments (both preflush and coinjection) to control fines migration.

In addition to the performance benefits, the synergistic effects of applying IOR techniques in combination are well documented. Hence, it can be attractive to combine LSWF with other IOR techniques to minimize the negative effects and enhance the positive consequence of LSWF. Skauge et al. (2011) confirmed that surfactant flooding under conditions of low salinity perform better than in cases where surfactants and low-salinity water are applied in isolation in terms of reduction of interfacial tension (IFT) and capillary number. Dang et al. (2016) reported the better performance of using secondary-mode LSW followed by LSWF combined with water-alternative-CO_2 flooding, rather than that of high-salinity WAG, standalone continuous LSWF and CO_2 flooding modes. This is because of the enhanced solubility of CO_2, ion exchange, carbonate mineral and clay distributions, and wettability alteration. Shiran and Skauge (2013) evaluated the synergistic effect on residual oil mobilization and EOR by combining LSWF with polymer injection in both secondary and tertiary modes. This takes advantage of fines migration in deep reservoirs away from injection well and prevents problems of fines migration in the near-wellbore region (Fig. 1.1).

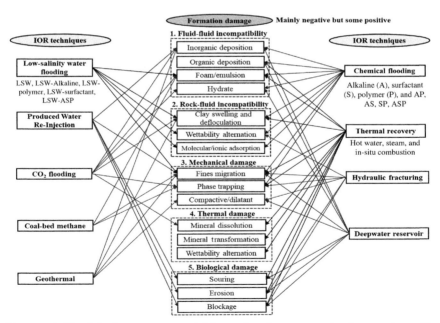

Figure 1.1 The links between potential formation damage mechanisms and specific EOR/IOR techniques and reservoir types.

1.4 CHEMICAL FLOODING

Among various techniques of EOR, chemical enhanced oil recovery (CEOR) has drawn increasing interest from oil companies. CEOR includes surfactant flooding (S), alkaline flooding (A), polymer flooding (P), alkaline-surfactant flooding (AS), alkaline-polymer (AP), surfactant-polymer (SP), and alkaline-surfactant-polymer (ASP). The performance of CEOR depends on the interaction efficiency among the injected chemicals, crude oil, formation water, reservoir conditions, and other stringent requirements, such as reservoir retention levels, compatibility, and thermal and aqueous stability of the injected chemicals.

The associated formation damage mechanisms with **surfactant flooding** include: (1) emulsion blockage and water-in-oil emulsion (Feng et al. 2011) tends to increase the viscosity in comparison to clean oil, produced by mixing surfactant and crude oil, which even with oil-in-water emulsions can improve the mobility ratio and sweep efficiency (Xu et al., 2011); (2) wettability alteration from oil-wet to water-wet; (3) phase trapping caused by the multiphase fluids transport with high capillary force (Nelson and Pope, 1978); (4) excessive adsorption and retention of surfactants leading to the decrease of porosity and permeability, and less efficiency of surfactant adsorption along oil-water interfaces for enhancing oil recovery (Hirasaki et al., 2011); (5) surfactant precipitation, viscous microemulsions, complexes, gels, and liquid crystals with increase of salinity that can damage permeability (Stellner and Scamehorn, 1986).

The damage issues caused by **polymer flooding** consist of: (1) polymer retention and plugging that reduce permeability caused by adsorption (Tang et al., 2001), mechanical trapping (Garti and Zour, 1997) and hydraulic retention (Bhardwaj et al., 2007); (2) incompatibility of polymer with formation fluids and rocks leading to both inorganic precipitation and organic deposition (Fletcher et al., 1992); (3) the mitigation of fines or sands caused by the strong adsorption of polymer onto clay particles (Borchardt, 1989). Consequently, before polymer is applied for IOR purposes, first, it should be subjected to a number of laboratory screening tests, including polymer injectivity, adsorption, and brine-compatibility tests. In addition, its microbial, mechanical, and chemical degradation characteristics should be established.

The formation damage mechanisms caused by **alkaline flooding** consist of: (1) migration and blockage of fine particles after alkaline injected fluids dissolve clays and other matrix minerals. Moreover, the necessity for high-salinity conditions to promote interactions between alkaline fluids and oil can lead to oil blockage of the flow paths (Ge et al., 2012); (2) the formation of carbonate scale, hydroxyl scale, silicate scale, or sulfate scale due to the incompatibility of alkaline with formation water, depending on formation water composition and temperature (Moghadasi et al., 2004; Sheng, 2011), which can result in significant damage to permeability; (3) precipitate formation due to the reactions among alkaline fluids and divalent or multivalent ions, which reduces permeability and also sometimes improves sweep efficiency by facilitating preferable flow of alkaline solutions into high-permeability zones (Sarem, 1974).

The synergistic effects of combinations of chemical flooding, such as, alkaline-surfactant flooding (AS), alkaline-polymer (AP), surfactant-polymer (SP), and alkaline-surfactant-polymer (ASP) are well known and documented. In addition to the benefits of combination flooding, one essential requirement is that the different chemicals must be compatible and stable in their mixture, otherwise, severe formation damage is likely to result significantly impairing the efficiency of chemical flooding (Sheng, 2016).

The potential formation damage mechanisms associated with **alkaline–surfactant (AS) flooding** include: (1) the addition of surfactant into the alkaline can generate more stable emulsions, which, on one hand, should carry more oil in the flowing water, but is adversely prone to block the pore-throats with the accumulation of emulsions (Rudin et al., 1994); (2) the detachment of sulfonate is likely to be enhanced and adversely lead to the damage of reservoirs, with the addition of alkaline as the charges of minerals become more negative (Hanna and Somasundaran, 1977).

Although the addition of polymer into alkaline should help improve mobility control by increasing the viscosity of displacing phase, formation damage can be induced during **alkaline–polymer (AP) flooding** by the increase of the alkaline concentration that can reduce polymer hydrolysis and polymer viscosity (Green and Willhite, 1998).

The addition of surfactants into polymer flooding can further reduce the interfacial tension and increase the viscosity of displacing phase by the surfactants forming chelation structures with polymers. In addition, as sacrificial agents, polymers react with the divalent ions on rock surfaces to reduce the loss of surfactants adsorption and precipitation to enhance the stability of the emulsions (Cui et al., 2011). However, during **surfactant–polymer (SP) flooding**, the emulsification effect of surfactant entering the oil phase, and the possibility of phase separation, may decrease the flowing efficiency of oil into wellbore (Pope et al., 1982).

The most likely formation damage mechanisms during **alkaline-surfactant–polymer (ASP) flooding** are fluid-fluid incompatibility and scale precipitation (Liu et al., 2007). Diverse techniques have been designed to prevent the problem of scaling, mainly including both chemical and physical antiscaling techniques (Xu et al., 2001).

1.5 THERMAL RECOVERY IN HEAVY OIL

The associated formation damage mechanisms with thermal recovery in heavy oil include: (1) migration of in-situ fines, kaolinite, detrital rock fragments, pyrobitumen, and other mobile particulates in clay-cemented clastic sandstones, and migration of dolomite or carbonate fines, and pyrobitumen in carbonate reservoirs (Bennion et al., 1995); (2) inevitably, large amounts of sands production along with the production of bitumen and heavy oil in less consolidated reservoirs (Tague 2000); (3) water-phase trapping occurs in heavy oil reservoirs with low permeability and porosity (Bennion et al., 1996); (4) gas-in-oil foam and water-in-oil emulsion with significant increase in apparent oil viscosity due to the entrainment of gas in solution (Chen et al., 2015); (5) clay welling and defloculation in an abrupt change of brine chemistry (Zhang et al., 2015a,b), such as fresh or low-salinity water, which can cause constriction, bridging, and blockage to damage permeability and well productivity; (6) wax and asphaltene deposition (Permadi et al., 2012): asphaltene coming out of solution as solid particles after destabilization of crude oil due to the reduction of temperature and pressure or by contact with precipitating agents, such as unsequestered hydrochloric acid, LPG, and carbon dioxide gas; the formation of crystalline wax caused by reduction in temperature; (7) biologically induced damage (Smith, 1995) including crude oil souring, blockage of bacteria-produced polyacharride with high molecular weight to reduce permeability, and mineral erosion due to the activities of bacteria onto rock surfaces; (8) thermally induced formation damage unique to the production of heavy oil by hot water, steam and in-situ combustion (Schembre and Koscel, 2005), including, transformation of kaolinite to water-sensitive clay, such as smectite in excess of 200°C, and dissolution of carbonate and silica due to increase of temperature, and reprecipitation of calcium, magnesium, and silicate solids caused by an abrupt change of temperature, and wettability alteration due to the decrease of adsorption of oil components as the temperature becomes higher.

1.6 PRODUCED-WATER-RE-INJECTION (PWRI)

Produced water compositions usually contain toxic pollutants, varying significantly from field to field, potentially posing a great threat to environment and increasing field oil and gas production costs. Produced water can contain contaminants of organic materials and inorganic materials, including: (1) dissolved and dispersed oil, (2) dissolved minerals, (3) treatment chemicals, (4) produced solids, (5) dissolved gases, and (6) bacteria and scales (Hansen and Davies, 1994). During production of oil/gas, the compounds contaminating the produced water may change over time, while other water enters the production zones, derived from water flooding or produced water re-injection (Sheng, 2013). For conventional oil and gas wells, as a well's production life advances, the volume of produced water typically increases associated with a decrease in produced volumes of oil and gas and an increase in waste-management costs; especially, after the breakthrough of injected water. The produced water in conventional oilfields usually contains emulsion, solution, suspension, particulates and adsorbed particles, and treatment chemicals, injected water, and bacterial contaminants (McCormack et al., 2001). The biggest difference of produced water in conventional gas fields is that usually no injected water is present in gas formations (Jacobs et al., 1992). In coalbed methane fields, a large amount of produced water is pumped out of coal seams at the early stage of production, and its produced volume decreases significantly as production advances (Nghiem et al., 2011). It is the uniqueness of requiring large amounts of fracturing fluids and sands for a large-scale fracture stimulation of unconventional reservoirs that distinguishes their produced water compositions from conventional reservoirs

(Yuan et al., 2015a; Yuan et al., 2016b). Hydraulic fracturing fluids, formation water, drilling and fracturing chemical additives (proppant, acids, surfactants, friction reducers, and others), and hydrocarbons are the common components of produced (flow back) water in such reservoirs (Ferrer and Thurman, 2015).

Approaches to produced water management for a specific well/field typically depends on many factors, including: cost, location, local law and feasible technologies. In general, the commonly applied methods for produced water management in oil and gas fields mainly include produced water minimization, produced water recycle/reuse and produced water disposal (Pichtel, 2016). Among them, produced water reinjection (PWRI) is an important IOR method with the potential to extend a reservoir's economic life, enhance oil recovery, increase water disposal, comply with national and local regulations, and minimize negative environment impacts. In the United States, most of offshore produced water is reinjected into the formations. However, the implementation of PWRI usually faces challenges with respect to safety, formation damage and injectivity, caused by low-quality water contaminated by clays, scale, bacteria, and oil droplets (Barkman and Davidson, 1972). When produced water is reinjected into a formation from an injector, the suspended particles tend to be deposited into the near-wellbore zones of the reservoir formations during the invasion process (internal filtration), and an external filter cake is formed on the walls of the wellbore (external cake filtration), resulting in reductions in the injectivity of the injector. The scale and extent of formation damage by PWRI is determined by the properties of rocks, such as pores and pore throat size, distribution and connectivity, as well as characteristics of injected produced water such as injection rate, temperature, pressure, suspended particle size, particle distribution, and surface charges (Yuan and Moghanloo, 2016a,b).

1.7 CO$_2$ FLOODING

The applications of CO$_2$ EOR projects are mostly accompanied with severe formation damage issues due to the incompatibility of injected fluids and reservoir fluids and injected fluids and formation minerals. The addition of CO$_2$ brings changes in oil composition and tends to decrease the solubility of asphaltenes in oil, and therefore acts as a precipitant for asphaltene. It does so by lowering the threshold of asphaltene precipitation (Gholoum et al., 2003), which can result in blockages of the reservoir pore throats, and damage to oil production by inducing reservoir rock wettability reversal with decrease in oil relative permeability (Novosad and Costain, 1990). CO_2 flooding can also result in the formation of carbonic acid, which lowers pH and increases Eh (activity of electrons) to dissolve quartz, feldspar, barite, anhydrite, mica, calcite (in some conditions), cements and some clays, including smectite, illite, and kaolinite (Chopping and Kaszuba, 2012; Miranda–Trevino and Cynthia, 2003). This leads to permeability damage through the movement of detached fines, precipitation of dissolved clays into pore throats, and the precipitation of calcite can be induced by the reactions between CO_2 and calcium ions (Ca^{2+}) in the formation water (Plummer and Busenberg, 1982).

1.8 HYDRAULIC FRACTURING IN SHALE FORMATIONS

Multistage fracturing in long horizontal wellbore sections has made shale oil and gas one of the most rapidly growing sources of worldwide energy supply (Yuan et al., 2015a,b; Yuan et al., 2017c; Zheng et al., 2016). However, the success and sustainability of oil and gas production from shale strongly depends on being able to avoid the reduction of fracture conductivity and matrix deliverability caused by formation damage mechanisms (Davudov et al., 2016). Major formation damage mechanisms during the production from shale reservoirs include the loss of fracture conductivity caused by proppant transport and its nonuniform placement (Liang et al., 2016), embedment and crushing (Kang et al., 2014), fine migration and plugging (Zhang et al., 2015a,b), gelling damage (Shaoul et al., 2011), and multiphase flow effects (Palisch et al., 2007). To solve the above damage problems, the following strategies can be applied, such as: foam-based fracturing fluids (Tong et al. 2017), and CO_2 foam stabilized fluids (Xiao et al., 2016), LPG–based fracturing fluids (Lestz et al., 2007; Zhao et al., 2017), low–density soft proppant (Jackson and Orekha, 2017), channel fracturing technique (Wang et al., 2018), and pumping design optimization (Bestaoui-Spurr and Hudson, 2017). Moreover, the severe damage of the reservoir matrix can be attributed to the pore shrinkage and connectivity loss caused by the decline of reservoir pressure (Davudov et al., 2017; Dong et al., 2010). In addition, in shale reservoirs, high capillary pressure tends to result in water trapping and significant damage to the relative permeability of oil and gas (Zitha et al., 2013).

Formation damage issues in shale reservoirs require that field operators pay close attention to the strategies capable of achieving sustainable production. Formation damage effects can be mitigated by optimizing fracturing design and operations, improving the choice of fracturing fluids and proppants, optimizing the online production strategy and conditions, and/or introducing chemical agents to remediate the damage caused to near-fracture and near-wellbore regions.

1.9 COAL-BED METHANE (CBM)

CBM is a natural gas extracted from coal formations. Dewatering is typically required to initiate CBM extraction, through which natural gas is gradually liberated from the coal matrix and fractures with the depletion of reservoir pressure (Bassett, 2009). However, various types of formation damage can be induced during the development of CBM by the physical-chemical-biological-thermal incompatibilities among clay particles, fluids, and coals, including fines migration, phase trapping, wettability alteration, and biological activities, and mechanical damage of the coal formations can also occur (Huang et al., 2015). During drilling and fracture stimulation of CBM wells, formation damage can be generated in the form of water blockage, clay swelling, and solid particles invasion, together with the leak-off of drilling and fracturing fluids (Lu et al., 2015). During the CBM-production phase, the main formation damage is the matrix shrinkage or swelling caused by gas desorption/adsorption, leading to the change of cleat aperture (Harpalani and Chen, 1995). Clay minerals in micropores can hydrate once in contact with water, leading to the blockage of flow paths (Zhang et al., 2016). In addition, the fluctuations or increase of production rates can lead to problems of fines migration and subsequent blockage of flow paths in coal seams (Han et al., 2015; Paul et al., 2014). During enhanced CBM recovery, the injection of CO_2-N_2 mixtures has been explored to diminish swelling-related formation damage caused by CO_2 (Vishal et al., 2015).

1.10 GEOTHERMAL RESERVOIRS

Geothermal energy provides a direct source of heat energy from within the Earth that can also be exploited for power generation with potential benefits associated with its reliability, sustainability, abundant reserve, and low environmental impacts (Glassley, 2010). In recent decades, the exploitation of geothermal energy has been extensively investigated and exploited in areas with high geothermal gradients, such as Geysers in California, USA, Rhine Graben in Soultz-sous-Forêts, France, Landau and Insheim, Germany, and Reykjavík, Iceland (Moore and Simmons, 2013).

The incompatibility among drilling fluids and formation minerals can lead to the shrinkage of wellbore due to clay swelling, and also, the reactions of invasion fluids with in-situ reservoir minerals can lead to severe scaling problems. Also, transporting particles along with leak-off fluids can result in permeability damage in geothermal reservoirs (Finger and Blankenship, 2010). The application of enhanced geothermal systems (EGS) is to improve the efficiency of extracting heat energy by pumping out injection water after it has become heated in deep high-temperature reservoirs (Markus et al., 2008). However, EGS has been reported to trigger earthquakes associated with some intensive applications of hydraulic fracturing (Deichmann et al., 2007). In addition, the migration of fines and clay swelling in rocks during water injection and recycling can lead to severe damage to reservoir permeability and consequential decline of well injectivity and productivity (You et al., 2016).

1.11 DEEPWATER RESERVOIRS

Deepwater and offshore oilfields contribute about 30% of total oil production over the past decade. In 2015, offshore production reached 29% of total global production, with a moderate decrease from 32% in 2005 (Manning, 2016). Deepwater oilfields under high-rates of depletion typically demonstrate rapid decline of reservoir pressure as production continues, leading to significant damage issues including: (1) severe reduction of permeability and well productivity (Yuan et al., 2016a; Wang et al., 2018); (2) sands production caused by excessive reservoir compaction due to the pressure drawdown and depletion in the reservoirs (Bianco, 1999); (3) asphaltene precipitation and damage in the near-wellbore zones of the reservoir formations, as the depletion of deepwater reservoirs progresses below the asphaltene onset pressure (Gonzalez et al., 2012). Frac-and-pack completions are often applied to deepwater wells, especially in unconsolidated formations for sand production control (Sanchez and Packing, 2007). However, the excessive decrease of pressure in the near-wellbore region can, in some cases, lead to the failure of well completion and sand control systems.

1.12 SUMMARY

Here, we combine theoretical knowledge with field practices to evaluate a wide spectrum of formation damage issues arising during enhanced oil/gas and geothermal energy recovery. The problems of formation damage in reservoirs caused by the changes to the chemical-thermal-mechanical-physical-biological environment induced during processes of enhanced oil/gas and geothermal recovery can lead to major negative consequences for geosystems. However, in some instances, formation damage can result in some benefits to enhanced oil/gas and geothermal recovery. The various potential formation damage issues associated with types of improved and enhanced petroleum and geothermal recovery methods, including low-salinity water flooding, chemical flooding, CO_2 flooding, thermal recovery in heavy oil, and hydraulic fracturing, are considered in terms of the types of reservoirs to which they are applied, such as sandstone, shale, coalbed methane, deepwater turbidite, and geothermal reservoirs. This information should provide insight to formation damage issues and aid the formulation of integrated and systematic designs to improve the efficiencies of improved and enhanced oil/gas and geothermal recovery techniques and strategies.

REFERENCES

Aksulu, H., Hamso, D., Strand, S., et al., 2012. The evaluation of low salinity enhanced oil recovery effects in sandstone: effects of temperature and pH gradient. Energy Fuels 26, 3497–3503.
Alagic, E., Skauge, A., 2010. Combined low salinity brine injection and surfactant flooding in mixed-wet sandstone cores. Energy Fuels 24 (06), 3551–3559.

Amorim, C.L.G., Lopes, R.T., Barroso, R.C., et al., 2007. Effect of clay—water interactions on clay swelling by X-ray diffraction. Nucl Instrum Methods Phys. Res. A 580, 768—770.

Arab, D., Pourafshary, P., 2013. Nanoparticles-assisted surface charge modification of the porous medium to treat colloidal particles migration induced by low salinity water flooding. Colloids Surf. A Physicochem Eng. Asp. 436, 803—814.

Austad, T. Rezaeidoust, A., Puntervold, T., 2008. Chemical mechanisms of low salinity water flooding in sandstone reservoirs. Paper SPE 129767 presented at SPE Improved Oil Recovery Symposium, Tulsa OK, USA.

Bassett, L., 2009. Successful strategies for dewatering wells using ESP's. Paper SPE-1254119 presented at SPE Eastern Regional Meeting, Charleston, West Virginia, USA.

Barkman, J.H., Davidson, D.H., 1972. Measuring water quality and predicting well impairment. J. Pet. Technol. 24, 865—873.

Bennion, D.B., Thomas, F.B., Bennion, D.W., 1995. Mechanisms of formation damage and permeability impairment associated with the drilling, completion and production of low API gravity oil reservoirs. SPE-30320 presented at SPE International Heavy Oil Symposium, 19—21 June, Calgary, Alberta, Canada.

Bennion, D.B., Thomas, F.B., Bietz, R.F., et al., 1996. Water and hydrocarbon phase trapping in porous media: diagnosis, prevention, and treatment. J. Can. Pet. Technol. 35 (10), 29—36.

Bestaoui-Spurr, N., Hudson, H., 2017. Ultra-light weight proppant and pumping design lead to greater conductive fracture area in unconventional reservoirs. SPE Oil and Gas India Conference and Exhibition, 4—6 April, Mumbai, India.

Bhardwaj, A., et al., 2007. Water retention and hydraulic conductivity of cross-linked polyacrylamiders in sandy soils. Soil Sci. Soc. Am. J. 71 (02), 406—412.

Bianco, L.C.B., 1999. Phenomena of sand production in non-consolidated sandstones; Ph.D. Thesis, The Pennsylvania State University, State College, PA, USA.

Borchardt, J.K., Young, B.M., Halliburton Co, 1985. Methods for stabilizing swelling clays or migrating fines in subterranean formations. US Patent 4,536,305.

Chen, Z., Sun, J., Wang, R., 2015. A pseudobubblepoint model and its simulation for foamy oil in porous media. SPE J. 20 (02), 239—247.

Chopping, C., Kaszuba, J.P., 2012. Supercritical carbon dioxide-brine-rock reactions in the Madison Limestone of Southwest Wyoming: an experimental investigation of a sulfur-rich natural carbon dioxide reservoir. Chem. Geol. 322, 223—236.

Civan, F., 2015. Reservoir Formation Damage. Gulf Professional Publishing.

Cui, Z., et al., 2011. Synthesis of N-(3-Oxapropanoxyl) dodecanamide and its application in surfactant-polymer flooding. J. Surfactants. Deterg. 14 (3), 317—324.

Dang, C., Nghiem, L., Nguyen, N., et al., 2016. Evaluation of CO_2 low salinity water-alternating-gas for enhanced oil recovery. J. Nat. Gas. Sci. Eng. 35, 237—258.

Darihim, N.M., Nassar, I., Salah, N., et al., 2013. A road map for integrating a Giant Field: case study. Paper SPE-164717 presented at North Africa Technical Conference and Exhibition, Cairo, Egypt.

Davudov, D., Moghanloo, R.G., Yuan, B., 2016. Impact of pore connectivity and topology on gas productivity in Barnett and Haynesville shale plays. Paper URTEC-2461331 presented at the Unconventional Resources Technology Conference, San Antonio, Texas, USA.

Davudov, D., Moghanloo, R.G., Lan, Y., et al., 2017. Investigation of shale pore compressibility impact on production with reservoir simulation. Paper SPE-185059 presented at SPE Unconventional Resources Conference, Calgary, Canada.

Deichmann, N., Mai, M., Bethmann, F., et al., 2007. Seismicity induced by water injection for geothermal reservoir stimulation 5 km below the city of Basel, Switzerland American Geophysical Union, American Geophysical Union, 53: 08.

Dong, J.J., Hsu, J.Y., Wu, W.J., 2010. Stress-dependence of the permeability and porosity of sandstone and shale from TCDP Hole-A. Int. J. Rock Mech. Min. Sci. 47 (7), 1141−1157.

Feng, X., Xiao, G., Wang, W., et al., 2011, Case study: numerical simulation of surfactant flooding in low permeability oil field, Paper SPE 145036 presented at the SPE Enhanced Oil Recovery Conference held in Kuala Lumpur, Malaysia.

Ferrer, I., Thurman, E.M., 2015. Chemical constituents and analytical approaches for hydraulic fracturing waters. Trends Environ. Analyt. Chem. 5, 18−25.

Finger, J., Blankenship, D., 2010. Handbook of Best Practices for Geothermal Drilling. Sandia National Laboratories, Albuquerque.

Fletcher, A., Lamb, S., Clifford, P., 1992. Formation damage from polymer solutions: factors governing injectivity. SPE Reserv. Eng. 7 (02), 237−246.

Garti, N., Zour, H., 1997. The effect of surfactants on the crystallization and polymorphic transformation of glutamic acid. J. Cryst. Growth. 172 (3), 486−498.

Ge, J., Feng, A., Zhang, G., 2012. Study of the Factors Influencing Alkaline Flooding in Heavy-Oil, Reservoirs. Energy Fuels 26, 2875−2882.

Gholoum, E.F., Oskui, G.P., Salman, M., 2003. Investigation of asphaltene precipitation onset conditions for Kuwaiti reservoirs. Middle East Oil Show. Society of Petroleum Engineers.

Glassley, W.E., 2010. Geothermal Energy: Renewable Energy and the Environment. CRC Press.

Gonzalez, D.L., Mahmoodaghdam, E., Lim, F.H., et al., 2012. Effects of gas additions to deepwater Gulf of Mexico reservoir oil: experimental investigation of asphaltene precipitation and deposition. SPE Annual Technical Conference and Exhibition, 8−10 October, San Antonio, Texas, USA.

Green, D.W., Willhite, G.P., 1998. Enhanced Oil Recovery. Society of Petroleum Engineers, Dallas.

Hanna, H.S., Somasundaran, P., 1977. Physic-chemical aspects of adsorption at solid/liquid interfaces, II: Mahogany Sulfonate/Berea sandstone, kaolinite. Improved Oil Recovery by Surfactant and Polymer Flooding. Academic Press, pp. 253−274.

Han, G., Ling, K., Wu, H., 2015. An experimental study of coal-fines migration in Coalbed-methane production wells. J. Nat. Gas. Sci. Eng. 26, 1542−1548.

Hansen, B.R., Davies, S.R., 1994. Review of potential technologies for the removal of dissolved components from produced water. Chem. Eng. Res. Des. 72, 176−188.

Harpalani, S., Chen, G., 1995. Influence of gas production induced volumetric strain on permeability of coal. Geotech. Geol. Eng. 15, 303−325.

Hirasaki, G., Miller, C.A., Puerto, M., 2011. Recent advances in surfactant EOR. SPE J 16 (4), 889−907.

Huang, W., Lei, M., Qui, Z., et al., 2015. Damage mechanism and protection measures of a coalbed methane reservoir in the Zhenghuang block. J. Nat. Gas. Sci. Eng. 26, 683−694.

Jackson, K., Orekha, O. 2017. Low density proppant in slickwater applications improves reservoir contact and fracture complexity—a permian basin case history. SPE Liquids-Rich Basins Conference-North America. Midland, Texas, USA.

Jacobs, R., Grant, R.O.H., Kwant, J., Marquenie, J.M., Mentzer, E., 1992. The composition of produced water from Shell operated oil and gas production in the North Sea. Produced Water. Springer, Berlin, pp. 13−21.

Kang, Y., Xu, C., You, L.J., 2014. Comprehensive evaluation of formation damage induced by working fluid loss in fractured tight gas reservoir. J. Nat. Gas. Sci. Eng. 18 (1), 353−359.

Khilar, K.C., Fogler, H.S., 1998. Migrations of Fines in Porous Media. Springer, Netherlands.

Lager, A., Webb, K.J., Black, C.J., et al., 2006. Low salinity oil recovery- an experimental investigation. International Symposium of the Society of Core Analysis, Trondheim, Norway.

Lake, W.L., Johns, R., Rossen, B., et al., 2014. Fundamentals of Enhanced Oil Recovery. Society of Petroleum Engineers.

Lestz, R.S., Wilson, L., Taylor, R.S., Funkhouser, G.P., Watkins, H., Attaway, D., 2007. Liquid petroleum gas fracturing fluids for unconventional gas reservoirs. J. Can. Pet. Technol. 46, 68.

Liang, F., Sayed, M., Al-Muntasheri, G.A., et al., 2016. A comprehensive review on proppant technologies. Petroleum 2 (1), 26−39.

Liu, D., et al., 2007. Study on scaling characteristics of strong base ASP flooding and the anti-scaling measures. Acta Petrolei. Sinica 28 (5), 139−141.

Lu, Y., Yang, Z., Li, X., et al., 2015. Problems and methods for optimization of hydraulic fracturing of deep coal beds in china. Chem Technol Fuels Oils 51, 41−48.

Manning, M., 2016. Offshore oil production in deepwater and ultra-deepwater is increasing. Today in Energy. US Energy Information Administration.

Markus, O.H., Ulrich, S., Florentin, L., et al., 2008. Characterisation of the basel 1 enhanced geothermal system. Geothermics 37 (5), 469−495.

McCormack, P., Jones, P., Hetheridge, M.J., 2001. Analysis of oilfield produced water and production chemicals by electrospay ionisation multi-stage mass spectrometry (ESI-MSn). Water Res. 35 (15), 3567−3578.

Miranda-Trevino, J.C., Cynthia, A.C., 2003. Kaolinite properties, structure and influence of metal retention on pH. Appl. Clay. Sci. 23 (1−4), 133−139.

Moghadasi, J., Jamialahmadi, M., Müller-Steinhagen, H., Sharif, A., 2004. Formation damage due to scale formation in porous media resulting from water injection, paper SPE 86524 presented at the SPE International Symposium and Exhibition on Formation Damage Control, 18−20 February, Lafayette, Louisiana.

Moore, J.N., Simmons, S.F., 2013. More power from below. Sci. 340 (6135), 933−934.

Nelson, R.C., Pope, G.A., 1978. Phase relationships in chemical flooding. SPE J. 18 (05), 325−338.

Nghiem, L.D., Ren, T., Aziz, N., Porter, I., Regmi, G., 2011. Treatment of coal seam gas produced water for beneficial use in Australia: a review of best practices. Desalination Water Treat. 32, 316−323.

Novosad, Z., Costain, T.G., 1990. Experimental and modeling studies of asphaltene equilibria for a reservoir under CO2 injection. SPE Annual Technical Conference and Exhibition, New Orleans, Louisiana.

Palisch, T.T., Duenckel, R.J., Bazan, L.W., 2007. Determining realistic fracture conductivity and understanding its impact on well performance-theory and field examples. Paper SPE-106301 presented at SPE Hydraulic Fracturing Technology Conference, College Station, Texas, USA.

Paul, M.R.S., Iyer, M.E., Timothy, N., 2014. An experimental study on characterizing coal bed methane (CBM) fines production and migration of mineral matter in coal beds. Energy Fuels 28 (2), 766−773.

Permadi, A.K., Naser, M.A., Mucharam, L., et al., 2012. Formation damage and permeability impairment associated with chemical and thermal treatments: future challenges in EOR applications.

Pichtel, J., 2016. Oil and gas production wastewater: soil contamination and pollution prevention. Appl. Environ. Soil Sci. 2016.

Plummer, L.N., Busenberg, E., 1982. The solubilities of calcite, aragonite and vaterite in CO_2-H_2O solutions between 0 and 90 C, and an evaluation of the aqueous model for the system $CaCO_3$-CO_2-H_2O. Geochim Cosmochim Acta. 46 1011−1040.

Pope, G.A., Tsaur, K., Schechter, R.S., 1982. The effect of several polymers on the phase behavior of micellar fluids. SPE J. 22 (6), 816−830.

Porter, K.E., 1989. An overview of formation damage. J. Pet. Tech. 41 (8), 780−786.

Sanchez, M., Packing, F., 2007. Fracturing for sand control. Middle East Asia Res. Rev. 37−49.

Schembre, J.M., Koscel, 2005. Mechanism of formation damage at elevated temperature. J Energy Resour. Technol. 127 (03), 171−180.

Shaoul, J., van Zelm, L., Pater, C.J., et al., 2011. Damage mechanisms in unconventional gas well stimulation-a new look at an old problem. SPE Prod. Oper. 26 (4), 388−400.

Smith, S.A., 1995. Monitoring and Remediation Wells: Problem Prevention, Maintenance, and Rehabilitation. CRC Press.

Robertson, E.P., 2010. Oil recovery increases by low salinity flooding: Minnelusa and Green River Formations. Paper SPE 132154 presented at the SPE Annual Technical Conference and Exhibition, Florence, Italy.

Rudin, J., Bernard, C., Wasan, D.T., 1994. Effect of added surfactant on interfacial tension and spontaneous emulsification in alkali/acidic oil systems. Ind. Eng. Chem. Res. 33, 150−1158.

Sarem, A.M., 1974. Secondary and tertiary recovery of oil by MCCF (mobility-controlled caustic flooding) process, paper SPE-4901 presented at the SPE−AIME 44th Annual California Regional Meeting, 4−5 April, San Francisco, USA.

Seccombe, J., Lager, A., Jerauld G.R., et al., 2010. Demonstration of low salinity EOR at interwell scale, Endicott Field, Alaska. Paper SPE 129692 presented at SPE/DOE Improved Oil Recovery Symposium, Tulsa, USA.

Sheng, J.J., 2011. Modern Chemical Enhanced Oil Recovery: Theory and Practice. Elsevier, Burlington, Massachusetts, USA.

Sheng, J.J., 2013. Enhanced Oil Recovery Field Case Studies. Gulf Professional Publishing.

Sheng, J.J., 2016. Formation damage in chemical enhanced oil recovery processes. Asia-Pac. J. Chem. Eng. 11, 826−835.

Shiran, B.S., Skauge, A., 2013. Enhanced oil recovery (EOR) by combined low salinity water/polymer flooding. Energy Fuels 27 (3), 1223−1235.

Skrettingland, K., Holt, T., Tweheyo, M.T., et al., 2010. Snorre low salinity water injection − core flooding experiments and single well field pilot. Paper SPE 129877 presented at the SPE Improved Oil Recovery Symposium, Tulsa, OK, USA.

Skauge, A. 2008. Microscopic diversion: a new EOR technique. 29th IEA Workshop & Symposium. Beijing China.

Skauge, A., Ghorbani, Z., Delshad, M., 2011. Simulation of combined low salinity brine and surfactant flooding. 16th European Symposium on Improved Oil Recovery, Cambridge, UK.

Sorbie, K.S., Collins I.R., 2010. A proposed pore-scale mechanism for how low salinity water flooding works. Paper SPE 129833 presented at SPE Improved Oil Recovery Symposium, Tulsa, Oklahoma, USA.

Stellner, K.L., Scamehorn, J.F., 1986. Surfactant precipitation in aqueous solutions containing mixtures of anionic and nonionic surfactants. J. Am. Oil. Chem. Soc. 63 (04), 566−574.

Tague, J.R., 2000. Overcoming formation damage in heavy oil fields: a comprehensive approach. Paper SPE-62546 presented at SPE/AAPG Western Regional Meeting, 19-22 June, Long Beach, California.

Tang, H., Meng, Y., Yang, X., 2001. A study of adsorption consumption of polyacrylamide on reservoir minerals. Oilfield Chem. 18 (4), 343−346.

Tong, S., Singh, R., Mohanty, K.K., 2017. Proppant Transport in Fractures with Foam-Based Fracturing Fluids. SPE Annual Technical Conference and Exhibition, San Antonio, Texas, USA.

Vishal, V., Singh, T.N., Ranjith, P.G., 2015. Influence of sorption time in CO2-ECBM process in Indian coals using coupled numerical simulation. Fuel 139, 51−58.

Wang, W., Yuan, B., Su, Y., 2018. A composite dual-porosity fractal model for channel-fractured horizontal wells. Eng. Appl. Comp. Fluid. Mech. 12 (01), 104−116.

Xiao, C., Balasubramanian, S.N., Clapp, L.W., 2016. Rheology of Supercritical CO 2 Foam Stabilized by Nanoparticles. In SPE Improved Oil Recovery Conference. SPE Improved Oil Recovery Conference, 11−13 April, Tulsa, Oklahoma, USA.

Xu, F., Guo, X., and Wang, W., et al., 2011. Case Study: numerical simulation of surfactant flooding in low permeability oil field, Paper SPE 145036 presented at the SPE Enhanced Oil Recovery Conference, 19−21 July, Kuala Lumpur, Malaysia.

Xu, L., He, Y., Xi, H., 2001. Study on the oxy-starch water treatment agents. Water. Purif. Technol. 20 (02), 27−29.

You, Z., Yang, Y., Badalyan, A., 2016. Mathematical modelling of fines migration in geothermal reservoirs. Geothermics 59, 123−133.

Yuan, B., Su, Y., Moghanloo, R.G., et al., 2015a. A new analytical multi-linear solution for gas flow toward fractured horizontal well with different fracture intensity. J. Nat. Gas. Sci. Eng. 23, 227−238.

Yuan, B., Wood, D.A., Yu, W., 2015b. Stimulation and hydraulic fracturing technology in natural gas reservoirs: theory and case study (2012−2015). J. Nat. Gas. Sci. Eng. 26, 1414−1421.

Yuan, B., Moghanloo, R.G., 2016a. Analytical solution of nanoparticles utilization to reduce fines migration in porous medium. SPE J. 21 (06), 2317−2332.

Yuan, B., Moghanloo, R.G., Shariff, E., et al., 2016b. Integrated investigation of dynamic drainage volume (DDV) and inflow performance relationship (Transient IPR) to optimize multi-stage fractured horizontal wells in shale oil. J. Energy Resour. Technol. 138 (5), 052901−052909.

Yuan, B., 2017a. Modeling Nanofluids Utilization to Control Fines Migration. Doctoral Dissertation. The University of Oklahoma, Norman, Oklahoma.

Yuan, B., Moghanloo, R.G., 2017b. Analytical modeling improved well performance by nanofluid pre-flush. Fuel 202, 380−394.

Yuan, B., Moghanloo, G.R., Zheng, D., 2017c. A novel integrated production analysis workflow for evaluation, optimization and predication in shale plays. Int. J. Coal Geol. 180, 18−28.

Yuan, B., Moghanloo, R.G., 2018a. Nanofluid treatment, an effective approach to improve performance of low salinity water flooding, J. Petrol. Sci. Eng., in press. doi:10.1016/j.petrol.2017.11.032.

Yuan, B., Moghanloo, R., Wang, W., 2018b. Using nanofluids to control fines migration for oil recovery: nanofluids co-injection or nanofluids pre-flush? − A comprehensive answer. Fuel 215, 474−483.

Yuan, B., Moghanloo, R.G., 2018c. Nanofluid pre-coating, an effective method to reduce fines migration in radial system saturated with mobile immiscible fluids. SPE J . Available from: https://doi.org/10.2118/189464-PASPE-189464-PA.

Zeinijahromi, A., Nguyen, T.K.P., Bedrikovetsky, P., 2013. Mathematical model for fines-migration-assisted waterflooding with induced formation damage. SPE J. 18 (03), 518–533.

Zhang, J., Wu, X., Zhang, K., et al., 2015a. Damage by swelling clay and experimental study of cyclic foam stimulation. Acta Geologica Sinica (Eng. Ed.) 89, 215–216.

Zhang, J., Ouyang, L., Zhu, D., 2015b. Experimental and numerical studies of reduced fracture conductivity due to proppant embedment in the shale reservoirs. J. Pet. Sci. Eng. 130 (1), 37–45.

Zhang, Y., Lebedev, M., Sarmadivaleh, M., et al., 2016. Swelling effect on coal micro structure and associated permeability reduction. Fuel 182, 568–576.

Zheng, D., Yuan, B., Moghanloo, R.G., et al., 2016. Analytical modeling dynamic drainage volume for transient flow towards multi-stage fractured wells in composite shale reservoirs. J. Pet. Sci. Eng. 149, 756–764.

Zhao, J., Chen, P., Liu, Y., Mao, Y., 2017. Development of an LPG fracturing fluid with improved temperature stability. J. Pet. Sci. Eng. in press. Available from: https://doi.org/10.1016/j.petrol.2017.10.060.

Zitha, P., Frequin, D., Bedrikovetsky, P., 2013. CT scan study of the leak-off of oil-based drilling fluids into saturated media. Paper SPE-165193 presented at SPE Formation Damage Conference and Exhibition Held in Njirwjik, Netherlands.

CHAPTER TWO

Low-Salinity Water Flooding: from Novel to Mature Technology

David A. Wood[1] and Bin Yuan[2]

[1]DWA Energy Limited, Lincoln, United Kingdom
[2]The University of Oklahoma, Norman, OK, United States

Contents

Formation Damage during Improved Oil Recovery.
DOI: https://doi.org/10.1016/B978-0-12-813782-6.00002-6

© 2018 Elsevier Inc.
All rights reserved.

2.1 INTRODUCTION

Low-salinity water flooding (LSWF) has become an established enhanced oil recovery (EOR) method (i.e., improving oil recovery by 5–38%, compared to conventional, high-salinity water flooding), as demonstrated in numerous experimental studies and field trials for both tertiary (residual oil) and secondary (initial water condition) modes of water flooding (Bernard, 1967; Hourshad and Jerauld, 2012; Behruz and Skauge, 2013). Early chemical injection studies of the 1960s and 1970s established that injecting water with low divalent ion concentrations tended to improve oil recovery. The pioneering work of Bernard (1967) identified some key attributes of the outcomes of LSWF: *"In general, it appears that water sensitive cores will produce more oil with a freshwater flood than with a brine flood. However, the fresh-water flood is accompanied by a lowering of permeability and the development of a relatively high pressure drop"*. That work also highlighted the complexity of the mechanisms at play. Although the technique has been applied successfully in many core experiments and field tests, it does not work in all formations being dependent on many factors; particularly, the initial wetting conditions, formation water composition, crude oil composition, and clay mineralogy. Also, LSFW involves multiple mechanisms impacting the reservoir, and there remains disagreement and lack of understanding as to which of these are dominant and which are merely effects rather than causes of incremental oil recovery associated with LSWF process.

2.2 ORIGINS OF LSWF AND IDENTIFICATION OF RESERVOIR MECHANISM DRIVING INCREMENTAL OIL RECOVERY

Changes to reservoir wettability caused by LSWF, and the impact of fines migration on associated oil recovery, were recognized by Tang and Morrow (1999). Several specific mechanisms have been proposed to explain how LSWF actually works. These include: the partial stripping of fines (Tang and Morrow, 1999); reduction of interfacial tension (IFT)

(McGuire et al., 2005); correlation of contact angles and IFT with salinity (Ashraf et al., 2010; Yousef et al., 2010); the electrical double layers associated with a thin film of saline formation water coating the clay minerals of the rock matrix changing thickness in double layer expansion (DLE) (Nasralla and Nasr-El-Din, 2012; Myint and Firoozabadi, 2015); and, multicomponent ionic exchange (MIE) with the sandstone reservoir becoming negatively charged (Fig. 2.1), partly due to Mg^{2+} exchange, with oil desorbing from the reservoir matrix due to repulsive charges (Lager et al., 2008). Austad et al. (2010) suggested that potentially several chemical mechanisms were at play in making LSWF effective.

There is a strong case to be made (Fig. 2.1) that both DLE and MIE are the key underlying mechanisms, which complement each other and not mutually exclusive, in enabling LSWF to incrementally improve oil recovery. Changes in wettability, formation water pH, and composition observed during LSWF are likely to be the effects of those underlying mechanisms. The complex and varied bonding of the polar elements of crude oil, some of which are displayed in Fig. 2.1, mean that there are potentially many types of ion exchange reactions involved depending on the composition of the oil, formation water, clay mineralogy, and fabric (e.g., availability of potentially mobile fines particles) in the reservoir matrix.

The first LSWF field test (Webb et al., 2004), on a single producing well in response to different water salinities injected into the reservoir, reported 25−50% reductions in residual oil saturation attributed to water flooding with low-salinity brine. That study concluded that as much as 50% additional oil could be produced from the well, if water with salinity of <4000 ppm was injected into the reservoir instead of sea water or higher salinity produced water. The significance of salinity thresholds was confirmed by multiwell field tests revealing that reservoirs flooded with water salinities of >7000 ppm showed the least incremental oil recovery (McGuire et al., 2005), which was supported by core studies (Zhang et al., 2007). Another pilot well test suggested that the improved incremental recovery was related to a reduction in the water cut of the produced fluids, rather than an increase in oil production rate (Seccombe et al., 2010).

Reservoir simulations applying salinity-dependent relative permeability and capillary pressure functions (Jerauld et al., 2006), the addition of a dual porosity model (Wu and Bai, 2009), and correlations of residual oil saturation, contact angle, and IFT (Al-adasani and Bai, 2012) have

Double-layer expansion (DLE) and multicomponent ion exchange (MIE) are key mechanisms enabled by low-salinity water flooding (LSWF) that mobilize oil bonded to the pore walls

Expansion of electrical double layers caused by LSWF

Thin saline fluid layers or films wet the surface of the rock matrix in oil reservoirs. That film <10nm thick has two interfaces (film/oil and film/rock) each with an electrical double layer. Double-layer expansion (DLE) induced by LSWF results in a thicker, more stable formation-water film resulting in a more water-wet state.

Some of the adsorption mechanisms bonding oil to clay minerals in porous reservoirs

In multi-ion exchange, divalent cations, particularly Ca^{2+} and Mg^{2+} replace monovalent cations. A process that is enhanced by expansion of the double layer during LSWF.

The electrical potentials at the two interfaces of the double layer are estimated by zeta potentials. ζ_3 and ζ_4 are more negatively charged than ζ_1 and ζ_2.

Figure 2.1 Schematic illustrating DLE and the diversity of adsorption mechanisms attaching the polar components of crude oil to the clay minerals and double layer of the thin film of formation water covering them; left side modified after Myint and Firoozabadi (2015); right side modified after Lager et al. (2008).

improved the accuracy of predictions of potential incremental production achievable with LSWF. Vledder et al. (2010) claimed from core experiments and field production results that LSWF causes desorption of petroleum heavy ends from the clays present on the pore wall, resulting in a more water-wet rock surface, a lower remaining oil saturation, and higher oil recovery. A reservoir simulation sensitivity study (Al-adasani et al., 2014) suggested that it is the initial and final wetting states of an oil reservoir which control oil recovery through an increase in the relative permeability oil. That study concluded that a decrease in IFT was shown to be the primary recovery effect in reservoirs which were strongly water-wet, but other factors dominated in other types of reservoir (i.e., lowering of capillary pressure in weak water-wet reservoirs; change of nonwetting phase to oil in weak oil–wet conditions).

The significance of polar bonding of crude oil to saline-water film covering the clay minerals of the rock matrix (Fig. 2.2), and the ability of LSWF to break those bonds to form mixed-wet fines that mobilize to the oil–water interface flowing through the pore space of a reservoir, was recognized by Tang and Morrow (1999).

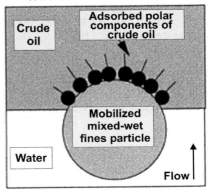

Figure 2.2 Polar components from crude play an important role in binding some oil to the pore walls and in the formation of wet-mixed fines some of which can be mobilized by changing flow conditions and fluid chemistry. *Modified after Tang & Morrow (1999).*

LSFW is also effective in carbonate reservoirs with a range of potential mechanisms proposed to explain its impacts on oil recovery, e.g., wettability (Al Shalabi et al., 2013), with IFT suggested as having a bigger impact on LSWF in carbonate reservoirs, with contact angle controlling final oil recovery (Al-adasani and Bai, 2012). Studies on chalk reservoirs have identified that wettability alterations are impacted by the concentration of calcium and magnesium ions in the presence of sulfate (i.e., the divalent ions: Ca^{2+}, Mg^{2+}, SO_4^{2-}) and the overall salinity level of the injected water (Zhang et al., 2006; Yousef et al., 2010; 2012; Hamouda and Gupta, 2017). Many offshore carbonate reservoirs (e.g., North Sea chalk reservoirs) have traditionally been flooded with sea water; so, LSWF core experiments often compare the impacts of injected fluids of various diluted brines with that of sea water. In doing so, Hamouda and Gupta (2017) found that brines diluted 1:10 versus sea water were more effective in terms of oil recovery from chalk reservoir than brines diluted 1:50. Also, the sulfate-rich brines performed the best, and were associated with an increase in pH of the fluids, changes in the Ca^{2+} and Mg^{2+} ion concentrations and pressure drop that suggested ion exchange and precipitation of magnesium minerals. Experiments conducted on oil-wet low-permeability limestones (Gandomkar and Rahimpour, 2015) suggested different outcomes for LSWF in secondary recovery mode versus tertiary recovery mode; with no benefits observed for the latter due to an increase in water relative permeability.

The uptake of LSWF in a wide range of reservoirs is now high for several reasons: (1) low capital expenditure and incremental operating costs for those reservoirs already developed with water injection facilities and wells; (2) ease of injection into most oil reservoirs; (3) high incremental recovery gains for light to medium gravity oil reservoirs; (4) reduction in the scaling and corrosion of wellbore tubulars and surface-water-handling facilities (Collins, 2011); and (5) potentially avoid reservoir souring.

2.3 FINES MIGRATION: DETACHMENT, TRANSPORT, AND REDEPOSITION

Khilar and Fogler (1984) identified that low-salinity water injected into oil-bearing sandstones could cause significant formation damage reducing permeability. They identified that this was due to the release then trapping in pore throats of clay particles. The release of the particles (fines) to form a colloidal solution was shown to be sensitive to salinity falling below certain threshold levels. Mechanism initiating the release of fines and how to control it continue to be a major focus of LSWF research.

Much of the early work on the water-composition sensitivity of fines migration was based on core-flooding experiments conducted on the Berea sandstone (Devonian, Ohio / West Virginia, USA). Kia et al. (1987) established that pH of the injected fluids played a role in the process of releasing fines from the pore walls. They demonstrated that significant permeability reductions occurred if LSW was injected with *pH* > 6, but that formation damage decreased as *pH* of the fluids decreased. Injecting LSW with pH less than about 4.8 resulted in no permeability reduction. They identified that the LSW pH was influencing the multivalent ion-exchange process at work, which was related to the surface charges of the mineral grains (clays and sand particles). The surface charges were very low at low *pH*, but increased as *pH* increased causing the electrostatic repulsion forces between the clay particles and the pore walls to increase. Kia et al. (1987) conceived a double layer model to predict the release of fines particles.

Ochi and Vernoux (1998) demonstrated that the combined effects of salinity and flow rate influenced formation damage in the Berea sandstone. Their core-flooding experiments' results identified a critical flow rate above which the permeability reduced related to hydrodynamic fines release. The critical flow rate increased with the fluid salinity. The hydrodynamic release of fines could, alone, result in a 50% reduction in permeability; but, the formation damage impacts related to salinity were shown to be greater. Increase in flow rate leads to a less significant drop in permeability than salinity decline, because the hydrodynamic changes cause just limited fines to mobilize with those particles quickly deposited in the pore throats.

Tang and Morrow (1999) established with core-flooding experiments on the Berea sandstone that adsorption of crude oil to the pore walls, mobile fines, and a water-oil formation fluid were all required for LSWF to increase oil recovery (Fig. 2.3). Their results suggested that oil recovery and wettability trends during LSWF required the heavy polar components of oil adsorb onto matrix grains on pore walls to generate mixed-wet fines.

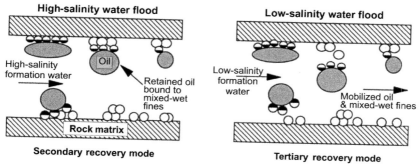

Figure 2.3 Mobilization of crude oil during water flooding of a high-water saturation reservoir with high-salinity and low-salinity injection water. *Modified after Tang & Morrow (1999).*

Mixed wettability is required for the adsorption of heavy polar oil components, as particles coated with formation water would not allow this (Salathiel, 1973). Some of the mixed-wet fines (i.e., oil adsorbed onto clay particles) are released (mechanical and chemical stripping) into the flowing pore fluids from the pore walls during LSWF. Once in the fluid system, the mixed-wet fines are expected migrate toward the oil–water interface (Muecke, 1979). A reduction in salinity was considered by Tang and Morrow (1999) to expand the electrical double layer in the aqueous phase between particles, which would ease the stripping of mixed-wet fines and thereby improve oil recovery.

Visualization models enabled Song and Kovscek (2016) to confirm that salinity of LSWF had to be below a critical level for clay swelling and fines migration to occur and significantly impact permeability. They surmized that the redeposition of fines in pore throats and clay bridging due to swellings is more likely to occur in the high–water-cut flow channels. Oil sweep of the reservoir would then be improved as the blocked flow channels would gradually divert injected water toward the higher-oil-cut flow channels.

Nguyen et al. (2013) developed a simulation to model the concept that fines migration could be induced by LSWF and exploited to positive effect to inhibit encroaching water in a depleting gas reservoir. In the simulation, involving two-phase immiscible compressible flow with a maximum particle retention function, fresh water was injected, into a watered-out production well, over a limited period to form a barrier fluid to decrease water encroachment up-dip into a gas reservoir with a strong water drive during the pressure-blowdown production phase (Fig. 2.4). The induced fines migration (pore plugging) formation damage was localized to the vicinity of the injection well. The impact is to reduce the phase permeability for water, but with very minor impacts on gas-phase permeability.

This technique leads, in the simulation at least, to a decrease in up-dip water cut, and prolonged the life of up-dip gas production wells, thereby increasing gas recovery by 15—18%. The volume of LSW injected should be carefully balanced to suit the reservoir characteristics and scale; too large a volume was found to increase water cut and decreased incremental gas recovery water. It was found for the gas reservoirs simulated that injecting banks of LSW of about 1.5—3% of the total reservoir pore volume, and then halting LSWF (after about 1.5 to 3 weeks in the reservoir volumes evaluated), lead to optimum gas recovery.

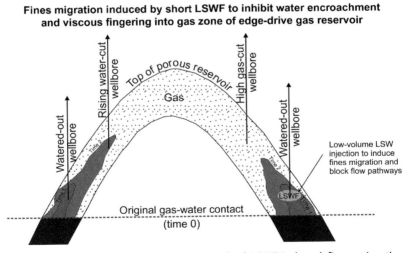

Fines migration induced by short LSWF to inhibit water encroachment and viscous fingering into gas zone of edge-drive gas reservoir

Figure 2.4 Schematic illustrating the potential of LSWF-induced fines migration to increase gas recovery by inhibiting water encroachment by targeted reduction in permeability in watered-out leg of a reservoir. *Modified after Nguyen et al. (2013).*

2.4 CLAY SWELLING, DETACHMENT, AND PORE BLOCKING LEADING TO REDUCTIONS IN PERMEABILITY AND POROSITY

In addition to fines migration, particle swelling and swelling-induced migration are the other main formation damage mechanisms which impact sandstone reservoirs subjected to LSWF. Mohan et al. (1993) conducted experiments on cores from the Upper Miocene Stevens sandstone from the Elk Hills oil field (California, USA) and compared the results with those from the Berea sandstone. The Stevens sandstone has similar porosity (18–20%) to the Berea sandstone, but lower permeability (<10 mD compared to >100 mD). Whereas, both sandstones have similar total clay contents (~8%), the Stevens sandstone contains a range of swelling (i.e., smectite, mixed-layer) and nonswelling clays (i.e., kaolinite and illite); whereas, the Berea sandstone clays are dominated by kaolinite and some illite (nonswelling).

Their LSWF experiments (Mohan et al.,1993) showed that the critical salt concentrations (CSCs) of chlorides of sodium, potassium, and calcium required to prevent formation damage in the form of permeability loss were higher than those required to prevent that effect in the Berea sandstone. They attributed this difference to the expansion of swelling clays in the Stevens sandstone. In contrast to the Berea sandstone, lowering the pH of the injection fluid had a limited effect in inhibiting the permeability reduction process in the Stevens sandstone (Fig. 2.5). These results suggest that water sensitivity of various sandstones to LSWF is dependent on the composition of the clays they contain, in addition to the total clay content and distribution of clays grains within the formation.

Clay swelling is a widely studied phenomenon in relation to formation damage and can have significant negative impacts on hydraulic fracture conductivity (Sanaei et al., 2016), as well as on the effectiveness of LSWF. The structural layers of clay minerals are deficient, to varying degrees, in positive charges due to cation substitution (Zhou et al. (1996). This means that interlayer cations are required to balance the negative charges. The distance between two structural layers in clay minerals (d-spacing) varies according to the exchangeable cation, the formation water composition, and the type of clay mineral. Salts that keep the d-spacing of clays low are considered to be good clay stabilizers or swelling inhibiters. Amorim et al. (2007) studied the effects of various salts to inhibit clay

Mechanisms causing permeability reduction in porous reservoirs related to changes in ionic conditions and fluid flow

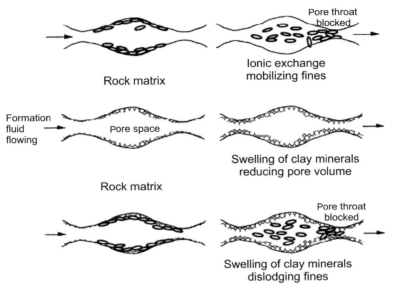

Figure 2.5 Schematic of permeability-reduction mechanisms in porous rocks related to fines migration, clay swelling, and changing ionic conditions. *Modified after Mohan et al. (1993).*

swelling on various clays, including samples from the Calumbi Formation (Sergipe-Alagoas Basin, NE Brazil). Their findings supported those of Mohan et al. (1993) in that $CaCl_2$ concentrations were shown to be most effective at inhibiting clay swelling. The CSC to inhibit clay swelling in the samples studied were: 0.5 M for NaCl, 0.4 M for KCl but only 0.2 M for $CaCl_2$.

To study clay detachment and pore blocking, Song and Kovscek (2016) constructed clay-functionalized, etched-silicon micromodels to visualize directly the mobilization of clay-mineral grains subjected to low-salinity conditions. Their results suggested that fines migration that preferentially blocks the high-water-cut flow channels is a mechanism that is much more effective in kaolinite-rich systems. They therefore concluded that LSWF is more likely to result in IOR in kaolinite-rich systems. On the other hand, in montmorillonite-rich systems, the mobilization of swollen montmorillonite from the pore linings degraded permeability more generally, but crucially did not appear to result directly in IOR under the LSWF conditions studied. The reduction of permeability by the blocking of pore throats with fines and swelled clays are both forms of formation damage that could be of benefit regarding EOR. However, these mechanisms work in subtly different ways, which seem to impact their potential IOR benefits that may vary from formation to formation.

2.5 SALINITY THRESHOLDS AND RESERVOIR HETEROGENEITY INFLUENCES ON PARTICLE DETACHMENT

The CSC and critical rate of salinity decrease (CRSD) (Khilar et al., 1983) are significant in terms of particle detachment thresholds in porous rocks. CSC is the salinity threshold below which fine particles, particularly clays, are released from matrix surfaces (pore linings). Khilar and Fogler (1984) identified that CSCs differ from sediment to sediment and are dependent on valence and size of the dissolved solubilized salt cations in the formation water. Blume et al. (2005) found that particle detachment in heterogeneous formations could be quite different from those observed in relatively clean quartz sandstones. They found that the quantity of particles released and the CSC could be an order of magnitude higher for heterogeneous formations, and significant particle detachment could also occur above the CSC in such formations. These findings bring into question the concept of all formations displaying a sharp CSC. This highlights the complex interactions between formation fluid salinity, cations present in those fluids, hydrodynamic forces, potentially-mobile-particle availability at the pore linings, and clay mineralogy of the particles on the matrix surfaces. All of these factors are likely to have an impact on particle release, and their relative impacts are difficult to distinguish in specific formations.

2.6 EXPLOITING PORE PLUGGING TO PREFERENTIALLY ENHANCE OIL RECOVERY

2.6.1 Quantifying and modeling fines migration and pore plugging

Fine particles in porous formations tend to exist in mechanical equilibrium balancing the drag, lift, electrostatic, and gravitational forces acting on them. By weakening the electrostatic forces, LSWF can, in certain conditions, disturb that equilibrium leading to fines particles being dislodged from the pore linings and dragged along with flowing formation water into pore throats (Fig. 2.6), some of which will become blocked by those dislodged particles (Khilar et al., 1983).

Bedrikovetsky et al. (2012) proposed a maximum retention function that could be used to model fines migration in porous media. This involves the calculation of the torque balance of forces to calculate the maximum concentration of particles that can remain attached to the pore walls. The mechanical equilibrium of a particle can be expressed as Eq. (2.1).

$$F_d(U)l_d + F_l(U)l_n = (F_e(\gamma) + F_g)l_n \tag{2.1}$$

where: γ is salt concentration, U is the Darcy velocity, and l_d and l_n represent levers for drag and normal forces, respectively. If the detaching forces (left-hand side Eq. (2.1): torque /drag F_d and lift F_l) are greater than the attaching forces (right-hand side Eq. (2.1): gravity F_g and electrostatic F_e) the hydrodynamics of the flowing formation water is likely to dislodge the particle (Fig. 2.7). LSWF can cause the electrostatic forces to reduce and mobilize the particle.

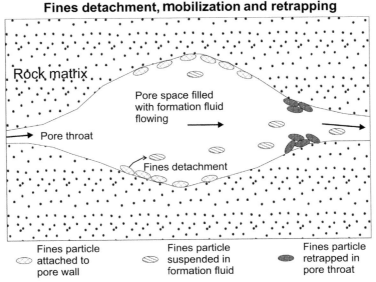

Figure 2.6 Schematic of fines distribution and mobilization within porous rock. *Modified after Khilar et al. (1983).*

Forces acting on a fines particle within a fluid-saturated pore volume that influence the maximum retention concentration of particles

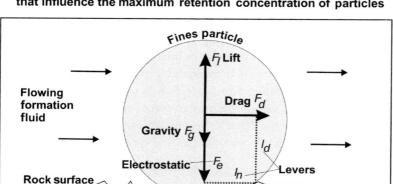

Figure 2.7 Forces acting on a fines particle attached to a pore wall (i.e., adsorbed) within a reservoir rock saturated with formation fluid. *Modified after Bedrikovetsky et al. (2012).*

Equation 2.1 can be used to calculate the dimensionless erosion number, ε, which is the ratio between detaching and attaching forces (Eq. (2.2)).

$$\varepsilon = \frac{F_d l_d + F_l l_n}{(F_e + F_g)l_n} \tag{2.2}$$

In-situ (nonswelling) fines will only be mobilized if the detachment forces exceed the attachment forces, i.e., where ε exceeds the electrostatic force. The maximum retention concentration of particles σ_a for specific flow and salinity conditions can then be expressed as Eq. (2.3).

$$\sigma_a = \sigma_{cr}(\varepsilon(U, \gamma)) \tag{2.3}$$

where: σ_{cr} is the maximum volumetric concentration of captured particles, U is formation fluid velocity, and γ is fluid salinity. This equation can be evaluated for γ_i (formation fluid salinity) and γ_o (and injected fluid salinity) to evaluate the impacts of LSWF on particle retention and calculate the concentration of fines released ($\Delta\sigma$). In a large reservoir volume, it is reasonable to assume that the concentration of strained particles (σ_s) that plug pore throats is approximately the same as the concentration of fines released ($\Delta\sigma$) during LSWF (Eq. (2.4)).

$$\sigma_s(\varepsilon(\gamma^0)) = \Delta\sigma(\varepsilon(\gamma^0)) \qquad (2.4)$$

Normalized permeability is then inversely proportional to the strained particle concentration, enabling the induced formation damage via LSWF to be quantified (Zeinijahromi et al., 2015a) using Eq. (2.5):

$$\frac{k_0}{k(\sigma_s)} = 1 + \beta\sigma_s(\varepsilon) \qquad (2.5)$$

where: k_0 is the initial undamaged permeability, $k(\sigma_s)$ is the damaged formation permeability, and β is the formation damage coefficient reflecting the degree of formation damage due to pore plugging. This approach to explain and model fines–assisted water flooding is supported by experimental studies (Lemon, 2011; Hussain et al., 2013).

2.6.2 LSWF to induce fines-migration-related formation damage

Zeinijahromi et al. (2015a) used these relationships to simulate the effects of induced formation damage applying LSWF to improve the sweep efficiency of edge–water drive reservoirs with in–situ movable fines. A small volume of LSW was injected into the downdip, watered-out wells, creating a low-permeability barrier to water encroachment. Slowing water encroachment and inhibiting its preferential flow relative to oil resulted in IOR of 3—5% in the reservoir simulated. The incremental oil recovery was attributed to the delay of water breakthrough in the up-dip producing wells and delaying their abandonment.

Zeinijahromi et al. (2015b) reported on the performance of long-term LSWF in the Zichebashskoe oil field (Russia) in reducing permeability by fines migration decreasing injected water mobility and increasing reservoir sweep. The 24-year production history of that field included only 7 years of LSWF. Reservoir simulation sensitivity analysis matching the field production history was used to explain why the incremental oil recovery and decrease in water cut due to LSWF was so small (i.e., just 4%). Three factors can be identified from the simulation analysis as possible explanation for this: (1) the oil-bearing formation had been extensively flooded during the primary production phase by high-salinity water, encroaching from the underlying aquifer, during the 17 years of production resulting in high water cuts of produced fluids. This change to the reservoir had occurred before LSWF began, which potentially impacted the availability of mobile fines once LSWF began; (2) LSWF injection directly into aquifer (as was the case in Zichebashskoe oil field) results in lower incremental oil recovery than LSWF injection directly into the oil zone. This second effect could be due to the usually high sweep efficiency of water injection directly into the aquifer; (3) reservoir heterogeneity limits the field-wide impacts of LSWF and fail to efficiently sweep the low-permeability flow channels.

Gamage & Thyne (2011) compared experimentally the impacts of LSWF conducted as either a secondary or tertiary recovery mode and as a two-phase (secondary plus tertiary) mode. They used cores from the Berea sandstone outcrops and the Minnelusa Donkey Creek sandstone reservoir (Permian, Powder River Basin Wyoming) and two crude oils from other Minnelusa oil fields (API gravity 29° and 34°; viscosity 11.5 and 8.0 cp). Their results showed increased oil recovery for LSWF in both secondary and tertiary recovery modes (2−8% OOIP) for the Berea sandstone, with the secondary modes at the top end of that oil recovery range. The Minnelusa tests showed no impacts on oil recovery in tertiary mode, but oil recovery increased significantly in secondary mode (10−22% OOIP) with both types of crude oils. pH increased in the effluent brine from the cores in all experiments, but less so for the Minnelusa cores.

The experience of LSWF at the Zichebashskoe oil field and the experiments conducted by Gamage & Thyne (2011) suggest that in planning LSWF IOR projects, the best results are likely to be achieved by commencing LSWF from the outset of field development in secondary recovery mode. The outcomes of the Zichebashskoe oil field suggest that it is also prudent to consider injecting LSW directly into the oil leg of the reservoir in some fields, and to carefully consider reservoir heterogeneities when selecting the locations for the injection wells.

2.6.3 Enhanced sweep efficiency by induced fines migration during LSW flooding

The issues associated with fines migration induced by chemical environments of low-salinity fluids have aroused significant debate, due to its positive and negative impacts. Fines migration may carry small amounts of residual along with detached oil-coated particles from rock grains, i.e., improve the displacement efficiency (Aksulu et al. 2012). In addition, the reduction of the effective permeability of the local water-phase in water-swept areas caused by the blockage of mobile fines can provide mobility control to enhance the sweep efficiency of the reservoir (Lemon, 2011;

Zeinijahromi et al., 2011, 2013). However, fines migration and its size-exclusion effects can also result in severe damage to reservoir permeability, which leads to declines of well injectivity in the case of injection wells, and productivity in case of production wells. During low–salinity waterflooding, the majority of injection pressure loss occurs in the vicinity of the wellbores. This is attributed to the high fluid flow velocities in these zones.

Therefore, understanding how to control or avoid fines migration in reservoirs is an important issue for LSWF. On the one hand, it is desirable to control fines migration to take advantages of its positive effects far from the wellbore. On the other, it is important to minimize its negative formation damage impacts near the wellbores. Here, we develop a mathematical framework for designing a nanofluid-slug, preflush to enhance well injectivity, while maintaining the mobility control assisted by fines migration that contributes to improving LSWF performance (both EOR and well injectivity).

During LSWF, the injected low-salinity fluid gradually sweeps out of the reservoir the in-situ fluids with higher salinity. In the low-salinity environment where smaller amounts of ions exist, according to the theory of Debye and Hückel (1923), the Debye-length (Double layer thickness in of the 1910 Gouy-Chapman theory, Greathouse et al., 1994) would increase. Therefore, the effects of fluid salinity can be reflected by changes to the inverse Debye-length, κ, m^{-1} (Elimelech et al. 1995), as shown in Eq. (2.6):

$$\kappa = 0.73 \times 10^{11} \sqrt{\sum \left| \frac{C_{mi,i}S_{wc} + C_{mi,j}(S_w - S_{wc})}{S_w} \right| z_i^2} \qquad (2.6)$$

where, C_{mi} is the molar i^{th} ion concentration in water phase (injected and initial conditions), moles/m^3; Z_i is a valence of ith ion. This relationship indicates that as the saturation of injected low-salinity water increases, the inverse Debye-length would decrease; thereby, the repulsive energy among the particles increases, and the bonding force among particles attenuates. The double electric layer repulsive energy V_{DLR} is described by the DLVO (Derjagin−Landau−Verwey−Overbeek) theory, as as expressed by Eq. (2.7):

$$V_{DLR} = \frac{128\pi r_{FP} n_\infty k_B T}{\kappa^2} \varsigma_{FP}\varsigma_{GS} e^{-\kappa h} \qquad (2.7)$$

For individual cylindrical shaped pores with water and oil two–phase flow, the maximum retention concentration of fine particles onto rock grains is expressed as Eq. (2.8) (Yuan and Moghanloo, 2016). This expression relates the increase of electric repulsive force to the low–salinity water's ability to decrease the maximum retention concentration.

$$\sigma_{cr}(S_w, r) = \left[1 - (\frac{\mu_w r_{FP}^2 \frac{qf_w}{2\pi r}}{2\phi S_w r p F_e y})^2\right] \phi S_w \left(1 - \phi_c\right) \tag{2.8}$$

Usually, the detachment/attachment of fines only leads to small changes of permeability, although the straining/plugging of fines into pore throats causes more significant formation damage. Hence, Yuan et al. (2018a) incorporated the effects of fines straining only into the damage of water–phase relative permeability. Based on the assumption of the instant straining of those detached fines caused by low–salinity water, the concentration of strained fines can be equal to the detached fines concentration (i.e., by subtracting the maximum retention concentration of fine particles under different conditions of low–salinity water saturation from the initially attached fines concentration). An implicit fractional flow relationship considering fines migration and subsequent straining during low–salinity waterflooding is expressed as Eq. (2.9):

$$f_w(S_w, x_D) =$$

$$\left(1 + \frac{K_{ro}\mu_w \left(1 + \beta_s \left(\sigma_{cr,initial}(0, S_{wc}) - \left[1 - \left(\frac{\mu_w r_{FP}^2 qf_w}{4\pi\phi r p F_e y}\right)^2 \frac{1}{x_D r_e^2}\right] \phi\left(1 - \phi_c\right)\right)\right)}{K_{nv}\mu_o}\right)^{-1}$$

$$\tag{2.9}$$

An iteration algorithm can calculate the fraction flow functions versus water saturation at different locations, as shown in Fig. 2.8. At the locations near the injection well, the calculated fractional flow velocities decline steeply, which means the decrease of water phase flowing velocities caused by the exaggerated fines migration close to the wellbores. In addition, the flood-front water saturation can be obtained by the method of Welge (1952) that increases due to the attenuation of fines migration as the injected water moves further toward to the production well.

Fig. 2.9 indicates that with an increase of low–salinity water saturation in the water-flooded reservoir, the average fluid salinity in reservoir

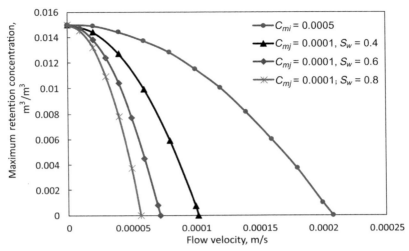

Figure 2.8 The effects of low-salinity water saturation on the maximum retention concentration of fine particles onto rock grains. C_{mi} refers to the initial ionic concentration in high-salinity water, i.e., 0.0005 m^3/m^3; C_{mj} refers to the ionic concentration in injected low-salinity water, 0.0001 m^3/m^3.

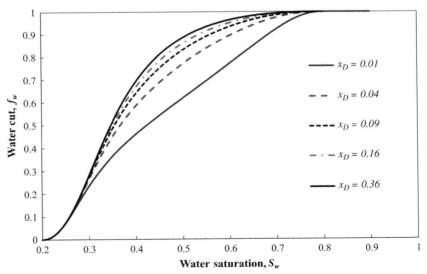

Figure 2.9 Fractional flow functions at different locations considering the effects of fines straining. x_D (dimensionless distance) specifies the different radial locations of asymmetric flowing system; it varies from 0.01 to 0.36.

decreases. As a result, the maximum retention concentration of fine particles onto rock grains also decreases, which means that fine particles are more likely to be dislodged as low-salinity waterflooding continues. In addition, with an increase in low-salinity water saturation, the sensitivity of fines detachment to the changes of flowing velocity becomes more significant. This phenomenon indicates that, in the late life of low-salinity waterflooding, field operators need to be aware that any abrupt changes of water injection and oil production rates can have significant impacts on fines detachment.

For cases of low-salinity waterflooding with fines migration effects, the mass-balance equation of fine particles and water flowing through the porous media, taking into account the detachment and straining effects of fine particles, can be expressed as Eq. (2.10) (pseudo-two-phase: water (solids only exist in water phase)/oil; three-component: water/oil/fines):

$$
\begin{cases}
\dfrac{\partial S_{wi}}{\partial t_D} + \left(\dfrac{\partial f_{wi}}{\partial S_{wi}}\right)_{x_D} \dfrac{\partial S_{wi}}{\partial x_D} + \left(\dfrac{\partial f_{wi}}{\partial x_D}\right)_{S_{wi}} = 0 \\[4mm]
\dfrac{q}{4\pi x_D} = \dfrac{k_0}{1 + \beta \sigma_s} \left(\dfrac{k_{nv}}{\mu_w} + \dfrac{k_{ro}}{\mu_o}\right) \dfrac{\partial p}{\partial x_D}
\end{cases}
\tag{2.10}
$$

The relationship expressed by equation 2.5x is a first-order, quasi-linear, partial differential equation. The method of characteristics (MOC) can be applied to solve this problem. For the equation 2.5x relationship, the characteristic directions and the variations of water saturation along the characteristic curves (lines) are expressed as Eq. (2.11):

$$
\frac{dx_D}{dt_D} = \left(\frac{\partial f_{wi}}{\partial S_w}\right)_{x_D} ; \quad \frac{dS_{wi}}{dt_D} = \left(\frac{\partial f_{wi}}{\partial x_D}\right)_{S_{wi}}
\tag{2.11}
$$

By combining the slopes of characteristic lines with the fraction flow function (Eq. (2.9)), changes in water saturation can be computed. As shown in Fig. 2.10, the effects of fines migration lead to the analytical solutions of low-salinity waterflooding being significantly different from conventional water flooding (i.e., with no changes of fluid salinity) expressed as a classical Buckley-Leverett (1942) problem. In the conventional waterflooding case, the characteristic velocity of the water-saturation wave with a specific water-saturation value remains constant, as it propagates toward the production well (i.e., the characteristic lines remain straight, and along those lines water saturation remains constant).

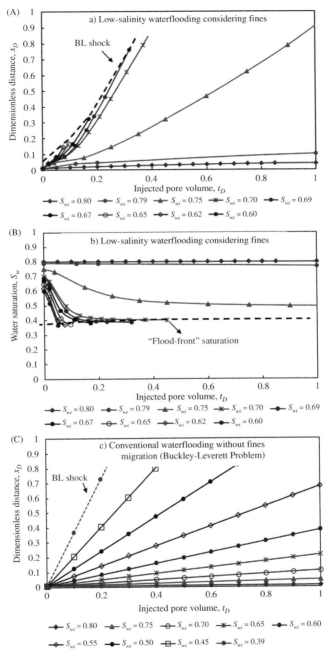

Figure 2.10 Distance-time diagrams and composition variations along characteristic lines. (A) shows distance-time diagram with characteristic lines representing the propagation of different water-saturation waves for low-salinity waterflooding. (B) shows variations of water saturation along different characteristic lines for low-salinity waterflooding. (C) shows distance-time diagram with characteristic lines for constant water-saturation values.

However, in the presence of fines migration effects, the characteristic lines are no longer straight; therefore, the water-saturation values, also, would not remain constant along those the curved characteristic lines. As the water-saturation wave with a specific injected water saturation at the inlet (injection well) moves toward the outlet (production well), the corresponding water saturation progressively decreases and the velocity at which that saturation wave moves (i.e., the slopes of characteristic lines) increase, as indicated in Fig. 2.10.

By drawing vertical lines with constant time (for example, $t_D = 0.2$) passing through the characteristic lines with varying water saturation, the water saturation distribution in a radial flow system can be derived. The comparisons of the water saturation profiles at specific times between conventional waterflooding and low-salinity waterflooding are illustrated in Fig. 2.11. A key effect of fines migration is to decrease the effective water-phase permeability, which can significantly slow down the movement of the injected water. This confirms the effectiveness of induced fines migration during low-salinity waterflooding to improve the mobility-control performance by slowing down the flow of injected water through the reservoir.

Nanoparticles can effectively enhance the attractive forces between fine particles and grain surfaces by changing the surface zeta potentials of fine particles, and significantly mitigating formation damage caused by

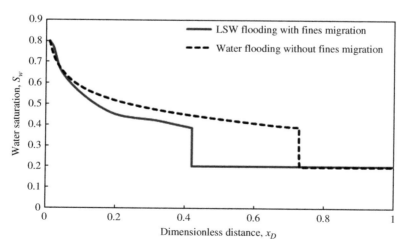

Figure 2.11 Water saturation profiles of conventional waterflooding cases and low-salinity waterflooding cases with fines migration effects at the same moment.

the clogging of fine particles into the pore-throats (Huang et al., 2008; Yuan and Moghanloo, 2016; Yuan, 2017a,b). In practice, it is, therefore, possible to improve the performance of LSWF by combining it with nanoparticles applications in multilayered radial flow systems.

In an example of multilayered radial flow system (Fig. 2.12), the ratio of permeability between two parallel layers is set as 2.0, and all the other properties are identical. During the injection of low–salinity water, the injection pressure for two layers is always kept identical. However, injection pressure increases with time due to progressive formation damage. The production pressure at the outlet (production well) is also kept constant. Even with the total injection rate held constant in the injection well, the fractional rates of water entering each layer continuously change over time, because of the changes of total fluid mobility within each layer.

The injection pressure loss at different times can be obtained by integrating the flowing pressure gradient from the injection well to the production well, as shown in Eq. (2.12):

$$
\left.
\begin{aligned}
\Delta p_1(t_D) &= q_1(t_D) \int_0^1 \frac{dx_D}{4\pi x_D \lambda_{t1}} \\[2ex]
\Delta p_2(t_D) &= q_2(t_D) \int_0^1 \frac{dx_D}{4\pi x_D \lambda_{t2}} \\[2ex]
\Delta p(t_D) &= q_t(t_D) \int_0^1 \frac{dx_D}{4\pi x_D \lambda_t}
\end{aligned}
\right\}
\xrightarrow{q_t(t_D) = Const.}
\frac{1}{\int_0^1 \frac{dx_D}{4\pi x_D \lambda_{t1}}} + \frac{1}{\int_0^1 \frac{dx_D}{4\pi x_D \lambda_{t2}}}
\quad (2.12)
$$

$$
= \frac{1}{\int_0^1 \frac{dx_D}{4\pi x_D \lambda_t}}
$$

Due to the layered heterogeneity, the fluid flowing mobility within each layer is different (Yuan et al. 2018b). Assuming constant injection rates for the sum of the layers, a harmonic mean of flowing mobility for the multilayered system can be derived from Eq. (2.12). As low–salinity waterflooding continues, the injection pressure to maintain a constant injection rate continues to increase due to the formation damage caused by fines migration. The injection pressure loss for the whole-layered system can be calculated using the harmonic mean of flowing mobility, as expressed by Eq. (2.13):

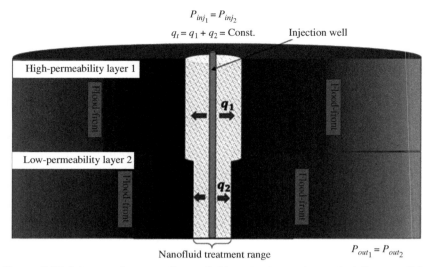

$$P_{inj_1} = P_{inj_2}$$
$$q_t = q_1 + q_2 = \text{Const.} \qquad \text{Injection well}$$

High-permeability layer 1

q_1

Low-permeability layer 2

q_2

Nanofluid treatment range $P_{out_1} = P_{out_2}$

Figure 2.12 Scheme of near-well nanofluid pretreatment to control fines particles during low-salinity waterflooding in two-layered heterogenous reservoirs.

$$\Delta p(t_D) = \frac{q_t(t_D)}{\displaystyle\int_0^1 \frac{dx_D}{4\pi x_D \lambda_{t1}} + \int_0^1 \frac{dx_D}{4\pi x_D \lambda_{t2}}} \tag{2.13}$$

The improvement of mobility control caused by fines migration/straining is expressed in Eq. (2.14) as the ratio (R) of the advancing locations of the flood fronts within each layer:

$$R = \frac{x_{fD1}}{x_{fD2}} = \frac{x_{fD10} + \left(\dfrac{\partial f_w}{\partial S_w}\right)_{S_{wf}} t_{D1}}{x_{fD20} + \left(\dfrac{\partial f_w}{\partial S_w}\right)_{S_{wf}} t_{D2}} \tag{2.14}$$

The evolution of water saturation profile along each layer at different moments is presented in Fig. 2.13. At the early period of low–salinity waterflooding, a larger percentage of injected water enters the high–permeability layer, and the flood-front along the high–permeability layer moves much faster. However, with the accumulation of low–salinity water in the high–permeability layer involving larger swept areas, more severe damage of permeability is induced by amounts of fines detachment and straining. This is a consequence of both higher flow rates and lower fluid salinity in the high–permeability layer. As a result, both the advancing

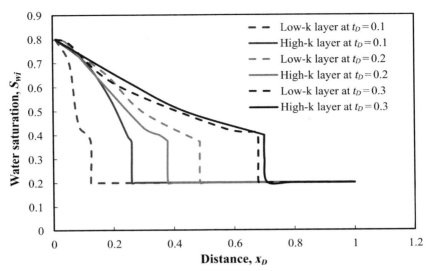

Figure 2.13 Water saturation profiles along each layer at different moments (t_D = 0.1, 0.2, and 0.3).

speed of the waterflood-front and the fraction of injection fluids flowing into the high-permeability layer decreases gradually, until it reaches the same as the fraction of injected fluids entering the low-permeability layer. As low-salinity waterflooding continues, the flow fractions of the injected water entering each layer and the movement of flood-front along each layer gradually become uniform. This uniformity of flood-front movement is attributed to the mobility control assisted by fines migration.

2.7 FACTORS INFLUENCING EOR IN SANDSTONE RESERVOIRS SUBJECTED TO LSWF

The concept of a CRC above which the effects of LSWF would not be realized was introduced by Khilar and Fogler (1984) as shown in Fig. 2.14.

Salehi et al. (2017) conducted an experimental study on a sandstone core from the Cheshme Khosh oil field (Iran). The system tested involved:

- Core with porosity 14.19% and permeability 20.8 mD;
- Formation water salinity $\sim$200,000 ppm;

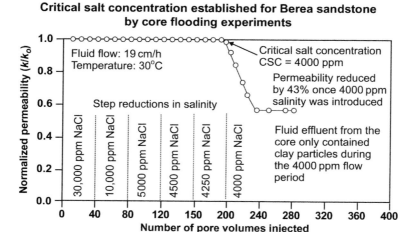

Figure 2.14 Sensitivity of Berea sandstone to salt concentration of injected fluids determined by core flooding experiments to reveal the CSC for that formation. K_o refers to initial permeability. *Adapted after Khilar and Fogler (1984).*

- Light crude oil with density 0.83 g/cm^3 and viscosity 3.3 cp; and
- Artificial brines made up with fresh water and NaCl to produce injection fluids with salinities varying from 30,000 to 200,000 ppm.

The study did not address fluids with salinities of <30,000 ppm (i.e., these are significantly above the CSC required to induce LSW sensitivity, Fig. 2.14) nor the influence of salts other than NaCl.

Despite the limitations (with respect to LSW conditions), Salehi et al. (2017) recorded a range of impacts related to a higher range of salinities for the injection water on oil recovery, reservoir pressure drop, reservoir permeability, IFT, viscosity, density, resistivity, *pH*, and relative permeability during the water–flooding process. Their results indicated that oil recovery increased as the injected water salinity increased (reaching 48% for 200,000 ppm water salinity). In addition, IFT decreased with increasing water salinity and both oil and water relative permeabilities curves moved toward higher water saturations as water salinity increased. This study highlights why historical many water floods conducted in secondary recovery mode have elected to use high-salinity (e.g., sea water) injection water.

Capillary forces play a significant role in limiting oil recovery efficiency during water flood displacement processes. Melrose and Brander (1974) identified that capillary forces are responsible for trapping between

10% and 50% of the oil phase within the interstices or pores of the rock. They identified using Eq. (2.15) that ultra-low values of the oil–water IFT are required to achieve improved recovery:

$$N_{ca} = \frac{\text{Viscous forces}}{\text{Capillary forces}} = \frac{\nu\mu}{\sigma\cos\theta} \qquad (2.15)$$

The combined effects of IFT (capillary forces) and viscosity ratio (viscous forces) on oil recovery are related to the capillary number (N_{ca}) (Melrose and Brander, 1974; Ayirala and Rao, 2004) where: N_{ca} is capillary number; ν is interstitial velocity of the displacing fluid (cm/s); μ is viscosity of the displacing fluid (water), σ is the IFT between oil and water (dyne/cm), and θ is the contact angle at the crude oil-formation water–porous rock interface.

As the capillary number (N_{ca}) increases, oil saturation (S_{or}) should decrease. This can be achieved by either lowering IFT and/or θ or increasing μ. Many studies that evaluate capillary number, essentially ignore the contact angle term Eq. (2.1), by setting $\cos\theta = 1.0$ ($\theta = 0°$), which simplistically assumes a perfect water-wet system. This was the assumption made by Salehi et al. (2017), observing in their experiments that S_{or} did indeed decrease with increasing capillary number and with increasing water salinity. Their capillary pressure curves also moved toward higher water saturations as salinity increased, indicating that S_{or} decreased with decreasing IFT. Plotting capillary pressure versus oil saturation can determine the irreducible water saturation and residual oil saturation of a formation.

With respect to the chemical mechanisms at play during LSWF, there has been some debate regarding the roles of MIE (e.g., Lager et al., 2008; Seccombe et al., 2010) and DLE (e.g., Nasralla and Nasr-El-Din, 2012, 2014; Xie et al., 2014). Charged formation surfaces in contact with formation water have an excess of ions close to that surface, which is referred to as the double layer. Pouryousefy et al. (2016) investigated, by combining simulation and experimental results, the relative roles in wettability change of MIE and DLE in sandstone reservoirs. They concluded that both mechanisms contribute to the oil recovery impacts of LSWF, which agree with the concept illustrated here in Fig. 2.1.

In MIE, different electrolyte concentrations to the formation brine are introduced by LSWF. This disturbs the equilibrium of the oil/water/rock system and variations in ionic concentration result in the substitution of divalent cations by the monovalent cations (Lager et al., 2008). This makes the system more water-wet and thereby increases the oil recovery factor. If the MIE mechanism is dominant, there should be an inverse correlation between oil recovery factors and divalent cation adsorption. The higher the divalent adsorption, the lower the oil recovery should be.

By injecting cores with solutions of different concentrations of both NaCl and CaCl$_2$, Pouryousefy et al. (2016) showed that the results did not match with those predicted by MIE. The expectations for MIE are that high concentration of NaCl should lead to higher oil recoveries than lower concentrations, and that all concentrations of CaCl$_2$ should lead to lower recoveries than NaCl with little variation in oil recovery due to different CaCl$_2$ concentrations. Experimental results showed that lower concentration of Ca^{2+} achieved higher oil recoveries than higher concentrations of Ca^{2+}. Also, although the oil recoveries with CaCl$_2$ were lower than for NaCl (consistent with MIE), the highest overall oil recovery was achieved with the lowest NaCl concentrations.

Based on these results, Pouryousefy et al. (2016) conclude that MIE can only be one of the mechanisms at play with LSWF and suggest that DLE, induced by LSWF (and perhaps in part by MIE) is also playing a role. The exact mechanism(s) by which this might occur remain unclear. Several possible chemical bonding mechanisms are likely involved, where polar components of crude oil are directly attached to the surface of the rock, or at least the thin water film covering its pore linings and its associated electrical double layers (Fig. 2.1). These include cation bridging, cation exchange, ligand bonding, anion bridging, which are all impacted by the electrolyte concentrations in the formation fluids. Myint and Firoozabadi (2015) have shown how some of the chemical mechanisms proposed by Austad et al. (2010) (e.g., organometallic bridges, hydrogen bonding, acid/base interactions) impact the expansion of the electrical double layer in sandstones and carbonates during LSFW.

2.8 RELATIONSHIPS BETWEEN OIL RECOVERY, SALINITY, AND WETTABILITY VARIABLES

The wettability of a porous rock refers to the ability of one fluid to spread across or adhere to the solid surface in presence of another immiscible fluid (oil and water being the two immiscible fluids of interest in IOR). The wetting properties determine the contact angle θ at which the oil–water interface intersects the surface of the porous media (Fig. 2.15), and that angle is used as the standard measure of wettability (Cieplak and Robbins, 1990).

The contact angle is a function of the interfacial energies between fluids and porous rock as defined by Eq. (2.16):

$$\sigma_{om} - \sigma_{wm} = \sigma_{ow} \cos\theta \qquad (2.16)$$

where,

σ_{om} = interfacial energy between oil and porous rock matrix (dyne/cm);

σ_{wm} = interfacial energy between formation water and porous rock matrix (dyne/cm);

σ_{ow} = IFT between oil and formation water (dyne/cm);

θ = contact angle at oil-formation water-porous rock interface measured through the water phase.

The IFT is the surface-free energy that exists between the two immiscible liquid phases (oil and formation water) in a crude oil-formation water-porous rock system. The effect of capillary forces at play involved in binding/trapping crude oil within the pores space of reservoir rocks are influenced by the IFT and contact angle. Those capillary forces are typically characterized by dimensionless number calculated as the capillary number (Eq. (2.1)).

A significant increase in the capillary number (about four to six orders of magnitude according to Ayirala and Rao, 2004) is required to significantly reduce residual oil saturation (i.e., improve oil recovery) in any

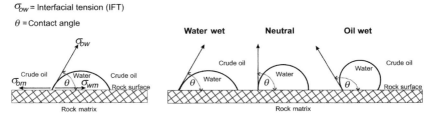

Figure 2.15 Metrics used to quantify wettability states in oil reservoirs.

IOR method. From Eq. (2.1), it is clear that the capillary number can be increased either by reducing the IFT(σ) or the contact angle (θ) (making the system more water wet) or increasing μ. Altering either IFT or θ involves inducing changes to the wettability of the crude oil-formation water-porous rock-fluids system. For LSWF to be effective at increasing oil recovery, a lowering of IFT or θ should be expected, although this may be an effect of the processes at work rather than the driving mechanism (Lager et al., 2008).

Barati-Harooni et al. (2016) conducted experiments and modeled the relationships between temperature, pressure, and synthetic formation water salinity and IFT of two carbonate oil reservoirs (Iran). Their results showed that the IFT data of reservoir A increased as the temperature, pressure, and salinity of synthetic formation water increased. On the other hand, IFT data of reservoir B increased as the pressure and salinity of synthetic formation water increased, but that IFT decreased as the temperature increased.

Electrokinetic effects in porous media, such as streaming potential and zeta potential (Fig. 2.1), are also related to its wettability (Jackson and Vinogradov, 2012). These effects can therefore be used as direct indicators to monitor changes in wettability throughout a reservoir or during IOR projects. When fluids pass through porous media, they carry mobile charges of double layers and an electric current is generated, which causes a potential difference to build up in the system, referred to as the "streaming potential". The streaming potential can be measured to map subsurface flow, detect subsurface flow patterns, and monitor wettability in oil reservoirs (Sadeqi-Moqadam et al. 2016).

The closest plane to the porous rock surface at which flow occurs is termed the shear plane or the slipping plane, and the electrical potential at this plane is called the zeta potential (Thanh and Sprik, 2015). The zeta potential depends upon several fluid and rock properties (e.g., electrolyte concentrations; mineralogy) and influences the degree of coupling between the electric flow and the fluid flow in porous media. The zeta potential can be determined by measuring the streaming potential.

Sadeqi-Moqadam et al. (2016) performed experiments on sand packs showing a spectrum of wetting conditions from completely water-wet to completely oil-wet and determined the zeta potential and streaming potential coupling coefficient. Their results showed good correlation between these electrokinetic parameters and wettability states. They also developed a bundle-of-tubes model to simulate and quantify wetting conditions in terms of these parameters by matching their experimental results. This approach offers a potentially useful method for monitoring wettability changes during dynamic reservoir water injection programs, making it possible to delineate progress and identify areas of formation damage (beneficial and detrimental) in LSWF projects.

2.9 WETTING MECHANISMS IN CARBONATES AND SANDSTONES

The distribution of oil and water in the porous system is linked to the wetting properties of the crude oil-formation water-rock matrix interface system, i.e., the contact between the rock surface and the two fluids, crude oil and formation fluid (a complex brine with varying salinity and ionic components). Referring to formation water as brine suggests to simplistic a composition for this key, but highly variable, component of the system.

The wetting properties of this system dictate, to an extent, how the two fluid phases in the system with flow-phase fluid flow the porous network of an oil reservoir, by influencing the capillary pressure, Pc, and the relative permeabilities of oil and water, kro, and krw, which in turn influence oil recovery (Jadhunandan and Morrow, 1995); systems that are slightly water wet tend to show the best incremental oil recovery through water flooding.

Austad (2013) makes the important distinction between water flooding as a secondary and tertiary recovery technique: injection of its own formation water into an oil reservoir is a secondary recovery process; injection of water with a different composition to the initial formation water may change wetting properties of the reservoir (for better or worse from an oil recovery process) and is therefore a tertiary recovery process. Wetting characteristics of carbonates and sandstones are quite distinct.

Oil recovery from carbonates is typically less than 30% due to low water wetness, the presence of natural fractures, low permeability, and inhomogeneous rock properties. This is due to the strong bond between the negatively charged carboxylic group, $-COO-$ (present in heavy components of crude oil) and the positively charged sites on carbonate surface (Fathi et al., 2011; Austad, 2013). Higher temperature carbonate reservoirs are more likely to be water-wet (Rao, 1996) as temperature tends to lower the acid number of the oil and lower its negatively charged components. Sulfate is the most effective ion in terms of its ability to alter the wetting properties in carbonates (Shariatpanahi et al., 2011); its presence increases the water-wetness of the system. Austad (2013) suggests that it is the combined effect of the presence of SO_4^{2-} ions and a decrease in the NaCl concentration of the injected fluid, which is important in altering wettability in carbonates and making LSWF effective. This combined effect is also temperature sensitive with 90°C to 110°C proposed as the optimum range, because although high temperatures are beneficial, they also lead to a reduction in sulfate ion concentrations in the formation fluids, due to sulfate mineral precipitation.

In contrast to carbonates, the surface of silicate minerals, particularly the clays, is negatively charged. The permanent negative charge of clay minerals enables them to act as cation exchangers with a variable affinity to exchange with specific cations (Austad, 2013); that affinity increases in the following order $Li^+ < Na^+ < K^+ < Mg^{2+} < Ca^{2+} < H^+$. As the cations adsorb onto the clay, the clay surfaces become more oil-wet, a process that is sensitive to the pH of the formation water (Madsen and Lind, 1998). As clays are normally not uniformly distributed throughout an oil reservoir, the clay-rich local areas may be less water-wet than the clay-poor zones.

Tang and Morrow (1999), in experiments on the Berea sandstone, established that adsorption from crude oil, the presence of potentially mobile fines, and initial water saturation were all requirements for an increase in oil recovery to result from a decrease in salinity of the injected water. Lager et al. (2008) concluded that pH induced IFT reduction or emulsification and fines migration were effects rather than the driving mechanisms of LSWF, which, based on their experiments with sandstone cores, they concluded was primarily cation exchange (i.e., MIE). However, they noted for LSWF to be effective, the system also needs to involve oil with polar components (i.e., acids and/or bases), the indigenous formation water needed to have meaningful concentrations of divalent cations, i.e., Ca^{2+}, Mg^{2+}, and salinity of the injection fluid should be between 1000 ppm and 2000 ppm, but could work with salinities up to about 5000 ppm. The effective salinity level of the LSW seems to depend on the divalent (mainly Ca^{2+}, Mg^{2+}) versus monovalent (Na^+) ionic composition of the injection fluid, consistent with the findings of Yildiz and Morrow (1996). In a MIE mechanism, it is the Ca^{2+} and Mg^{2+} ions which are adsorbed from the LSW by the rock matrix until it is fully saturated in those ions which lead to the incremental oil recovery benefits (Lager et al., 2008). If the rock matrix is already saturated in Ca^{2+} and Mg^{2+} ions before LSWF begins, the MIE mechanism will not be effective.

Morrow et al. (1994) reported on spontaneous imbibition experiments to characterize wettability effects of fluids in porous media. Buckley and Liu (1998) distinguished four mechanisms by which crude oil components may adsorb on high-energy mineral surfaces within a porous rock, and the effect of different oil compositions in this regard. The four mechanisms are:

- **Polar interactions**: These prevail in oil-wet systems where there is no water film between oil and the bulk of the mineral grains
- **Surface precipitation**: If the oil is a poor solvent for the asphaltenes (e.g., high API/low density), its wetting alteration capabilities tend to be enhanced
- **Acid/base interactions**: In the presence of water, both the solid and oil interfaces become charged, either behaving as acids (giving up a proton and becoming negatively charged) and bases (gaining a proton and a positive charge). For monovalent salt solutions at low concentrations, pH of the fluid impacts these interactions
- **Ion-binding**: Divalent or multivalent ions can at both oil and solid water interfaces or bridge between them, with calcium ions playing a key role in wettability impacts

Buckley and Liu (1998) also identified three different properties of crude oils that in combination influenced their ability to alter the wettability of a reservoir: oil gravity (API), acid number, and base number. This highlights the role which crude oil composition plays in the effectiveness of LSWF in a particular reservoir.

2.10 EXAMPLE FIELD FIELD-SCALE TESTS AND OUTCOMES OF LSWF

In contrast to core water-flood experiments, relatively little multi-well, multiyear, field-wide production data specifically focused on evaluating the effects of LSWF are published. Robertson (2007) evaluated water-flood data from public records for several oil fields from the Powder River basin of Wyoming (USA) which were flooded with LSW (salinity ~ 1000 ppm), the injection water coming from the Madison limestone and Fox Hills sandstone. Three Minnelusa oil fields (Moran, North Semlek, and West Semlek) provided sufficient variation in formation and injection water salinities to compare the LSWF outcomes in terms of salinity of the formation and injection water and the reservoir characteristics. The analysis of production performance of those fields (Fig. 2.16) indicated that oil recovery with LSWF flooding was greater than for other fields in the same reservoir using higher-salinity injection water. Also, field recovery trends for the fields matched the predictions of core experiments and single-well tests.

Vledder et al. (2010) reported the results of a 10-year field-wide LSWF secondary recovery project for the Omar Field (Syria), which were supported by experiments on sandstone cores from the reservoir. The project resulted in $10-15\%$ of incremental oil recovery. A change in wettability from an oil-wet system to a water-wet system over the field production life, results in a step change in water-cut consistent with the change in wettability.

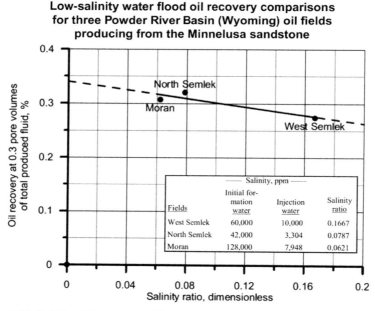

Figure 2.16 Relationship between oil recovery and salinity ratio for Minnelusa sandstone LSWF in three oil fields in the Powder River Basin, Wyoming (USA). *Modified after Robertson (2007).*

A LSWF field trial, in tertiary recovery mode, was planned (based on earlier core-flood test results) and conducted between 2007 and 2009 in the Endicott oil field (Alaska) in a single reservoir zone where the water cut was stabilized at 95% following saline water injection (Seccombe et al., 2010, Lager et al., 2011). This trial involved a single injector-producer well pair 1040 ft apart, with the producer monitored for changes in water cut and ionic compositional changes. LSW injection began in June 2008 and was detected in the producer well 3 months later, and that LSW breakthrough coincided with water cut dropping from 95% to 92%. Over the course of about 1 year, 1.3 pore volumes of LSW were injected in the field area leading to incremental oil recovery of 10% for the pore volume swept by the trial. The tertiary mode LSWF was expected to reduce the residual oil saturation from 41% to 28% with no adverse production outcomes.

Not all core-flooding experiments and field tests have shown LSWF to be of benefit. Skrettingland et al. (2011) conducted core-flooding experiments and a single-well chemical tracer pilot study involving three reservoirs (upper and lower Statfjord and Lunde formations) in the Snorre field, offshore Norway. The Statfjord formations demonstrated incremental oil recovery from the core experiments of only about 2% for seawater flooding and LSWF; whereas, the Lunde formation showed no improved recovery. The field pilot test was carried out on the Upper Statfjord formation, but recorded no change in oil saturation. It was concluded that the initial wetting conditions of these reservoirs were probably unsuitable for LSWF.

On the back of successful LSWF test outcomes at Endicott oil field and other fields, BP in 2011 decided to apply its own proprietary brand of the technique to some large-scale field developments (e.g., Mad Dog Phase 2 and Clair Ridge) as early stage secondary-recovery mode projects. In 2016 BP, ConocoPhillips, Chevron, and Shell initiated production at about 120,000 bopd at the largest offshore LSWF project, to date, in Clair Ridge oil field (offshore West of Shetland, UK). Clair Ridge is expected to produce some 42 million barrels of additional oil, contributing to a total of 640 million barrels of expected total oil recovery (from about 8 billion barrels of oil in-place), at a relatively low cost (BP, 2015). The £4.5- billion development includes around USD$120 million for the desalination facilities to supply LSW for "waterflooding" from sea water.

2.11 POTENTIAL TO COMBINE LSWF WITH OTHER IOR MECHANISMS

There is significant potential to combine LSWF with other IOR technologies at it works to complement and enhance several other EOR methods in the form of hybrid processes. However, formation damage mechanisms are also more complex in terms of potential causes and effects.

2.11.1 LSWF combined with surfactants

Skauge et al. (2011) reported the results for core-flood experiments for the Berea sandstone and simulation models involving LSW combined with surfactant flooding. The concept behind this combination of IOR methods (Alagic and Skauge,2010) is that some of the discontinuous oil that is potentially mobilized by LSWF is likely to be retrapped in the reservoir at the given capillary pressure. By also injecting surfactant, the capillary pressure in the swept reservoir should be further reduced, and retrapping of mobilized oil should be reduced.

Wettability alteration was assumed to be the main LSWF mechanism at work in the developed simulation models, although the authors recognized that other factors were also probably influencing reservoir behavior. Nevertheless, the simplified models built on two different simulation software packages developed salinity dependent oil and water capillary pressure and relative permeability curves, which successfully matched oil recovery and differential pressure from the core-flood experiments with salinity change. Skauge et al. (2011) concluded that surfactant flooding at low salinity produced better than expected results in terms of reduction of IFT and capillary number.

2.11.2 LSWF combined with polymers

Shiran and Skauge (2013) evaluated the synergistic effect on residual oil mobilization and final oil recovery of combining LSWF with polymer injection and, based on Berea sandstone core experiments, whether this worked better in secondary or tertiary recovery modes. The concept is to enhance LSWF with the benefits of polymer flooding that involves the addition of large molecular weight, water-soluble polymers, and linked polymer solutions to injection water to increase the viscosity of displacing fluid and improve the adverse mobility ratio. The wettability of the reservoir is known to influence the effectiveness of polymer injection (Chiappa et al., 1999). This approach is expected to reduce the viscous fingering effect of water flow within the reservoir and achieve better sweep efficiency from the displacement process.

The results showed, consistent with other studies mentioned, that LSWF is more efficient if commenced in secondary recovery mode. Synergistic benefits were recorded by combining LSWF with polymer (and nanosized polymer) injection. These were most effective when the process commenced at initial reservoir oil saturation, achieving more than 50% reduction in residual oil saturation after the secondary recovery mode waterflood, even using low (300 ppm) concentrations of polymer. Shiran and Skauge (2013) attributed the beneficial effects to improved banking of low-salinity mobilized oil with only a slight change in mobility ratio.

2.11.3 LSWF combined with CO_2 water-alternating gas injection

Through experimental core-flooding studies of wettability impacts during conventional high-salinity water flooding (formation water followed by sea water) followed by CO_2 flooding on North Sea reservoirs at temperatures ranging from $50°C$ to $130°C$, Fjelde and Asen (2010) have demonstrated alterations to more water-wet conditions. After the third-phase water-alternating gas (WAG) cycle residual oil saturation were reduced to between 3% and 5%, demonstrating the effectiveness of the reservoir sweep achieved by this combination. Zolfaghari et al. (2013) conducted core-flood experiments. Based on a series of core-flood experiments combining LSW with CO_2 in a WAG injection system, they reported an incremental oil recovery of up to 18% OOIP.

Dang et al. (2016) evaluated the merits of combining LSW with CO_2 injection in a hybrid CO_2-LSWAG process using 1D and full-field simulation models combining ion exchange and reservoir geochemistry. A secondary-mode LSW followed by tertiary-mode CO_2-LSWAG model outperformed high-salinity WAG, standalone LSWF, and continuous CO_2 flooding models. Sensitivity cases for the models were used to identify the effects of solubility of CO_2 in various injected-water salinities, dissolution of carbonate minerals, ion exchange, wettability alteration, and clay distribution. CO_2-LSWAG was simulated at full-field scale for the North Sea Brugge oil field sandstone reservoir (Peters et al., 2009) yielding between 4.5% and 9% of incremental OOIP compared to a high–salinity WAG process.

Field–scale simulation indicated that the WAG ratio has a large effect on the ultimate oil recovery with a WAG ratio of 1:2 resulting in the highest oil recovery for the Brugge oil field case. The longer the CO_2-LSWAG cycling is applied, the greater the benefit in terms of oil recovery. Also, the shorter the water injection period involved in each WAG cycle, the greater the ultimate oil recovery. Of course, these latter two recovery benefits come at the cost of additional CO_2 required for injection. It is important to design the WAG parameters carefully (Batruny and Babadagli, 2015), on a field-by-field basis taking into account historical water injection, if appropriate.

The success of CO_2-LSWAG is also reported by Dang et al. (2016) who identify a number of factors that are likely to have variable impacts for each oil field and reservoir. These factors require careful evaluation and analysis with pilot well tests and field trials in addition to simulations. These factors include:

- type and quantity of clay minerals
- initial reservoir wettability condition
- reservoir heterogeneity
- nonsilicate reservoir mineralogy (e.g., calcite and dolomite)
- composition of formation water and injected brine
- reservoir pressure and temperature for achieving CO_2 miscible condition, and
- WAG parameters

2.11.4 LSWF combined with nanofluid treatments

On the one hand, the problem of fines migration induced by low–salinity water can improve mobility control as already described. However, the straining effects of fines also bring significant damage to formation permeability and the subsequent increase of injection pressure. This places strain on the management of surface facilities with loss of economic profits.

In the radial flow system, the majority of pressure loss is attributed to the tremendously large flow in close vicinity to the wellbores. Because of the differences in flowing velocities through each layer, the distributions of maximum retention concentration of fines are also not identical for each layer. Yuan and Moghanloo, (2017b, 2018c) introduced different slug sizes of nanofluids to retain fines in the near–wellbore zones, i.e., 0.10 and 0.20 of the flowing system for each layer, respectively, as indicated in Fig. 2.12.

After nanofluid–slug treatments are injected into each layer, the movement of injected fluids becomes accelerated. However, the increase of injection pressure loss could be adequately controlled due to the prevention of fines migration/detachment near the wellbores. In addition, despite the changes of the moving velocity of the flood front through each layer, the ratio of front-location along each layer, R (Eq. (2.14)), shows no significant changes after nanofluid treatment. This suggests that nanofluid pretreatment for the near–wellbore region does not negatively affect the improvement of sweep efficiency attributed to fines migration for layered heterogeneous reservoirs. Indeed, it can help mitigate the increase of injection pressure loss and thereby maintain well injectivity. The only potentially negative consequence of nanofluid treatment is that it may advance the breakthrough of injected water by accelerating the movement of injected water within the nanofluid–treated region (Fig. 2.17).

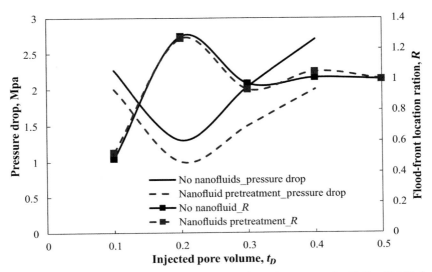

Figure 2.17 Injection pressure drop and flood-front location ratio (R, Eq. (2.14)) for cases with and without nanofluid pretreatment.

2.12 CONCLUSIONS

- Many experimental studies and several field tests have demonstrated the benefits of the LSWF method in terms of increased oil recovery factors and lower water cuts. However, the technique does not work in all sandstone reservoirs and requires quite specific conditions to work in carbonate reservoirs.
- Considering a wide range of recent studies, tests, and trials, it is apparent that there are multiple factors and driving mechanisms at play in LSWF. These include electrical DLE, MEI, the presence of calcium and magnesium ions in the injected water, and multiwet fines migration. The net effect of the combined mechanisms is a desorption of petroleum heavy ends from the clays present on the pore wall, resulting in a more water-wet rock surface, a lower remaining oil saturation, and higher oil recovery.
- Some of the impact of the multiple mechanisms at work in LSWF lead to both positive and negative formation damage outcomes (e.g., fines migration).
- The induced fines migration by low-salinity water delays water breakthrough, improves sweep efficient by fines migration assisted mobility control, and leads to uniform allocation of injected fluid entering each layer. However, fines migration can also lead to troublesome formation damage resulting in a reduction in permeability and the decline of well injectivity.
- Established correlations between residual oil saturation and the wettability-related attributes, including contact angle, IFT, and capillary pressure as functions of salinity help to improve the accuracy of reservoir simulation predictions associated with LSWF. Recently developed electrokinetic techniques offer potential to monitor the dynamic changes in wettability as LSWF progress through streaming potential and zeta potential measurements.
- As uncertainty exists concerning the main mechanisms at work in specific reservoirs subjected to LSWF, there is the risk of it being ineffective or less effective than expected. Consequently, the predictions of core-flooding experiments and simulation analysis may be inaccurate, due to incomplete or incorrect assumptions, meaning that pilot well testing and/or field trials are typically required to confirm its performance.

- Core experiments and simulations have demonstrated that synergies can be achieved by combining LSWF with several other IOR techniques.
- Nanofluid treatment prior to LSWF can help control fines migration adjacent to injection wells, and significantly maintain well injectivity. Although nanofluid treatment has the capacity to advance the break-through of injected water, it has minimal negative effects on the mobility control assisted by fines migration in layered heterogeneous reservoirs.

NOMENCLATURE

$C_{mi,i}$	Molar i^{th} ion concentration in water at initial conditions, moles/m^3
$C_{mi,j}$	Molar i^{th} ion concentration in injected water, moles/m^3
f_w	Fractional flow function
S_w	Water saturation, decimal
S_{wc}	Connate water saturation, decimal
Z_i	Valence of ith ion
n_∞	Bulk number density of ions, 6.022×10^{25} number/m^3
κ	Inverse Debye-length, m^{-1}
k_B	The Boltzmann's constant, 1.381×10^{-21} J/K
l	Levers (l_d for drag forces; l_n for normal forces)
r_{FP}	Size of fine particles, m
$\varsigma_{FP}, \varsigma_{NP}, \varsigma_{GS}$	Zeta potentials for fine particles, nanoparticles, and grain surfaces, mV
k_{rw}	Relative permeability of water phase
k_{ro}	Relative permeability of oil phase
k_0	Initial undamaged permeability
$k(\sigma_s)$	Damaged formation permeability
h	The surface-to-surface separation length, m
n_∞	Bulk number density of ions, 6.022×10^{25} number/m^3
γ	The ratio between the drag and electrostatic force
ϕ	Porosity of sand pack
ϕ_c	The percentage of internal cake thickness, about 0.10
μ	Viscosity of fluid
μ_w	Water viscosity, Pa.s
μ_o	Oil viscosity, Pa.s
ε	Ratio between detaching and attaching forces
F	Forces (drag, lift, gravity, electrostatic)
F_e	Electrostatic forces, N
β	Formation damage coefficient reflecting the degree of formation damage due to pore plugging

β_s	Formation damage coefficient for straining
q	Injection rate per formation height, m^2/s
T	Absolute temperature of reservoir, K
p	The flowing pressure at different locations, Pa
Δp	Pressure drop, MPa
$\Delta p_{1,2}$	Pressure drop along different layers 1 and 2, MPa
λ_t	Total flow mobility (both water- and oil-phase)
$\lambda_{t1,2}$	Total flow mobility within different layers 1 and 2
r	Radial location of flowing system, m
r_p	Pore radius, m
r_e	Outer radius of radial flowing system, m
r_{NP}	Radius of nanoparticles, m
S_{or}	Oil saturation
N_{ca}	Capillary number
θ	Contact angle at the crude oil-formation water-porous rock interface
σ	Interfacial tension (IFT) between oil and water
σ_{ow}	Interfacial tension (IFT) between oil and formation water
σ_{om}	Interfacial energy between oil and porous rock matrix
σ_{wm}	Interfacial energy between formation water and porous rock matrix
$\sigma_{cr,initial}(0, S_{wc})$	Critical retention concentration of fines at initial condition with connate water saturation Swc, m^3/m^3
σ_a	Maximum retention concentration of particles at specific flow and salinity conditions
σ_{cr}	Maximum volumetric concentration of captured particles
σ_{cr}	Maximum retention concentration of fine particles, m^3/m^3
σ_s	Strained fine particles concentration, m^3/m^3
$\Delta\sigma$	Concentration of fine particles released
xf_D	Locations of water-saturation front
t_D	Dimensionless time or injected pore volume
γ	Salt concentration; γ_i is formation fluid salinity; γ_o injected fluid salinity
v	interstitial velocity of the displacing fluid
U	Darcy velocity
x_D	Dimensionless distance

REFERENCES

Aksulu, H., Hamso, D., Strand, S., et al., 2012. The evaluation of low salinity enhanced oil recovery effects in sandstone: effects of temperature and pH gradient. Energy Fuels 26, 3497—3503.

Al-adasani, A., Bai, B., 2012. Investigating low-salinity water flooding recovery mechanism(s) in carbonate reservoirs. In: Proceedings of the 2012 SPE EOR Conference at Oil and Gas West Asia, Muscat, Oman, 16—18 April, SPE155560.

Al-adasani, A., Bai, B., Wu, Y.-S., Salehi, S., 2014. Studying low-salinity waterflooding recovery effects in sandstone reservoirs. J. Pet. Sci. Eng. 120, 39—51.

Alagic, E., Skauge, A., 2010. Combined Low Salinity Brine Injection and Surfactant Flooding in Mixed — Wet Sandstone Cores. Energy Fuels 2010. Available from: https://doi.org/10.1021/ef1000908.

Al Shalabi, E.W., Sepehrnoori, K., Delshad, M., 2013. Mechanisms behind low salinity water flooding in carbonate reservoirs. SPE-165339-MS. SPE Western Regional &

AAPG Pacific Section Meeting 2013 Joint Technical Conference, 19-25 April, Monterey, California, USA.

Amorim, C.L.G., Lopes, R.T., Barroso, R.C., Queiroz, J.C., Alves, D.B., Perez, C.A., et al., 2007. Effect of clay—water interactions on clay swelling by X-ray diffraction. Nucl. Instrum. Methods Phys. Res. A 580, 768—770.

Ashraf, A., Hadia, N.J., Torsaeter, O., 2010. Laboratory investigation of low salinity waterflooding as secondary recovery process: effect of wettability. In: Proceedings of the SPE Oil and Gas India Conference and Exhibition, Mumbai, India, 20—22 January, SPE129012. Available from: https://doi.org/10.2118/129012-MS.

Austad, T., 2013. Water based EOR in carbonates and sandstones: new chemical understanding of the EOR potential using "Smart Water". (Chapter 13). In: Sheng, J. (Ed.), Enhanced Oil Recovery Field Case Studies. Elsevier, pp. 301—335.

Austad, T., Rezaeidoust, A., Puntervold, T., 2010. Chemical mechanism of low salinity waterflooding in sandstone reservoirs. Paper SPE 129767 presented at the SPE Improved Oil Recovery Symposium, 24-28 April, Tulsa, Oklahoma, USA.

Ayirala, S.C., Rao, D.N., 2004. Multiphase flow and wettability effects of surfactants in porous media. Colloids Surf A Physicochem Eng Asp. 241, 313—322.

Barati-Harooni, A., Soleymanzadeha, A., Tatar, A., Najafi-Marghmaleki, A., Samadi, S.-J., Yari, A., et al., 2016. Experimental and modeling studies on the effects of temperature, pressure and brine salinity on interfacial tension in live oil-brine systems. J. Mol. Liq. 219, 985—993.

Batruny, P., Babadagli, T., 2015. Effect of water flooding history on the efficiency of fully miscible tertiary solvent injection and optimal design of water-alternating-gas process. J. Pet. Sci. Eng. 130, 114—122.

Bedrikovetsky, P., Zeinijahromi, A., Siqueira, F., Furtado, C., deSouza, A., 2012. Particle detachment under velocity alternation during suspension. Transp. Porous Media 91 (1), 173—197.

Behruz, S.S., Skauge, A., 2013. Enhanced oil recovery (EOR) by combined low salinity water/polymer flooding. Energy Fuels 27 (3), 1223—1235.

Bernard, G.G. 1967. Effect of floodwater salinity on recovery of oil from cores containing clays, Paper SPE-1725-MS presented at SPE California Regional Meeting. Los Angeles, 26-27 October, Los Angeles, California.

Blume, T., Weisbrod, N., Selker, J.S., 2005. On the critical salt concentrations for particle detachment in homogeneous sand and heterogeneous Hanford sediments. Geoderma 124, 121—132.

BP, 2015. Low salinity water brings award for BP. http://www.bp.com/en/global/corporate/bp-magazine/innovations/offshore-technology-award-for-clair-ridge.html.

Buckley, J.S., Liu, 1998. Some mechanisms of crude oil/brine/solid interactions. J. Pet. Sci. Eng. 20, 155—160.

Buckley, S.E., Leverett, M.C., 1942. Mechanism of fluid displacements in sands. Trans AIME 146, 107—116.

Chiappa, L., Mennella, A., Lockhart, T.P., Burrafato, G., 1999. Polymer adsorption at the brine-rock interface: the role of electrostatic interactions and wettability. J. Pet. Sci. Eng. 24, 113—122.

Cieplak, M., Robbins, M.O., 1990. Influence of contact angle on the quasi-static fluid invasion of porous media. The American Physical Society. Phys. Rev. B. 41 (16), 11508—11521.

Collins, I.R., 2011. Holistic benefits of low salinity waterflooding, 16th European Symposium on Improved Oil Recovery, Cambridge, U.K., April. (extended abstract) DOI: 10.3997/2214-4609.201404797.

Dang, C., Nghiem, L., Nguyen, N., Chen, Z., Nguyen, Q., 2016. Evaluation of CO_2 low salinity water-alternating-gas for enhanced oil recovery. J. Nat. Gas. Sci. Eng. 35, 237−258.

Debye, P., Hückel, E., 1923. The theory of electrolytes. I. Lowering of freezing point and related phenomena. Phys Z. 24, 185−206.

Elimelech, M., Gregory, J., Jia, X., Williams, R.A., 1995. Particle Deposition & Aggregation: xiii-xv. Butterworth-Heinemann, Woburn.

Fathi, S.J., Austad, T., Strand, S., 2011. Effect of water-extractable carboxylic acids in crude oil on wettability in carbonates. Energy Fuels 25, 2587−2592.

Fjelde, I., Asen, S.M., 2010. Wettability alteration during water flooding and carbon dioxide flooding of reservoir chalk rocks. In: Paper SPE 130992, SPE EUROPEC/EAGE Annual Conference and Exhibition, Barcelona, Spain.

Gandomkar, A., Rahimpour, M.R., 2015. Investigation of low-salinity waterflooding in secondary and tertiary enhanced oil recovery in limestone reservoirs. Energy Fuels 29, 7781−7792.

Greathouse, J.A., Feller, S.E., McQuarrie, D.A., 1994. The Modified Gouy-Chapman Theory: Comparisons between Electrical Double Layer Models of Clay Swelling, 10. American Chemical Society., Langmuir, pp. 2125−2130 (7).

Hamouda, A.A., Gupta, S., 2017. Enhancing oil recovery from chalk reservoirs by a low-salinity water flooding mechanism and fluid/rock interactions. Energies 10 (576), 16.

Hourshad, M., Jerauld, G., 2012. Mechanistic modeling of the benefit of combining polymer with low salinity water for enhanced oil recovery. Paper SPE-153161-MS presented at SPE Improved Oil Recovery Symposium, 14−18 April, Tulsa, Oklahoma, USA.

Huang, T., Crews, J., Willingham, J.R., 2008. Using nanoparticles technology to control fine migration, Paper SPE-115384-MS presented at SPE Annual Technical Conference and Exhibition, 21−24 September, Denver, Colorado, USA.

Hussain, F., Zeinijahromi, A., Bedrikovetsky, P., Cinar, Y., Badalyan, A., Carageorgos, T., 2013. An experimental study of improved oil recovery through fines-assisted waterflooding. J. Pet. Sci. Eng. 109, 187−197.

Jackson, M.D., Vinogradov, J., 2012. Impact of wettability on laboratory measurements of streaming potential in carbonates. Colloids Surf. Physicochem. Eng. Asp. 393, 86−95.

Jadhunandan, P.P., Morrow, N.R., 1995. Effect of wettability on waterflood recovery for crude-oil/brine/rock systems. SPE. Reserv. Eng, 40−46.

Jerauld, G.R., Lin, C.Y., Webb, K.J., Seccombe, J.C., 2006. Modeling low-salinity waterflooding. In: Proceedings of the SPE Annual Technical Conference and Exhibition, San Antonio, TX, USA, 24−27 September, SPE102239. https://doi.org/10.2118/102239-MS.

Khilar, D.C., Fogler, H.S.J., 1984. The existence of a critical salt concentration for particle release. J. Colloid Interface Sci. 101, 214−224.

Khilar, K.C., Fogler, H.S., Ahluwalia, J.S., 1983. Sandstone water sensitivity: existence of a critical rate of salinity decrease for particle capture. Chem. Engr. Sci. 38 (5), 789−800.

Kia, S.F., Fogler, H.S., Reed, M.G., 1987. Effect of pH on colloidally induced fines migration. J. Colloid Interface Sci. 118 (1), 158−168.

Lager, A., Webb, K.J. and Seccombe, J.C., 2011. Low salinity waterflood, Endicott, Alaska: Geochemical study & field evidence of multicomponent ion exchange. 16th European Symposium on Improved Oil Recovery, 12−14 April.

Lager, Arnaud, Webb, K.J., Black, C.J.J., Singleton, M., Sorbie, K.S., 2008. Low salinity oil recovery − an experimental investigation. Petrophysics 49 (1), 28−35.

Lemon, P., 2011. Effects of injected-water salinity on waterflood sweep efficiency through induced fines migration. J. Can. Pet. Technol. 50, 9−10.

Madsen, L., Lind, I., 1998. Adsorption of carboxylic acids on reservoir minerals from organic and aqueous phase. SPE. Reserv. Eval. Eng, February 47−51.

McGuire, P.L.L., Chatham, J.R.R., Paskvan, F.K.K., Sommer, D.M.M., Carini, F.H.H., B P exploration, 2005. Low salinity oil recovery: an exciting new EOR opportunity for Alaska's North slope. In: SPE Western Regional Meeting, pp. 1e15. Available from: https://doi.org/10.2118/93903-MS.

Melrose, J.C., Brander, C.F., 1974. Role of capillary forces in determining microscopic displacement efficiency for oil recovery by waterflooding. J. Can. Pet. Technol. 13 (4), 54−61.

Mohan, K., Vaidya, R.N., Reed, M.G., Scott, F.H., 1993. Water sensitivity of sandstones containing swelling and non-swelling clays, Colloids Surf A Physicochem. Eng. Asp, 73. p. 237e254. Available from: http://dx.doi.org/10.1016/0927-7757(93)80019-B.

Morrow, N.R., Ma, S., Zhou, X., Zhang, X., 1994. Characterization of wettability from spontaneous imbibition measurements, Paper CIM 94-47 presented at the 1994 Petr. Soc. of CIM Ann. Tech. Meeting and AOSTRA 1994 Ann. Tech. Conf., Calgary, June 12−15.

Muecke, T.W., 1979. Formation fines and factors controlling their movement in porous media. J. Pet. Technol. 144−150.

Myint, P.C., Firoozabadi, A., 2015. Thin liquid films in improved oil recovery from low-salinity brine. Curr. Opin. Colloid Interface Sci. 20, 105−114.

Nasralla, R.A., Nasr-El-Din, H.A., 2012. Double-layer expansion: is it a primary mechanism of improved oil recovery by low-salinity waterflooding?, SPE-154334. In: SPE Improved Oil Recovery Symposium, Society of Petroleum Engineers, Tulsa, Oklahoma, USA, 2012.

Nasralla, R.A., Nasr-El-Din, H.A., 2014. Impact of cation type and concentration in injected brine on oil recovery in sandstone reservoirs. J. Pet. Sci. Eng. 122, 384−395.

Nguyen, T.K.P., Zeinijahromi, A., Bedrikovetsky, P., 2013. Fines-migration-assisted improved gas recovery during gas field depletion. J. Pet. Sci. Eng. 109, 26−37.

Ochi, J., Vernoux, J.F., 1998. Permeability decrease in sandstone reservoirs by fluid injection: hydrodynamic and chemical effects. J Hydrol 208 (1998), 237−248.

Peters, L., Arts, R., Brouwer, G., Geel, C., 2009. Results of the Brugge benchmark study for flooding optimisation and history matching. Paper SPE 119094 Presented at the SPE Reservoir Simulation Symposium, e4.

Pouryousefy, E., Xie, Q., Saeedi, A., 2016. Effect of multi-component ions exchange on low salinity EOR: coupled geochemical simulation study. Petroleum 2, 215−224.

Rao, D.N., 1996. Wettability effects in thermal recovery operations. SPE/DOE Improved Oil Recovery Symposium, 21−24 April.

Robertson, E.P., 2007. Low-salinity waterflooding to improve oil recovery - Historical field evidence. Paper SPE 109965 presented at the 2007 SPE Annual Technical Conference and Exhibition, 11−14 Nov., 9 pages.

Sadeqi-Moqadam, M., Riahi, S., Bahramian, A., 2016. Monitoring wettability alteration of porous media by streaming potential measurements: Experimental and modelling investigation. Colloids Surf A: Physicochem. Eng. Asp. 497, 182−193.

Salathiel, R.A., 1973. Oil recovery by surface film drainage in mixed-wettability rocks. JPT. 1216−1224.

Salehi, M.M., Omidvar, P., Naeimi, F., 2017. Salinity of injection water and its impact on oil recovery absolute permeability, residual oil saturation, interfacial tension and capillary pressure. Egy. J. Pet. 26, 301−312.

Sanaei, A., Shakiba, M., Varavei, A., Sepehrnoori, K., 2016. Modeling clay swelling induced conductivity damage in hydraulic fractures. SPE-180211. SPE Low Perm Symposium, 5−6 May, Denver, Colorado, USA. 14 pages.

Seccombe, J., Lager, A., Jerauld, G., Jhaveri, B., Buikema, T., Bassler, S., et al., 2010. Demonstration of low-salinity EOR at inter well scale, Endicott Field, Alaska. In: Proceedings of the SPE Improved Oil Recovery Symposium, Tulsa, OK, USA, 24–28 April, SPE129692. 12 pages ISBN:978-1-55563-289.

Shariatpanahi, S.F., Strand, S., Austad, T., 2011. Initial wetting properties of carbonate oil reservoirs: effect of the temperature and presence of sulfate in formation water. Energy Fuels 25 (7), 3021–3028.

Shiran, B.S., Skauge, A., 2013. Enhanced oil recovery (EOR) by combined low salinity water/polymer flooding. Energy Fuels 27 (3), 1223–1235.

Skauge, A., Ghorbani, Z., Delshad, M., 2011. Simulation of combined low salinity brine and surfactant flooding. 16th European Symposium on Improved Oil Recovery Cambridge, UK, 12–14 April 2011, 13 pages.

Skrettingland, K., Holt, T., Tweheyo, M.T., Skjevrak, I., 2011. Snorre low salinity water injection - Core flooding experiments and single well field pilot. Paper SPE129877 presented at the 2010 SPE Improved oil recovery symposium 22–26 April.

Song, W., Kovscek, A., 2016. Direct visualization of pore-scale fines migration and formation damage during low-salinity waterflooding. J. Nat. Gas. Sci. Eng. 34, 1276–1283.

Tang, G., Morrow, N.R., 1999. Influence of brine composition and fines migration on crude oil/brine/rock interactions and oil recovery. J. Pet. Sci. Eng. 24 (2), 99–111. Available from: http://dx.doi.org/10.1016/S0920-4105(99)00034-0.

Thanh, L.D., Sprik, R., 2015. Zeta potential measurement using streaming potential in porous media. VNU J. Sci: Mat–Phys 31 (4), 56–65.

Vledder, P., Fonseca, J.C., Wells, T., Gonzalez, I. and Ligthelm, D., 2010. Low salinity water flooding: Proof of wettability alteration on a field wide scale. Paper SPE 129564 presented at the 2010 SPE Improved Oil Recovery Symposium, 24–28 April. 10 pages.

Webb, K.J., Black, C.J.J., Al-Ajeel, H., 2004. Low salinity oil recovery—log-inject-log. In: Proceedings of the SPE/DOE Symposium on Improved Oil Recovery, Tulsa, OK, USA, 17–21 April, SPE89379-MS. Available from: https://doi.org/10.2118/89379-ms.

Welge, H.J., 1952. A Simplified method for computing oil recovery by gas or water drive. J. Pet. Technol. 4 (04), 91–98.

Wu, Y., Bai, B., 2009. Efficient simulation of low-salinity waterflooding in porous and fractured reservoirs. In: Proceedings of the 2009 SPE Reservoir Simulation Symposium, The Woodlands, TX, USA, 2–4 February, SPE118830. Available from: https://doi.org/10.2118/118830-ms.

Xie, Q., Liu, Y., Wu, J., Liu, Q., 2014. Ions tuning water flooding experiments and interpretation by thermodynamics of wettability. J. Pet. Sci. Eng. 124, 350–358.

Yildiz, H.O., Morrow, N.R., 1996. Effect of brine composition on recovery waterflooding of Moutray crude oil by. J. Pet. Sci. Eng. 14, 159–168.

Yousef, A.A., Al-Saleh, S., Al-kaabi, A., Al-Jawfi, M., 2010. Laboratory investigation of novel oil recovery method for carbonate reservoirs. In: Proceedings of the Canadian Unconventional Reservoirs & International Petroleum Conference, Calgary, Alberta, Canada, 19–21 October, CSUG/SPE137634. Available from: https://doi.org/10.2118/137634-PA.

Yousef, A.A.; Al-Saleh, S.; Al-Jawfi, M.S. 2012. The impact of the injection water chemistry on oil recovery from carbonate reservoirs. In Proceedings of the SPE EOR Conference at Oil and Gas West Asia, Muscat, Oman, 16–18 April 2012.

Yuan, B., 2017a. Modeling Nanofluids Utilization to Control Fines Migration. The University of Oklahoma, Norman, Oklahoma.

Yuan, B., Moghanloo, 2017b. Analytical modeling improved well performance by nanofluid pre-flush. Fuel 202, 380–394.

Yuan, B., Moghanloo, R.G., 2016. Analytical solution of nanoparticles utilization to reduce fines migration in porous medium. SPE. J. 21 (06), 2317–2332.

Yuan, B., Moghanloo, R.G., 2018a. Nanofluid treatment, an effective approach to improve performance of low salinity water flooding. J. Petrol. Sci. Eng. in press. Available from: https://doi.org/10.1016/j.petrol.2017.11.032.

Yuan, B., Moghanloo, R.G., Wang, W., 2018b. Using nanofluids to control fines migration for oil recovery: nanofluids co-injection or nanofluids pre-flush? A comprehensive answer. Fuel 215, 474−483.

Yuan, B., Moghanloo, R.G., 2018c. Nanofluid pre-coating, an effective method to reduce fines migration in radial systems saturated with mobile immiscible fluids. SPE. J. Available from: https://doi.org/10.2118/189464-PASPE-189464-PA.

Zeinijahromi, A., Lemon, P., Bedrikovetsky, P., 2011. Effects of induced fines migration on water cut during waterflooding. J. Pet. Sci. Eng. 78 (3−4), 609−617.

Zeinijahromi, A., Nguyen, T.K.P., Bedrikovetsky, P., 2013. Mathematical model for fines-migration-assisted waterflooding with induced formation damage. SPE. J. 18 (03), 518−533.

Zeinijahromi, A., Al-Jassasi, H., Begg, S., Bedrikovetsky, P., 2015a. Improving sweep efficiency of edge-water drive reservoirs using induced formation damage. J. Pet. Sci. Eng. 130, 123−129.

Zeinijahromi, A., Ahmetgareev, V., Ibatullin, R., Bedrikovetsky, P., 2015b. Sensitivity study of low salinity water injection in Zichebashskoe Oilfield. J. Pet. Gas. Eng. 6, 10−21.

Zhang, P., Tweheyo, M.T., Austad, T., 2006. Wettability alteration and improved oil recovery in chalk: the effect of calcium in the presence of sulfate. Energy Fuels 20, 2056−2062.

Zhang, Y., Xie, X., Morrow, N.R., 2007. Waterflood performance by injection of brine with different salinity for reservoir cores. In: Proceedings of the SPE Annual Technical Conference and Exhibition, Anaheim, CA, USA, 11−14 November, SPE109849. Available from: https://doi.org/10.2118/109849-ms.

Zhou, Z., Gunter, W.D., Kadatz, B., Cameron, S., 1996. Effect of clay swelling on reservoir quality. J. Can. Pet. Technol. 35 (07), 18−23.

Zolfaghari, H., Zebarjadi, A., Shahrokhi, O., Ghazanfari, M.H., 2013. An experimental study of CO_2 low salinity WAG injection in sandstone heavy oil reservoirs. Iranian. J. Oil Gas Sci. Technol. 2 (3), 37−47.

CHAPTER THREE

Formation Damage by Fines Migration: Mathematical and Laboratory Modeling, Field Cases

Thomas Russell, Larissa Chequer, Sara Borazjani, Zhenjiang You, Abbas Zeinijahromi and Pavel Bedrikovetsky
The University of Adelaide, Adelaide, SA, Australia

Contents

Formation Damage during Improved Oil Recovery.
DOI: https://doi.org/10.1016/B978-0-12-813782-6.00003-8

© 2018 Elsevier Inc.
All rights reserved.

3.1 INTRODUCTION

Migration of natural reservoir fines is one of the main formation damage mechanisms during waterflooding and enhanced oil recovery (EOR) operations (Civan, 2014). Formation damage is induced by mobilization, migration, and straining of particles which are initially attached to the rock surface. Fig. 3.1 shows fine particles attached to grains, detached particles, and those strained in thin pores.

The most common fines in subterranean reservoirs are clays (e.g., kaolinite, illite, chlorite). Initially, fine particles coat the rock surface, so their detachment causes a small increase of permeability. However, straining of particles in thin pores reduces the number of available flow paths in the porous media, causing significant permeability decline. Fig. 3.2 shows a Scanning Electron Microscope (SEM) image of kaolinite leaflets on a grain surface. The platelet shape of the kaolinite particles means that although detachment of kaolinite will have a small effect on the porosity, detached particles can strain in even large pore throats (Al-Yaseri et al., 2017; 2016). Sarkar and Sharma 1990 report typical permeability decrease for fines migration during single-phase flow of $100-1000$ times, although the presence of residual oil decreases this number by $10-100$.

Fig. 3.3 shows the primary forces acting on an attached particle. The drag and lifting force will act to detach the particle, while the electrostatic and gravitational force will act against particle mobilisation. It is commonly assumed that at the moment of detachment, the particle rotates around a neighboring particle or an asperity on the grain surface, so the

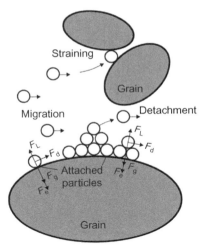

Figure 3.1 Schematic representation of particle detachment from grains, migration in carrier water, and straining in thin pores.

Figure 3.2 SEM photograph of kaolinite leaflets attached to the grain surface.

torque balance is the condition of the particle mechanical equilibrium (Bradford et al., 2013; Das et al., 1994; Schechter, 1992):

$$F_d(U, r_s)l_d(r_s) = \left[F_e(r_s, \gamma, pH, T) - F_L(U, r_s) + F_g(r_s)\right]l_n, \qquad (3.1)$$

where F_d, F_e, F_L, and F_g are drag, electrostatic, lifting, and gravitational forces, γ is the ionic strength of water, further called salinity, T is the temperature, r_s is the particle radius, and l_d and l_n are the tangential and normal lever arms, respectively.

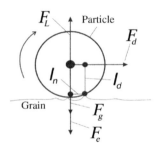

Figure 3.3 Torque balance of electrostatic, gravity, lift, and drag forces (F_e, F_g, F_L, and F_d) and lever arms l_n and l_d for attached particles.

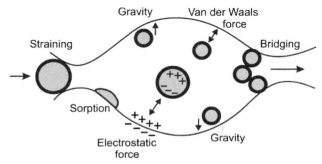

Figure 3.4 Fine-particle-capture mechanisms in a single pore.

Fig. 3.4 shows different mechanisms of particle capture by porous rock: size exclusion in thin pores, bridging, electrostatic attraction, gravity segregation, and diffusion into dead-end pores and stagnant flow zones.

The traditional colloidal flow model assumes simultaneous attachment and detachment of colloidal particles in porous media (Bradford et al., 2003; Bradford and Torkzaban, 2008; Sharma and Yortsos, 1987; Tufenkji, 2007):

$$\frac{\partial \sigma_a}{\partial t} = \lambda c U - k_{\text{det}} \sigma_a, \tag{3.2}$$

where the attachment rate is proportional to the advective flux of suspended particles cU, and the detachment rate is proportional to the attached concentration. Here, c and σ_a are the suspended and attached concentrations, U is the Darcy's velocity, λ is the filtration coefficient, which is the particle capture probability per unitary length of its trajectory, and k_{det} is the kinetic detachment coefficient, which is the reciprocal of the reference detachment time.

Kinetic Eq. (3.2) is analogous to the equation for nonequilibrium adsorption, where adsorption and desorption occur simultaneously. Equilibrium sorption corresponds to an equality of the adsorbed and desorbed rates. The rate expression for nonequilibrium adsorption reflects interface mass transfer that is proportional to the difference between Gibbs potentials of suspended and adsorbed matter. However, the work of the dissipative drag force F_d that is present in the condition of mechanical equilibrium Eq. (3.1) cannot be included into an energy potential, resulting in velocity-dependent attached concentration.

Kinetics Eq. (3.2) does not reflect the mechanical equilibrium condition expressed by Eq. (3.1). The filtration function can be derived by averaging of microscale flow in a pore, although the detachment coefficient is purely phenomenological and cannot be derived from microscopic physics.

On the contrary to the abovementioned shortcomings of Eq. (3.2), the modified particle detachment model follows from the torque balance condition of mechanical equilibrium Eq. (3.1). Consider an ensemble of particles on the rock surface, and apply the flow with given velocity U, ionic strength γ, pH, temperature T, etc. For each particle, Eq. (3.1) makes it possible to define whether the particle is attached to the rock or lifted by the detaching torques. The particles submitted to detaching torques, which exceed the attaching torque defined by the maximum electrostatic attraction, are mobilized. It makes the attached concentration of remaining fines a function of velocity U, ionic strength, pH, temperature T, etc., which is called the maximum retention function (Bedrikovetsky et al., 2011a; 2012; You et al., 2016; Yuan et al., 2016).

The interface mass transfer by particles between the solid and fluid becomes:

$$\frac{\partial \sigma_a}{\partial t} = \begin{cases} \lambda c U, & \sigma_a < \sigma_{cr}(U, \gamma) \\ \sigma_a = \sigma_{cr}(U, \gamma), & \sigma_a = \sigma_{cr}(U, \gamma) \end{cases}, \tag{3.3}$$

i.e., the particles attach until reaching the maximum retention value, and afterward the attached concentration remains equal to the maximum retention function. Continuous particle detachment during flows with decreasing salinity or increasing velocity and pH corresponds to the second line of Eq. (3.3), where time variations of U, γ, pH, and T result in a decrease of the maximum retention function.

The above fines-lifting mechanisms are supported by the typical skin histories for production wells in the Campos Basin (Brazil), presented in

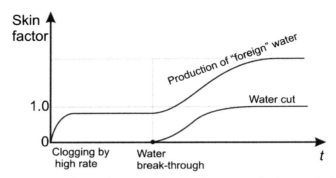

Figure 3.5 Increase of skin factor at the production well due to high-pressure gradient, after switching the well on, and due to the arrival of "foreign" water.

Fig. 3.5 (Bedrikovetsky et al., 2011b). Switching wells on yields the creation of large pressure gradients in the vicinity of the wellbore, which decrease with time and stabilize during the pressure wave propagation into the formation. Equation (3.3) shows that the mobilized particles are strained preferentially where the flow velocity is high, i.e., near to the wellbore causing maximum damage. It corresponds to steep growth of the skin factor. Pressure gradient stabilization leads to a stabilized skin factor. The skin remains constant until the appearance of water in the produced fluid. This could be either injected or aquifer water. The particle detachment due to difference in compositions of formation and breakthrough waters causes growing skin simultaneously with increasing water-cut (Fig. 3.5). Skin growth during this second stage is substantially higher when the injected (foreign) water has a smaller salinity than the formation water.

The rate Eq. (3.3) with mass balance of attached, suspended, and strained particles, and modified Darcy's law accounting for permeability damage by strained fines form a closed system of equations. Recently, the analytical and numerical solutions of this system of equations have been applied to well productivity and injectivity damage for single-phase and two-phase flows.

Single-phase steady-state inflow well performance with fines migration after stabilization shows zones of complete fines detachment, partial detachment, and nonmobilized fines. It allows determining the size of fines mobilization zone (Bedrikovetsky et al., 2012). Steady-state inflow with fines migration occurs also during the intermediate production state, where the decrease of fines mobilization due to spreading of pressure wave is compensated by the increase of the spreading area (Zeinijahromi et al., 2012a; 2012b). All stages of inflow well performance with fines lifting, migration, and straining are described by the exact solution for nonsteady-state transient flow (Marquez et al., 2014).

Exact solutions for fines-migration during high-rate water injection show steep well skin growth at the beginning of injection, and its relatively fast stabilization during pressure-wave spreading and pressure gradient decrease (Bedrikovetsky and Caruso, 2014). Here, the fines are lifted by the drag force during high-velocity flow. Lifting of fines due to low-salinity water injection and weakening of electrostatic force is described by the analytical model derived by Yang and Bedrikovetsky (2017).

Rosenbrand et al. (2015) applied the notion of a maximum retention function for fines detachment in geothermal reservoirs. Yuan and Shapiro (2011) and Zeinijahromi et al. (2013) applied it for 3D modeling of fines-assisted low-salinity waterflooding. Guo et al. (2015), Huang et al. (2017), and Zhu et al. (2017) applied it for coal beds.

Yuan et al. (2016) and Yuan and Moghanloo (2017) applied the modified model with the maximum retention function for fines fixing by nanoparticles, both for the problems of well productivity and injectivity. The authors derived exact solutions for different regimes of injection and production (Yuan et al., 2018a), accounting for preflush and simultaneous injection, corresponding to laboratory tests and field exploitation (Yuan et al., 2018b).

Mobilized particles in porous media are commonly considered to be transported in suspension in the carrier fluid. This would imply that the advective velocity of the suspended particles would be equal to that of the carrier fluid. Thus, the stabilization of permeability, which will occur after the arrival at the outlet of the fines mobilized at the inlet, will occur after the injection of one pore volume. Several authors have presented a contrary theory, where particles may exhibit near-surface motion, where their velocity is significantly smaller than that of the carrier fluid (Bradford et al., 2011; Yuan and Shapiro, 2010). Calculations of particle movement near the pore wall using the Navier-Stokes equations confirm that these particles will travel with a significantly reduced speed. (Sefrioui et al., 2013). Despite these results, most mathematical formulations of fines migration assume an equality of particle and water velocities (Bradford and Torkzaban, 2008).

Laboratory studies of fines transport in cores show that mobilized fines role or slide 100—1000 times slower than the injected water velocity

(Oliveira et al., 2014). The mathematical model assumes that fast particles move with the water velocity, so slow fines drift adds only one unknown parameter to be determined from the laboratory tests. It is the so-called delay drift factor α, which is the ratio between slow fines and water velocities (Yang et al., 2016). The model exhibits an excellent agreement with the laboratory data. Therefore, the drift fines velocity is equal to αU, which must be used in mass balance and straining kinetics equations.

Russell et al. (2017) undertook a systematic laboratory study of the effect of kaolinite contents in the rock on· fines-migration formation-damage. Nonmonotonic permeability decrease versus decreasing injected salinity has been observed. The results indicated that only $0.2-1.6\%$ of the kaolinite is movable by low–salinity and fresh-water injection.

Two-phase flow with varying salinity and fines detachment occurs during displacement of oil by low–salinity (smart) waterflooding. Saturation effects on fines detachment and transport are particle flow with phase velocity of water (Yuan and Shapiro, 2011) and saturation-dependency of the maximum retention function (Zeinijahromi et al., 2013). Lemon et al. (2011) and Zeinijahromi et al. (2011) introduced fines-migration formation-damage into the Dietz model for waterflooding, for displacement regimes with a given rate and pressure drop, respectively. The analytical model shows significant sweep enhancement due to plugging of the swept zone.

Laboratory studies of low–salinity waterflooding in Berea sandstones was undertaken by Hussain et al. (2013), and Zeinijahromi et al. (2016). So-called double coreflood consists of oil displacement by formation-water and resaturation of the core with further low–salinity water injection. The Welge's and Johnson-Bossler-Naumann (JBN) methods are applied for the case of low–salinity water injection, yielding two pairs of relative permeability — for oil-formation water and for oil-injected water. No fines release and ionic exchange occur during displacement by formation water, so this case is a base case, to compare other cases with. Application of Welge's and JBN methods to laboratory coreflood data resulted in nonmonotonic water relative permeability as a function of saturation. This unusual effect is explained by continuous sweep of new rock surfaces by continuous water saturation increase that yields continuous release of fines and growing permeability damage.

Mapping of basic equations for fines-assisted low–salinity waterflooding on the polymer option of black-oil equations allows for 3D numerical modeling using commercial reservoir simulators (Zeinijahromi et al., 2013). The method was applied for optimization of injected water compositions for smart waterflooding, and also for technologies of produced water reduction by injection of small banks of low–salinity/fresh water.

Regarding water-production control, injection of small portions of fresh water into watered-out producer-wells before their abandonment, injection of small portion of fresh water above the water-oil contact (WOC) or water-gas contact (WGC) before the production, or huff-n-puff with a small portion of fresh water can decrease water-cut by 10–20% during a 4–8 month period, in oil and gas wells.

Injection of small fresh-water banks into watered-out gas or oil wells just before their abandonment yields significant deceleration of encroaching water fingers, leading to the prolongation of the lifetime of up-dip producer wells and improved sweep efficiency (Zeinijahromi et al., 2015; 2013). Other methods of deliberate fines-migration stimulation with consequent induction of permeability damage by small fresh-water bank injection include huff-n-puff (Zeinijahromi and Bedrikovetsky, 2016) and injection just above the WGC or WOC to reduce water coning (Zeinijahromi et al., 2015).

To summarize, the abovementioned results on fines migration in two-phase flow, the fines-assisted low-salinity waterflooding is a mobility-control EOR method, on the contrary to low-salinity waterflood with wettability alteration and S_{or} reduction. Permeability decrease due to low-salinity water in already swept areas can yield a significant increase in the reservoir sweep efficiency. Injection of small volume banks of low-salinity (fresh) water during commingled oil and water production can significantly decrease water cut.

Yuan and Moghanloo (2017) and Yuan et al. (2018c) derived analytical models for waterflooding with nanoparticles accounting for fines release. Fines detachment is described by the nanoparticle-concentration-dependent maximum retention function. The splitting technique is used for exact integration of two-phase systems with fines migration.

The current chapter presents exact solutions for 1D coreflooding with stepwise increasing velocity and stepwise decreasing salinity with matching laboratory data and tuning model parameters. The derived analytical model for inflow performance of production wells with fines migration uses the model parameters, retrieved from the laboratory tests, for field-scale predictions. The same data are used for injectivity decline prediction during low-salinity water injection using the analytical model. Exact solutions are also obtained for fines-migration systems with delay in fines release. The main result is that the laboratory data can be matched with equal precision by the slow-particle model and by the delay-fines-release model. The obtained parameters are also used for field-scale prediction of injectivity decline during low-salinity water injection. Exact solutions for 1D low-salinity waterflooding with a delay in fines release and fines straining are derived using the splitting mapping, separating two-phase flow equations from the system of equations expressing fines release, migration, and straining.

The structure of the text is as follows. Section 3.2 presents governing equations for single-phase colloidal flow in porous media with particle detachment modeled by the maximum retention function. Section 3.3 derives an exact x,t-solution for fines migration under high injection rates and matches the laboratory coreflood data. Section 3.4 derives an exact r,t-solution for well inflow performance with fines migration; well behavior is predicted using the model coefficients obtained by the tuning of the laboratory coreflood data. Section 3.5 derives an exact x,t- and r,t-solutions for fines migration, mobilized by low-salinity water; x,t-solution matches the laboratory coreflood data and tunes the model coefficients, which is used by well behavior prediction. Section 3.6 derives an exact x,t and semianalytical r,t-solutions for single-phase fines migration accounting for the delay in fines release by low-salinity water. Section 3.7 derives an exact r,t-solution for two-phase fines migration during low-salinity water injection into oilfield; the section discusses how the solution can be applied for well injectivity in 3D reservoir simulation.

3.2 GOVERNING EQUATIONS FOR FLOW WITH FINES MIGRATION

In this section, we present the torque balance condition for mechanical equilibrium of fine particles attached to the rock surface (Section 3.2.1) and use it further to derive the maximum retention concentration as a function of flow velocity and salinity (Section 3.2.2). The mass balance equation for suspended, attached and strained particles, rate expressions for straining and attachment, and the maximum retention function form a closed system of equations describing suspension/colloidal flow in porous media with particle detachment (Section 3.2.3).

3.2.1 Torque balance of forces acting on particle

Attached particles present in porous media are adhered either on the grains comprising the rock matrix or an internal filter cake formed by other attached particles. Particle detachment depends on the mechanical equilibrium of forces acting on the particle. Drag, electrostatic, lift, and gravitational are the most important forces acting on the attached particles. Drag and lift forces act to detach particles from the surface; whereas, electrostatic and gravitational forces act to maintain the particle attached to the surface, considering the particles sitting on grains in the porous media (Muecke, 1979; Sarkar and Sharma, 1990).

Detachment of a particle attached to the rock can occur through one of three primary mechanisms: horizontal translation along the rock surface, translation vertically away from the surface, or rotation around the rock asperity of neighboring particles. These equilibria can be evaluated by summing either the acting forces tangential or normal to the surface, or summing the torques generated by each acting force. By Newton's second law of motion, the balance among these forces is such that any of these sums equals zero will give a mechanical equilibrium condition (i.e., the threshold between attachment and detachment). Analysis of these forces for fine particle detachment in porous media has shown that particle rotation, or rolling, is significantly more likely than translation (Sharma et al., 1992). This simplifies the quantitative prediction of detachment by reducing the three equations outlined above to simply the torque balance.

The torque balance of detaching and attaching forces acting on the particle is expressed as Eq. (3.1) (Bradford et al., 2013). The normal forces are responsible for the deformation of the particle in contact with the grain surface. If the particle rotates around a point of contact with the surface, the lever arm of the normal forces l_n is assumed to be equal to the radius of the contact area deformation by the normal force. Thus, l_n can be calculated from Hertz's theory as (Derjaguin et al., 1975; Schechter, 1992):

$$l_n^3 = \frac{F_e r_s}{4K}, \quad K \equiv \frac{4}{3\left(\frac{1-\nu_1^2}{E_1} + \frac{1-\nu_2^2}{E_2}\right)}, \tag{3.4}$$

where K is the composite Young's modulus, ν is the Poisson's ratio, E is the Young's modulus, and the subscripts 1 and 2 refer to the particle and the grain, respectively. Once the normal lever arm is calculated, the drag lever arm can be calculated from the geometrical relationship between the particle radius and the normal lever arm, $l_d = \sqrt{r_s^2 - l_n^2}$. For Young's modulus typical for silica and kaolinite, corresponding to more frequent cases of attached fines in sandstone rocks, the contact deformation radius is very small (Kalantariasl and Bedrikovetsky, 2013), so:

$$r_s >> l_n, \quad l_d \cong r_s, \tag{3.5}$$

Eq. (3.5) shows that in this case, the electrostatic force significantly exceeds drag under the mechanical equilibrium.

Assuming the particle to be spherical, the gravitational force can be expressed as:

$$F_g = \frac{4}{3}\pi r_s^3 \Delta \rho g, \tag{3.6}$$

where $\Delta \rho$ is the difference of densities between the particle and the water, and g is the gravitational acceleration.

Drag and lift forces are the hydrodynamic forces exerting on the particle and mainly dependent on the carrier fluid velocity. Expressions for lift and drag forces are:

$$F_L = \chi r_s^3 \sqrt{\frac{\rho \mu u^3}{r_p^3}}, \tag{3.7}$$

$$F_d = \frac{\varpi \pi \mu r_s^2 u}{r_p}, \tag{3.8}$$

respectively, where χ is the lift coefficient, which Kang et al. (2004) reported to be equal to 89.5 while Altmann and Ripperger (1997) reported a value of 1190, ρ is the fluid density, μ is the fluid viscosity, ω is the drag coefficient, and r_p is the pore size. The geometrical shape coefficients χ and ω can be calculated using computational fluid dynamics modeling (CFD).

Eqs. (3.7) and (3.8) show that both of the hydrodynamic forces increase with increasing fluid velocity, which will favor the conditions for particle detachment from the surface.

The electrostatic force is calculated from the extended DLVO (Derjaguin–Landau–Verwey–Overbeek) theory that takes into account the interacting energies between the particles and the grain surface. The total energy is the sum of the London–Van der Waals, Electrical Double Layer, and Born Repulsive potentials (Derjaguin and Landau, 1941; Elimelech et al., 2013; Gregory, 1981; Hogg et al., 1966; Israelachvili, 2011; Verwey et al., 1999).

$$V = V_{LVW} + V_{EDL} + V_{BR}, \tag{3.9}$$

where V is the total energy, V_{LVW} is the London-Van der Waals potential, V_{EDL} is the Electrical Double Layer potential, and V_{BR} is the Born Repulsive potential.

The London-Van der Waals potential is the attraction between two closely separated surfaces and will mostly be attractive for the interaction between particles and grains in natural porous media. The force arises from the spontaneous electrical and magnetic polarizations, giving a fluctuating electromagnetic field within the medium. The classical approach to evaluating the London-Van der Waals interaction between two macroscopic bodies is obtained by pairwise summation of all the relevant intermolecular interactions. Retarded London-Van der Waals energy for sphere-plate interaction is given by the following formula (Gregory (1981)):

$$V_{LVW} = -\frac{A_{132}r_s}{6h}\left[1 - \frac{5.32h}{\lambda_w}\ln\left(1 + \frac{\lambda_w}{5.32h}\right)\right], \tag{3.10}$$

where A_{132} is the Hamaker constant, λ_w is the characteristic wavelength of interaction, and h is the surface-to-surface separation distance.

The expression for the Hamaker constant has the form (Israelachvili, 2011):

$$\begin{aligned}
A_{132} = &\frac{3}{4}k_BT\left(\frac{\varepsilon_1 - \varepsilon_3}{\varepsilon_1 + \varepsilon_3}\right)\left(\frac{\varepsilon_2 - \varepsilon_3}{\varepsilon_2 + \varepsilon_3}\right) \\
&+ \frac{3h\nu_e}{8\sqrt{2}}\frac{\left(n_1^2 - n_3^2\right)\left(n_2^2 - n_3^2\right)}{\left(n_1^2 + n_3^2\right)^{1/2}\left(n_2^2 + n_3^2\right)\left\{\left(n_1^2 + n_3^2\right)^{1/2} + \left(n_2^2 + n_3^2\right)\right\}},
\end{aligned} \tag{3.11}$$

where $k_B = 1.381 \times 10^{-23}$ J/K is Boltzmann constant, ε_1, ε_2, and ε_3 are static dielectric constants of the particle, grain, and fluid, respectively, n_1, n_2, and n_3 are the refractive indices of particle, grain, and fluid, respectively, and $\nu_e = 3.0 \times 10^{15}$ s^{-1} is the constant value of absorption frequency.

The Born repulsive potential is a short-range repulsion that originates from the strong repulsive forces between the atoms as their electron shells interpenetrate each other. An expression for the Born repulsion potential between a sphere and a plate was developed by Ruckenstein and Prieve (1976) as:

$$V_{BR} = \frac{A_{132}}{7560} \left(\frac{\sigma_{LJ}}{r_s}\right)^6 \left[\frac{8+Z}{(2+Z)^7} + \frac{6-Z}{Z^7}\right]; \quad Z = \frac{h}{r_s}, \quad (3.12)$$

where $\sigma_{LJ} = 0.5$ nm is the atomic collision diameter adopted from Elimelech et al. (2013).

The Electrical Double Layer energy can be attractive or repulsive if the colloids interacting have the opposite or similar charges, respectively. The charge on the particle surface affects the ion distribution in the neighborhood of the particle, creating an electric double layer around the surface. The overlap of two electric double layers results in a net potential energy of interaction. Considering the case of interaction between clay and sand in the porous medium, the Electrical Double Layer will generally be repulsive due to the similar charges of the two materials. The smaller is the surface-to-surface distance, the higher is the potential energy of the Electrical Double Layer.

Several expressions for Electrical Double Layer energy are presented in the literature (Elimelech et al., 2013). Gregory (1975) proposed the following expression to calculate the interaction between a spherical particle and a surface in the porous matrix:

$$V_{EDL} = \frac{128\pi r_s n_\infty k_B T}{\kappa^2} \psi_s \psi_b e^{-\kappa h}$$

$$\kappa = \sqrt{\frac{e^2 \sum n_{i0} z_i^2}{\varepsilon_0 \varepsilon_3 k_B T}} \quad (3.13)$$

$$\psi_s = \tanh\left(\frac{ze\zeta_s}{4k_B T}\right), \psi_g = \tanh\left(\frac{ze\zeta_g}{4k_B T}\right),$$

where κ is the inverse Debye length, n_∞ is the bulk density of ions, $e = 1.602 \times 10^{-19}$ C is the elementary electric charge, n_{i0} is the concentration of ions i in bulk solution, z is the valence of symmetrical electrolyte solution, $\varepsilon_0 = 8.854 \times 10^{-12}$ F/m is the dielectric permittivity of vacuum, ε_3 is the dielectric constant of the fluid, ψ_s and ψ_g are the reduced zeta potentials for the particle and grain, and ζ_s and ζ_g are the zeta potentials for the particle and grain.

The value taken for the electrostatic force depends primarily on the form of the total potential energy profile. Two typical forms of these

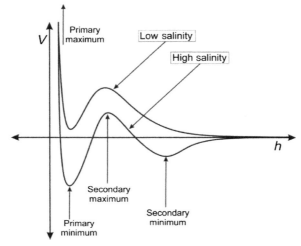

Figure 3.6 Typical forms for total potential energy profile for high- and low-salinity water (V: potential energy of interaction, h: particle-grain separation distance).

profiles are presented in Fig. 3.6. The total force acting on the attached particle is opposite to the gradient of the total energy:

$$F_e = -\frac{\partial V}{\partial h}, \tag{3.14}$$

where F_e is the total electrostatic force, and V is the total interaction energy.

In equilibrium, the system will tend to a state in which the acting force is zero. These positions are either at maxima, minima, or at infinite separation distance. Maxima are unstable equilibrium points because slight variations in the separation distance will result in forces driving the particle away from the maximum. Minima, however, are stable points, because the energy profile on either side of the minima will drive particles back toward the minimum. Thus, when a particle is attached, it is considered to either be in the primary or the secondary minimum. When attempting to leave either of these minima (increasing the separation distance), the particles will encounter a positive potential energy gradient, and hence a negative, or attractive, force. The maximum force experienced during this passage, located at the inflection point on the potential energy profile, is taken as the electrostatic force used in the torque balance equation, outlined above.

Which of the two minima should be used is still a matter of some contention. Some authors suggest that the electrostatic force experienced when leaving the primary minimum is too great for any other forces to detach the particle; hence, all particles that can detach must have resided in the secondary minimum. On the other hand, as outlined in the low-salinity curve in Fig. 3.6, some potential energy profiles either do not have secondary minima, or they are too small to prevent particles from freely leaving due to Brownian motion (Bradford et al., 2013).

Calculations for gravitational and lift forces indicate that their order of magnitude range is between 10^{-14} and 10^{-13}, although drag and electrostatic forces range between 10^{-11} and 10^{-8} (Bedrikovetsky et al., 2012; Kalantariasl and Bedrikovetsky, 2013). Therefore, gravitational and lift forces are disregarded and only drag and electrostatic forces are considered to be relevant to determine conditions for particle detachment. Calculations of the torque balance will only take into account drag and electrostatic forces and thus, Eq. (3.1) simplifies to the following form:

$$F_d(U, r_s) = F_e(r_s, \gamma, pH, T)l(r_s, F_e), \qquad (3.15)$$

Favorable conditions for particle mobilization happen when the drag torque exceeds the electrostatic torque. If the particle is attached to the pore surface, this means that it is in a stable equilibrium, where the net force on the particle is zero.

The equations for lever arms, drag, and electrostatic forces provide an understanding of which parameters can lead to favorable conditions for particle detachment during waterflooding. The drag force is mainly a function of velocity and particle size. The higher the carrier fluid velocity and particle size, the higher the drag force, increasing the likelihood of particle detachment. The electrostatic force is primarily affected by water salinity, pH, and temperature. The attractive London-Van der Waals potential and repulsive Born potential are functions of the Hamaker constant, which is a function of the dielectric constant and refractive index for particle, grain, and fluid, both are temperature dependent (Israelachvili, 2011). An increase in the temperature will increase repulsion between the particle and surface. The repulsive electrical double layer is a function of the inverse Debye length and zeta potentials. The first is inversely proportional to the water salinity and pH, meaning that lower salinity and higher pH will lead to higher repulsion between particle and surface. The second is a function of pH and temperature, higher pH and temperature will also lead to stronger repulsion. Thus, high pH

and temperatures and low-salinity water are favorable conditions for particle mobilization (You et al., 2014).

3.2.2 Using the torque balance to derive expressions for the maximum retention function

The principle of particle mobilization in porous media due to perturbation in the torque balance leads to the development of a macroscale mathematical model for attached particle concentration. Such an expression can be used to model suspension transport in porous media with particle detachment given changes to the particle mechanical equilibrium (Bedrikovetsky et al., 2011a).

The classical approach to modeling particle detachment is to implement a first-order rate equation to capture the detachment kinetics. Several authors have, however, demonstrated that changes to flow rate or salinity result in seemingly instantaneous changes to the attached concentration. A more thorough discussion of the appropriateness of particle detachment kinetics will be presented in Section 3.6.

Nonetheless, regardless of whether kinetics is deemed important in the formulation of particle detachment, an equilibrium function is required which describes the final attached concentration. Such a function should recognize the changes to attachment conditions governed by the dependencies of the acting forces outlined above. This function will be referred to here as the maximum retention function.

The classical model to describe particle attachment and detachment accounts for particle detachment kinetics yielding to an asymptotical stabilization for retention concentration and permeability when time tends to infinity, Eq. (3.2). Nonetheless, the fines mobilization and permeability increase after sharp increase in flow rate or shocks of salinity alteration (Jaiswal et al., 2011; Khilar and Fogler, 1998; Ochi and Vernoux, 1998; Oliveira et al., 2014; Sarkar and Sharma, 1990). As the traditional model fails to reflect the prompt permeability and particle release observed in many laboratory experiments, alternative models capable of modeling the instant particle release have to be applied to calculate the maximum retention concentration.

In the following sections, two models for the maximum retention function will be presented. The first assumes that mono-sized particles form a multilayer internal cake within pores. The second recognizes a distribution in particle sizes, but restricts particle attachment on the pore wall to a single layer.

3.2.2.1 Internal filter cake of multilayer with mono-sized fine particles

This model assumes that the drag, lift, gravitational, and electrostatic forces describe the mechanical equilibrium for attached particles as outlined in the previous section. Here, all attached particles are assumed to have the same size and electrostatic properties. The pore space is modeled as a bundle of square, parallel capillaries with the same size. An attached particle with acting forces and corresponding lever arms is shown in Fig. 3.7.

Particles attach to the pore walls in discrete layers. As more particle layers appear, the cross-sectional area within the pore that is available for flow decreases. Thus, the flow velocity increases, increasing the drag and lifting forces. This is captured in the equations for these two forces with an inverse dependency on the pore size (Eqs. (3.7 and 3.8)). Therefore, as more layers of particles appear, the condition for mechanical equilibrium shifts further toward detachment. At some critical internal cake thickness, h_c, any additional layers of attached particles would be unstable and detach. This cake thickness, when upscaled, defines the maximum concentration of particles that can be attached for given values of total fluid velocity, fluid salinity, pH, etc.

The assumption of spherical particles, accompanied by the assumption that the particles will arrange evenly on the pore space, makes it possible to approximate the lever arm ratio l_d/l_n as $\sqrt{3}$. Given this assumption, the torque balance can be evaluated as:

$$F_e + \frac{4\pi r_s^3}{3}\Delta\rho g - \chi r_s^3 \sqrt{\frac{\rho\mu U^3}{(H-2h_c)^3}} = \frac{\sqrt{3}\omega\pi\mu r_s^2 U}{H-2h_c}. \qquad (3.16)$$

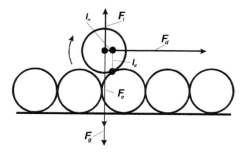

Figure 3.7 Mechanical equilibrium of fines on the rock surface: drag, lift, electrostatic, and gravitational forces exerting on the particle at the cake surface (Bedrikovetsky et al., 2011a).

For rectangular pores, the critical retention function can be related to the critical cake thickness as per:

$$\sigma_{cr} = \left[1 - \left(1 - \frac{h_c}{H}\right)^2\right](1 - \phi_c)\phi. \tag{3.17}$$

where ϕ_c is the porosity of the internal filter cake formed by attached particles. This parameter is often unknown and must be determined empirically.

Removing the gravitational and lift forces from the torque balance and substituting the resulting expression into Eq. (3.17) yield the expression for the critical retention function:

$$\sigma_{cr}(U, \gamma, pH, T) = \left[1 - \left(\frac{1}{2} + \frac{\sqrt{3}\omega\pi\mu r_s^2 U}{2F_e(\gamma, pH, T)H}\right)^2\right](1 - \phi_c)\phi. \tag{3.18}$$

The above equation provides the qualitative descriptions of particle detachment as outlined above. This equation predicts a decrease in σ_{cr} both with an increase in fluid velocity and a decrease in fluid salinity, which will decrease the electrostatic force. Both of these dependencies are monotonic. Typical forms of the critical retention function plotted against the fluid velocity and fluid salinity are presented in Fig. 3.8. Figs. 3.8A and B demonstrate the monotonic decline of σ_{cr} both with increases to velocity and decreases in salinity. Both forms also show the highly nonlinear dependency of the critical retention function on the injection conditions.

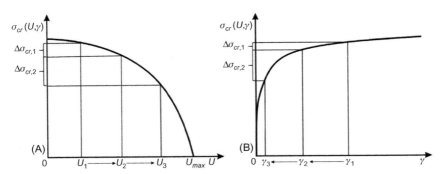

Figure 3.8 The critical retention function as a function of (A) the fluid velocity and (B) the fluid salinity (σ_{cr}: critical retention function, U: fluid velocity, γ: fluid salinity, $\Delta\sigma_{cr}$: detached particle concentration).

For the rock with pore sizes distributed as $f_s(r_s)$ and pores distributed by sizes as $f_p(r_p)$, the following estimate of the maximum retention function is:

$$\langle \sigma_{cr} \rangle (U, \gamma, pH, T) = \int_0^\infty \int_0^\infty \left[1 - \left(\frac{1}{2} + \frac{\sqrt{3}\omega\pi\mu r_s^2 U}{2F_e(\gamma, pH, T)H} \right)^2 \right]$$

$$(1 - \phi_c)\phi f_p(r_p) f_s(r_s) dr_s dr_p. \tag{3.19}$$

3.2.2.2 Monolayer of multisized fine particles

The monolayer model restricts each pore to contain at most a single layer of attached particles. For each of these particles, the torque balance can be applied to determine individually whether each particle will detach. The nuance in this model is that particles are no longer assumed to be of the same size.

The equations for the acting forces (Eqs. (3.6, 3.7, 3.8, and 3.14)) show some form of particle size dependence. A particle with larger size experiences larger drag and lifting forces as well as larger electrostatic and gravitational forces. Thus, both the detaching and attaching torques increase with increasing particle size. When evaluated, it can be shown that the particle size has a stronger influence on the detaching torque than the attaching torque. Fig. 3.9 shows how the critical (minimum) radius decreases as velocity increases for different Hamaker constants. The attaching electrostatic force increases as the Hamaker constant increases.

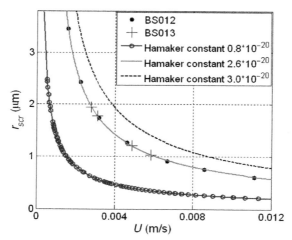

Figure 3.9 The minimum size of attached particles that can be detached by flux with velocity U (r_{scr}: critical detached particle size).

Thus, the higher the Hamaker constant, the larger the particle removed under a given velocity U. Therefore, the curve $r_{scr}(U)$ shifts to the right as the Hamaker constant increases.

Fig. 3.9 also shows that the experimental points obtained by Ochi and Vernoux (1998), are located on the theoretical curve with high precision.

Thus, larger particles are more likely to detach than smaller particles. The explicit form of the torque balance allows the definition of the critical particle size:

$$r_{scr} = r_s(U, \gamma, pH, T). \tag{3.20}$$

All particles greater than the critical size will detach, and all particles smaller will remain attached.

The maximum retention function can then be calculated as:

$$\sigma_{cr}(U, \gamma, pH, T) = \int_0^{r_{scr}(U, \gamma, pH, T)} \sigma_{aI} f(r_s) dr_s, \tag{3.21}$$

where σ_{aI} is the initial attached particle concentration. This can be visualized from the particle size distribution as is shown in Fig. 3.10. Here, the maximum retention function is determined from the area under the curve for all particle sizes smaller than the critical particle size. When the velocity increases, or the salinity decreases, the critical particle size will decrease, and correspondingly, the critical retention function will decrease.

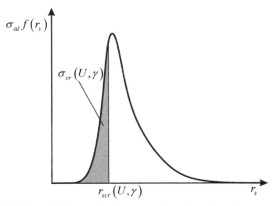

Figure 3.10 Calculation of the maximum retention function for a monolayer of sized distributed particles (σ_{aI}: initial attached particle concentration, $f(r_s)$: attached particle size probability density function, σ_{cr}: critical retention function, r_s: particle size, r_{scr}: critical detached particle size).

The final form of the critical retention function versus velocity and salinity will not only roughly follow those shown in Fig. 3.8, but will also be highly dependent on the form of the particle size distribution.

3.2.3 Single-phase equations for fines transport

Mathematical modeling of fines migration typically revolves around a continuity equation for the suspended particle concentration. The form of this equation is as follows:

Rate of accumulation = Divergence of advective flux
$$\qquad\qquad\qquad - \text{Rate of detachment} - \text{Rate of straining.}$$

Suspended particle transport is given by the sum of the advective and diffusive flux. For the purposes of modeling fines migration in petroleum reservoirs, diffusive flux is typically negligible and thus is removed for convenience. Thus, the continuity equation can be expressed as:

$$\frac{\partial \phi c}{\partial t} = -U \frac{\partial c}{\partial x} - \frac{\partial \sigma_a}{\partial t} - \frac{\partial \sigma_s}{\partial t}. \qquad (3.22)$$

This is more commonly given as:

$$\frac{\partial}{\partial t}[\phi c + \sigma_a + \sigma_s] + U \frac{\partial c}{\partial x} = 0, \qquad (3.23)$$

where c, σ_a, and σ_s are the concentrations of suspended, attached, and strained particles, respectively, ϕ is the porosity of the porous media, U is the fluid flow velocity, x is the distance, and t is the time. The porosity term arises because the suspended concentration is defined as the volume of particles in suspension per pore volume, while the attached and strained concentrations are defined similarly but for the bulk rock volume.

For Eq. (3.23) to be valid, it is necessary here to assume that during the transport, detachment, and capture, the particle and fluid volumes are additive (Amagat's law), and c, σ_a, and σ_s are the volumetric concentrations. Eq. (3.23) is also valid where the mass concentrations of all particle species are negligibly small if compared with the mass of water.

The straining rate is assumed to be proportional to particle advection flux, cU (Bedrikovetsky, 2008; Herzig et al., 1970):

$$\frac{\partial \sigma_s}{\partial t} = \lambda(\sigma_s)cU, \qquad (3.24)$$

where λ is the filtration coefficient for straining.

The typical form assumed for the filtration coefficient follows from the principles of Langmuir adsorption curves (Wang et al., 2017; Yang and Balhoff, 2017). Assuming that there are a fixed and finite number of vacancies available for particle straining, the concentration of these vacancies is denoted σ_m. When there are no strained particles, the filtration coefficient has some initial value λ_0. As particles begin to strain, the likelihood of suspended particles encountering a vacancy decreases; therefore, the filtration coefficient decreases. The filtration coefficient then takes the following form:

$$\lambda(\sigma) = \lambda_0 \left(1 - \frac{\sigma_s}{\sigma_m} \right). \tag{3.25}$$

When the strained concentration is significantly smaller than the number of vacancies, the filtration coefficient will only be weakly dependent on the strained concentration. In this case, it is often assumed that the filtration coefficient is constant altogether, and so Eq. (3.24) simplifies to:

$$\frac{\partial \sigma_s}{\partial t} = \lambda c U. \tag{3.26}$$

The result of the analysis of particle detachment presented in the previous section is the maximum retention function:

$$\sigma_a = \sigma_{cr}(U, \gamma), \tag{3.27}$$

which characterizes the changes to the attached concentration based on shifts in the torque balance. When equilibrium between the attached concentration and the maximum retention function is reached, no particle detachment occurs. As discussed earlier, the finiteness of the detachment rate is still a matter of debate. What is known is that once the attached concentration has reached the value of the maximum retention function, the detachment rate will be zero.

The mass balance equation for salt neglecting the diffusion/dispersion is:

$$\phi \frac{\partial \gamma}{\partial t} + U \frac{\partial \gamma}{\partial x} = 0. \tag{3.28}$$

With given expressions for the straining and detachment rates, a system of equations can be produced that allows solving for the three particle concentrations and salinity. The remaining component of modeling fines migration is to determine the effect of the in-situ particles on the permeability of the rock.

Assuming that the rock permeability is a function of both the strained and attached concentrations, a Taylor's series expansion can be used to derive:

$$\frac{k_0}{k(\sigma_a, \sigma_s)} = 1 + \beta_a \sigma_a + \beta_s \sigma_s + O(\sigma^2), \tag{3.29}$$

where the small magnitude of both concentrations typically allows neglecting all terms of order σ^2 and above.

Substituting Eq. (3.29) into Darcy's law for flow in porous media produces the modified Darcy's law accounting for permeability decline due to both strained and attached particles (Bedrikovetsky et al., 2011a; Pang and Sharma, 1997):

$$U = -\frac{k}{\mu(c)(1 + \beta_s \sigma_s + \beta_a \sigma_a)}\frac{\partial p}{\partial x}. \tag{3.30}$$

Note that the fluid viscosity is a function of the suspended particle concentration. In almost all applications of fines migration, this dependency is small and thus is neglected. The use of Darcy's law requires the assumption of steady state, laminar, and incompressible flow.

Usually, it is also assumed that the coating of grains by attached particles causes insignificant permeability damage compared to that caused by the straining of particles in the pore throats. Therefore, $\beta_a << \beta_s$, and the modified Darcy's law can be expressed simply as:

$$U = -\frac{k}{\mu(1 + \beta_s \sigma_s)}\frac{\partial p}{\partial x}. \tag{3.31}$$

The mass balance equation, combined with expressions for the straining and detachment rates and the modified Darcy's law, presents a closed mathematical system capable of modeling the process of fines migration in porous media and its effect on rock permeability.

3.3 FINES MIGRATION RESULTING FROM HIGH FLUID VELOCITIES

In Section 3.2 it was shown that increasing fluid velocity can result in particle detachment through an increase in the hydrodynamic drag force. In the following section, the mathematical and laboratory modeling of fines migration induced by high fluid velocities will be presented. In Section 3.3.1 the concept of slow particle migration and the drift delay factor will be introduced. Following this, a reformulation of the governing Eqs. (3.23, 3.26, and 3.31) alongside the appropriate initial and boundary conditions for the problem of fines migration under high velocity will be given. The exact solution for this problem for 1D linear flow will be presented in Section 3.3.2 which will be used to analyze laboratory data in Section 3.3.3.

3.3.1 Formulation of mathematical model

The primary assumptions of the model follow from those presented in Section 3.2. These being the additivity of particle concentrations, sufficiently small strained concentration to provide a constant filtration coefficient, neglecting permeability decline due to attached particles, and steady-state, incompressible, laminar flow.

In the formulation of the mass balance equation for fine particles, the advective flux of suspended particles is proportional to their velocity U_s. This parameter is typically set to the value of the fluid velocity as the suspended particles experience advective flux due to the transport of the bulk fluid through the porous media (Khilar and Fogler, 1983, 1998; Sharma and Yortsos, 1987). Long stabilization times in experimental studies of fines migration (Oliveira et al., 2014; Yang et al., 2016) suggest that the particle velocity is significantly smaller than that of the carrier fluid. Rolling of particles along the pore surface as well as limited accessibility to small pores provide physical justification for the distinction between the two velocities. As both of these effects will reduce the particle velocity, it is typically assumed that the particle velocity is strictly smaller than the fluid velocity.

Following the decoupling of particle and fluid velocities, their ratio is introduced as a dimensionless parameter:

$$\alpha = \frac{U_s}{U}. \tag{3.32}$$

This parameter will be referred to as the drift delay factor α, and assumed to be constant for all particles during any period of injection. Although the fluid velocity is known under test conditions, the particle velocity is unknown; hence, the drift delay factor will be an unknown parameter of the model.

The system of equations governing fines migration can now be presented. Firstly, the mass balance equation for suspended, strained, and attached particles where the suspended particles are now transported by the reduced velocity $U_s = \alpha U$:

$$\frac{\partial}{\partial t}(\phi c + \sigma_s + \sigma_a) + \alpha U \frac{\partial c}{\partial x} = 0. \tag{3.33}$$

Similarly, as the rate of straining is proportional to particle flux, the relationship for the straining rate becomes:

$$\frac{\partial \sigma_s}{\partial t} = \lambda c \alpha U. \tag{3.34}$$

The attached particle concentration is expressed by the maximum retention function, which will vary with the bulk fluid velocity:

$$\sigma_a = \sigma_{cr}(U). \tag{3.35}$$

Finally, the effect of strained particles is incorporated into the modified Darcy's law:

$$U = -\frac{k_0}{\mu(1 + \beta\sigma_s)}\frac{\partial p}{\partial x}. \tag{3.36}$$

The four Eqs. (3.33−3.36) constitute a closed system of equations in the four unknowns c, σ_a, σ_s, and p which describe the detachment, migration, and straining of fine particles in porous media.

During a coreflood with velocity altered from U_0 to U_1 with $U_1 > U_0$, the critical retention function will decrease by an amount $\Delta\sigma = \sigma_{cr}(U_0) - \sigma_{cr}(U_1)$. The detached concentration will immediately enter the colloidal suspension. These conditions give the initial state of the suspended and attached particle concentrations:

$$t = 0{:}c = \frac{\Delta\sigma}{\phi}, \sigma_a = \sigma_{cr}(U_1). \tag{3.37}$$

As the process of straining is considered to be irreversible, any particles that were strained prior to the injection period under consideration (i.e., prior to time $t = 0$) can simply be considered as a part of the initial permeability k_0. Doing so allows defining the initial condition for the strained particle concentration as follows:

$$t = 0{:}\sigma_s = 0. \tag{3.38}$$

The boundary conditions correspond to the absence of suspended particles in the injected fluid:

$$x = 0{:}c = 0. \tag{3.39}$$

Integration of Eq. (3.34) at the core inlet using boundary condition (Eq. (3.39)) allows calculating the strained concentration at the inlet:

$$x = 0: \sigma_s = 0. \tag{3.40}$$

Before attempting to solve the above system of equations, first the following dimensionless parameters are introduced to produce dimensionless relationships:

$$X = \frac{x}{L}, T = \frac{\int_0^t U(\gamma)d\gamma}{\phi L}, C = \frac{c}{\Delta\sigma}, S_s = \frac{\sigma_s}{\phi\Delta\sigma}, S_a = \frac{\sigma_a}{\phi\Delta\sigma}, \Lambda = \lambda L, P = \frac{pk_0}{U\mu L}, \tag{3.41}$$

where L is the core length, ϕ is the rock porosity, $\Delta\sigma = \sigma_{cr}(U_0) - \sigma_{cr}(U_1)$ is the total detached concentration when the fluid velocity is increased from U_0 to U_1, and μ is the fluid viscosity.

Substituting these parameters into the system of Eqs. (3.33−3.36) yields, for the mass balance:

$$\frac{\partial}{\partial T}(C + S_s + S_a) + \alpha\frac{\partial C}{\partial X} = 0. \tag{3.42}$$

The kinetics of particle straining is now given by:

$$\frac{\partial S_s}{\partial T} = \Lambda\alpha C. \tag{3.43}$$

The attached concentration is similarly expressed as:

$$S_a = S_{cr}(U). \tag{3.44}$$

And the dimensionless modified Darcy's law becomes:

$$1 = -\frac{1}{1 + \beta\phi\Delta\sigma S_s}\frac{\partial P}{\partial X}. \tag{3.45}$$

In dimensionless coordinates, the initial and boundary conditions become, respectively:

$$T = 0: C = 1, S_s = 0, S_a = S_{cr}(U_1), \tag{3.46}$$

$$X = 0: C = 0, S_s = 0. \tag{3.47}$$

In the following section, the solution for the dimensionless Eqs. (3.42−3.45) subject to initial and boundary conditions (Eqs. (3.46 and 3.47)) is presented.

3.3.2 Exact analytical solution for 1D problem

Substitution of the equation for straining rate (Eq. (3.43)) into the mass balance Eq. (3.42) and accounting for the fact that the attached concentration is constant after the initial detachment yields a first-order hyperbolic equation:

$$\frac{\partial C}{\partial T} + \alpha \frac{\partial C}{\partial X} = -\Lambda \alpha C. \tag{3.48}$$

Here, the filtration coefficient is considered to be constant.

Eq. (3.48) can be solved separately for the two regions behind and ahead of the suspended particle front using the method of characteristics.

Ahead of the particle front $(X > \alpha T)$, along parametric curves given by:

$$\frac{dX}{dT} = \alpha, \tag{3.49}$$

Eq. (3.48) reduces to the ordinary differential equation:

$$\frac{dC}{dT} = -\Lambda \alpha C. \tag{3.50}$$

Integrating Eq. (3.50) subject to initial condition (Eq. (3.46)) yields the solution for the suspended concentration ahead of the particle front:

$$C(X, T) = e^{-\alpha \Lambda T}. \tag{3.51}$$

Similarly, behind the concentration front $(X < \alpha T)$ Eq. (3.48) reduces to the ordinary differential equation:

$$\frac{dC}{dX} = -\Lambda C, \tag{3.52}$$

Along parametric curves given by:

$$\frac{dT}{dX} = \frac{1}{\alpha}, \tag{3.53}$$

Using the boundary condition (Eq. (3.47)) corresponding to the absence of suspended particles in the injected fluid, Eq. (3.52) can be integrated to yield zero suspended concentration behind the particle front, $C(X, T) = 0$.

Integrating Eq. (3.43) using separation of variables accounting for the derived solution for $C(X, T)$ enables determination of the strained concentration $S_s(X, T)$.

Table 3.1 Solutions for suspended and strained concentrations, and impedance ahead and behind the particle front

Line	Term	Zone	Solution
1	$C(X,T)$	$X > \alpha T$	$e^{-\alpha \Lambda T}$
2		$X < \alpha T$	0
3	$S_s(X,T)$	$X > \alpha T$	$1 - e^{-\alpha \Lambda T}$
4		$X < \alpha T$	$1 - e^{-\Lambda X}$
5	$J(T)$	$T < \frac{1}{\alpha}$	$1 + \beta \phi \Delta \sigma_{cr} \left[1 - \frac{1}{\Lambda} - \left(1 - \frac{1}{\Lambda} - \alpha T \right) e^{-\alpha \Lambda T} \right]$
6		$T > \frac{1}{\alpha}$	$1 + \beta \phi \Delta \sigma_{cr} \left[1 - \frac{1}{\Lambda} - \frac{e^{-\Lambda}}{\Lambda} \right]$

The dimensionless pressure drop $J(T)$, referred to as the impedance, can be solved by integrating Eq. (3.45) directly using separation of variables:

$$J(T) = \int_0^1 \left(-\frac{\partial P}{\partial X} \right) dX = 1 + \beta \phi \Delta \sigma \int_0^1 S_s(X, T) dX. \qquad (3.54)$$

Substituting the expression for the strained concentration and integrating make it possible to express the impedance explicitly as:

$$J(T) = \begin{cases} 1 + \beta \phi \Delta \sigma_{cr} \left[1 - \dfrac{1}{\Lambda} - \left(1 - \dfrac{1}{\Lambda} - \alpha T \right) e^{-\alpha \Lambda T} \right], & T < \dfrac{1}{\alpha} \\[2ex] 1 + \beta \phi \Delta \sigma_{cr} \left(1 - \dfrac{1}{\Lambda} + \dfrac{e^{-\Lambda}}{\Lambda} \right), & T \geq \dfrac{1}{\alpha} \end{cases} \qquad (3.55)$$

The two solutions correspond to times before and after the passing of the suspended particle front. The latter of the two solutions is independent of time, as after time $1/\alpha$, the core no longer contains any suspended particles. Table 3.1 summarizes the solutions for suspended and strained particle concentrations, and impedance.

3.3.3 Qualitative analysis of the solution

Fig. 3.11A shows the X-T plane with the two regions in which the solution is presented. Figs. 3.11B and C show the profiles of the suspended and strained particle concentrations across the core. The profiles are taken at the following moments: initially $(T = 0)$, and before and after permeability stabilization (T_a and T_b, respectively).

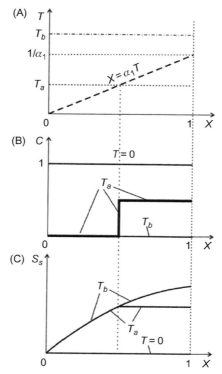

Figure 3.11 Solution space (A) and concentration profiles for the suspended (B) and strained (C) concentrations for the derived analytical solution for fines migration (X: dimensionless linear coordinate, T: dimensionless time, C: dimensionless suspended concentration, S_s: dimensionless strained concentration).

All detached particles move through the core with velocity α, which defines the slope of the characteristics along which Eq. (3.48) is solved.

Ahead of the particle concentration front $(X > \alpha T)$, the suspended particles are uniformly distributed along the core. This follows from the fact that each particle moves with the same velocity and has the same probability of capture via straining. This uniform distribution decreases monotonically with time as the particles are captured in small pore throats. As a result of the uniform distribution of C, and the constant straining rate, the strained concentration ahead of the concentration front is similarly independent of X.

Behind the concentration front $(X < \alpha T)$, all suspended particles have either been captured, or transported further into the core. As such, the

Table 3.2 Rock properties

Initial permeability (mD)	Porosity (%)	Diameter (mm)	Length (mm)	Pore volume (mL)
235	38.54	38.04	46.00	20.14

suspended concentration is zero, and the strained concentration is constant with time. As the core outlet is defined at point $X = 1$, the time of stabilization, where the core contains no suspended particles, will be at $T = 1/\alpha$. At this time, the distribution of strained particles, and consequently the impedance, will be constant with time. By Eqs. (3.43 and 3.45), both the strained concentration and the impedance grow monotonically with time until stabilization.

3.3.4 Analysis of laboratory data

A laboratory coreflood on an artificially prepared sand-kaolinite core was performed to simulate permeability decline due to fines migration at high velocities. The core was comprised of 10% kaolinite and 90% sand by weight. The properties of the artificial core are presented in Table 3.2.

The core was compacted in a core holder to create a reproducible sample with stable permeability. Pressure drop across the core and outlet suspended concentration were measured during the injection period. A full description of the experimental design has been presented by Russell et al. (2017).

The core was flooded with a sodium chloride solution with ionic strength of 0.01 M. Experimental results with the fitted model for a single injection cycle are shown in Fig. 3.12. The injection rate prior to this injection cycle was 40 mL/min (Superficial velocity, $U = 5.869 \times 10^{-4}$ m/s). Increases to velocity prior to this point resulted in negligible permeability decline. The data shown demonstrates the response of the core to increasing the injection rate to 50 mL/min ($U = 7.336 \times 10^{-4}$ m/s). The result of changing the velocity was a reduction in permeability from 235 mD to 219 mD. The results are presented in the form of the dimensionless pressure drop, or the impedance, and the accumulated outlet concentration, which is defined as:

$$
C_{acc} = \alpha \int_0^T C(1, \gamma) d\gamma = \begin{cases} -\dfrac{1}{\Lambda}(\exp(-\alpha\Lambda T) - 1), & T < 1/\alpha \\[2mm] -\dfrac{1}{\Lambda}(\exp(-\Lambda) - 1), & T > 1/\alpha \end{cases}. \quad (3.56)
$$

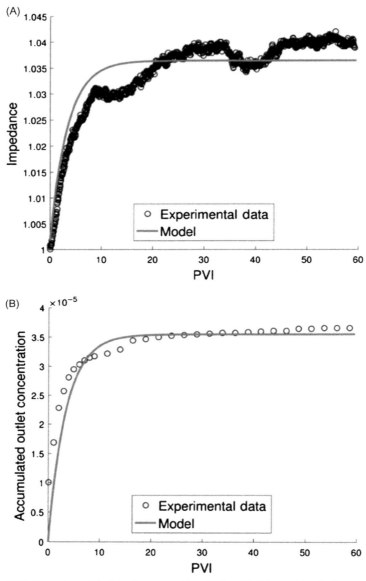

Figure 3.12 Experimental data from coreflood of artificial sand-kaolinite core with the model for velocity-induced fines migration, (A) core impedance, (B) accumulated concentration of suspended particles at the outlet (PVI: number of pore volumes injected).

Table 3.3 Tuned fines-migration parameters (drift delay factor α, concentration of released particles $\Delta\sigma$, filtration coefficient λ, and formation damage coefficient β) from the coreflood data

Parameter	Value
Drift delay factor, α	0.003267
Total detached concentration, $\Delta\sigma$	0.003144
Filtration coefficient, λ (1/m)	1927
Formation damage coefficient, β	28.97
Coefficient of Determination, R^2	0.7693

The data were fitted using the model outlined above. Both datasets were fitted simultaneously using a genetic algorithm least-squared fitting procedure implemented in MATLAB (Mathworks, 2010). Table 3.3 presents the four parameters obtained by fitting the experimental data.

The fitting in Fig. 3.12 shows good agreement between the experimental data and the theoretical model. The stabilization time of this experiment highly exceeds the time to inject a single-pore volume and this feature is captured in the value of the drift delay factor in Table 3.3, which is significantly smaller than one. The fitting suggests that the particles move at a substantially lower velocity than that of the bulk fluid.

Introducing the drift delay factor results in a system of equations capable of modeling fines migration induced by high velocities. The system of equations allows for the analytical solution which is presented here. The analytical solution shows good agreement with the laboratory data and the fitted parameters demonstrate the importance of the new model parameter (i.e., the drift delay factor).

In the following section, it is shown how modeling of fines migration under high velocities can be used to predict the in-flow performance of a production well.

3.4 PRODUCTIVITY DECLINE DUE TO FINES MIGRATION

In the previous section, fines migration induced by high velocities was introduced and a solution was presented for the case of linear, incompressible flow. In the current section, this formulation will be extended to radial coordinates to calculate the impedance for a production well.

Consider flow with fines mobilization, where the attached fine particle concentration is given by the maximum retention function $\sigma_{cr}(U)$. The analytical form of the maximum retention function for a bundle of capillaries of equal size has been derived from the torque balance (Bedrikovetsky et al., 2011a; 2012):

$$\sigma_a = \sigma_{cr}(U) = \sigma_0 \left[1 - \left(\frac{U}{U_m} \right)^2 \right]. \tag{3.57}$$

The attached concentration in the porous media will begin at some initial value, σ_{ai}. For sufficiently small fluid velocities, the critical retention function may lie above this initial value. It follows that for velocities smaller than some velocity, U_i:

$$\sigma_a = \sigma_{ai}. \tag{3.58}$$

That is, the attached concentration remains at its initial value.

For axisymmetric flow during production in a petroleum reservoir, the fluid velocity decreases as the distance from the wellbore increases. Therefore, at some distance from the wellbore, r_i, the fluid velocity will be equal to U_i. No particles will detach at any points further from the wellbore than this distance. The reservoir can thus be divided into two zones, the damaged zone, $r_w < r < r_i$, and the undamaged zone, $r_i < r < r_e$. Here, r_w and r_e are the wellbore and drainage radii, respectively.

In the following formulation, the effect of fines migration on a producing well are calculated. The assumption of fluid incompressibility will only be used for the damaged zone.

3.4.1 Mathematical formulation

First, consider the fluid flow within the damaged zone. Within this zone, the fluid is assumed to be incompressible. The fluid velocity is calculated as:

$$U = \frac{q}{2\pi r}. \tag{3.59}$$

The mass balance equation for suspended, strained, and attached particles in radial coordinates is:

$$r \frac{\partial}{\partial t} (\phi c + \sigma_a + \sigma_s) - \frac{\partial}{\partial r} (rcU) = 0. \tag{3.60}$$

The rate of the particle capture by straining is given by the kinetics equation:

$$\frac{\partial \sigma_s}{\partial t} = \lambda c |U|. \tag{3.61}$$

Darcy's law accounts for permeability damage due to straining:

$$U = -\frac{k}{\mu(1 + \beta \sigma_s)} \frac{\partial p}{\partial r}. \tag{3.62}$$

As particle detachment is instantaneous, the initial condition for the suspended concentration is the difference between the attached concentration and the critical retention function:

$$t = 0 : c(r, 0) = \begin{cases} \dfrac{\sigma_{ai}}{\phi}, & r_w < r < r_m \\[2ex] \dfrac{\sigma_{ai} - \sigma_a\left(\dfrac{q}{2\pi r}\right)}{\phi}, & r_m < r < r_i \\[2ex] 0 & r > r_i \end{cases} \tag{3.63}$$

where

$$r_i = \frac{q}{2\pi |U_i|}. \tag{3.64}$$

As before, the strained concentration begins at zero:

$$t = 0 : \sigma_s = 0. \tag{3.65}$$

The boundary condition for this system is given at the boundary between the damaged and undamaged zones:

$$r = r_i : c(r_i, t) = 0, p = p_i. \tag{3.66}$$

Introducing the following dimensionless variables:

$$X = \left(\frac{r}{r_e}\right)^2, T = \frac{qt}{\pi \phi r_e^2}, \Lambda = \lambda r_e, S_a = \frac{\sigma_a}{\phi \sigma_{ai}}, S_{cr} = \frac{\sigma_{cr}}{\phi \sigma_{ai}} S_s = \frac{\sigma_s}{\phi \sigma_{ai}},$$
$$C = \frac{c}{\sigma_{ai}}, P = \frac{2\pi kp}{\mu q(t)}, \tag{3.67}$$

makes it possible to express the system of Eqs. (3.60—3.62) in dimension-less form:

$$\frac{\partial C}{\partial T} - \frac{\partial C}{\partial X} = -\frac{\partial S_s}{\partial T},$$ (3.68)

$$\frac{\partial S_s}{\partial T} = \Lambda \frac{C}{2\sqrt{X}},$$ (3.69)

$$\frac{2X}{1 + \beta\phi\sigma_{ai}S_s}\frac{\partial P}{\partial X} = 1.$$ (3.70)

For the undamaged zone, $r_i < r < r_e$, the assumption of incompressibility is no longer used. Although it would be preferred to solve in both zones for a compressible fluid for accuracy, the impact of both fluid compressibility and formation damage due to fines migration complicates the derivation of an exact solution. Here, the diffusivity equation determines the pressure distribution within the reservoir:

$$\frac{\partial p}{\partial t} = \frac{k}{\phi\mu c_p}\frac{1}{r}\frac{\partial}{\partial r}\left(r\frac{\partial p}{\partial r}\right),$$ (3.71)

where c_p is the compressibility of the fluid.

The initial condition corresponds to the reservoir pressure:

$$t = 0: p = p_{res}.$$ (3.72)

The outer boundary condition is:

$$r = r_e: \frac{\partial p}{\partial r} = 0.$$ (3.73)

The inner boundary condition is given by Darcy's law:

$$r = r_i: \frac{\partial p}{\partial r} = \frac{q\mu}{2\pi kr_i}.$$ (3.74)

3.4.2 Analytical solution

All particle detachment occurs on the line $T = 0$ and propagates through the X-T plane along characteristics of slope -1. The equations for these characteristics are:

$$X = X_0 - T,$$ (3.75)

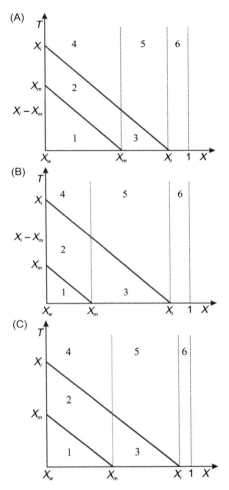

Figure 3.13 Solution zones for the three cases: (A) $X_i < 2X_m$, (B) $X_i > 2X_m$, and (C) $X_i = 2X_m$ (X: dimensionless spatial coordinate, T: dimensionless time, X_i: spatial coordinate of damaged radius (outside of which, no particles are detached), X_m: spatial coordinate of the maximum detachment radius (inside of which, all attached particles are detached)).

where X_0 denotes the starting point on the X-axis. The solution space in Fig. 3.13 is split into multiple zones. The solution for the suspended and strained concentrations is presented below for each zone.

Along the characteristic lines, Eq. (3.68) reduces to the ODE:

$$\frac{dC}{dT} = -\frac{\Lambda C}{2\sqrt{X_0 - T}}. \tag{3.76}$$

Zone 1 ($X_w < X + T < X_m$) contains the zone where the entirety of the initial attached concentration is detached. As such, the suspension concentration in this zone is initially constant with X.

The initial condition for the suspended concentration in zone 1 is given by the first line of Eq. (3.63).

Integrating Eq. (3.76) by separation of variables yields:

$$C(X, T) = e^{A(\sqrt{X} - \sqrt{X+T})}.$$
(3.77)

Integrating Eq. (3.69) yields the strained concentration distribution in this zone:

$$S_s(X, T) = \frac{1}{A\sqrt{X}} \left[A\sqrt{X} + 1 - \left(A\sqrt{X+T} + 1 \right) e^{A(\sqrt{X} + \sqrt{X+T})} \right].$$
(3.78)

Zones 2 ($X_w < X < X_m$, $X_m < X + T < X_i$) and 3 ($X_m < X < X_i$, $X + T < X_i$) both contain suspended particles detached on the X-axis between points X_m and X_i such that the initial suspended concentration is governed by the critical retention function. The initial value of the suspended concentration will be a function of X for these zones.

For both of these zones, Eq. (3.76) is integrated by separation of variables using the middle line of Eq. (3.63) as the initial condition. This yields:

$$C(X, T) = \left(S_{ai} - S_{cr} \left(\frac{q}{2\pi r_i \sqrt{X+T}} \right) \right) e^{A(\sqrt{X} - \sqrt{X+T})}.$$
(3.79)

The strained concentration is derived by integrating Eq. (3.69) as an Ordinary Differential Equation (ODE). Zone 2 will inherit the particles strained in zone 1. The solution (Eq. (3.78)) thus acts as an initial condition for solving the straining kinetics Eq. (3.69). The solution is:

$$S_s(X, T) = \frac{1}{A\sqrt{X}} \left[A\sqrt{X} + 1 - \left(A\sqrt{X_m} + 1 \right) e^{A(\sqrt{X} - \sqrt{X_m})} \right]$$
$$+ \frac{A}{2\sqrt{X}} e^{A\sqrt{X}} \int_{X_m - X}^{T} \left[S_{ai} - S_a \left(\frac{q}{2\pi r_i \sqrt{X+T}} \right) \right] e^{-A\sqrt{X+T}} dT.$$
(3.80)

The first term here corresponds to the strained particles inherited from zone 1. The second term is the straining of particles detached in zone 3. Substituting the expression for the maximum retention function (Eq. (3.57)) enables the evaluation of the integral:

$$S_s(X, T) = \frac{1}{\Lambda\sqrt{X}}\left[\Lambda\sqrt{X} + 1 - \left(\Lambda\sqrt{X_m} + 1\right)e^{\Lambda\left(\sqrt{X} - \sqrt{X_m}\right)}\right]$$

$$+ \frac{\sigma_{ai} - \sigma_0}{\phi\sigma_{ai}\Lambda\sqrt{X}}\left[\left(\Lambda\sqrt{X_m} + 1\right)e^{\Lambda\left(\sqrt{X} - \sqrt{X_m}\right)} - \left(\Lambda\sqrt{X+T} + 1\right)e^{\Lambda\left(\sqrt{X} - \sqrt{X+T}\right)}\right]$$

$$+ \frac{\Lambda\sigma_0}{8\pi^2 r_i^2 U_m^2 \phi\sigma_{ai}\sqrt{X}}e^{\Lambda X}\int_{X_m - X}^{T}\frac{q(T)^2}{X+T}e^{-\Lambda\sqrt{X+T}}dT.$$

$$(3.81)$$

The solution is presented here for an arbitrary injection rate, $q(T)$.

The solution for zone 3 is similar, with the exception that the initial condition corresponds to the absence of strained particles along the line $T = 0$.

$$S_s(X, T) = \frac{\sigma_{ai} - \sigma_0}{\phi\sigma_{ai}\Lambda\sqrt{X}}e^{\Lambda\sqrt{X}}\left[\Lambda\sqrt{X} + 1 - \left(\Lambda\sqrt{X+T} + 1\right)e^{\Lambda\left(\sqrt{X} - \sqrt{X+T}\right)}\right]$$

$$+ \frac{\Lambda\sigma_0}{8\pi^2 r_i^2 U_m^2 \phi\sigma_{ai}\sqrt{X}}e^{\Lambda X}\int_0^{T}\frac{q(T)^2}{X+T}e^{-\Lambda\sqrt{X+T}}dT.$$

$$(3.82)$$

All characteristics comprising zone 4 ($X_w < X < X_m$, $X + T > X_i$) and zone 5 ($X_m < X < X_i$, $X + T > X_i$) begin on the X-axis at X values greater than X_i. As such, the suspended concentration is zero in these zones.

The strained concentration in zone 4 consists of particles strained in zones 1 and 2. The strained particle concentration in zone 5 is equal to the total concentration of particles that strained in zone 3. As such, the strained concentration for these two zones will be given by:

$$S_s(X, T) = S_s(X, X_i - X),$$
$$(3.83)$$

and will be independent of time.

Zone 6 lies entirely outside of the damaged zone. As such, there are no suspended or strained particles in this zone.

$$C(X, T) = S_s(X, T) = 0.$$
$$(3.84)$$

Table 3.4 presents the summary for all the solutions derived in this section.

Table 3.4 Analytical solutions for attached particle concentration S_a, suspended particle concentration C, and strained particle concentration S_s

Line	Term	Zone	Solution
1	$S_a(X,T)$	$X_w < X < X_m$	1
2		$X_m < X < X_i$	$\dfrac{\sigma_0}{\sigma_{ai}}\left[1 - \dfrac{1}{X}\left(\dfrac{qr_e}{2\pi U_m}\right)^2\right]$
3	$C(X,T)$	$X_i < X$	0
4		1	$e^{A(\sqrt{X}-\sqrt{X+T})}$
5		2	$\left(1 - S_\sigma\left(\dfrac{q}{2\pi r_f\sqrt{X+T}}\right)\right)e^{A(\sqrt{X}-\sqrt{X+T})}$
6		3	$\left(1 - S_\sigma\left(\dfrac{q}{2\pi r_f\sqrt{X+T}}\right)\right)e^{A(\sqrt{X}-\sqrt{X+T})}$
7		4	0
8		5	0
9		6	0
10	$S_s(X,T)$	1	$\dfrac{1}{A\sqrt{X}}\left[A\sqrt{X}+1-\left(A\sqrt{X+T}+1\right)e^{A(\sqrt{X}+\sqrt{X+T})}\right]$
11		2	$\dfrac{1}{A\sqrt{X}}\left[A\sqrt{X}+1-\left(A\sqrt{X_m}+1\right)e^{A(\sqrt{X}-\sqrt{X_m})}\right]$ $+ \dfrac{\sigma_{ai}-\sigma_0}{\phi\sigma_{ai}A\sqrt{X}}\left[\begin{array}{l}\left(A\sqrt{X_m}+1\right)e^{A(\sqrt{X}-\sqrt{X_m})}\\ -\left(A\sqrt{X+T}+1\right)e^{A(\sqrt{X}-\sqrt{X+T})}\end{array}\right]$ $+ \dfrac{A\sigma_0}{8\pi^2 r_f^2 U_m^2 \phi\sigma_{ai}\sqrt{X}}e^{AX}\int^T\int_{X_m-X}\dfrac{q(T)^2}{X+T}e^{-A\sqrt{X+T}}\,dT$

3

$$\frac{\sigma_{ai}-\sigma_0}{\phi\sigma_{ai}\Lambda\sqrt{X}}e^{\Lambda\sqrt{X}}\left[\Lambda\sqrt{X}+1-\left(\Lambda\sqrt{X+T}+1\right)e^{\Lambda(\sqrt{X}-\sqrt{X+T})}\right]$$
$$+\frac{\Lambda\sigma_0}{8\pi^2 r_i^2 U_m^2 \phi\sigma_{ai}\sqrt{X}}e^{\Lambda X}\int_0^T \frac{q(T)^2}{X+T}e^{-\Lambda\sqrt{X+T}}dT$$

4

$$\frac{1}{\Lambda\sqrt{X}}\left[\Lambda\sqrt{X}+1-(\Lambda\sqrt{X_m}+1)e^{\Lambda(\sqrt{X}-\sqrt{X_m})}\right]$$
$$+\frac{\sigma_{ai}-\sigma_0}{\phi\sigma_{ai}\Lambda\sqrt{X}}\left[(\Lambda\sqrt{X_m}+1)e^{\Lambda(\sqrt{X}-\sqrt{X_m})}-(\Lambda\sqrt{X_i}+1)e^{\Lambda(\sqrt{X}-\sqrt{X_i})}\right]$$
$$+\frac{\Lambda\sigma_0}{8\pi^2 r_i^2 U_m^2 \phi\sigma_{ai}\sqrt{X}X_i}e^{\Lambda X-\Lambda\sqrt{X_i}}\int_{X_m-X}^{X_i-X}q(T=X_i-X)^2\,dT$$

5

$$\frac{\sigma_{ai}-\sigma_0}{\phi\sigma_{ai}\Lambda\sqrt{X}}e^{\Lambda\sqrt{X}}\left[\Lambda\sqrt{X}+1-(\Lambda\sqrt{X_i}+1)e^{\Lambda(\sqrt{X}-\sqrt{X_i})}\right]$$
$$+\frac{\Lambda\sigma_0}{8\pi^2 r_i^2 U_m^2 \phi\sigma_{ai}\sqrt{X}X_i}e^{\Lambda X-\Lambda\sqrt{X_i}}\int_0^{X_i-X}q(T=X_i-X)^2\,dT$$

6

$$0$$

12

13

14

15

3.4.3 Calculation of impedance

The next step is to calculate the impedance of the well. The impedance is defined as the normalized inverse of the well index.

$$J(T) = \frac{II(0)}{II(T)}, \tag{3.85}$$

where the well index is the flow rate per unit pressure drop:

$$II(T) = \frac{q}{\Delta P}. \tag{3.86}$$

At $T = 0$ there are no suspended particles. Therefore, the initial rate follows from:

$$q_0 = \frac{2\pi k}{\mu} r \frac{\partial p}{\partial r}. \tag{3.87}$$

The initial pressure drop is given by:

$$\Delta p_0 = -\frac{q_0 \mu}{4\pi k} \ln(X_w). \tag{3.88}$$

The dimensionless initial pressure drop is therefore:

$$\Delta P_0 = -\frac{1}{2} \ln(X_w). \tag{3.89}$$

It will from this point be assumed that the production rate from the well is constant. From Eq. (3.85), it then follows that the impedance in the damaged zone is:

$$J_D = \frac{\Delta P(T)}{\Delta P_0} = 1 - \frac{\beta \phi S_{ai}}{\ln(X_w)} \int_{X_w}^{1} \frac{S_s}{X} dX. \tag{3.90}$$

The total well index is the sum of that in damaged zone and that in undamaged zone:

$$J = J_D + J_{UD} = \frac{\Delta P(T) + P_e - P_i}{\Delta P_0}. \tag{3.91}$$

An example calculation of impedance is shown in Fig. 3.14 with field data taken from an oil-producing well. The associated critical retention function is given in Fig. 3.15. The fitted impedance curve using the above model (black curve) shows good agreement with the data.

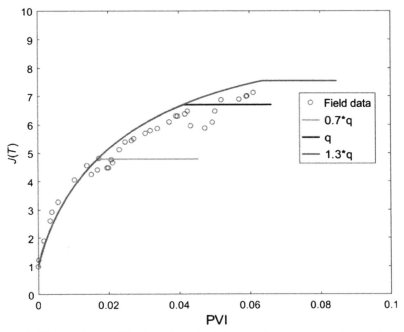

Figure 3.14 Impedance J(T) plotted against the number of pore volumes injected (PVI) for three theoretical curves and field data from an oil-production well.

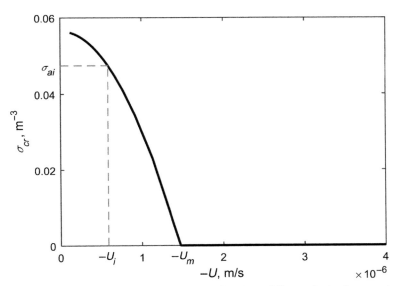

Figure 3.15 Maximum retention function as a function of flow velocity (σ_{cr}: maximum retention function, U: flow velocity).

Two additional curves are shown to demonstrate the effect of production rate. Larger rates create higher fluid velocities in the reservoir which consequently result in more fines detachment and formation damage.

3.5 FINES DETACHMENT AND MIGRATION AT LOW SALINITY

Weakening of the electrostatic force between particle grains during low-salinity waterflooding will lead to particle mobilization, as discussed in Section 3.2. Formation damage during low-salinity waterflooding due to fines mobilization, migration, and straining in thin pores has been widely presented in the literature (Civan, 2014; Sarkar and Sharma, 1990). Significant pressure-drop increase and fines appearance in the core outlet have been observed during numerous corefloods with injection of low-salinity or fresh water (Morrow and Buckley, 2011).

This section presents a mathematical model for fines migration during low-salinity waterflooding. Different from the model for particle detachment due to velocity alteration discussed in Section 3.3, in the case for particle detachment due to salinity alteration particles only detach upon the arrival of the injected low-salinity water. Therefore, it is not possible to assume that particles detach as soon as the salinity of the injected water changes and an equation that describes the transport of salt must be included.

3.5.1 1D analytical solution with instant fines detachment

During low-salinity waterflooding, the colloidal-suspension flux is assumed to be incompressible, so the carrier water flux $U(t)$ is constant (Barenblatt et al., 1989; Lake, 2010). All particle concentrations are small enough not to affect the volumetric water balance. Porosity is considered to be constant. Particles drift with velocity αU that is significantly lower than the carrier water velocity U, i.e., $\alpha << 1$, as stated in Section 3.3 (Oliveira et al., 2014; Yuan and Shapiro, 2011). Particle and salt diffusion are ignored.

To take into account the water salinity alteration, it is necessary to introduce the mass balance equation for salt transport:

$$\phi \frac{\partial \gamma}{\partial t} + U \frac{\partial \gamma}{\partial x} = 0, \tag{3.92}$$

where γ is the salinity.

Including the mass balance for solute concentration in the system of Eqs. (3.33–3.36) presented in Section 3.3 allows for a full system of equations with five equations and five unknowns, c, σ_a, σ_s, γ, and p.

Let us introduce the following dimensionless parameters and the full system of equations for single-phase flow with varying salinity accounting for fines mobilization due to salinity change in the dimensionless form.

$$X \to \frac{x}{L}, \quad T \to \frac{Ut}{\phi L}, \quad C \to \frac{c}{\Delta\sigma}, \quad S_a \to \frac{\sigma_a}{\phi\Delta\sigma}, \quad S_s \to \frac{\sigma_s}{\phi\Delta\sigma},$$

$$\Delta\sigma = S_{cr}(\gamma_I) - S_{cr}(\gamma_J), \quad \Lambda \to \lambda L, \quad P \to \frac{k_0}{U\mu L}p, \quad \Gamma = \frac{\gamma - \gamma_J}{\gamma_I - \gamma_J}, \qquad (3.93)$$

$$\frac{\partial}{\partial T}(C + S_s + S_a) + \alpha\frac{\partial C}{\partial X} = 0, \qquad (3.94)$$

$$\frac{\partial S_s}{\partial T} = \alpha\Lambda C, \qquad (3.95)$$

$$S_a = S_{cr}(\Gamma), \qquad (3.96)$$

$$\frac{\partial\Gamma}{\partial T} + \frac{\partial\Gamma}{\partial X} = 0, \qquad (3.97)$$

$$1 = -\frac{1}{1 + \beta\Delta\sigma\phi S_s}\frac{\partial P}{\partial X}, \qquad (3.98)$$

where Γ is the dimensionless parameter for salinity.

Initial conditions correspond to an absence of suspended and strained particles, salinity of the formation water, and an initial attached concentration given by the value of the maximum retention function for the reservoir conditions

$$T = 0 : C = 0, \quad \Gamma = 1, \quad S_a = S_{cr}(\gamma_I), \quad S_s = 0. \qquad (3.99)$$

Boundary conditions correspond to injection of particle-free water with given salinity:

$$X = 0 : C = 0, \quad \Gamma = 0. \qquad (3.100)$$

The five Eqs. (3.94–3.98) subject to initial and boundary conditions (Eqs. (3.99 and 3.100)) determine unknowns C, S_a, S_s, Γ, and P. The salt transport Eq. (3.97) separates from the rest of the system, and as such can be solved separately. Pressure $P(X,T)$ is determined by integration of

Eq. (3.98) after solving the system of Eqs. (3.94–3.96). The exact solution for 1D flow with salinity alteration and fines migration is derived below. The solution is based on the mass-balance condition on a shock front, and on the method of characteristics (Polyanin and Zaitsev, 2011; Polyanin and Manzhirov, 2007).

Eqs. (3.96 and 3.97) for attached concentration and mass balance of salt transport both separate from the rest of system (Eqs. (3.94–3.98)) and can be solved immediately using the Method of Characteristics subject to the initial (Eq. (3.99)) and boundary (Eq. (3.100)) conditions. It follows that the salinity is equal to the formation water salinity ahead of the salinity front and equal to injected water salinity behind the front. This is shown in the dimensionless form in Eq. (3.101). The solution for the attached concentration corresponds to the critical retention function at the two salinities behind and ahead of the shock (Eq. (3.102)):

$$\Gamma = \begin{cases} 1, & X > T \\ 0, & X < T \end{cases}, \tag{3.101}$$

$$S_a = \begin{cases} S_{cr}(\gamma_I), & X > T \\ S_{cr}(\gamma_J), & X < T \end{cases}. \tag{3.102}$$

In all zones, the salinity values are constant, so the time derivative of attached concentration is equal to zero. Substituting the straining rate (Eq. (3.95)) into Eq. (3.94) yields:

$$\frac{\partial C}{\partial T} + \alpha \frac{\partial C}{\partial X} = -\alpha \Lambda C. \tag{3.103}$$

The above Partial Differential Equation (PDE) is solved using the method of characteristics in three different zones: ahead of the salinity front $(X > T)$, between the salinity front and the suspended particle front $(\alpha T < X < T)$, and behind the suspended particle front $(X < \alpha T)$. Fig. 3.16A shows the three different zones indicated by the numbers 0, 1 and 2, respectively.

For $X > T$, zone 0, along the parametric curves given by:

$$\frac{dX}{dT} = \alpha, \tag{3.104}$$

Eq. (3.103) reduces to the ordinary differential equation:

$$\frac{dC}{dT} = -\alpha \Lambda C. \tag{3.105}$$

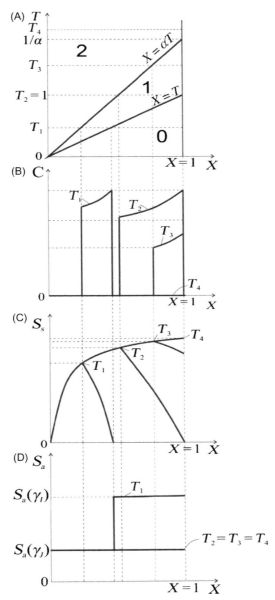

Figure 3.16 Exact solution of low-salinity coreflooding with fines migration: (A) concentration front in (X,T)-plane; (B) suspended concentration profiles in four moments $k = 1,2,3,4$; (C) profiles of strained concentration in four moments; (D) profiles (X: dimensionless linear coordinate, T: dimensionless time, C: dimensionless suspended concentration, S_s: dimensionless strained concentration, S_a: dimensionless attached concentration).

Solving Eq. (3.105) by separation of variables, accounting for initial condition (Eq. (3.99)), yields $C(X, T) = 0$ ahead of the salinity front.

To find a solution between the suspended particle and the salinity front, zone $(\alpha T < X < T)$, it is necessary to calculate the boundary condition for suspended particle concentration behind the front $X = T$. Consider the mass balance condition on the salinity front (Bedrikovetsky, 1993; Lake, 2010):

$$[C + S_s + S_a]D = \alpha[C], \tag{3.106}$$

where D is the shock front velocity, and jump of the quantity A is the difference between A-values ahead and behind the front: $[A] = A^+ - A^-$ (Polyanin and Zaitsev, 2011; Polyanin and Manzhirov, 2007). As it follows from the kinetics Eq. (3.95), the strained concentration is continuous across the shock. The values of the attached concentration ahead and behind the shock follow from the solution (Eq. (3.102)), and the value of the suspended concentration ahead of the shock follows from the solution to Eq. (3.103) for zone 0:

$$C^+ = 0, \quad S_a^+ = S_{cr}(\gamma_I), \quad S_a^- = S_{cr}(\gamma_J), \tag{3.107}$$

and as the shock is being evaluated across the line $X = T$, the shock velocity equals one.

Substituting Eq. (3.107) into Eq. (3.106) yields the suspended concentration behind the salinity front:

$$C = \frac{1}{(1 - \alpha)\phi}. \tag{3.108}$$

Solving Eq. (3.105) by separation of variables, accounting for the boundary condition for suspended particle concentration behind the front $X = T$, yields:

$$C(X, T) = \frac{1}{(1 - \alpha)\phi} \exp\left(-\alpha\Lambda\left(\frac{T - X}{1 - \alpha}\right)\right). \tag{3.109}$$

Using the method of characteristics for Eq. (3.103) along with the boundary condition (Eq. (3.100)) yields a zero suspended concentration $C(X, T) = 0$ behind the suspended particle front $X = \alpha T$, zone 2.

The strained concentration is obtained from Eq. (3.95) by integration of the suspended concentration in time. Table 3.5 presents a summary of the solutions for salt transport, and attached, suspended, and strained concentrations in the different zones.

Table 3.5 Formulae for salinity Γ, attached concentration S_a, suspended concentration C, strained concentration S_s, and impedance J during low-salinity coreflooding in the linear (X,T) analytical model

Line	Term	Zones	Solution
1	$\Gamma(X,T)$	$X > T$	1
2		$X < T$	0
3	$S_a(X,T)$	$X > T$	$S_\sigma\left(\gamma_I\right)$
4		$X < T$	$S_\sigma\left(\gamma_J\right)$
5	$C(X,T)$	$X > T$	0
6		$\alpha T < X < T$	$\frac{1}{(1-\alpha)\phi}\exp\left(-\alpha\Lambda\left(\frac{T-X}{1-\alpha}\right)\right)$
7	$S_s(X,T)$	$X < \alpha T$	0
8		$X > T$	0
9		$\alpha T < X < T$	$\frac{1}{\phi}\left[1 - \exp\left(-\alpha\Lambda\left(\frac{T-X}{1-\alpha}\right)\right)\right]$
10		$X < \alpha T$	$\frac{1}{\phi}\left[1 - \exp\left(-\Lambda X\right)\right]$
11	$J(T)$	$T < 1$	$1 + \beta\Delta\sigma\left[T + \frac{1}{\Lambda}(\exp(-\alpha\Lambda T)-1) + \frac{(1-\alpha)}{\alpha\Lambda}(\exp(-\alpha\Lambda T)-1)\right]$
12		$1 < T < \frac{1}{\alpha}$	$1 + \beta\Delta\sigma\left[1 + \frac{1}{\Lambda}(\exp(-\alpha\Lambda T)-1) + \frac{(1-\alpha)}{\alpha\Lambda}\left(\exp(-\alpha\Lambda T)-\exp\left(\alpha\Lambda\left(\frac{1-T}{1-\alpha}\right)\right)\right)\right]$
13		$T > \frac{1}{\alpha}$	$1 + \beta\Delta\sigma\left[1 + \frac{1}{\Lambda}(\exp(-\Lambda)-1)\right]$

Dimensionless pressure drop across the core J, called impedance, is calculated from Darcy's law (Eq. (3.98)):

$$J(T) = P(0, T) - P(1, T) = 1 + \beta\phi\Delta\sigma \int_0^1 S_s(X, T)dX. \qquad (3.110)$$

Substituting the expression for strained concentration into Eq. (3.110) and integrating in X yield:

$$J(T) = \begin{cases} 1+\beta\Delta\sigma\left[T+\dfrac{1}{\Lambda}(\exp(-\alpha\Lambda T)-1)+\dfrac{(1-\alpha)}{\alpha\Lambda}(\exp(-\alpha\Lambda T)-1)\right], T<1 \\[2em] 1+\beta\Delta\sigma\left[1+\dfrac{1}{\Lambda}(\exp(-\alpha\Lambda T)-1)+\dfrac{(1-\alpha)}{\alpha\Lambda}\right. \\[1em] \left.\left(\exp(-\alpha\Lambda T)-\exp\left(\alpha\Lambda\left(\dfrac{1-T}{1-\alpha}\right)\right)\right)\right], \qquad 1<T<\dfrac{1}{\alpha} \\[2em] 1+\beta\Delta\sigma\left[1+\dfrac{1}{\Lambda}(\exp(-\Lambda)-1)\right], \qquad T>\dfrac{1}{\alpha} \end{cases}$$

$$(3.111)$$

3.5.1.1 Qualitative analysis of the model

Figs. 3.16B, C, and D present profiles of suspended, strained, and attached concentrations, respectively, in four different moments. Moment T_1 is before the salinity front arrival at the outlet $(X=1)$, $T_2 = 1$ PVI is the exact moment when the injected salinity reaches the outlet, T_3 is before the particle front arrival at the outlet, and T_4 is at some point afterward.

First, consider the behavior of profiles in Figs. 3.16B and C. Consider two points $X_1 < X_2$ at the moment T_1 in zone 1. Suspended particles at point X_1 are submitted to deep bed filtration longer than those at point X_2; the released concentration C^- behind the salinity front is constant. Therefore, suspended concentration C at point X_1 is lower than that at point X_2. Suspended concentration monotonically increases along the core in zone 1.

Straining at point X_1 has occurred for longer than at point X_2, so the strained concentration at point X_1 is higher than that at point X_2. It follows that the strained concentration monotonically decreases along the core in zone 1.

Now, consider two points $X_1 < X_2$ at the moment T_3 in zone 2. There are no suspended particles in zone 2, so the strained profile is constant with time. Strained concentration in zone 2 remains the same as that reached at the moment of the passing of the particle front. In zone 1, particle straining at point X_2 has been occurring for longer than at point X_1, so the strained concentration at point X_2 is higher than at point X_1. The envelope profile in Fig. 3.16C increases.

3.5.2 Tuning experimental data

Three Berea sandstone cores have been used for coreflooding with water injection subjected to piecewise decreasing salinity. Table 3.6 presents the core properties. The sodium chloride solutions were prepared with Milli-Q water, which is ultrapure deionized water, filtered with a 0.22 μm filter. The tests start by injecting high–salinity water, which has 0.6 M (35,000 ppm) of NaCl. Then low–salinity water with 0.035 M (2000 ppm) of NaCl was injected. Finally, fresh water was injected. In test 3, low-salinity water with 0.018 M (1000 ppm) of NaCl was injected before fresh water.

These studies on fines mobilization due to low-salinity waterflooding and consequent permeability damage were carried out using a permeability apparatus. The laboratory setup consisted of a core-holder where the core was placed inside a 1.5" Viton sleeve. An overburden pressure of 1000 psi was applied. The solutions were injected into the core through a pump that delivered a constant flow rate of 0.5 mL/min. The pressure transducer measured the pressure drop across the overall core. The effluent was collected in glass sampling tubes located in a sampling collector for each pore volume produced.

After placing the core in the core-holder, injection of high-salinity water was carried out until the pressure drop stabilized, allowing for the absolute permeability to be calculated. Injection of low-salinity and fresh water follow. The pressure drop across the core was measured during the entire experiment for each core and particle concentration was

Table 3.6 Rock properties

Test	Absolute permeability (mD)	Porosity (%)	Pore volume (mL)	Core diameter (mm)	Core length (mm)
1	50	18	9.97	36.14	52.71
2	35	17	9.48	36.04	52.26
3	87	21	11,99	37.67	51.08

determined for each sample which allowed calculating the accumulated concentration in the outlet over time. Figs. 3.17—3.19 present the results of three corefloods.

The model contains four parameters: the drift delay factor α, the filtration coefficient λ, the concentration of released particles $\Delta\sigma$, and the formation damage coefficient β. The parameters are determined by the least squares method, i.e., by minimization of the total deviation between the modeling and laboratory curves for accumulated particle concentration and impedance. The reflective trust region algorithm (Coleman and Li, 1996) was applied to solve the optimization problem using the software MATLAB (Mathworks, 2010).

The initial value for the drift delay factor α is determined from the stabilization time for impedance, which is equal to $1/\alpha$. Accumulated particle breakthrough concentration is calculated from the exact solution:

$$c_{acc} = \alpha\Delta\sigma \int_0^T C(1,y)dy$$

$$= \begin{cases} 0, & T<1 \\ -a(\exp(-b(T-1))-1), & 1<T<1/\alpha, \quad a=\dfrac{\Delta\sigma}{\phi\Lambda}, \ b=\dfrac{\alpha\Lambda}{(1-\alpha)}, \\ -a\left(\exp\left(-b\left(\dfrac{1-\alpha}{\alpha}\right)\right)-1\right), & T>1/\alpha \end{cases}$$

$$(3.112)$$

where a and b are fitting parameters.

Fitting the curve for experimental accumulative particle breakthrough concentration with exponential formula (Eq. (3.112)) enables the calculation of the coefficients a and b. The initial filtration coefficient Λ from Eq. (3.112) for known α is determined from the expression for b. The initial value for the concentration of released particles $\Delta\sigma$ is determined for the known filtration coefficient Λ from the expression for a. The initial value for the formation damage coefficient β is determined by fitting the impedance data by the analytical formula (Eq. (3.111)). Afterward, the initial values are used in a minimization algorithm for model adjustment. Table 3.7 presents the tuned values of the four parameters. The coefficient of determination R^2 in all cases exceeds 0.95, demonstrating a high agreement between the experimental data and the theoretical model. The obtained values of the model coefficients α, λ, $\Delta\sigma$, and β in

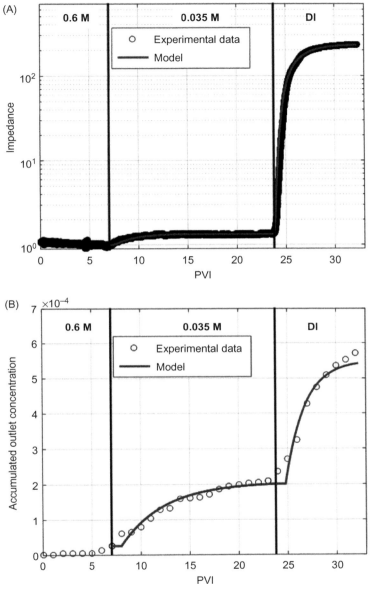

Figure 3.17 Laboratory data and the fitted analytical model: (A) impedance $J(T)$; (B) accumulated breakthrough concentration − Test 1 (PVI: number of pore volumes injected).

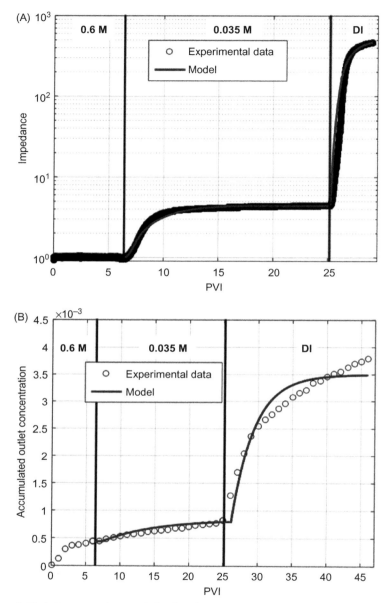

Figure 3.18 Laboratory data and the fitted analytical model: (A) impedance $J(T)$; (B) accumulated breakthrough concentration — Test 2 (PVI: number of pore volumes injected).

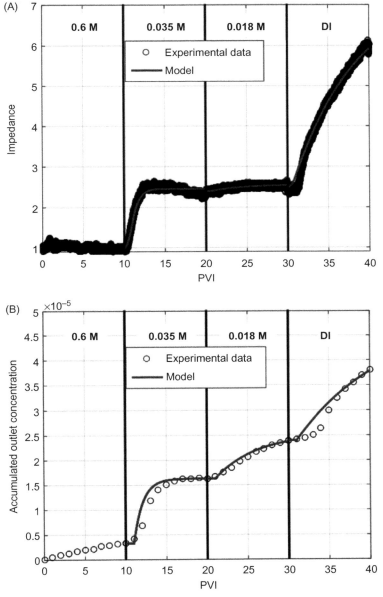

Figure 3.19 Laboratory data and the fitted analytical model: (A) impedance $J(T)$; (B) accumulated breakthrough concentration – Test 3 (PVI: number of pore volumes injected).

Table 3.7 Tuning the fines-migration parameters (drift delay factor α, concentration of released particles $\Delta\sigma$, filtration coefficient λ, and formation damage coefficient β) from the coreflood data

Test	γ_J	A	$\Delta\sigma$ (ppm)	λ (1/m)	β	R^2
1	0.035 M	0.11	78.6	46.04	6696	0.96
	DI	0.05	598.5	180.23	290670	0.98
2	0.035 M	0.08	101.6	28.60	70663	0.97
	DI	0.05	2600	109.07	76852	0.96
3	0.035 M	0.2	9.8	70.47	205050	0.97
	0.018 M	0.06	5.9	61.34	12281	0.98
	DI	0.03	15.3	56.96	191230	0.98

Table 3.7 belong to common intervals reported in the literature (Oliveira et al., 2014; Vaz et al., 2017; Yuan and Shapiro, 2011; Zeinijahromi et al., 2013).

3.5.3 Well injectivity decline during low-salinity water injection

Injectivity impairment is one of the main challenges of waterflooding projects (Civan, 2014). Injectivity decline is commonly associated with the capture of injected foreign particles and natural reservoir fines, and also due to the formation of an external filter cake (Kalantariasl and Bedrikovetsky, 2013; Pang and Sharma, 1997). The reliable prediction of well injectivity highly affects planning and design of waterflooding operations. Usually, reliable well behavior prediction involves laboratory-based mathematical modeling. This section presents a mathematical model for injectivity decline due to fines migration during low–salinity waterflooding. The assumptions of the model are the same as those formulated in Section 3.5.1.

The governing system consists of a mass balance for suspended, attached, and strained particles, kinetics rate for straining, expression for maximum retention function, mass balance for salt, and Darcy's law for axisymmetric flow, accounting for permeability damage due to fines straining. Assuming incompressibility of the carrier fluid, the velocity profile at any distance from the well follows:

$$U(r) = \frac{q}{2\pi r}, \qquad (3.113)$$

where r is the radial coordinate, and q is the injection rate per unit of formation thickness.

Inserting Eq. (3.113) in the basic governing system mentioned above yields the following system of equations in radial coordinates:

$$\frac{\partial}{\partial t}(\phi c + \sigma_s + \sigma_a) + \frac{\alpha q}{2\pi r}\frac{\partial c}{\partial r} = 0, \tag{3.114}$$

$$\frac{\partial \sigma_s}{\partial t} = \frac{\alpha \lambda c q}{2\pi r}, \tag{3.115}$$

$$\sigma_a = \sigma_{cr}\left(\frac{q}{2\pi r}, \gamma\right), \tag{3.116}$$

$$\phi\frac{\partial \gamma}{\partial t} + \frac{q}{2\pi r}\frac{\partial \gamma}{\partial r} = 0, \tag{3.117}$$

$$\frac{q}{2\pi r} = -\frac{k}{\mu(1 + \beta\sigma)}\frac{\partial p}{\partial r}. \tag{3.118}$$

Introducing the following parameters:

$$T \to \frac{qt}{\pi r_e^2 \phi}, X \to \left(\frac{r}{r_e}\right)^2, S_s \to \frac{\sigma_s}{\phi}, S_a \to \frac{\sigma_a}{\phi}, S_{cr} \to \frac{\sigma_{cr}}{\phi}, \Lambda \to \lambda r_w, P \to \frac{\pi k p}{q\mu}, \Gamma = \frac{\gamma - \gamma_J}{\gamma_I - \gamma_J}, \tag{3.119}$$

where r_e is the drainage radius, X is the dimensionless radial coordinate, and r_w is the wellbore radius. This transforms the system (Eqs. (3.114−3.118)) into the following dimensionless form:

$$\frac{\partial(C + S_s + S_a)}{\partial T} + \alpha\frac{\partial C}{\partial X} = 0, \tag{3.120}$$

$$\frac{\partial S_s}{\partial T} = \frac{\alpha \Lambda C}{2\sqrt{X}\sqrt{X_w}}, \tag{3.121}$$

$$S_a = S_{cr}\left(\frac{q}{2\pi r_e\sqrt{X}}, \Gamma\right), \tag{3.122}$$

$$\frac{\partial \Gamma}{\partial T} + \frac{\partial \Gamma}{\partial X} = 0, \tag{3.123}$$

$$\frac{1}{X} = -\frac{2}{(1 + \beta\phi S_s)}\frac{\partial P}{\partial X}, \tag{3.124}$$

Mass balance Eq. (3.120) for axisymmetric flow is the same as the one for linear flow, given by Eq. (3.42). Appearance of $\sqrt{X}$ in the

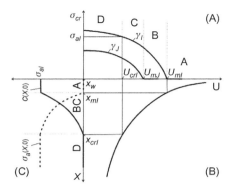

Figure 3.20 Graphical calculations of (A) maximum retention functions for initial and injected salinity in the areas A, B, C, and D; (B) velocity profile $U(X)$; (C) initial concentrations for suspended and attached particle concentration (σ_{cr}: maximum retention function, σ_{al}: initial attached concentration, γ: fluid salinity, U: fluid velocity, X: dimensionless spatial coordinate, σ_a: attached concentration,. c: suspended concentration).

denominator in Eq. (3.121) corresponds to the radius–dependency of velocity for axisymmetric flow, as expressed by Eq. (3.113). The change in the expression for Darcy's law, when compared with linear-flow given by Eq. (3.45), is due to the radius-dependency of velocity and of co-ordinate X (Eq. (3.124)).

Fig. 3.20A shows two arbitrary maximum retention functions for initial and injected salinities and the initial attached particle concentration (S_{al}). As the injected salinity is lower than the initial salinity, the maximum retention function for injected salinity lies below the initial salinity. U_{mI} and U_{mJ} are the maximum velocities, above which no further particle detachment is expected. As the electrostatic force decreases with salinity, the critical velocity for particle detachment will be also smaller and thus $U_{mI} > U_{mJ}$.

Initially, the drag force exceeds the maximum electrostatic force at high velocities $U > U_{mI}$, where the maximum retained concentration is equal to zero (Bedrikovetsky et al., 2011a; 2012). For the fine particles attached with initial concentration S_{aI}, the drag force does not exceed the maximum electrostatic force for low velocities $U < U_{crI}$, where no attached particles are removed by the flow. The attached concentration is equal to the maximum retention concentration S_{cr} for velocities varying from U_{crI} to U_{mI}.

The incompressibility assumption implies that the velocity profile $U(X)$ (Fig. 3.20B), given by Eq. (3.113), is established instantly from the very beginning of injection, at $t = +0$. The excess attached particles over the maximum retention concentration are removed. In some region near to the wellbore $X_w < X < X_{mI}$, the velocity is sufficient to detach all particles. At some distance far enough away from the wellbore, $X > X_{crI}$, no particles are mobilized at all. These distances are defined by:

$$X_w = \left(\frac{r_w}{r_e}\right)^2, \quad X_{mI} = \left(\frac{q}{2\pi U_{mI} r_e}\right)^2, \quad X_{mJ} = \left(\frac{q}{2\pi U_{mJ} r_e}\right)^2,$$

$$X_{crI} = \left(\frac{q}{2\pi U_{crI} r_e}\right)^2, \quad S_{cr}(U_{crI}, \gamma_I) = S_{aI} ,$$

(3.125)

where U_{mI} is the maximum velocity for which no particles can be held on the grain surface for initial water salinity, U_{mJ} is the maximum velocity for which no particles can be held on the grain surface for injected water salinity, U_{crI} is the critical particle detachment velocity for initial water salinity, X_{mI} is dimensionless radius where water velocity is equal to U_{mI}, X_{mJ} is dimensionless radius where the fluid velocity is equal to U_{mJ}, X_{crI} is the dimensionless radius corresponding to velocity U_{crI}, where initial attached concentration is equal to the critical value; X_w is the dimensionless radial coordinate for $r = r_w$. The relationship $U = U(X)$ translates typical velocities U_{mI} and U_{crI} in Eq. (3.125) into positions where those velocities are reached at X_{mI} and X_{crI}.

Fine particles detached due to high velocity near to the wellbore are removed from the attached state into suspension at the beginning of injection ($t = +0$). Therefore, initial suspension is constant near to the wellbore at $X_w < X < X_{mI}$, and zero outside the fines lifting zone for $X > X_{crI}$. The remaining attached concentration is equal to maximum retention function in the intermediate zone $X_{mI} < X < X_{crI}$. The initial conditions for the system (Eq. (3.120−3.124)) are:

$$T = 0: S_a(X, 0) = \begin{cases} 0, & X_w < X < X_{mI} \\ \dfrac{1}{\phi} S_{cr}\left(\dfrac{q}{2\pi r_e \sqrt{X}}, \gamma_I\right), & X_{mI} < X < X_{crI} , \\ S_{aI}/\phi, & X > X_{crI} \end{cases}$$

$$c(X, 0) = \begin{cases} S_{aI}/\phi, & X_w < X < X_{mI} \\ \dfrac{1}{\phi}\left(S_{aI} - S_{cr}\left(\dfrac{q}{2\pi r_e \sqrt{X}}, \gamma_I\right)\right), & X_{mI} < X < X_{crI} \\ 0, & X > X_{crI} \end{cases}$$

, (3.126)

$$\Gamma = 1$$

Fig. 3.20C shows the initial conditions (Eq. (3.126)). The sum of profiles of initial suspended and attached concentrations is equal to S_{aI}, which corresponds to the removal of the excess of the initial attached concentration S_{aI} over the maximum retention concentration into the suspension. Fig. 3.20C also shows those intervals in axis X, where $T = 0$; the corresponding areas are A, B, C and D, respectively.

The boundary condition is set at the wellbore:

$$X = X_w: c = 0, \ \Gamma = 0. \tag{3.127}$$

The radial system is composed of five Eqs. (3.120–3.124) in five variables (c, S_a, S_s, Γ, and P) and therefore presents a closed system. Eq. (3.123) for salt transport separates from Eqs. (3.120–3.122, and 3.124) and can be solved directly by the method of characteristics subject to the initial (Eq. (3.126)) and boundary (Eq. (3.127)) conditions outlined above. The solution corresponds to the initial salinity γ_I ahead of salinity front $X - X_w = T$, and the injected salinity γ_J behind this front. The solution for salt transport is presented in the dimensionless form as follows:

$$\Gamma = \begin{cases} 1, & X - X_w > T \\ 0, & X - X_w < T \end{cases}. \tag{3.128}$$

Initial conditions for the attached concentration (Eq. (3.122)) remain constant with time ahead of the salinity front in zones A, B, C, and D, i.e.,

$$S_a(X, T) = \begin{cases} 0, & X_w < X < X_{mI}, \ T < X - X_w \\ \dfrac{1}{\phi} S_{cr}\left(\dfrac{q}{2\pi r_e \sqrt{x}}, \gamma_I\right), & X_{mI} < X < X_{crI}, \ T < X - X_w \\ S_{aI}/\phi, & X > X_{crI}, \ T < X - X_w \end{cases}. \tag{3.129}$$

Fig. 3.20A shows the maximum retention curve behind the salinity front for $\gamma = \gamma_J$. The attached concentration behind the salinity front is zero for $X_w < X < X_{mJ}$ in areas A and B, and equal to maximum retention function for $X > X_{mJ}$ in areas C and D:

$$S_a(X, T) = \begin{cases} 0, & X_w < X < X_{mJ}, \ T > X - X_w \\ \dfrac{1}{\phi} S_{cr}\left(\dfrac{q}{2\pi r_e \sqrt{X}}, \gamma_J\right), & X_{mJ} < X < \infty, \ T > X - X_w \end{cases} .$$

(3.130)

Therefore, the attached concentration corresponds to the value of the maximum retention function of γ_I ahead of salinity front X-$X_w = T$, and of γ_J behind this front. The attached concentrations are steady state, which cancels S_a in the accumulative term of the mass balance Eq. (3.120). Substituting the straining rate (Eq. (3.121)) into Eq. (3.120) yields:

$$\frac{\partial c}{\partial T} + \alpha \frac{\partial c}{\partial X} = -\frac{\alpha \Lambda c}{2\sqrt{X}\sqrt{X_w}}.$$

(3.131)

In zones 0, 1, and 3 (Fig. 3.20B), Eq. (3.131) subject to initial conditions (Eq. (3.126)) is solved by the method of characteristics. Here, time T is set as the parameter along the characteristic lines:

$$\frac{dX}{dT} = \alpha, \quad X - X_w - X_0 = \alpha(T - T_0), \quad \frac{dc}{dT} = -\frac{\alpha \Lambda c}{2\sqrt{X}\sqrt{X_w}}, \quad (3.132)$$

where X_0 corresponds to the intersection of the characteristic line with the X-axis. Formula (Eq. (3.126)) for the suspended concentration for three X-intervals provides initial conditions for the ordinary differential Eqs. (3.132) in zones 3, 1, and 0, respectively.

Separation of variables in Eq. (3.132) yields explicit formulae for suspended concentration. In zone 0, the initial suspended concentration is zero. The solution along characteristic lines yields zero suspended concentration in the overall zone 0.

In zone 3, the initial suspended concentration c is a constant given by Eq. (3.126). The solution $c(X, T)$ along characteristics is given by the formula listed in the tenth row of Table 3.8. The solution is independent of X in zone 3. The solution in zone 1 with X-distributed initial suspended concentration given by formula (Eq. (3.126)) is presented in the formula given in the ninth row of Table 3.8.

Table 3.8 Formulae for salinity and particle concentrations in the axisymmetric analytical model during low-salinity water injection

Line	Term	Zones	Solution
1	$\Gamma(X,T)$	$X > X_w + T$	1
2		$X < X_w + T$	0
3	$S_a(X,T)$	A	0
4		$X > X_w + T$ B, C	$\frac{1}{\phi} S_\sigma\left(\frac{q}{2\pi r_e \sqrt{X}}, \gamma_1\right)$
5		D	S_{al}/ϕ
6		$X > X_w + T$ A, B	0
7		$X < X_w + T$ C, D	$\frac{1}{\phi} S_\sigma\left(\frac{q}{2\pi r_e \sqrt{X}}, \gamma_1\right)$
8	$c(X,T)$	0	0
9		1	1
10		3	$\frac{1}{\phi}\left(S_{al} - S_\sigma\left(\frac{q}{2\pi r_e \sqrt{X}}, \gamma_1\right)\right)\exp\left(-\frac{A}{\sqrt{X_w}}\left(\sqrt{X} - \sqrt{X - \alpha T}\right)\right)$
11		4	$S_{al}/\phi\,\exp\left(-\frac{A}{\sqrt{X_w}}\left(\sqrt{X} - \sqrt{X - \alpha T}\right)\right)$
12		5	$\left[S_{al}/\phi\,\exp\left(-\frac{A}{\sqrt{X_w}}\left(\sqrt{X} - \sqrt{(1-\alpha)X + \alpha X_w}\right)\right) + S_{al}/\phi\,\exp\left(-\frac{A}{\sqrt{X_w}}\left(\sqrt{X} - \sqrt{(1-\alpha)X + \alpha X_w}\right)\right)\right]\exp\left(-\frac{A}{\sqrt{X_w}}\left(\sqrt{X} - \sqrt{\frac{X - \alpha T}{1-\alpha}}\right)\right)$
13		6	$\left[\frac{1}{(1-\alpha)\phi} S_\sigma\left(\frac{q}{2\pi r_e \sqrt{X}}, \gamma_1\right) + \frac{1}{\phi}\left(S_{al} - S_\sigma\left(\frac{q}{2\pi r_e \sqrt{X}}, \gamma_1\right)\right)\right]\exp\left(-\frac{A}{\sqrt{X_w}}\left(\sqrt{X} - \sqrt{(1-\alpha)X + \alpha X_w}\right)\right)\exp\left(-\frac{A}{\sqrt{X_w}}\left(\sqrt{X} - \sqrt{\frac{X - \alpha T}{1-\alpha}}\right)\right)$
14		7	$\left[\frac{1}{(1-\alpha)\phi} S_\sigma\left(\frac{q}{2\pi r_e \sqrt{X}}, \gamma_1\right) + \frac{1}{\phi}\left(S_{al} - S_\sigma\left(\frac{q}{2\pi r_e \sqrt{X}}, \gamma_J\right)\right)\right]\exp\left(-\frac{A}{\sqrt{X_w}}\left(\sqrt{X} - \sqrt{(1-\alpha)X + \alpha X_w}\right)\right)\exp\left(-\frac{A}{\sqrt{X_w}}\left(\sqrt{X} - \sqrt{\frac{X - \alpha T}{1-\alpha}}\right)\right)$
15		8	$\left[\frac{1}{(1-\alpha)\phi}\left(S_\sigma\left(\frac{q}{2\pi r_e \sqrt{X}}, \gamma_1\right) - S_\sigma\left(\frac{q}{2\pi r_e \sqrt{X}}, \gamma_J\right)\right) + \frac{1}{\phi}\left(S_{al} - S_\sigma\left(\frac{q}{2\pi r_e \sqrt{X}}, \gamma_J\right)\right)\right]\exp\left(-\frac{A}{\sqrt{X_w}}\left(\sqrt{X} - \sqrt{(1-\alpha)X + \alpha X_w}\right)\right)\exp\left(-\frac{A}{\sqrt{X_w}}\left(\sqrt{X} - \sqrt{\frac{X - \alpha T}{1-\alpha}}\right)\right)$
16		9	$\frac{1}{(1-\alpha)\phi}\left[S_{al} - S_\sigma\left(\frac{q}{2\pi r_e \sqrt{X}}, \gamma_J\right)\right]\exp\left(-\frac{A}{\sqrt{X_w}}\left(\sqrt{X} - \sqrt{\frac{X - \alpha T}{1-\alpha}}\right)\right)$
17		2	0

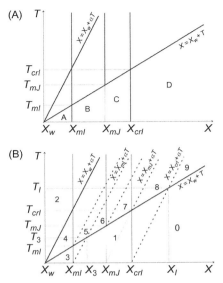

Figure 3.21 (A) Particle and (B) front trajectories in different flow zones in (X,T) plane (X: dimensionless spatial coordinate, T: dimensionless time).

Suspended concentration ahead of the salinity front is obtained by substitution of the trajectory coordinate $X = X_w + T$ into formulae for c (X,T) in zones 3, 1, and 0:

$$c^+(X_w + T, T)$$

$$= \begin{cases} S_{aI}/\phi \exp\left(-\dfrac{\Lambda}{\sqrt{X_w}}\left(\sqrt{X} - \sqrt{(1-\alpha)X + \alpha X_w}\right)\right), & T < T_1 \\ \dfrac{1}{\phi}\left(S_{aI} - S_{cr}\left(\dfrac{q}{2\pi r_e \sqrt{X}}, \gamma_I\right)\right)\exp\left(-\dfrac{\Lambda}{\sqrt{X_w}}\left(\sqrt{X} - \sqrt{(1-\alpha)X + \alpha X_w}\right)\right), & T_1 < T < T_3 \\ 0, & T > T_3 \end{cases}$$

$$(3.133)$$

Fig. 3.21A shows the moments T_1 and T_3 of arrival of the boundaries between zones 3, 1, and 0 to the salinity front:

$$T_3 = \frac{X_{mJ} - X_w}{1 - \alpha}, \quad T_1 = \frac{X_{crI} - X_w}{1 - \alpha}. \qquad (3.134)$$

Next, the suspension concentration released over the salinity front $X = X_w + T$ is calculated. i.e., $c^-(X, X-X_w) - c^+(X, X-X_w) = -[c]$. Kinetics Eq. (3.121) yields the continuous strained concentration. The salinity front speed is equal to unity. Particle mass balance on the salinity front follows from condition (Eq. (3.106)):

$$[\sigma_a] = (\alpha - 1)[c]. \tag{3.135}$$

According to Eq. (3.122), the attached concentrations ahead and behind the salinity front are:

$$S_a^+ = \begin{cases} S_{cr}\left(\dfrac{q}{2\pi r_e \sqrt{X_w + T}}, \gamma_I\right), & U > U_{crI} \\ S_{aI}, & U < U_{crI} \end{cases}, \quad S_a^- = S_{cr}\left(\dfrac{q}{2\pi r_e \sqrt{X_w + T}}, \gamma_J\right). \tag{3.136}$$

The jump in attached concentration across the salinity front $X = X_w + T$ is:

$$S_a^+(X, X_w + T) - S_a^-(X, X_w + T)$$

$$= \begin{cases} 0, & X_w < X < X_{mI} \\[2mm] \dfrac{1}{\phi} S_{cr}\left(\dfrac{q}{2\pi r_e \sqrt{X}}, \gamma_I\right), & X_{mI} < X < X_{mJ} \\[2mm] \dfrac{1}{\phi}\left[S_{cr}\left(\dfrac{q}{2\pi r_e \sqrt{X}}, \gamma_I\right) - S_{cr}\left(\dfrac{q}{2\pi r_e \sqrt{X}}, \gamma_J\right)\right], & X_{mJ} < X < X_{crI} \\[2mm] \dfrac{1}{\phi}\left[S_{aI} - S_{cr}\left(\dfrac{q}{2\pi r_e \sqrt{X}}, \gamma_J\right)\right], & X_{crI} < X < \infty \end{cases} \tag{3.137}$$

Substituting Eq. (3.137) into the mass balance condition on the salinity front (Eq. (3.135)) yields the expression for the suspended concentration c released by the salinity front

$$c^-(X_w + T) - c^+(X_w + T)$$

$$= \left[S_a^+\left(\dfrac{q}{2\pi r_e \sqrt{X_w + T}}, \gamma_I\right) - S_a^-\left(\dfrac{q}{2\pi r_e \sqrt{X_w + T}}, \gamma_J\right)\right](1-\alpha)^{-1}, \tag{3.138}$$

$$\bar{c}\left(X_w+T,T\right)$$

$$=\begin{cases}
S_{aI}/\phi\,\exp\left(-\dfrac{\Lambda}{\sqrt{X_w}}\left(\sqrt{X}-\sqrt{(1-\alpha)X+\alpha X_w}\right)\right), & T<X_{mI}-X_w \\[2em]
\dfrac{1}{(1-\alpha)\phi}S_{cr}\left(\dfrac{q}{2\pi r_e\sqrt{X}},\gamma_I\right) \\[1.5em]
\quad +S_{aI}/\phi\,\exp\left(-\dfrac{\Lambda}{\sqrt{X_w}}\left(\sqrt{X}-\sqrt{(1-\alpha)X+\alpha X_w}\right)\right), & X_{mI}-X_w<T<T_{mI} \\[2em]
\dfrac{1}{(1-\alpha)\phi}S_{cr}\left(\dfrac{q}{2\pi r_e\sqrt{X}},\gamma_I\right)+\dfrac{1}{\phi}\left(S_{aI}-S_{cr}\left(\dfrac{q}{2\pi r_e\sqrt{X}},\gamma_I\right)\right) \\[1.5em]
\quad \exp\left(-\dfrac{\Lambda}{\sqrt{X_w}}\left(\sqrt{X}-\sqrt{(1-\alpha)X+\alpha X_w}\right)\right), & T_{mI}<T<X_{mJ}-X_w \\[2em]
\dfrac{1}{(1-\alpha)\phi}\left[S_{cr}\left(\dfrac{q}{2\pi r_e\sqrt{X}},\gamma_I\right)-S_{cr}\left(\dfrac{q}{2\pi r_e\sqrt{X}},\gamma_J\right)\right]+\dfrac{1}{\phi}\left(S_{aI}-S_{cr}\left(\dfrac{q}{2\pi r_e\sqrt{X}},\gamma_I\right)\right) \\[1.5em]
\quad \exp\left(-\dfrac{\Lambda}{\sqrt{X_w}}\left(\sqrt{X}-\sqrt{(1-\alpha)X+\alpha X_w}\right)\right), & X_{mJ}-X_w<T<X_{crI}-X_w \\[2em]
\dfrac{1}{(1-\alpha)\phi}\left[S_{aI}-S_{cr}\left(\dfrac{q}{2\pi r_e\sqrt{X}},\gamma_J\right)\right]+\dfrac{1}{\phi}\left(S_{aI}-S_{cr}\left(\dfrac{q}{2\pi r_e\sqrt{X}},\gamma_I\right)\right) \\[1.5em]
\quad \exp\left(-\dfrac{\Lambda}{\sqrt{X_w}}\left(\sqrt{X}-\sqrt{(1-\alpha)X+\alpha X_w}\right)\right), & X_{crI}-X_w<T<T_{crI} \\[2em]
\dfrac{1}{(1-\alpha)\phi}\left[S_{aI}-S_{cr}\left(\dfrac{q}{2\pi r_e\sqrt{X}},\gamma_J\right)\right], & T>T_{crI}
\end{cases},$$

$$\tag{3.139}$$

where,

$$T_{mJ}=\frac{X_{mJ}-X_w}{1-\alpha},\ T_{mI}=\frac{X_{mI}-X_w}{1-\alpha},\ T_{crI}=\frac{X_{crI}-X_w}{1-\alpha}. \tag{3.140}$$

Eq. (3.139) is the boundary condition for the solution of Eq. (3.132) in zones 4–9 in Fig. 3.21B. Applying the method of characteristics yields explicit formulae for the suspended concentration in zones 4 to 9 (rows 11 to 16 in Table 3.8).

Solving in zone 2, the parameter along the characteristic lines is X. The suspended concentration is equal to zero (row 17 in Table 3.8).

Strained concentration in all zones is obtained from Eq. (3.121) by integration of suspended concentration in time.

Dimensionless pressure drop across the core is calculated from Darcy's law (Eq. (3.124)):

$$\Delta P(T) = P(0, T) - P(1, T) = \frac{1}{2} \int_{X_w}^{1} \frac{1 + \beta\phi S_s(X, T)}{X} dX$$

$$= -\frac{1}{2} \ln X_w + \frac{\beta\phi}{2} \int_{X_w}^{1} \frac{S_s(X, T)}{X} dX. \tag{3.141}$$

3.5.3.1 Qualitative analysis of the model

The evolution of attached, suspended and strained particle profiles during low-salinity water injection with fines migration is now analyzed. The trajectories of salinity and particle fronts in areas A, BC, and D are presented in Fig. 3.21A. Fig. 3.21B shows the different flow zones.

The initially mobilized fines move with the particle speed α and are captured by straining. The fines removed from area A move in zone 3. Movement of fines from zone 3 continues into zones 4 and 5 after by-passing of zone 3 by the salinity front; the fines propagation continues with speed α.

The fines initially detached in area BC also move with the particle speed α in zone 1; the time decrement in the formula for suspended concentration in zone 1 is the same as that in zone 3. Zone 1 disappears at moment T_1. The fine particle propagation from zone 1 continues into zones 6, 7, and 8 after by-passing of zone 1 by the salinity front. The particle speed and straining probability remain the same.

No particles are mobilized in area D; thus, there is no suspended particles in zone 0.

The maximum retention curve for $\gamma = \gamma_I$ lies above the curve for $\gamma = \gamma_J$, so the salinity front releases the attached fines. As it follows from particle mass balance conditions at the salinity front, given by formula (Eq. (3.106)), the released suspension concentration is equal to vertical distance between the curve $S_a{}^+$, given by first formula of Eq. (3.136), and the maximum retention curve for $\gamma = \gamma_J$, times $(\alpha\text{-}1)^{-1}$.

There are no attached particles in area A, both maximum retention concentrations for $\gamma = \gamma_I$ and γ_J are zero, so no particles are released during the salinity front propagation in area A. All suspended particles that move into zone 4 are mobilized in A.

The salinity front enters the area B at the moment T_{mI}. The attached concentration ahead of the front is not zero; however, the maximum retention function for high velocities $U > U_m$ is zero; thus, all attached particles are removed by the salinity front. The particle release in zone B corresponds to a vertical jump in the maximum retention curve at $\gamma = \gamma_I$ to zero at $\gamma = \gamma_J$ (Fig. 3.20A). The suspended particles that move in zone 5 are those mobilized in A and released by the salinity front.

At the moment T_{mJ}, the salinity front enters zone C. The particle release in zone C corresponds to a jump in the maximum retention curve at $\gamma = \gamma_I$ to some nonzero interval at $\gamma = \gamma_J$. Some attached fines remain behind the salinity front. The suspended particles that move in zones 6 and 7 are those mobilized in area B and released by the salinity front.

The salinity front enters the area D at the moment T_{crI}. The attached concentration ahead of the front is equal to S_{aI}. The particle release in zone D corresponds to a jump from horizontal line S_{aI} to some nonzero-interval of the curve $\gamma = \gamma_J$. The suspended particles that move in zone 8 are those mobilized in area C and released by the salinity front.

At the moment T_1, the salinity front by-passes zone 1. There are no suspended particles ahead of the front. All the suspended particles that move in zone 9 are those released by the salinity front.

Figs. 3.22B, C, and D show the evolution of the profiles for suspended, strained and attached concentrations at different moments. The profiles are taken at the above defined moments $T = 0$, T_{mI}, T_3, T_{mJ}, T_{crI}, T_1.

3.5.3.2 Injectivity decline prediction

The analytical model is now used to analyze well injectivity decline. Typical values for the parameters drift delay factor $\alpha = 0.05$, filtration coefficient $\lambda = 10$ 1/m, and formation damage coefficient $\beta = 1500$ are applied. A quadratic formula for the maximum retention function is used, which corresponds to a model for mono-sized fine particles attached in square pores, as derived in Bedrikovetsky et al. (2011a):

$$\sigma_{cr}(U, \gamma) = \begin{cases} \sigma_{aI}\left(1 - \left(\frac{U}{U_m(\gamma)}\right)^2\right), & U < U_m(\gamma) \\ 0, & U > U_m(\gamma) \end{cases} \qquad (3.142)$$

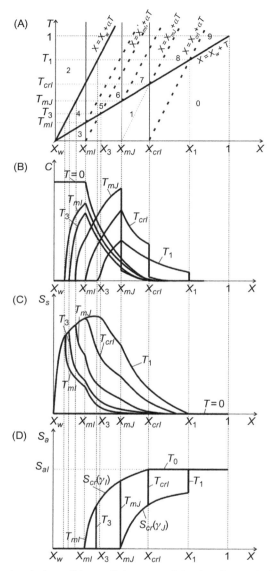

Figure 3.22 Exact solution of low-salinity water injection in vertical well with fines migration: (A) concentration front in (X,T)-plane; (B) suspended concentration profiles in four moments $k = 1, 2, 3, 4$; (C) profiles of strained concentration in the same four moments (X: dimensionless spatial coordinate, T: dimensionless time, C: dimensionless suspended concentration, S_s: dimensionless strained concentration, S_a: dimensionless attached concentration, S_{cr}: dimensionless maximum retention function, γ: fluid salinity).

where $U_m(\gamma)$ is the maximum velocity for which no particles can be held on the grain surface.

Fig. 3.23A shows impedance curves for different injection rates that vary from 50 to 200 bbl/day/m and under the formation water salinity that is equal to that of the high-salinity water injected. After injection of

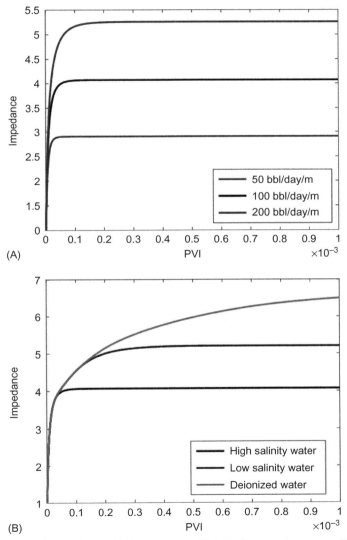

Figure 3.23 Well impedance $J(T)$ increases as (A) injection rate increases; (B) salinity decreases (PVI: number of pore volumes injected).

10^{-3} PVI, injectivity decreases from 2.5 to 5.1 times. The higher is the rate, the higher is the velocity and hence the greater is the detached concentration. Well index impairment is highly sensitive to injection rate.

Fig. 3.23B shows impedance curves for injection rate of 100 bbl/day/m. Arbitrary values for U_{mI} and U_{mJ} for low-salinity water and fresh water are applied. After injection of 10^{-3} PVI, injectivity decreases from 3.8 to 7.2 times. The injectivity damage after injection of formation water is induced by fine particles, mobilized near to the wellbore by excessive drag force under high velocities. Well index impairment is also highly sensitive to injection-water salinity.

Clearly from Figs. 3.23A and 3.23B, the effects of particle detachment and straining due to velocity happens much faster than the effects due to salinity alteration. Stabilization time for impedance in the first case happens almost instantly. This behavior can be explained by the fact that velocity alteration detaches particles at once and only near the wellbore where velocity is higher; whereas, salinity alteration causes particles to detach throughout the reservoir, but only when the salinity front reaches that point.

3.5.4 Field cases

This section presents three field cases where the effects of low-salinity waterflooding are observed. The model outlined in Section 3.5.3 is used to adjust the model parameters to produce good agreement between the field data and the model. This provides a means of evaluating the appropriateness of the model in capturing the effects of fines migration at the field-scale. The results are presented in Fig. 3.24.

The first example of injectivity decline due to low-salinity waterflooding comes from the Ventura Oil Field, located in the north of the city of Ventura, California, United States (Fig. 3.24A). The field was operated at the time the data were recorded by Shell. Extensive water quality control and monitoring to remove solids was performed to ensure that all injectivity decline was due to low-salinity water. The initial injected water salinity was equal to 0.35 M (20000 ppm); afterward, the water salinity was decreased to 0.08 M (5000 ppm). Initial injectivity loss of 23% was observed with total loss being equal to 50% at the end of 6 months (Barkman et al., 1975).

A similar effect was observed in the West Delta Block 73 field, located 27 miles offshore of Grand Isle, Lousiana, United States (Fig. 3.24B).

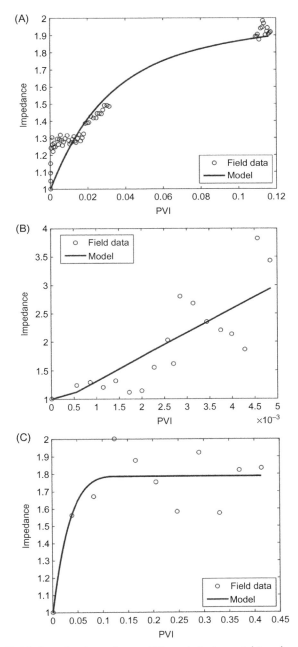

Figure 3.24 Field data for impedance $J(T)$ and their matching by the analytical model: (A) Ventura Oil Field; (B) West Delta Block 73; (C) Pervomaiskoye Oil Field (PVI: number of pore volumes injected).

This field was operated at the time the data were recorded by Exxon. Seawater was used as the injection fluid; however, it contained a lower salt concentration due to fresh water discharge from the Mississippi River. Coreflooding experiments using seawater showed significant permeability decline and the presence of suspended fines in the effluent indicating that permeability decline was due to fines migration. Qualitative analysis indicated that all intervals with permeability lower than 500 mD and clay content higher than 5% would experience significant permeability decline (Ogletree and Overly, 1977).

The Pervomaiskoye Oil Field in Republic of Tatarstan is another example of injectivity decline due to low-salinity waterflooding (Fig. 3.24C). Water was taken from the Kama River, which was the nearest water source. After 40 years of waterflooding and about two PVI injected, 30.1% of the injected water was low-salinity. High-salinity water, which was the formation water, had about 252,738 ppm of salt, while low-salinity water from the river had 848 ppm of salt. Coreflooding experiments showed fines release and formation damage when reducing water salinity (Akhmetgareev and Khisamov, 2015).

In all three field cases, the model was capable of reproducing the impedance growth caused by detachment and migration of fines lifted by low-salinity water injection. The obtained parameters and the R^2 coefficients are presented in Table 3.9.

3.6 EFFECTS OF NONEQUILIBRIUM/DELAY IN PARTICLE DETACHMENT ON FINES MIGRATION

In all previous discussions on particle detachment, changes to fluid properties were assumed to have an immediate effect on the detachment

Table 3.9 Tuning the fines-migration parameters (drift delay factor α, filtration coefficient λ, and formation damage coefficient β) from data collected from three oil fields subjected to low-salinity waterflooding

Field	α	λ (1/m)	β	R^2
Ventura Oil Field	2.7×10^{-7}	18.08	177.9	0.86
West Delta Block 73	1.4×10^{-7}	95.89	75190	0.66
Pervomaiskoye Oil Field	3.4×10^{-7}	57.58	3316	0.78

criteria for attached particles. In the following section, it will be proposed that changes to the fluid salinity may not result in an immediate detachment of particles. A nonequilibrium relation between the attached concentration and the critical retention function will be introduced to model this additional phenomenon.

3.6.1 Introduction of a delay in detachment

First, it is necessary to revisit the criteria for detachment outlined in Section 3.2. The primary forces acting on the particle are the hydrodynamic drag force and the electrostatic force. Increasing the velocity will increase the former; whereas, decreasing the salinity will decrease the latter. Both changes will result in shifting the mechanical equilibrium of attached particles toward detachment. Under the assumption of fluid incompressibility, changes to injection rate will naturally result in an immediate change to the acting hydrodynamic force. Consequently, the assertion that the attached concentration can instantaneously be determined from the critical retention function holds:

$$\sigma_a(x, t) = \sigma_{cr}(U). \qquad (3.143)$$

Changes to salinity are potentially more complex. DLVO theory states that the electrostatic force is a function of the type and concentration of ions in the confined, interparticle region between the attached particle and the grain surface. Ions in this region will form an equilibrium with the ions in the bulk solution; however, the ionic concentration in these two regions will generally not be equal. The difference arises due to the tendency of charged ionic species to surround oppositely charged surfaces (Israelachvili, 2011). Previous formulations of fines detachment due to salinity have relied on the assumption that changes to the bulk salinity recorded at any point in the core immediately translate into changes in the salinity in the interparticle region for any attached particles. The transfer rate of ions between the bulk solution and the interparticle region will however be finite, and the resulting delay between changes to salinity and the electrostatic force may have significant impacts on particle detachment.

Due to the proximity of the particle to the attached grain, advective flux is likely to be low, and ion transfer will be dominated by diffusion. Mahani et al. (2015) investigated the similar problem of ion diffusion between an oil droplet and clay and demonstrated that the exceedingly long stabilization times could not be justified by typical ranges of diffusion coefficients found in the literature. To account for the atypically slow rate of diffusion, these authors proposed the use of the Nernst-Planck (NP) equation for diffusion of charged ionic species among charged surfaces (Joekar–Niasar and Mahani, 2016; Zheng and Wei, 2011). The electrostatic influence of the charged surfaces coupled with the electrostatic gradient induced by the ionic concentration gradient will oppose the ion transfer and hence decrease the net ionic transfer rate with the bulk solution.

The NP equation has so far not been applied to incorporate delayed detachment explicitly into models for fines migration. Instead, authors have opted for simplified relationships between the attached concentration and the critical retention function.

Suppose that the attached concentration is governed by the critical retention function, but their equilibrium will occur only after some time τ. This can be expressed as:

$$\sigma_a(x, t + \tau) = \sigma_{cr}(U, \gamma(x, t)). \tag{3.144}$$

By taking a Taylor's series expansion of the attached concentration at time t, around time $t + \tau$ and discarding all but the first two terms in the expansion, the following expression can be derived:

$$\tau \frac{\partial \sigma_a(x, t)}{\partial t} = \sigma_{cr}(U, \gamma(x, t)) - \sigma_a(x, t). \tag{3.145}$$

The parameter τ is referred to as the delay time. This equation will replace Eq. (3.35) in the system of Eqs. (3.33–3.36, and 3.92) to incorporate the effect of nonequilibrium or delayed detachment into the system for fines migration.

3.6.2 Exact solution for 1D problem accounting for delay with detachment

The solution for the system of Eqs. (3.33, 3.34, 3.36, and 3.92) coupled with the new nonequilibrium Eq. (3.145) for particle detachment will mostly follow the solution procedure outlined in Section 3.5.1.

First, let us introduce the dimensionless delay factor:

$$\varepsilon = \frac{U\tau}{\phi L}. \tag{3.146}$$

which can be understood phenomenologically as the number of pore volumes that need to be injected to establish equilibrium between the attached concentration and the critical retention function. Using this parameter, Eq. (3.145) can be expressed in dimensionless form:

$$\varepsilon \frac{\partial S_a}{\partial T} = S_{cr}(U, \Gamma(X, T)) - S_a(X, T). \tag{3.147}$$

Eq. (3.147), alongside Eqs. (3.94, 3.95, 3.97, and 3.98) outlined in Section 3.5, form the dimensionless system of equations to be solved.

As before, the equation for salt transport can be solved using the method of characteristics to obtain:

$$\Gamma = \begin{cases} 1, & X > T \\ 0, & X < T \end{cases}. \tag{3.148}$$

Using this result, and integrating Eq. (3.147) as an ordinary differential equation allows deriving the attached concentration:

$$S_a = \begin{cases} S_{a1}, & X > T \\ S_{a0} + (S_{a1} - S_{a0})e^{\frac{X-T}{\varepsilon}}, & X < T \end{cases}. \tag{3.149}$$

where S_{a1} and S_{a0} are the values of the critical retention function when the dimensionless salinity is 1 and 0 respectively. The primary difference in the solution procedure for the suspended concentration is the continuity of the solution. In Section 3.5, the suspended concentration experienced a shock along the salinity front ($X = T$). A mass balance condition was applied across the shock to determine the initial condition used to solve for C behind this shock. It can be shown that when using nonequilibrium particle detachment, the suspended concentration is continuous across the salinity front.

The mass balance condition evaluated across the salinity shock is (Bedrikovetsky, 1993; Lake, 2010):

$$[C + S_a + S_s]D = \alpha[C], \tag{3.150}$$

where D is the slope of the salinity front and is equal to unity in this case.

Given that changes to C will be finite, Eq. (3.95) indicates that the strained concentration is continuous:

$$[S_s] = 0. \tag{3.151}$$

Eq. (3.149) shows that:

$$[S_a] = 0. \tag{3.152}$$

Thus, the mass balance condition indicates that:

$$[C] = \alpha[C]. \tag{3.153}$$

To satisfy the condition that α can be less that one, this becomes:

$$[C] = 0. \tag{3.154}$$

Thus, the suspended concentration is continuous, which differs from the solution presented in Section 3.5 for equilibrium particle detachment.

The solution of $C = 0$ for all $X > T$ follows in the same manner as presented previously. As the suspended concentration is zero ahead of the concentration front, and given that C is continuous across this front, the suspended concentration must also be zero behind it. This gives the initial condition for the second region:

$$T = X : C = 0. \tag{3.155}$$

Substituting the equation for the straining rate (Eq. (3.95)) and the solution for the attached concentration (Eq. (3.149)) into the dimensionless mass balance equation presents a first-order hyperbolic differential equation in C that can be solved using the initial condition (Eq. (3.155)):

$$\frac{\partial C}{\partial T} + \alpha \frac{\partial C}{\partial X} = -\alpha \Lambda C - \frac{\partial S_a}{\partial T}. \tag{3.156}$$

Between the particle and salinity fronts $(\alpha T < X < T)$, the PDE (Eq. (3.156)) reduces to an ordinary differential equation:

$$\frac{dC}{dT} = -\alpha \Lambda C - \frac{\partial S_a}{\partial T}, \tag{3.157}$$

along parametric curves given by,

$$\frac{dX}{dT} = \alpha. \tag{3.158}$$

From the solution for the attached concentration, it follows that for all $X < T$:

$$\frac{\partial S_a}{\partial T} = -\frac{(S_{a1} - S_{a0})}{\varepsilon} e^{\frac{X-T}{\varepsilon}}. \tag{3.159}$$

To integrate the right-hand side of Eq. (3.159), all terms must be expressed in the variable T. As such, the equation for the parametric curves (Eq. (3.158)) is solved to yield:

$$X - X_0 = \alpha(T - T_0). \tag{3.160}$$

As all of these curves originate on the line $X = T$, so $X_0 = T_0$, yielding:

$$X = \alpha T + T_0(1 - \alpha). \tag{3.161}$$

Eq. (3.161) can be used in combination with Eq. (3.159) in the ODE (Eq. (3.157)) to solve directly for the suspended concentration.

Behind the particle front, the PDE (Eq. (3.156)) is reduced to:

$$\frac{dC}{dX} = -\Lambda C - \frac{1}{\alpha}\frac{\partial S_a}{\partial T}. \tag{3.162}$$

which can be solved similarly using the expression for the change in the attached concentration (Eq. (3.159)).

The final solution for the suspended concentration in all regions is presented in Table 3.10. The solutions for the strained concentration and impedance are derived using the solution for the suspended concentration and by integrating Eqs. (3.95) and (3.98), respectively. The complete analytical expressions for these terms are given in Table 3.10.

3.6.3 Analysis of laboratory data and tuning of the model coefficients

Applications of this model to practical problems follows a similar workflow to that outlined for the models given in previous sections. First and foremost, laboratory testing should be completed, and the model should be used to fit the model parameters. An example of this is presented alongside a brief methodology to aid further work.

In typical experiments on fines migration, two experimental datasets are used to treat the data; the pressure drop and the outlet particle concentration. For the case of model tuning, as before, these are transformed for the convenience and accuracy of fitting.

Under the condition of a constant injection rate, the pressure drop can be represented as the impedance:

$$J = \frac{\Delta p}{\Delta p_0}. \tag{3.163}$$

Table 3.10 Analytical solutions for attached particle concentration S_a, suspended particle concentration S_σ, suspended particle concentration C, strained particle concentration S_s, and impedance J, for nonequilibrium delay in particle detachment

Line	Term	Zone	Solution
1	$S_a(X, T)$	$X < T$	S_{a1}
2		$X > T$	$S_{a0} + e^{\frac{X-T}{\varepsilon}}(S_{a1} - S_{a0})$
3	$C(X, T)$	$X < \alpha T$	$\dfrac{S_{a1} - S_{a0}}{\alpha\Lambda\varepsilon + \alpha - 1}\left(e^{\frac{X-T}{\varepsilon}} - e^{\frac{X-T}{\varepsilon} - \Lambda X}\right)$
4		$\alpha T < X < T$	$\dfrac{S_{a1} - S_{a0}}{\alpha\Lambda\varepsilon + \alpha - 1}\left(e^{\frac{X-T}{\varepsilon}} - e^{-\alpha\Lambda\left(\frac{X-T}{\alpha-1}\right)}\right)$
5		$X > T$	0
6	$S_s(X, T)$	$X < \alpha T$	$\dfrac{\alpha\Lambda(S_{a1} - S_{a0})}{\alpha\Lambda\varepsilon + \alpha - 1}\left(\left(\dfrac{1-\alpha}{\alpha\Lambda}\right)e^{-\Lambda X} - \varepsilon e^{\frac{X-T}{\varepsilon}} + \varepsilon\left(e^{-\frac{T}{\varepsilon} + X\left(\frac{1}{\alpha\varepsilon} - \Lambda\right)} - e^{-\Lambda X}\right) + \varepsilon - \left(\dfrac{1-\alpha}{\alpha\Lambda}\right)\right)$
7		$\alpha T < X < T$	$\dfrac{\alpha\Lambda(S_{a1} - S_{a0})}{\alpha\Lambda\varepsilon + \alpha - 1}\left(-\varepsilon e^{\frac{X-T}{\varepsilon}} + \left(\dfrac{1-\alpha}{\alpha\Lambda}\right)e^{\frac{\alpha\Lambda(X-T)}{1-\alpha}} + \varepsilon - \left(\dfrac{1-\alpha}{\alpha\Lambda}\right)\right)$
8		$X > T$	0
9	$J(T)$	$T < 1$	$1 + \beta\phi\Delta\sigma_a\alpha\Lambda\left(\dfrac{S_{a1} - S_{a0}}{\alpha\Lambda\varepsilon + \alpha - 1}\right)\left(\dfrac{\alpha - 1}{(\alpha\Lambda)^2}e^{-\alpha\Lambda T} + \varepsilon^2 e^{-\frac{T}{\varepsilon}} \right.$ $+ \left(\dfrac{\varepsilon}{\frac{1}{\alpha\varepsilon} - \Lambda}\right)\left(e^{-\alpha\Lambda T} - e^{-\frac{T}{\varepsilon}}\right)$ $\left. + \dfrac{\varepsilon}{\Lambda}e^{-\alpha\Lambda T} + T\left(\varepsilon - \dfrac{1-\alpha}{\alpha\Lambda}\right) - \dfrac{\alpha - 1}{(\alpha\Lambda)^2} - \dfrac{\varepsilon}{\Lambda} - \varepsilon^2 \right)$

$1 < T < \frac{1}{\alpha}$

10

$$1 + \beta\phi\Delta\sigma_a\alpha\Lambda\left(\frac{S_{a1}-S_{a0}}{\alpha\Lambda\varepsilon+\alpha-1}\right)\left(\begin{array}{l}\left(\frac{1-\alpha}{\alpha\Lambda}\right)^2 e^{\frac{\alpha\Lambda(1-T)}{1-\alpha}} + \frac{\alpha-1}{(\alpha\Lambda)^2}e^{-\alpha\Lambda T} - \varepsilon^2\left(e^{\frac{1-T}{\varepsilon}} - e^{-\frac{T}{\varepsilon}}\right) \\[2mm] + \frac{\varepsilon}{\frac{1}{\alpha\varepsilon}-\Lambda}\left(e^{-\alpha\Lambda T} - e^{-\frac{T}{\varepsilon}}\right) + \frac{\varepsilon}{\Lambda}e^{-\alpha\Lambda T} \\[2mm] -\frac{\alpha-1}{\alpha\Lambda^2} - \frac{\varepsilon}{\Lambda} + \varepsilon - \frac{1-\alpha}{\alpha\Lambda}\end{array}\right)$$

$T > \frac{1}{\alpha}$

11

$$1 + \beta\phi\Delta\sigma_a\alpha\Lambda\left(\frac{S_{a1}-S_{a0}}{\alpha\Lambda\varepsilon+\alpha-1}\right)\left(\begin{array}{l}\frac{\alpha-1}{\alpha\Lambda^2}\left(e^{-\Lambda}-1\right) - \varepsilon^2\left(e^{\frac{1-T}{\varepsilon}} - e^{-\frac{T}{\varepsilon}}\right) \\[2mm] + \frac{\varepsilon}{\frac{1}{\alpha\varepsilon}-\Lambda}\left(e^{-\frac{T}{\varepsilon}+\left(\frac{1}{\alpha\varepsilon}-\Lambda\right)} - e^{-\frac{T}{\varepsilon}}\right) \\[2mm] + \frac{\varepsilon}{\Lambda}\left(e^{-\Lambda}-1\right) + \varepsilon - \frac{1-\alpha}{\alpha\Lambda}\end{array}\right)$$

And the outlet concentration is treated as the accumulated outlet concentration:

$$C_{acc} = \alpha \int C(1, \gamma) d\gamma. \qquad (3.164)$$

These two datasets can be tuned using any optimization algorithm.

To demonstrate the applicability of the model, one experiment on artificial sandstone cores is presented. The cores comprised of 5% kaolinite and 95% chemically-washed silica sand. The methodology for the test can be found in the work of Russell et al. (2017).

The five model parameters α, ε, Λ, β, and $\Delta\sigma$ are tuned here using a least-squared genetic algorithm implemented in MATLAB. The experimental data and tuned model results are presented below in Fig. 3.25 and the resulting model parameters are given in Table 3.11.

The model fit shows good agreement with the experimental data, achieving a value for R^2 higher than 0.95. Note that although the drift delay factor is not equal to one, modeling fines migration without the detachment delay factor would result in a prediction of permeability stabilization after the injection of less than 10 pore volumes. The delay in detachment accounts for the remainder of the lengthy stabilization time.

3.6.4 Semianalytical model for axisymmetric flow

As in Section 3.5.3, the equations for fines migration can be extended to axisymmetric radial flow for the purpose of predicting well behavior. With the inclusion of delayed detachment, the system of Eqs. (3.114–3.118) is modified to include a nonequilibrium expression.

The problem of delayed detachment is more nuanced in radial coordinates. In corefloods, changes to salinity or injection rate are controlled; thus, to study the underlying processes, only one of these variables is changed during any one injection cycle. This allows the modeling to be done with one of the analytical solutions presented previously (Section 3.3 for changes to velocity, and Sections 3.5 and 3.6.2 for changes to salinity). In radial flow, the fluid velocity varies with distance from the wellbore. With low-salinity injection, particle detachment will result from both changes to velocity and fluid salinity. In Section 3.6.1, it was reasoned that the delay in detachment was only appropriate when particle detachment was induced by changes to salinity. As such, detachment during radial injection should comprise of both instant and delayed detachment.

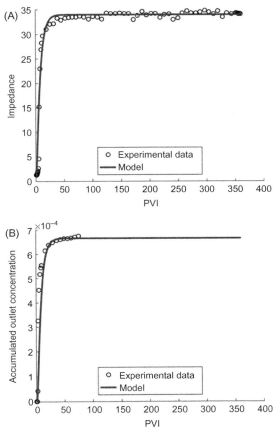

Figure 3.25 Experimental data from coreflood of artificial sand-kaolinite core with the fines migration model with delayed detachment, (A) core impedance, (B) accumulated concentration of suspended particles at the outlet (PVI: number of pore volumes injected).

Table 3.11 Tuned model parameters from the tuning of experimental data or coreflood data presented in Fig. 3.25 for nonequilibrium delay in particle detachment

Parameter	Value
Drift delay factor, α	0.1061
Delay time, τ (s)	1126
Filtration coefficient, λ $(1/m)$	125.6
Formation damage coefficient, β	57,771
Total detached concentration, $\Delta\sigma$	0.00361
Coefficient of Determination, R^2	0.9649

To incorporate this, we propose that the attached concentration is governed by:

$$\sigma_a(x, t) = \sigma_{cr}(U(x, t), \gamma'(x, t)). \tag{3.165}$$

where γ' is a new variable, referred to as the pseudo-salinity, which is related to the bulk fluid salinity by:

$$\gamma'(x, t + \tau) = \gamma(x, t). \tag{3.166}$$

This parallels the initial introduction of delayed detachment given in Section 3.6.1. It follows that using a Taylor's series expansion, a relationship can be derived for the pseudo-salinity:

$$\tau \frac{\partial \gamma'}{\partial t} = \gamma - \gamma'. \tag{3.167}$$

The Eqs. (3.165) and (3.167) satisfy the condition that changes to the fluid velocity result in an immediate change in the attached concentration, while changes to the fluid salinity have a delayed effect according to a nonequilibrium relation.

The full system of equations can then be given:

$$\frac{\partial}{\partial t}(\phi c + \sigma_s + \sigma_a) + \frac{\alpha q}{2\pi r} \frac{\partial c}{\partial r} = 0, \tag{3.168}$$

$$\frac{\partial \sigma_s}{\partial t} = \frac{\alpha \lambda c q}{2\pi r}, \tag{3.169}$$

$$\sigma_a = \sigma_{cr}\left(\frac{q}{2\pi r}, \gamma'\right), \tag{3.170}$$

$$\phi \frac{\partial \gamma}{\partial t} + \frac{q}{2\pi r} \frac{\partial \gamma}{\partial r} = 0, \tag{3.171}$$

$$\frac{q}{2\pi r} = -\frac{k}{\mu(1 + \beta \sigma_s)} \frac{\partial p}{\partial r}. \tag{3.172}$$

Using the dimensionless coordinates outlined in Section 3.5,

$$T \rightarrow \frac{qt}{\pi r_e^2 \phi}, X \rightarrow \left(\frac{r}{r_e}\right)^2, \ S_s \rightarrow \frac{\sigma_s}{\phi}, S_a \rightarrow \frac{\sigma_a}{\phi}, \ S_{cr} \rightarrow \frac{\sigma_{cr}}{\phi}, \Lambda \rightarrow \lambda r_w, P \rightarrow \frac{\pi k p}{q\mu},$$

$$\Gamma = \frac{\gamma - \gamma_J}{\gamma_I - \gamma_J}, \varepsilon \rightarrow \frac{q\tau}{\pi r_e^2 \phi}.$$

$$\tag{3.173}$$

The system of equations becomes:

$$\frac{\partial(C + S_s + S_a)}{\partial T} + \alpha\frac{\partial C}{\partial X} = 0, \tag{3.174}$$

$$\frac{\partial S_s}{\partial T} = \frac{\alpha\Lambda C}{2\sqrt{X}\sqrt{X_w}}, \tag{3.175}$$

$$S_a = S_{cr}(X, \Gamma'), \tag{3.176}$$

$$\frac{\partial \Gamma}{\partial T} + \frac{\partial \Gamma}{\partial X} = 0, \tag{3.177}$$

$$\varepsilon\frac{\partial \Gamma'}{\partial T} = \Gamma - \Gamma', \tag{3.178}$$

$$\frac{1}{X} = -\frac{2}{(1 + \beta\phi S_s)}\frac{\partial P}{\partial X}. \tag{3.179}$$

The initial conditions for this system correspond to high–salinity fluid, and an absence of strained particles:

$$T = 0{:}\Gamma = 1, S_s = 0. \tag{3.180}$$

The attached concentration begins at some initial value S_{aI} for all X. However, as the fluid is assumed to be incompressible, any particle detachment due to velocity at high salinity will occur instantaneously. Hence, the initial attached concentration can be given as:

$$S_a(X, T = 0) = \begin{cases} S_{aI}, & S_{cr}(X, \Gamma' = 1) > S_{aI} \\ S_{cr}(X, \Gamma' = 1), & S_{cr}(X, \Gamma' = 1) < S_{aI} \end{cases}, \tag{3.181}$$

The initial value of suspended concentration will be given by:

$$C(X, T = 0) = S_{aI} - S_a(X, T = 0). \tag{3.182}$$

The boundary condition is given by the injection of low–salinity water with an absence of suspended particles:

$$X = 0{:}\Gamma = 0, C = 0. \tag{3.183}$$

The solution procedure is presented for an arbitrary form of the maximum retention function. First, the salt mass balance Eq. (3.177) is solved using the method of characteristics to yield:

$$\Gamma = \begin{cases} 1, & X > X_w + T \\ 0, & X < X_w + T \end{cases}. \tag{3.184}$$

Next, the pseudo-salinity is solved by integrating Eq. (3.178) by separation of variables:

$$\Gamma' = \begin{cases} 1, & X > X_w + T \\ e^{\frac{X-T}{\varepsilon}}, & X < X_w + T \end{cases}.$$ (3.185)

The suspended concentration is then solved using the mass balance Eq. (3.174) by substituting the expressions for the straining rate and the attached concentration:

$$\frac{\partial C}{\partial T} + \alpha \frac{\partial C}{\partial X} = -\frac{\alpha \Lambda C}{2\sqrt{X}\sqrt{X_w}} - \frac{\partial S_{cr}(X, \Gamma)}{\partial T}.$$ (3.186)

As before, this equation is solved separately for the three solution regions. For all points ahead of the particle concentration front, along parametric curves given by:

$$\frac{dX}{dT} = \alpha.$$ (3.187)

Eq. (3.186) reduces to the ODE:

$$\frac{dC}{dT} = -\frac{\alpha \Lambda C}{2\sqrt{X}\sqrt{X_w}} - \frac{\partial S_{cr}(X, \Gamma)}{\partial T}.$$ (3.188)

For the region ahead of the salinity front, salinity is constant, so the critical retention function does not change with time. Using the initial condition (3.182), this equation can be integrated to yield:

$$C(X, T) = e^{-\Lambda\left(\sqrt{X} - \sqrt{X - \alpha T}\right)}(S_{aI} - S_a(T = 0)).$$ (3.189)

Following the reasoning presented in Section 3.6.2, it can be shown that the suspended concentration is continuous across the salinity shock. As such, the solution (Eq. (3.189)) is used as an initial condition for the second region, between the salinity and concentration shock.

Integrating Eq. (3.188) using separation of variables with the initial condition:

$$C(X = T, T) = e^{-\Lambda\left(\sqrt{X} - \sqrt{X - \alpha T}\right)}(S_{aI} - S_a(T = 0)),$$ (3.190)

yields:

$$C(X, T) = e^{\left[A\left(\sqrt{X+\alpha(X-T)}-2\sqrt{X}+\sqrt{X-\alpha T}\right)\right]}(S_{aI} - S_a(T=0))$$
$$- e^{\left[-A\left(\sqrt{X}-\sqrt{X-\alpha T}\right)\right]} \int_0^T e^{\left[A\left(\sqrt{\alpha T+T_0(1-\alpha)}-\sqrt{T_0(1-\alpha)}\right)\right]} \frac{\partial S_{cr}(T)}{\partial T} dT.$$

$$(3.191)$$

Note that the integral term contains the constant T_0 defined by the solution of Eq. (3.187):

$$X - X_w - X_0 = \alpha(T - T_0),$$

$$(3.192)$$

where on the salinity shock $X_w + X_0 = T_0$ and so $X = \alpha T + T_0(1-\alpha)$.

This constant is used to eliminate X from the integral such that the integrand is a function of T only. For cases where the integral can be evaluated explicitly, the constant can be substituted following integration.

For the region behind the concentration shock, the original PDE (Eq. (3.186)) is reduced to the ODE:

$$\frac{dC}{dX} = -\frac{\Lambda C}{2\sqrt{X}\sqrt{X_w}} - \frac{1}{\alpha}\frac{\partial S_a}{\partial T}.$$

$$(3.193)$$

Along the parametric curves given by:

$$\frac{dT}{dX} = \frac{1}{\alpha}.$$

$$(3.194)$$

Integrating Eq. (3.193) subject to the boundary condition (Eq. (3.183)) yields:

$$C(X, T) = -\frac{e^{-A\left(\sqrt{X}-\sqrt{X_w}\right)}}{\alpha} \int_{X_w}^X e^{A\left(\sqrt{X}-\sqrt{X_w}\right)} \frac{\partial S_{cr}(X)}{\partial T} dX.$$

$$(3.195)$$

The solution here is clearly presented in an implicit form due to the requirement of a continuous function $S_{cr}(U,\gamma)$ of which there is yet no accepted general form. This requirement complicates generating results, and for particular cases of the critical retention function will lead to a need for numerically solving the integrals in Eqs. (3.191 and 3.195).

The strained concentration is solved by integrating Eq. (3.175) as an ordinary differential equation. The dimensionless pressure drop is then calculated by integrating Eq. (3.179). The results are summarized in Table 3.12.

Table 3.12 Solution for suspended concentration during uniaxial injection of low-salinity water. Presented for arbitrary critical retention function $S_{cr}(U, \Gamma)$

Line	Term	Zone	Solution
1	$\Gamma(X, T)$	$X > X_w + T$	1
2		$X < X_w + T$	0
3	$\Gamma'(X, T)$	$X > X_w + T$	1
4		$X < X_w + T$	$e^{\frac{X-T}{\varepsilon}}$
5		$X > X_w + T$	$e^{-\Lambda\left(\sqrt{X}-\sqrt{X-\alpha T}\right)}\left(S_{al} - S_a(T=0)\right)$
6	$C(X, T)$	$X_w + \alpha T < X < X_w + T$	$e^{\left[\Lambda\left(\sqrt{X+\alpha(X-T)}-2\sqrt{X}+\sqrt{X-\alpha T}\right)\right]}\left(S_{al} - S_a(T=0)\right)$ $-e^{\left[-\Lambda\left(\sqrt{X}-\sqrt{X-\alpha T}\right)\right]}\int_0^T e^{\left[\Lambda\left(\sqrt{\alpha T + T_0(1-\alpha)}-\sqrt{T_0(1-\alpha)}\right)\right]}\frac{\partial S_{cr}(T)}{\partial T}dT$
7		$X < X_w + \alpha T$	$-\frac{e^{-\Lambda\left(\sqrt{X}-\sqrt{X_w}\right)}}{\alpha}\int_{X_w}^{X} e^{\Lambda\left(\sqrt{X}-\sqrt{X_w}\right)}\frac{\partial S_a(T)}{\partial T}dX$

A simplified form of the critical retention function can be used to demonstrate the effect of the delay in detachment on well injectivity. For example, suppose that the maximum velocity present in the critical retention function:

$$\sigma_{cr}(U, \gamma) = \begin{cases} \sigma_{aI}\left(1 - \left(\frac{U}{U_m(\gamma)}\right)^2\right), & U < U_m(\gamma), \\ 0, & U > U_m(\gamma) \end{cases} \quad (3.196)$$

decreases linearly with decreases in salinity. As such, the critical retention function will be lower for lower salinities, as predicted by the qualitative analysis of the torque balance outlined in Section 3.2. The resulting form of the critical retention function in dimensionless coordinates is:

$$S_{cr}\left(U, \gamma'\right) = a\left(1 - \frac{d}{X(U_{m0} + (U_{m1} - U_{m0})\gamma')^2}\right), \quad (3.197)$$

where,

$$a = \frac{\sigma_0}{\phi}, \quad d = \left(\frac{q}{2\pi r_e}\right)^2, \sigma_0 = \sigma_{cr}(U \to 0, \gamma). \quad (3.198)$$

This form of the critical retention function provides a simple but effective means to predict injectivity decline during low-salinity water injection with delayed detachment.

3.6.5 Prediction of injection well behavior

A calculation of injectivity is performed to demonstrate the behavior of the model, primarily reflecting the effect of the delay factor. The results are shown in Fig. 3.26. The values of the other parameters used in these calculations are given in Table 3.13.

Fig. 3.26 clearly demonstrates the impact of the delay factor τ. Increasing the delay factor results in a longer stabilization time for the impedance, but does not change the stabilized value.

The computation of impedance varies significantly with the maximum retention function applied. To illustrate the effect of including a delay in detachment, a simplified σ_{cr} model is presented. To model injectivity decline for the purposes of industrial applications, it is possible that a more rigorous model for particle detachment would be required.

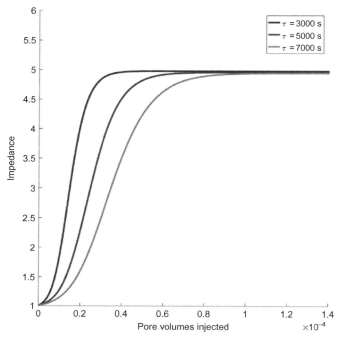

Figure 3.26 Injection well impedance plotted against the number of PVI for three values of the detachment delay factor τ.

Table 3.13 Reservoir properties and model parameters used for the sensitivity study presented in Fig. 3.26

Variable	Value
Injection rate, q $[\mathrm{m^2.s^{-1}}]$ (bbl.day^{-1}.m^{-1})	$\left[1.84 \times 10^{-4}\right]$ (100)
Reservoir properties	
Porosity, ϕ	0.2
Wellbore radius, r_w	0.1
Drainage radius, r_e	1000
Fines migration parameters	
Drift delay factor, α	0.05
Filtration coefficient, λ $[\mathrm{m^{-1}}]$	1000
Formation damage coefficient, β	2000
Maximum retention function parameters	
Maximum velocity (High Salinity), U_{m1} $[\mathrm{m.s^{-1}}]$	10^{-2}
Maximum velocity (Low Salinity), U_{m0} $[\mathrm{m.s^{-1}}]$	10^{-3}
σ_0	0.8×10^{-3}

3.7 TWO-PHASE FINES MIGRATION DURING LOW-SALINITY WATERFLOOD: ANALYTICAL MODELING

All discussions so far of formation damage due to fines migration have been solely in a single-phase flow environment. Most scenarios of fluid flow in porous media, especially in the petroleum industry, contain more than one mobile phase. Low-salinity water is used primarily to increase the oil-recovery in petroleum reservoirs. As such, in most cases of low-salinity water injection, where fines migration is most prominent, both mobile oil and water phases are present.

The two primary characteristics of two-phase flow are the relative volume of each of the phases in the pore space, or the saturation:

$$s = \frac{V_{water}}{V_{oil}},\qquad(3.199)$$

and the relative permeability for each phase:

$$k_{ri} = \frac{k_i}{k},\qquad(3.200)$$

where k_{ri} denotes the relative permeability for phase i, k_i denotes the absolute permeability for phase i, and k is the rock permeability as before. In this section, subscripts w and o will be used for water and oil, respectively.

As suspended colloids are typically contained in the water phase only, the effect of particle straining is suspected to influence only the water flux. Thus, the effect of particle straining will not only be a net reduction in fluid flux, but also a shift in the relative flow of the two mobile phases.

In the following discussion, a mathematical model is presented which incorporates the effects of fines migration into two-phase flow.

3.7.1 Fines migration in two-phase flow

Ignoring capillary effects on fluid transport, the mass balance equation for an incompressible, immiscible water phase is:

$$\phi\,\frac{\partial s}{\partial t} + \frac{1}{r}\frac{\partial(rfU)}{\partial r} = 0,\qquad(3.201)$$

where the fractional flow function f is:

$$f(s, \gamma, \sigma_s) = \frac{\frac{k_{rw}(s,\gamma,\sigma_s)}{\mu_w}}{\frac{k_{rw}(s,\gamma,\sigma_s)}{\mu_w} + \frac{k_{ro}(s,\gamma)}{\mu_o}}. \qquad (3.202)$$

The effect of fines straining on the aqueous phase flow is expressed as (Coronado and Díaz-Viera, 2017; Zeinijahromi et al., 2013):

$$k_{rw}(s, \gamma, \sigma_s) = \frac{k_{rw}(s, \gamma)}{1 + \beta \sigma_s}. \qquad (3.203)$$

Eq. (3.204) provides the mass conservation law for the salt, which is assumed to only exist within the aqueous phase. Here, both the capillary effects and advective dispersion are ignored (Omekeh et al., 2013):

$$\phi \frac{\partial(\gamma s)}{\partial t} + \frac{1}{r} \frac{\partial (r\gamma f U)}{\partial r} = 0. \qquad (3.204)$$

In an oil–water system, a portion of the surface area of the porous space is occupied by water, and the remainder by oil. It is assumed that fines detachment is only possible where the rock surface is coated by water, as the salt transfer that governs the electrostatic force (Section 3.2) occurs only in the aqueous phase.

The overall attached fine particle concentration can be expressed as the total of the particles attached to the water-accessible surface and those attached to the oil-accessible surfaces.

Fines detachment here is modeled without a delay in detachment. Thus, the attached concentration is given by (Borazjani et al., 2017; Mohammadmoradi et al., 2017):

$$\sigma_a = \sigma_a(\gamma) \frac{A_w(s, \theta)}{A} + \sigma_{aI} \frac{A - A_w(s, \theta)}{A}, \qquad (3.205)$$

where A_w is the area of the pore surface covered by water and A is the total pore surface area.

The conservation law for suspended, attached, and strained fines, similarly ignoring capillary effects and advective dispersion, is:

$$\frac{\partial}{\partial t}(\phi s c + \sigma_a + \sigma_s) + \frac{1}{r} \frac{\partial}{\partial r}(r c f U) = 0. \qquad (3.206)$$

The kinetic rate of fines straining is proportional to the advective flux of suspended fines:

$$\frac{\partial \sigma_s}{\partial t} = \lambda(\gamma, \sigma_s) cfU. \tag{3.207}$$

A modified Darcy's law accounts for permeability damage due to particle straining is given as:

$$\frac{q}{2\pi r} = -k\left(\frac{k_{rw}(s, \gamma, \sigma_s)}{\mu_w} + \frac{k_{ro}(s, \gamma)}{\mu_o}\right)\frac{\partial p_w}{\partial r} - \frac{k\, k_{ro}(s, \gamma)}{\sqrt{k/\phi}\mu_o}\frac{\partial\left(\sigma_{WO}(\gamma)\cos\theta J\right)}{\partial r}. \tag{3.208}$$

$$\frac{q}{2\pi r} = -k\left(\frac{k_{rw}(s, \gamma, \sigma_s)}{\mu_w} + \frac{k_{ro}(s, \gamma)}{\mu_o}\right)\frac{\partial p_w}{\partial r}. \tag{3.209}$$

Introducing the following dimensionless variables:

$$T = \frac{qt}{\pi r_e^2 \phi}, X = \left(\frac{r}{r_e}\right)^2, \Lambda = \lambda r_e, S_a = \frac{\sigma_a}{\phi}, S_s = \frac{\sigma_s}{\phi}, P_w = \frac{k\pi p_w}{q\mu_o}, \tag{3.210}$$

transforms the system of Eqs. (3.201, 3.204−3.207, and 3.209) in the following dimensionless form:

$$\frac{\partial s}{\partial T} + \frac{\partial f}{\partial X} = 0, \tag{3.211}$$

$$\frac{\partial(\gamma s)}{\partial T} + \frac{\partial(\gamma f)}{\partial X} = 0, \tag{3.212}$$

$$S_a = S_{cr}\left(\frac{q}{2\pi r_e\sqrt{X}}, \gamma\right), \tag{3.213}$$

$$\frac{\partial}{\partial T}(sc + S_a + S_s) + \frac{\partial(cf)}{\partial X} = 0, \tag{3.214}$$

$$\frac{\partial S_s}{\partial T} = \frac{1}{2\sqrt{X}}\Lambda(\gamma, \sigma_s)cf, \tag{3.215}$$

$$\frac{\partial P_w}{\partial X} = \frac{-1}{2\left(\mu_o\frac{k_{rw}(s, \gamma, \sigma_s)}{\mu_w} + k_{ro}(s, \gamma)\right)2X}, \tag{3.216}$$

The result is a system of six equations which determines six unknowns: saturation s, salinity γ, attached fines concentration S_a, strained fines concentration fines S_s, suspended fines concentration c, and water pressure P_W. The equation for water pressure separates from the system. As particle detachment is modeled in equilibrium conditions here, the detachment rate will be zero; hence, S_a will only provide initial and boundary conditions, as in Section 3.5. Thus, the system has four Eqs. (3.211, 3.212, 3.214, and 3.215) with four unknowns, s, γ, S_s, and c.

The initial conditions of this system are:

$$T=0 : S_a = \begin{cases} 0 & X_w < X < X_{mI} \\ S_{cr}\left(\dfrac{q}{2\pi r_e \sqrt{X}}, \gamma_I\right) & X_{mI} < X < X_{crI}, \\ S_{aI} & X > X_{crI} \end{cases} \qquad (3.217)$$

$$T=0 : c(X,0) = \begin{cases} S_{aI} & X_w < X < X_{mI} \\ \left(S_{aI} - S_{cr}\left(\dfrac{q}{2\pi r_e \sqrt{X}}, \gamma_I\right)\right) & X_{mI} < X < X_{crI}, \\ 0 & X > X_{crI} \end{cases} \qquad (3.218)$$

$$T=0 : \gamma = \gamma_I, s = s_I, \qquad (3.219)$$

$$X = X_w : c = 0, \gamma = \gamma_J, f = 1. \qquad (3.220)$$

3.7.2 Splitting method for integration of two-phase systems

Based on works by Pires et al. (2006), Shen (2016), and Borazjani et al. (2017), this section presents the splitting technique for a hyperbolic system of equations.

In order to split, the stream-function $\varphi(X,T)$ is introduced, assuming that the solution of the parameters $s(X,T)$, $c(X,T)$, $S_s(X,T)$, $S_a(X,T)$, and $\gamma(X,T)$ is already known:

$$\varphi(X,T) = \int_{(0,0)}^{(X,T)} f dT - s dX. \qquad (3.221)$$

It follows from Eq. (3.211) that s and f are the partial derivatives of the stream-function $\varphi(X,T)$:

$$s = -\frac{\partial \varphi}{\partial X}, \quad f = \frac{\partial \varphi}{\partial T}, \qquad (3.222)$$

This function is independent of the integration path that links point (X,T) with the origin $(0,0)$ (Courant and Friedrichs, 1976).

The corresponding differential form of Eq. (3.221) results in:

$$d\varphi = fdT - sdX, \tag{3.223}$$

and the lines of constant stream-function are stream-lines of the flow:

$$\frac{dX}{dT} = \frac{f}{s}. \tag{3.224}$$

Time derivative, dT can be found from Eq. (3.223) as:

$$dT = \frac{d\varphi}{f} + \frac{sdX}{f}. \tag{3.225}$$

The equality of second mixed derivatives of function $T = T(X,\varphi)$ in Eq. (3.225) yields:

$$\frac{\partial F(U,\gamma)}{\partial \varphi} + \frac{\partial U}{\partial X} = 0. \tag{3.226}$$

where,

$$U = \frac{1}{f(s,\gamma)}, \quad F(U,\gamma) = -\frac{s}{f(s,\gamma)}. \tag{3.227}$$

Eq. (3.226) is the transformed form of Eq. (3.211) in (X,φ) plane.

Using Green's theorem over any arbitrary domain ϖ with a continuous boundary to Eq. (3.212) and applying to Eq. (3.223):

$$0 = \oint_{\partial\varpi} (\gamma f)dT - (\gamma s)dX = \oint_{\partial\varpi} \gamma(fdT - sdX) = \oint_{\partial\varpi} \gamma d\varphi = \iint_{\varpi} \left(\frac{\partial\gamma}{\partial X}\right) dXd\varphi. \tag{3.228}$$

transforms the salt transport equation into (X,φ)-coordinates.

$$\frac{\partial\gamma}{\partial X} = 0. \tag{3.229}$$

In the same way, applying Green's theorem to Eqs. (3.214) and (3.215) transforms the fines transport and straining rate equations into the new coordinate system:

$$\frac{\partial c}{\partial X} + \frac{\partial S_s}{\partial \varphi} = 0, \tag{3.230}$$

$$\frac{\partial S_s}{\partial \varphi} = \frac{\Lambda c}{2\sqrt{X}}. \tag{3.231}$$

So the transformation of the original system (Eqs. (3.211−3.215)) into (X,φ)-coordinates has the form (Eqs. (3.226, 3.229, 3.230, and 3.231). System (Eqs. (3.229−3.231)) and Eq. (3.226) are called the auxiliary system and the lifting equation, respectively. The auxiliary system (Eqs. (3.229−3.231)) splits from the lifting Eq. (3.226). A detailed derivation of the splitting technique and its application to colloidal flow in porous media is provided by Borazjani and Bedrikovetsky (2017).

Substituting a trajectory $\varphi(X)$ and $T(X)$ into the flux (Eq. (3.223)) and taking the corresponding derivatives show the relationship between rarefaction and shock wave speeds in planes (X,φ) and (X,T):

$$V = fD - s. \qquad (3.232)$$

Here, V and D are the wave speeds in (X,φ) and (X,T) coordinates, respectively.

The initial and boundary conditions translate to:

$$\varphi = -s_I X : S_a = \begin{cases} 0 & X_w < X < X_{mI} \\ S_{cr}\left(\dfrac{q}{2\pi r_e\sqrt{X}}, \gamma_I\right) & X_{mI} < X < X_{crI}, \\ S_{aI} & X > X_{crI} \end{cases} \qquad (3.233)$$

$$\varphi = -s_I X : c = \begin{cases} S_{aI} & X_w < X < X_{mI} \\ \left(S_{aI} - S_{cr}\left(\dfrac{q}{2\pi r_e\sqrt{X}}, \gamma_I\right)\right) & X_{mI} < X < X_{crI}, \\ 0 & X > X_{crI} \end{cases} \qquad (3.234)$$

$$\varphi = -s_I X : \gamma = \gamma_I, F = -\infty, \qquad (3.235)$$

$$X = X_w : c = 0, \gamma = \gamma_J, U = 1. \qquad (3.236)$$

By using the splitting procedure, the original $(4) \times (4)$ system of quasi-linear hyperbolic Eqs. (3.211, 3.212, 3.214, and 3.215) is reduced to a single equation for the saturation (Eq. (3.226)), which separates from the remaining three Eqs. (3.229−3.231) for the salinity and the suspended and strained particle concentrations. This new $(3) \times (3)$ system is called the auxiliary system.

3.7.3 Exact solution for the auxiliary system

As performed in Section 3.5, the salt and suspended particle concentration can be solved from Eqs. (3.230 and 3.231) using the method of

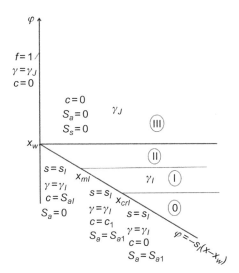

Figure 3.27 The solutions of the auxiliary system (φ: stream function, X: dimensionless spatial coordinate, s: water saturation, γ: fluid salinity, c: suspended particle concentration, S_a: dimensionless attached concentration).

characteristics alongside the initial (Eq. (3.235)) and boundary conditions (Eq. (3.236)). The strained concentration can be derived by integration of Eqs. (3.230 and 3.231) as an ODE. Fig. 3.27 presents the solution of the auxiliary problem (Eqs. (3.229–3.231)) in (X, φ) plane.

The final solution for the auxiliary system is given in Table 3.14.

3.7.4 Lifting equation

The remaining calculations involve solving Eq. (3.226) to "lift" the solution presented in Table 3.14. The lifting problem was solved numerically using the computer code presented by Shampine and Thompson (2001) and is available from http://faculty.smu.edu/shampine/current.html. This code implements the second-order Richtmyer's two-step variant of the Lax–Wendroff method.

3.7.5 Inverse mapping

To obtain the solution of the problem (Eqs. (3.211–3.220)), let us consider the inverse transformation of an independent variable in the solution of auxiliary and lifting problems:

$$(X, \varphi) \rightarrow (X, T). \tag{3.237}$$

Table 3.14 Analytical solutions for salt transport γ, suspended particle concentration c, and strained particle concentration S_s in (X, φ) coordinate for the auxiliary system

Zone	Term	Domain	Solution
III	$\gamma(X, \varphi)$	$\varphi > 0$	γ_I
0, I, II		$-s_I X < \varphi < 0$	γ_I
0		$\varphi < \varphi(X_{crl})$	0
I	$c(X, \varphi)$	$\varphi(X_{crl}) < \varphi < \varphi(X_{ml})$	$S_{al} - S_{cr}\left(\frac{q}{2\pi r_e \sqrt{x}}, \gamma_I\right) e^{-\Lambda\left(\sqrt{x} - \sqrt{\frac{\varphi}{-\gamma} + x_w}\right)}$
II		$\varphi(X_{ml}) < \varphi < 0$	$S_{al}\, e^{-\Lambda\left(\sqrt{x} - \sqrt{\frac{\varphi}{-\gamma} + x_w}\right)}$
III		$\varphi > 0$	0
0		$\varphi < \varphi(X_{crl})$	0
I		$\varphi(X_{crl}) < \varphi < \varphi(X_{ml})$	$\frac{1}{2\sqrt{X}}\left[\Lambda S_{al}\left(\varphi + s^I(X - X_w)\right)\left(\dfrac{e^{\Lambda\sqrt{X_w - \frac{\varphi}{s^I}}}\; e^{-\Lambda\sqrt{X}\sqrt{X_w - \frac{\varphi}{s^I}}}}{\Lambda} - \dfrac{e^{-\Lambda\sqrt{x}}}{\Lambda^2} \right) + 2 S_{cr} s^I\left(-e^{\Lambda\sqrt{X}}\left(\dfrac{e^{-\Lambda X}}{\Lambda} - \dfrac{e^{-\Lambda\sqrt{X}}}{\Lambda^2} \right) \right) \right]$
II	$S_s(X, \varphi)$	$\varphi(X_{ml}) < \varphi < 0$	$-\dfrac{\Lambda S_{al}}{2\sqrt{X}\Lambda^2}\left(s^I\left(2 - 2\Lambda\sqrt{X_w - \frac{\varphi}{s^I}}\right) e^{\Lambda\sqrt{X_w - \frac{\varphi}{s^I}} - \sqrt{X}} - s^I\left(2 - 2\Lambda\sqrt{X}\right) \right)$
III		$\varphi > 0$	0

$$T = \int_{0,0}^{X,\varphi} \left(\frac{1}{f(X,\varphi)} d\varphi + \frac{s(X,\varphi)}{f(X,\varphi)} dX \right). \tag{3.238}$$

Fig. 3.28 presents the trajectories of saturation and concentration in the (X, T)-plane.

An example calculation of the Impedance from Eq. (3.216) is given in Fig. 3.29.

Relative phase permeability is given by Corey's formulae:

$$
\begin{aligned}
k_{ro} &= k_{rowi} \left(\frac{1-s_{or}-s}{1-s_{or}-s_I} \right)^{n_o} \\
k_{rw} &= k_{rwor} \left(\frac{s-s_I}{1-s_{or}-s_I} \right)^{n_w} /(1 + \beta S_s)
\end{aligned}
\tag{3.239}
$$

where the values of endpoint saturations and relative permeability are presented in Table 3.15 for injected and reservoir salinities.

The impedance initially rises when the injection fluid is high–salinity water, but then declines gradually as more of the reservoir fills with the less viscous injected water. When low–salinity water is used, fines detachment causes the impedance to rise substantially.

3.7.6 Implementation of fines migration using reservoir simulators

The explicit expressions for impedance derived here, i.e., those for well productivity in Section 3.4 and well injectivity in Sections 3.5.3, 3.6.4, and 3.7.3, can be implemented in 3D reservoir simulators.

The extent of formation damage zones during fines migration in production and injection wells typically does not exceed 1−3 m (Nunes et al., 2010). Therefore, this damage can be accounted for in the skin factor. The main commercial 3D reservoir simulators contain well options, where skin can be expressed by a table or formula, yielding the well boundary condition for the damage-free flow in the reservoir. An example of matching the analytical model for near-well formation damage with the damage-free waterflooding is presented by Bedrikovetsky et al. (2011c). Implementation of Shapiro's model for menisci dynamics and fines lifting by capillary forces is an important next development of fines migration in two–phase flows (Shapiro 2015, 2016).

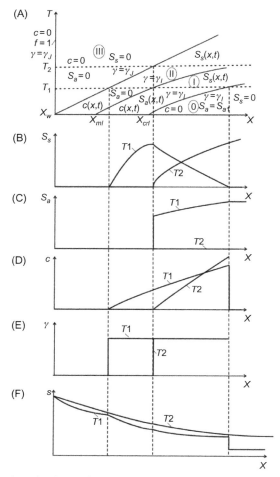

Figure 3.28 Analytical solution for injection of low-salinity water in (X, T) coordinates applying inverse transformation of the independent variables: (A) trajectories of saturation and concentration waves in the (X, T)-plane along with typical zones 0, I, II, and III; (B) strained concentration profiles at two different times; (C) attached concentration profiles at two different times; (D) suspended concentration at two different times, (E) salinity profiles, and (F) saturation profiles (X: dimensionless spatial coordinate, T: dimensionless time, S_s: dimensionless strained concentration, S_a: dimensionless attached concentration, c: dimensionless suspended concentration, γ: fluid salinity, s: water saturation).

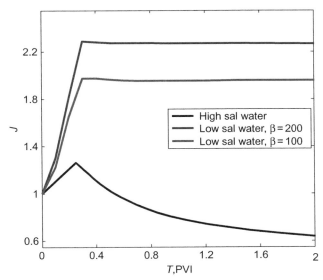

Figure 3.29 Impedance plotted against the number of PVI for high-salinity water injection, and two cases of low-salinity water injection with different formation damage coefficients (J: impedance, PVI: number of pore volumes injected).

Table 3.15 Impedance calculation during low-salinity and high-salinity water injection applying inverse mapping

Case	s_l	k_{rwor}	s_{or}	k_{rwoi}	n_w	n_o	β
High–salinity water	0.2	0.5	0.3	1	3	4	0
Low–salinity water	0.2	0.5	0.3	1	3	4	100, 200

3.8 CONCLUSIONS

The mathematical formulations and exact solutions presented here provide a means for petroleum engineers to quantify and predict formation damage due to fines migration. The main two causes of fines migration are high fluid velocities and low fluid salinities. Equations for both of these cases, as well as descriptions of the mechanisms for damage, have been presented in the prceeding sections. Finally, a brief extension of these works into the two–phase environment has been presented alongside a methodology for predicting injectivity decline.

All formulations presented here are still reliant on several limitations and assumptions. Of the most relevance to application in the petroleum industry are the homogeneity of the reservoir and the simplicity of the reservoir fluid. The spatial distributions of permeability and porosity are already recognized as critical for proper simulation of fluid flow in petroleum reservoirs. The model parameters characterising the susceptibility of rocks to damage by fines migration will similarly vary substantially within reservoirs. Quantifying the spatial variability of these parameters will most likely come from correlations with porosity and permeability. Such correlations are at the moment unclear and remain a limiting factor for the practical application of fines migration modeling on the field scale.

Fluid compressibility and the validity of Darcy's law are two other assumptions that currently limit the range of application of fines migration. Extension of the models to include these effects would allow the prediction of fines migration in the presence of complex fluids such as in natural gas reservoirs.

Despite the possibilities for extensions to the presented models, the existing literature provides petroleum engineers with sufficient tools to make informed decisions in relation to formation damage due to fines migration.

NOMENCLATURE

A	Total pore surface area, L^2
A_{132}	Hamaker constant
c	Suspended particle concentration, L^{-3}
C	Dimensionless suspended particle concentration
D	Shock front velocity in coordinates (X, T)
e	Elementary electric charge, IT
E	Young's modulus, $ML^{-1}T^{-2}$
F_d	Drag force, MLT^{-2}
F_e	Electrostatic force, MLT^{-2}
F_g	Gravitational force, MLT^{-2}
F_l	Lift force, MLT^{-2}
f	Fractional flow
h	Surface-to-surface separation distance, L
h_c	Critical internal cake thickness, L
II	Injectivity index
J	Impedance
J_D	Impedance in the damage zone
J_{UD}	Impedance in undamaged zone
k_{det}	Kinetic detachment coefficient

k	Permeability, L^2
k_B	Boltzmann constant, $L^2MT^{-2}\Theta^{-1}$
k_i	Absolute permeability for phase i, L^2
k_{ri}	Relative permeability for phase i
K	Composite Young's modulus, $ML^{-1}T^{-2}$
L	Core length, L
l_n	Normal lever arm, L
l_d	Drag lever arm, L
n_{i0}	Concentration of i ions in bulk solution, L^{-3}
n_1	Refractive index of particle
n_2	Refractive index of grain
n_3	Refractive index of fluid
n_∞	Bulk density of ions
p	Pressure, $MT^{-2}L^{-1}$
P	Dimensionless pressure
q	Flow rate per unit reservoir thickness, L^2T^{-1}
r	Radial coordinate, L
r_p	Pore size, L
r_s	Particle radius, L
r_e	Drainage radius, L
s	Saturation
S_a	Dimensionless attached particle concentration
S_s	Dimensionless strained particle concentration
ΔS_{cr}	Dimensionless mobilized concentration of detached particle with salinity alteration
t	Time, T
T	Temperature, Θ
T	Dimensionless time
T_0	Intersection of characteristic line and the T-axis
U	Darcy's velocity, LT^{-1}
U_s	Particle velocity, LT^{-1}
V	Total energy, ML^2T^{-2}
V_{LVW}	London-Van der Waals potential, ML^2T^{-2}
V_{EDL}	Electrical Double Layer potential, ML^2T^{-2}
V_{BR}	Born Repulsive potential, ML^2T^{-2}
x	Linear coordinate, L
X	Dimensionless linear coordinate
z	Valence symmetrical electrolyte solution

Greek symbols

α	Drift delay factor
β_a	Formation damage coefficient for attached particles
β_s	Formation damage coefficient for strained particles
γ	Salinity, L^{-3}
Γ	Dimenssionless salinity
ε	Dimensionless delay time
ε_0	Dielectric permittivity of vacuum, MT^{-2}

ε_1 Dielectric constant of the particle, MT^{-2}
ε_2 Dielectric constant of the grain, MT^{-2}
ε_1 Dielectric constant of the fluid, MT^{-2}
ζ_s Zeta potential for particle, $ML^2T^{-3}I^{-1}$
ζ_g Zeta potential for grain, $ML^2T^{-3}I^{-1}$
κ Inverse Debye length, L
λ Filtration coefficient for re-attachment, L^{-1}
λ_w Characteristic wavelength of interaction, L
λ_0 Initial filtration coefficient, L^{-1}
Λ Dimensionless filtration coefficient for re-attachment
Λ_s Dimensionless filtration coefficient for straining
μ Dynamic viscosity, $ML^{-1}T^{-1}$
ν Poison's ratio
ν_e Constant value of absorption frequency, T^{-1}
σ_{LJ} Atomic collision diameter, L
σ_{cr} Critical retention function, L^{-3}
$\Delta\sigma$ Mobilized concentration of detached particles with salinity decrease, L^{-3}
σ_a Concentration of attached particles, L^{-3}
σ_s Concentration of strained particles, L^{-3}
σ_m Number of vacancies available for particle straining, L^{-3}
τ Delay time of particle release, T
φ Lagrangian coordinate (potential)
ϕ Porosity
ϕ_c Porosity of the internal cake
χ Lift coefficient
Ψ_s Reduced zeta potential for particle
Ψ_g Reduced zeta potential for grain
ω Drag coefficient

Super/Subscripts

0 Initial value or condition
$+$ Ahead of the front
$-$ Behind the front
cr Critical, for radius, retention concentration, velocity, and salinity
I Initial, for pressure, salinity, and concentrations
J Injected, for salinity
m Minimum, for velocity
mI Minimum under the initial water salinity, for radius and velocity
mJ Minimum under the injected salinity, for radius and velocity
o Oil phase
w Water phase, for relative permeability, and area of pore surface

REFERENCES

Akhmetgareev, V., Khisamov, R., 2015. 40 years of low-salinity waterflooding in Pervomaiskoye field, Russia: incremental oil. SPE European Formation Damage Conference and Exhibition. Society of Petroleum Engineers.
Altmann, J., Ripperger, S., 1997. Particle deposition and layer formation at the crossflow microfiltration. J. Memb. Sci. 124 (1), 119−128.

Al-Yaseri, A., Al Mukainah, H., Lebedev, M., Barifcani, A., Iglauer, S., 2016. Impact of fines and rock wettability on reservoir formation damage. Geophys. Prospect. 64 (4), 860—874.

Al-Yaseri, A.Z., Zhang, Y., Ghasemiziarani, M., Sarmadivaleh, M., Lebedev, M., Roshan, H., et al., 2017. Permeability evolution in sandstone due to CO_2 injection. Energy. Fuels. 13 (11), 12390—12398.

Barenblatt, G.I., Entov, V.M., Ryzhik, V.M., 1989. Theory of Fluid Flows through Natural Rocks. Kluwer Acadmeic Publishers, Dordrecht.

Barkman, J., Abrams, A., Darley, H., Hill, H., 1975. An oil-coating process to stabilize clays in fresh waterflooding operations (includes associated paper 6405). J. Pet. Technol. 27 (09), 1053—1059.

Bedrikovetsky, P., 1993. Mathematical Theory of Oil and Gas Recovery. Kluwer Academic Publishers, Dordrecht, The Netherlands.

Bedrikovetsky, P., 2008. Upscaling of stochastic micro model for suspension transport in porous media. Trans. Porous Media 75 (3), 335—369.

Bedrikovetsky, P., Caruso, N., 2014. Analytical model for fines migration during water injection. Trans. Porous Media 101 (2), 161—189.

Bedrikovetsky, P., Siqueira, F.D., Furtado, C.A., Souza, A.L.S., 2011a. Modified particle detachment model for colloidal transport in porous media. Trans. Porous Media 86 (2), 353—383.

Bedrikovetsky, P.G., Vaz Jr, A., Machado, F.A., Zeinijahromi, A., Borazjani, S., 2011b. Well productivity decline due to fines migration and production:(Analytical model for the regime of strained particles accumulation). SPE European Formation Damage Conference. Society of Petroleum Engineers.

Bedrikovetsky, P.G., Nguyen, T.K., Hage, A., Ciccarelli, J.R., ab Wahab, M., Chang, G., et al., 2011c. Taking advantage of injectivity decline for improved recovery during waterflood with horizontal wells. J. Pet. Sci. Eng. 78 (2), 288—303.

Bedrikovetsky, P., Zeinijahromi, A., Siqueira, F.D., Furtado, C.A., de Souza, A.L.S., 2012. Particle detachment under velocity alternation during suspension transport in porous media. Trans. Porous Media 91 (1), 173—197.

Borazjani, S., Bedrikovetsky, P., 2017. Exact solutions for two-phase colloidal-suspension transport in porous media. Appl Math Model 44, 296—320.

Borazjani, S., Behr, A., Genolet, L., Van Der Net, A., Bedrikovetsky, P., 2017. Effects of fines migration on low-salinity waterflooding: analytical modelling. Trans Porous Media 116 (1), 213—249.

Bradford, S.A., Torkzaban, S., 2008. Colloid transport and retention in unsaturated porous media: a review of interface-, collector-, and pore-scale processes and models. Vadose Zone J 7 (2), 667—681.

Bradford, S.A., Simunek, J., Bettahar, M., van Genuchten, M.T., Yates, S.R., 2003. Modeling colloid attachment, straining, and exclusion in saturated porous media. Environ. Sci. Technol. 37 (10), 2242—2250.

Bradford, S.A., Torkzaban, S., Wiegmann, A., 2011. Pore-scale simulations to determine the applied hydrodynamic torque and colloid immobilization. Vadose Zone J 10 (1), 252—261.

Bradford, S.A., Torkzaban, S., Shapiro, A., 2013. A theoretical analysis of colloid attachment and straining in chemically heterogeneous porous media. Langmuir 29 (23), 6944—6952.

Civan, F., 2014. Reservoir Formation Damage. Gulf Professional Publishing, Burlington, MA, USA.

Coleman, T.F., Li, Y.Y., 1996. An interior trust region approach for nonlinear minimization subject to bounds. SIAM J Optimiz 6 (2), 418—445.

Coronado, M., Díaz-Viera, M.A., 2017. Modeling fines migration and permeability loss caused by low salinity in porous media. J. Pet. Sci. Eng. 150, 355–365.

Courant, R., Friedrichs, K.O., 1976. Supersonic Flow and Shock Waves. Interscience Publisher Ltd, London.

Das, S.K., Schechter, R.S., Sharma, M.M., 1994. The role of surface roughness and contact deformation on the hydrodynamic detachment of particles from surfaces. J. Colloid. Interface. Sci. 164 (1), 63–77.

Derjaguin, B., Landau, L., 1941. Theory of the stability of strongly charged lyophobic sols and of the adhesion of strongly charged particles in solutions of electrolytes. Acta Physicochim. URSS 14 (6), 633–662.

Derjaguin, B.V., Muller, V.M., Toporov, Y.P., 1975. Effect of contact deformations on the adhesion of particles. J. Colloid. Interface. Sci. 53 (2), 314–326.

Elimelech, M., Gregory, J., Jia, X., 2013. Particle Deposition and Aggregation: Measurement, Modelling and Simulation. Butterworth-Heinemann.

Gregory, J., 1975. Interaction of unequal double layers at constant charge. J. Colloid. Interface. Sci. 51 (1), 44–51.

Gregory, J., 1981. Approximate expressions for retarded van der Waals interactions. J. Colloid. Interface. Sci. 83 (1), 138–145.

Guo, Z., Hussain, F., Cinar, Y., 2015. Permeability variation associated with fines production from anthracite coal during water injection. Int. J. Coal. Geol. 147, 46–57.

Herzig, J., Leclerc, D., Goff, P.L., 1970. Flow of suspensions through porous media—application to deep filtration. J. Indus. Eng. Chem. 62 (5), 8–35.

Hogg, R., Healy, T.W., Fuerstenau, D., 1966. Mutual coagulation of colloidal dispersions. J. Chem. Soc. Faraday Tra 62, 1638–1651.

Huang, F., Kang, Y., You, Z., You, L., Xu, C., 2017. Critical conditions for massive fines detachment induced by single-phase flow in coalbed methane reservoirs: modeling and experiments. Energy. Fuels. 31 (7), 6782–6793.

Hussain, F., Zeinijahromi, A., Bedrikovetsky, P., Badalyan, A., Carageorgos, T., Cinar, Y., 2013. An experimental study of improved oil recovery through fines-assisted water-flooding. J. Pet. Sci. Eng. 109, 187–197.

Israelachvili, J.N., 2011. Intermolecular and Surface Forces: Revised Third Edition. Academic Press, Burlington, USA.

Jaiswal, D.K., Kumar, A., Yadav, R.R., 2011. Analytical solution to the one-dimensional advection-diffusion equation with temporally dependent coefficients. J. Water. Res. Protect. 3 (01), 76.

Joekar-Niasar, V., Mahani, H., 2016. Nonmonotonic pressure field induced by ionic diffusion in charged thin films. J. Indus. Eng. Chem. Res. 55 (21), 6227–6235.

Kalantariasl, A., Bedrikovetsky, P., 2013. Stabilization of external filter cake by colloidal forces in a "well–reservoir" system. J. Indus. Eng. Chem. Res. 53 (2), 930–944.

Kang, S.-T., Subramani, A., Hoek, E.M., Deshusses, M.A., Matsumoto, M.R., 2004. Direct observation of biofouling in cross-flow microfiltration: mechanisms of deposition and release. J. Memb. Sci. 244 (1), 151–165.

Khilar, K.C., Fogler, H.S., 1983. Water sensitivity of sandstones. SPE. J. 23 (01), 55–64.

Khilar, K.C., Fogler, H.S., 1998. Migrations of Fines in Porous Media. Kluwer Academic Publishers, Dordrecht.

Lake, L.W., 2010. Enhanced Oil Recovery. Society of Petroleum Engineers, Richardson, TX.

Lemon, P., Zeinijahromi, A., Bedrikovetsky, P., Shahin, I., 2011. Effects of injected-water salinity on waterflood sweep efficiency through induced fines migration. J. Can. Pet. Technol. 50 (9/10), 82–94.

Mahani, H., Berg, S., Ilic, D., Bartels, W.B., Joekar-Niasar, V., 2015. Kinetics of low-salinity-flooding effect. SPE. J. 20 (1), 8–20.

Marquez, M., Williams, W., Knobles, M.M., Bedrikovetsky, P., You, Z., 2014. Fines migration in fractured wells: integrating modeling with field and laboratory data. SPE. Pro. Ope. 29 (04), 309−322.

Mohammadmoradi, P., Taheri, S., Kantzas, A., 2017. Interfacial areas in Athabasca oil sands. Energy Fuels 31 (8), 8131−8145.

Morrow, N., Buckley, J., 2011. Improved oil recovery by low-salinity waterflooding. J. Pet. Technol. 63 (05), 106−112.

Muecke, T.W., 1979. Formation fines and factors controlling their movement in porous rocks. J. Pet. Technol. 32 (2), 144−150.

Nunes, M., Bedrikovetsky, P., Newbery, B., Paiva, R., Furtado, C., De Souza, A., 2010. Theoretical definition of formation damage zone with applications to well stimulation. J. Energy. Res. Technol. 132 (3), 033101.

Ochi, J., Vernoux, J.-F., 1998. Permeability decrease in sandstone reservoirs by fluid injection: hydrodynamic and chemical effects. J. Hydrol. 208 (3), 237−248.

Ogletree, J., Overly, R., 1977. Sea-water and subsurface-water injection in West Delta Block 73 waterflood operations. J. Pet. Technol. 29 (06), 623−628.

Oliveira, M.A., Vaz, A.S., Siqueira, F.D., Yang, Y., You, Z., Bedrikovetsky, P., 2014. Slow migration of mobilised fines during flow in reservoir rocks: laboratory study. J Pet Sci Eng 122, 534−541.

Omekeh, A.V., Evje, S., Friis, H.A., 2013. Modeling of low salinity effects in sandstone oil rocks. Int. J. Numer. Anal. Model., Ser. B 4 (2), 95−128.

Pang, S., Sharma, M., 1997. A model for predicting injectivity decline in water-injection wells. SPE Form Eval 12 (03), 194−201.

Pires, A.P., Bedrikovetsky, P.G., Shapiro, A.A., 2006. A splitting technique for analytical modelling of two-phase multicomponent flow in porous media. J. Pet. Sci. Eng. 51 (1), 54−67.

Polyanin, A., Zaitsev, V., 2011. Handbook of Nonlinear Partial Differential Equations. CRC press, Hoboken, USA.

Polyanin, A.D., Manzhirov, A.V., 2007. Handbook of Mathematics for Engineers and Scientists. CRC Press, Boca Raton, USA.

Rosenbrand, E., Kjøller, C., Riis, J.F., Kets, F., Fabricius, I.L., 2015. Different effects of temperature and salinity on permeability reduction by fines migration in Berea sandstone. Geothermics 53, 225−235.

Ruckenstein, E., Prieve, D.C., 1976. Adsorption and desorption of particles and their chromatographic separation. AIChE J 22 (2), 276−283.

Russell, T., Pham, D., Neishaboor, M.T., Badalyan, A., Behr, A., Genolet, L., et al., 2017. Effects of kaolinite in rocks on fines migration. J. Nat. Gas. Sci. Eng. 45, 243−255.

Sarkar, A.K., Sharma, M.M., 1990. Fines migration in two-phase flow. J Pet Technol 42 (05), 646−652.

Schechter, R.S., 1992. Oil Well Stimulation. Prentice Hall, NJ.

Sefrioui, N., Ahmadi, A., Omari, A., Bertin, H., 2013. Numerical simulation of retention and release of colloids in porous media at the pore scale. Colloids Surf A: Physicochem. Eng. Asp. 427, 33−40.

Shampine, L.F., Thompson, S., 2001. Solving ddes in matlab. Appl. Num. Math. 37 (4), 441−458.

Shapiro, A.A., 2015. Two-phase immiscible flows in porous media: The mesoscopic Maxwell-Stefan approach. Transport Porous Media 107 (2), 335−363.

Shapiro, A.A., 2016. Mechanics of the separating surface for a two-phase co-current flow in a porous media. Transport Porous Media 112 (2), 489−517.

Sharma, M., Yortsos, Y., 1987. Fines migration in porous media. AIChE. J. 33 (10), 1654−1662.

Sharma, M.M., Chamoun, H., Sarma, D.S.R., Schechter, R.S., 1992. Factors controlling the hydrodynamic detachment of particles from surfaces. J. Colloid. Interface. Sci. 149 (1), 121–134.

Shen, W., 2016. On the Cauchy problems for polymer flooding with gravitation. Journal of differential equations. 261 (1), 627–653.

Tufenkji, N., 2007. Colloid and microbe migration in granular experiments: a discussion of modelling methods. In: Frimmel, F.H., von der Kammer, F., Flemming, F.C. (Eds.), Colloidal Transport in Porous Media. Springer-Verlag, Berlin.

Vaz, A., Bedrikovetsky, P., Fernandes, P., Badalyan, A., Carageorgos, T., 2017. Determining model parameters for non-linear deep-bed filtration using laboratory pressure measurements. J. Pet. Sci. Eng. 151, 421–433.

Verwey, E.J.W., Overbeek, J.T.G., Overbeek, J.T.G., 1999. Theory of the Stability of Lyophobic Colloids. Dover Publications, Mineola, New York, USA.

Wang, J., Liu, Hq, Zhang, Hl, Sepehrnoori, K., 2017. Simulation of deformable pre-formed particle gel propagation in porous media. AIChE. J. 63 (10), 4628–4641.

Yang, H., Balhoff, M.T., 2017. Pore-network modeling of particle retention in porous media. AIChE. J. 63 (7), 3118–3131.

Yang, Y., Bedrikovetsky, P., 2017. Exact solutions for nonlinear high retention-concentration fines migration. Trans. Porous. Media. 119 (2), 351–372.

Yang, Y., Siqueira, F.D., Vaz, A.S., You, Z., Bedrikovetsky, P., 2016. Slow migration of detached fine particles over rock surface in porous media. J. Nat. Gas. Sci. Eng. 34, 1159–1173.

You, Z., Badalyan, A., Yang, Y., Bedrikovetsky, P., Hand, M., 2014. Laboratory study of fines migration in geothermal reservoirs. Geothermics (submitted for publication).

You, Z., Yang, Y., Badalyan, A., Bedrikovetsky, P., Hand, M., 2016. Mathematical modelling of fines migration in geothermal reservoirs. Geothermics 59, 123–133.

Yuan, B., Moghanloo, R.G., 2017. Analytical model of well injectivity improvement using nanofluid preflush. Fuel 202, 380–394.

Yuan, B., Moghanloo, R.G., Zheng, D., 2016. Analytical evaluation of nanoparticle application to mitigate fines migration in porous media. SPE. J. 21 (06), 2317–2332.

Yuan, B., Moghanloo, R., Wang, W., 2018a. Using nanofluids to control fines migration for oil recovery: nanofluids co-injection or nanofluids preflush? – a comprehensive answer. Fuel 215, 474–48312.

Yuan, B., Moghanloo, R., 2018b. Nanofluid treatment, an effective approach to improve performance of low salinity water flooding. J. Pet. Sci. Eng. in press Available from: https://doi.org/10.1016/j.petrol.2017.11.032.

Yuan, B., Moghanloo, R.G., 2018c. Nanofluid pre-coating, an effective method to reduce fines migration in radial systems saturated with mobile immiscible fluids. SPE J., SPE-189464-PA. https://doi.org/10.2118/189464-PA

Yuan, H., Shapiro, A.A., 2010. Modeling non-Fickian transport and hyperexponential deposition for deep bed filtration. Chem. Eng. J. 162 (3), 974–988.

Yuan, H., Shapiro, A.A., 2011. Induced migration of fines during waterflooding in communicating layer-cake reservoirs. J. Pet. Sci. Eng. 78 (3), 618–626.

Zeinijahromi, A., Bedrikovetsky, P., 2016. Water production control using low-salinity water injection. SPE Asia Pacific Oil & Gas Conference and Exhibition. Society of Petroleum Engineers, Perth, Australia.

Zeinijahromi, A., Lemon, P., Bedrikovetsky, P., 2011. Effects of induced fines migration on water cut during waterflooding. J. Pet. Sci. Eng. 78 (3), 609–617.

Zeinijahromi, A., Vaz, A., Bedrikovetsky, P., 2012a. Well impairment by fines migration in gas fields. J. Pet. Sci. Eng. 88, 125–135.

Zeinijahromi, A., Vaz, A., Bedrikovetsky, P., Borazjani, S., 2012b. Effects of fines migration on well productivity during steady state production. J. Porous. Media. 15 (7), 665–679.

Zeinijahromi, A., Nguyen, T.K.P., Bedrikovetsky, P., 2013. Mathematical model for fines-migration-assisted waterflooding with induced formation damage. SPE. J. 18 (03), 518−533.

Zeinijahromi, A., Al-Jassasi, H., Begg, S., Bedrikovetski, P., 2015. Improving sweep efficiency of edge-water drive reservoirs using induced formation damage. J. Pet. Sci. Eng. 130, 123−129.

Zeinijahromi, A., Farajzadeh, R., Bruining, J.H., Bedrikovetsky, P., 2016. Effect of fines migration on oil-water relative permeability during two-phase flow in porous media. Fuel 176, 222−236.

Zheng, Q., Wei, G.-W., 2011. Poisson−Boltzmann−Nernst−Planck model. J. Chem. Phy. 134 (19), 194101.

Zhu, S.-Y., Peng, X.-L., Du, Z.-M., Wang, C.-W., Deng, P., Mo, F., et al., 2017. Modeling of coal fine migration during cbm production in high-rank coal. Trans. Porous. Media. 118 (1), 65−83.

CHAPTER FOUR

Using Nanofluids to Control Fines Migration in Porous Systems

Bin Yuan[1,2] and Rouzbeh G. Moghanloo[1]

[1]The University of Oklahoma, Norman, OK, United States
[2]Coven Energy Technology Research Institute, Qingdao, Shandong, China

Contents

4.1 INTRODUCTION

Nanofluid that contains nanoparticles can exhibit unique electrical, magnetic, and chemical properties. Recently, nanofluids are widely reported in diverse applications associated with the oil and gas industry (Achinta and Belhaj, 2016), including oilfield exploration, reservoir characterization, drilling and completion, and enhanced oil recovery (EOR), etc.

Formation Damage during Improved Oil Recovery.
DOI: https://doi.org/10.1016/B978-0-12-813782-6.00004-X
© 2018 Elsevier Inc.
All rights reserved.

The types of nanoparticles involved mainly include Al_2O_3, MgO, ZrO_2, CeO_2, TiO_2, SiO_2, ZnO, and Fe_2O_3. Song and Marcus (2007) proposed hyperpolarized silicon nanoparticles for taking images of hydrocarbon reserves. Nanosensor and nanoidentification techniques were proposed to identify the physical and chemical properties, and mechanical characteristics of both fluids and rocks (Jahagirdar, 2008; Abousleiman et al., 2009; Kapusta et al., 2011; Berlin et al., 2011). Nanoparticles are now also used as additives in drilling and completion fluids for clay stabilization (McDonald, 2012), fluid-loss control (Huang et al., 2008; Contreras et al., 2014), viscosity alternation (Gurluk et al., 2013), wellbore stability (Zhang et al., 2015), decrease of drag and torque friction (Sharma et al., 2012), cementation (Van Zanten et al., 2010; Santra et al., 2012; Pang et al., 2014). Moreover, nanofluids have been extensively applied to enhance oil recovery through multiple mechanisms, including wettability alteration (Crews and Gomaa, 2012; Li et al., 2013, 2014), interfacial tension (IFT) reduction (Moghadam and Azizian, 2014), enhancing emulsion and foam stability (Adkins et al., 2007; Gonzenbach et al., 2007; Aminzadeh et al., 2012; Prigiobbe et al., 2016), and channels plugging (Ju et al., 2006, 2009; Ogolo et al., 2012).

Various approaches have been developed to control formation fines migration and to remove the formation damage caused by formation fines plugging in the near-wellbore region. A variety of special-designed clay-control agents have been applied to minimize the damage effects of fines migration in high clay-content wells (Jaramillo et al., 2010). Different organic & inorganic acid systems were developed to remove the formation fines that plugged reservoir pores, gravel packs, and sand control screens under different downhole conditions (Hibbeler et al., 2003, Huang et al., 2002). Recently, the introduction of nanoparticles to control fines migration were proposed (Huang et al., 2008; Ahmadi et al., 2011; Assef et al., 2014; Yuan et al., 2015; Yuan and Moghanloo, 2016). Laboratory experiments have confirmed that nanoparticles with extremely high surface areas are suitable to fixate mobile fines by decreasing the double-layer repulsive forces between fine particles and rock grains effectively (Huang et al., 2008). Yuan et al. (2015, 2017a,b,c,d) presented a series of analytical solutions to characterize nanoparticle/fines migration in porous media saturated with only single-phase water and quantified the positive contributions of nanoparticles treatment (both pre-flush and co-injection) in mitigating problems of fines migration. Followed by, Yuan et al. (2018a,b,c) further developed a mobility-control method attributed to the alteration of water-phase permeability caused by fines migration & plugging, and also introduced the nanofluid-slug treatment to enhance the injectivity of low-salinity water.

4.2 LABORATORY PROOF AND FIELD CASES

Yuan et al. (2018b) reviewed both the experimental findings and field application examples related to nanofluid application for fines migration control. Among them, Huang et al. (2015) delivered laboratory testing of nanoparticles to stabilize both expandable (Bentonite) and non-expandable clays (Illite) in sand packs. Pressure drops along each tube packed with a mixture of sands and clays were measured with separate transducers. To compare the clay-stabilizing capability of nanoparticles with commercial clay stabilizer for the control of expandable clays, another two sand packs, one without clays (Bentonite or Illite) and one with bentonite, but using 2% bv CS-38 for clay stabilization, were presented. CS-38 is a type of liquid commercial clay stabilizer with polyquat amines as active components. In Fig. 4.1, the decreased values of pressure drop as 5% bw KCl water flows through the pack containing magnesium oxide nanoparticles and clay stabilizer confirms the effectiveness of magnesium oxide nanoparticles in stabilizing both expandable and nonexpandable clays and keeps the clay particles at their original locations.

Huang et al. (2010) also reported a field example of deepwater well treated with nanoparticles-coated proppant for a frac-packing operation in Gulf of Mexico. Before the utilization of nanoparticles, the

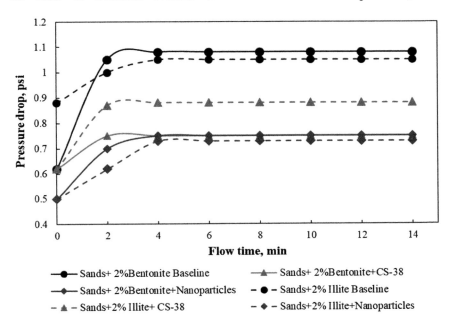

Figure 4.1 Comparison of pressure drop along sand packs with Bentonite or Illite clays at 10 mL/min of 5%bw KCL (Huang et al., 2015; Yuan et al., 2018).

production rate of the target well had declined by 5300BOPD oil and 4000 MCF gas with even flowing tubing pressure (FTP) decrease by 1900 psi, due to the severe problem of fines migration and plugging the reservoirs. In 2008, a frac-packing treatment was introduced with approximate 97,000 lb 20/40 mesh proppant, pre-coated with 0.1% (lb/lb) nanoparticles. After the nanoparticles-coated frac-packing treatment, the well production rates recovered with zero fines migration. Three months after that treatment, the production increased to 3200 BOPD oil and 2700 MCF gas with FTP 2300 psi. This field application example provides very good evidence to confirm the effectiveness of nanoparticles to fixate formation fines migration for the purpose of maintaining well productivity.

For the purpose of enhanced oil recovery, Arab et al. (2013) investigated the nanoparticles treatment during low-salinity waterflooding to mitigate the induced fines migration. In their experimental work, five types of metal oxide nanoparticles, Al_2O_3, MgO, CuO, SiO_2 and ZnO were selected to find the best type of nanoparticles to control fines migration by comparing the effluent fines concentrations. Also, coreflood test were performed using Berea sandstone cores to investigate the removal of permeability damage using nanoparticles. Those results confirmed the efficiency of soaking the medium with nanofluids prior to low-salinity waterflooding to mitigate the formation damage consequence induced by fines migration.

4.3 NANOPARTICLES TRANSPORT IN POROUS MEDIA: ADSORPTION, STRAINING, AND DETACHMENT BEHAVIORS

A comprehensive study of nanoparticles adsorption/detachment behaviors is essential to provide a foundation to illustrate the numerous benefits of nanoparticles applications. Li et al. (2015) performed experiments to study the transport phenomenon of hydrophilic silica nanoparticles and their effects in damaging core permeability. To quantify hydrophilic nanoparticles adsorption, straining, and detachment behaviors, and associated formation damage effects, the effluent-nanoparticle concentrations and pressure drop across the cores were used to estimate nanoparticles adsorption and retention behavior, as well as nanoparticles detachment behaviors during the postflush of the brine (Yuan et al., 2017a,c).

As nanoparticles pass through porous medium, they are adsorbed and strained at the stagnant points on the pore-throat surfaces, which can be confirmed by the reduction of effluent-nanoparticle concentrations from the injection source (Zhang et al., 2013; Li et al., 2015). The mass-balance equation of nanoparticles flowing through permeable media, considering their deposition onto rock grains and straining into pore-throats, can be written as,

$$\frac{\partial C_{NP}}{\partial x_D} + \frac{\partial C_{NP}}{\partial t_D} + \frac{1}{\phi}\frac{\partial \sigma_{NP}}{\partial t_D} + \frac{1}{\phi}\frac{\partial S_{NP}}{\partial t_D} = 0 \qquad (4.1)$$

where, $x_D = \frac{x}{L}$, $t_D = \frac{q_{inj}t}{\phi AL}$.

Nanoparticles straining rate can be expressed by classical filtration kinetics (Gruedes et al., 2006; Massoudieh and Ginn, 2010). Until nanoparticles adsorption reaches the maximum retention concentration (Yuan, 2017a), the classic particle-capture kinetics can be applied to quantify the transient attachment rates of nanoparticles (Vafai, 2005).

$$\frac{\partial S_{NP}}{\partial t_D} = \lambda_s C_{NP}\phi L, \qquad \frac{\partial \sigma_{NP}}{\partial t_D} = \lambda_{ad} C_{NP}\phi L \qquad (4.2)$$

when, $\sigma_{NP} < \sigma_{NP,max}$, $\sigma_{NP,max1} = \left[1 - (\frac{\mu r_{NP}^2 U}{2\phi r_p F_{e,max}\gamma})^2\right]\phi(1 - S_{or})$.

During the stage of brine postflush, there are no changes of fluid salinity. Hence, it can be concluded that any changes of nanoparticles retention concentration are only attributed to the decrease of average fluid density, because of the changes of nanoparticle concentrations. The average fluid density is expressed as a weighted average of nanoparticles and carrier water density:

$$\rho = \rho_w(1 - C_{NP}) + \rho_{NP}C_{NP} = \rho_w + C_{NP}(\rho_{NP} - \rho_w) \qquad (4.3)$$

The maximum retention concentration of nanoparticles becomes a function of injected nanoparticles concentration. The detachment of nanoparticles occurs instantly along with the abrupt changes of flowing nanoparticles concentration. Thus, the mass-balance equation of nanoparticles during the postflush of brine could be expressed as follows:

$$\frac{\partial C_{NP}}{\partial x_D} + \left(1 + \frac{1}{\phi}\frac{\sigma_{NP,max1} - \sigma_{NP,max2}}{C_{NP}}\right)\frac{\partial C_{NP}}{\partial t_D} = 0 \qquad (4.4)$$

The modified Darcy's flow equation is then applied by considering the damage of core permeability caused by nanoparticles adsorption and straining effects (Sharma, 1987):

$$U = \frac{k_0}{L\mu(1 + \beta_a \sigma_{NP} + \beta_s S_{NP})} \frac{dp}{dx_D} \tag{4.5}$$

As the injected nanoparticles concentration increases, the nanoparticles adsorption quantities can be enhanced (the solid line in Fig. 4.2A). The detachment of reversible adsorbed nanoparticles occurs during the postflush of brine. The amounts of reversible nanoparticles adsorption also increase along with the increase of the injected nanoparticles concentration (the dashed line in Fig. 4.2A). In addition, for the cases of different nanofluid injection concentration, the percentages of reversible adsorption remain approximately the same, i.e., about 30% of the total amount of nanoparticles adsorption. Fig. 4.2 bindicates that the nanoparticles adsorption and straining rates vary as functions of injected nanoparticles concentration. The higher the injected nanoparticles concentration is, the larger the nanoparticles adsorption and straining rates would be. In addition, the rates of nanoparticles adsorption are usually larger than nanoparticles straining rates. Fig. 4.2C summarizes the formation damage coefficients caused by both: nanoparticles adsorption and straining. It explains the reasons why pressure drop increases during nanofluid injection. In contrast to the relationship between nanoparticles adsorption and straining rates, the formation damage effects of nanoparticles straining are much larger than that of nanoparticles adsorption. That is to say, the formation damage caused by nanoparticles straining dominates the increase of pressure drop.

4.4 EFFECTIVENESS OF NANOPARTICLES UTILIZATION TO MITIGATE FINES MIGRATION IN WATER FLOW

The positive contributions of nanoparticles to mitigate fines migration can be characterized by the enhancement of the maximum retention concentration of fine particles onto rock grains through two reactions (Yuan and Moghanloo, 2016): (1) adsorption of nanoparticles onto the fines/grain surfaces; and (2) increased concentration of fines attachment onto pore surfaces via reducing the surface potential between grains and fines. Yuan et al. (2018a,b,c) proposed two different approaches of nanoparticles utilization to effectively enhance the capability of porous

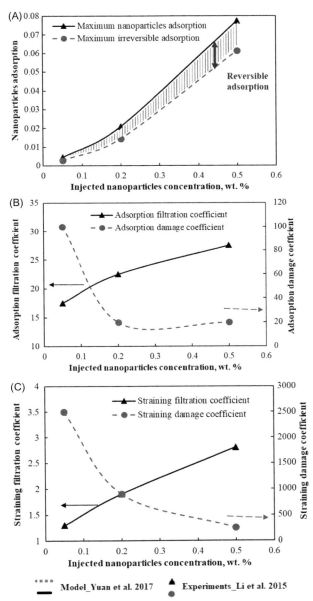

Figure 4.2 Effects of injected nanoparticles concentrations on irreversible and reversible adsorption of nanoparticles, nanoparticles adsorption and straining rates, and formation damage caused by nanoparticles adsorption and straining (revised after Yuan et al., 2017c).

medium to capture unsettled fines: (1) coinjection of nanoparticles with fines suspension into porous medium; and (2) precoating medium with nanoparticles prior to fines invasion.

4.4.1 Approach I: Coinjection of nanoparticles and fines into porous media

Usually, the surface potential of fines is less than that of rock grains, which results in the stronger attractive force between nanoparticles and fines than that between nanoparticles and rock grains (Yuan et al., 2016 and 2018b). Hence, during the co-injection of a mixture of nanoparticles and fines, nanoparticles should be assumed to be preferentially adsorbed onto the surfaces of mobile fines rather than the surfaces of rock grains. As a result, the nanoparticles adsorption onto fines would alter the surface potential of fines, which consequently enhance the attractive force between fines and rock grains. This is why the co-injection of nanoparticles could help increase the maximum retention concentration of fines onto rock grains.

As shown in Fig. 4.3a, the attachment of fine particles onto rock grains has already reached the maximum retention concentration of fines initially (Eq. 4.6a), which is determined by the surface charge of fine particles and rock grains. However, still excessive unattached fines are left in flowing suspension through porous medium. As nanoparticles arrive, the already attached fine particles onto rock grains would not be the immediate targets to capture nanoparticles; instead, the remaining unattached fines in the carrier suspension are more prone to first host the arriving nanoparticles. Continuously, the more nanoparticle adsorption, the larger the decrease of the surface potential of fines; thus, the attractive electrostatic forces between unattached fines and rock grains could be further enlarged. As a result, more attachment of initially unattached fine particles onto the pore surface instantaneously occurs following the adsorption of nanoparticles onto their surfaces. The n-nanoparticles-fine complex would be attached onto the rock grain surfaces continuously until the attachment of fine particles with nanoparticles adsorption onto rock grains reaches a new maximum limit (Eq. 4.6b), which is controlled by the maximum amounts of nanoparticles adsorbed onto suspended fines.

The stepwise reaction process of nanoparticles adsorption and subsequent n-nanoparticles-fine complex attachment would continuously repeat, and thus more unattached fines can be retained by rock grains with the increase of nanoparticles adsorption onto their surfaces (Yuan et al., 2018b).

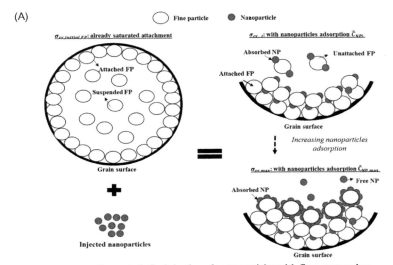

Scenario I: Co-injection of nanoparticles with fines suspension

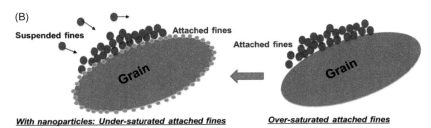

Scenario II: Nanoparticle treated pores prior to fines injection

Figure 4.3 Mutual interactions among nanoparticles, fines, and rock grains indicating various physical mechanisms by which nanoparticles control fines migration. (A). Scenario I: Co-injection of nanoparticles with fines suspension (B). Scenario II: Nanoparticle treated pores prior to fines injection (revised after Yuan et al., 2018b).

However, the adsorption capacity of nanoparticles onto fines is not infinite, which means beyond the maximum limit of nanoparticle adsorption (as a function of salinity, pH, temperature, and et al.) onto fine particles, the surface potential of fine particles does not change further. At that moment, the surface potential of fines can be assumed to become very close to that of adsorbed nanoparticles. As a result, the maximum retention concentration of fine particles onto rock grains also reaches the ultimate maximum attachment concentration value, which is determined by the surface charges of both nanoparticles and rock grains as shown in Eq. (4.6c) (Yuan et al., 2016, 2018b).

Without nanoparticles utilization to enhance fines attachment:

$$\sigma_{cr,initial} = \left[1 - \left(\frac{\mu_w r_{FP}^2 U}{2\phi(1-S_{or})r_P F_{ei}\gamma}\right)^2\right]\phi(1-S_{or}) \qquad (4.6a)$$

With arbitrary amounts of nanoparticles adsorption onto fine particles as the coinjection continues:

$$\sigma_{cr,i} = \left[1 - \left(\frac{\mu_w r_{FP}^2 U}{2\phi(1-S_{or})r_P\gamma\left(F_{ei} + \frac{128\pi r_{FP}n_\infty k_B T}{\kappa}e^{-\kappa h}\left[\frac{K_{NP}C_{NP}}{1+K_{NP}C_{NP}}(\varsigma_{FP}-\varsigma_{NP})\varsigma_{GS}\right]\right)}\right)^2\right]\phi(1-S_{or})$$

$$(4.6b)$$

With the maximum nanoparticles adsorption onto fine particles finally:

$$\sigma_{cr,max} = \left[1 - \left(\frac{\mu_w r_{FP}^2 U}{2\phi(1-S_{or})r_P\gamma(F_{ei} + \frac{128\pi r_{FP}n_\infty k_B T}{\kappa}e^{-\kappa h}(\varsigma_{FP}-\varsigma_{NP})\varsigma_{GS}}\right)^2\right]\phi(1-S_{or}) \quad (4.6c)$$

As the coinjection of nanoparticles and fines continues, the transport phenomenon of nanoparticles and fines passing through porous media was analyzed by Yuan et al. (2016; 2017a,b). For both nanoparticles and fines concentrations profiles along 1-D permeable medium in Fig. 4.4A, there are four different-state regions progressing from (Yuan et al., 2018b): (1) injection-condition region; (2) spreading-wave region; (3) constant-state region; to (4) initial-condition region. Meanwhile, a genuine shock "nanoparticle absorption front" connecting the constant-state region and the initial-condition region occurs to maintain the physical integrity of the analytical solutions (Moghanloo, 2012; Noh and Lake, 2004). The existence of coinjected nanoparticles leads to a "concave" profile in Fig. 4.4A that indicates the reduced amounts of mobile fines in the carrier fluid. Also, in Fig. 4.4B, the performance of nanoparticles to control fines migration are quantified as the coinjection continues. Prior to the breakthrough of the injected mixture, the reduced amounts of effluent fines (differences between dashed line and solid line) increase as time progresses. Following the arrival of the injection condition, the attachment ability of fine particles onto rock grains has already reached the maximum limit; meaning that, from that point in time, no more fines control by nanoparticles can be realized.

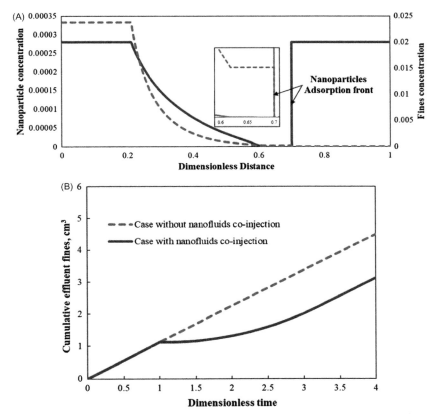

Figure 4.4 Nanoparticles and fines concentration profiles and cumulative effluent fines quantities with and without nanoparticles coinjection (Yuan et al., 2016).

4.4.2 Approach II: Precoat porous media with nanofluids prior to fines invasion

As illustrated in Fig. 4.3B, before the introduction of new fine particles into porous medium, a nanofluid slug is injected to coat the surfaces of rock grains and thereby modify the surface potential of the rock grains. As a result, the attractive forces between mobile fines and rock grains are enhanced, which could help mitigate the further movement of newly injected fines later. To characterize the enhanced attachment capacity of fines onto pore surfaces, the maximum retention concentration of fines onto rock grains is also applied, which can be also enlarged by the contributions of pre-coated nanoparticles onto rock grains, see Eq. 4.7a–c (Yuan et al., 2016, 2018b).

Without nanoparticles pretreatment:

$$\sigma_{cr,initial} = \left[1 - \left(\frac{\mu_w r_{FP}^2 U}{2\phi(1-S_{or})r_P F_{ei}\gamma}\right)^2\right]\phi(1-S_{or}) \tag{4.7a}$$

With arbitrary amounts of nanoparticles usage to coat rock grains:

$$\sigma_{cr,j} = \left[1 - \left(\frac{\mu_w r_{FP}^2 U}{2\phi(1-S_{or})r_P\gamma\left(F_{ei}+\frac{128\pi r_{FP}n_\infty k_B T}{\kappa}e^{-\kappa h}\left[\frac{K_{NP}C_{NP}}{1+K_{NP}C_{NP}}(\varsigma_{GS}-\varsigma_{NP})\varsigma_{FP}\right]\right)}\right)^2\right]$$
$$\times \phi(1-S_{or}) \tag{4.7b}$$

With maximum amounts of nanoparticles to cover the surfaces of rock grains:

$$\sigma_{cr,max} = \left[1 - \left(\frac{\mu_w r_{FP}^2 U}{2\phi(1-S_{or})r_P\gamma(F_{ei}+\frac{128\pi r_{FP}n_\infty k_B T}{\kappa}e^{-\kappa h}(\varsigma_{GS}-\varsigma_{NP})\varsigma_{FP})}\right)^2\right]$$
$$\times \phi(1-S_{or}) \tag{4.7c}$$

Yuan and Moghanloo (2017b) developed analytical solutions (Eq. (4.8)) to explain the experimental results of Arab and Pourafshary (2013) and evaluated the effectiveness of nanofluid pretreatment to prevent fines migration and the associated damage to the porous medium's permeability. Fig. 4.5 presents the effluent history of fines concentration with and without different types of nanofluids to treat mobile fines. Inferred from Fig. 4.5, Al_2O_3-based nanoparticle is confirmed as the best type of nanoparticle to reduce fines migration from Arab's laboratory experimental results. By applying Eq. (4.8) to match with the experimental results in Fig. 4.5, the fines attachment and straining rates can be enhanced most significantly with Al_2O_3-based nanoparticles, which indicates the optimal type of nanoparticles to control fines migration (Yuan et al., 2018b):

$$\frac{C_{FP,eff}}{C_{FP,inj}} = \exp\left(-(\lambda_a+\lambda_s)L\left(x_D-(t_D-x_D)\frac{L(\lambda_a+\lambda_s)+1}{L(\lambda_a+\lambda_s)(t_{cr}-1)}+\frac{1}{(\lambda_a+\lambda_s)L}\right)\right) \tag{4.8}$$

As shown in Fig. 4.6, before the breakthrough of nanoparticles, about $t_{D1} = 1.0$, there are no changes to effluent-fines concentration, which remains the same as the initial condition ($C_{FP,initial} = 0.02 \text{ m}^3/m^3$).

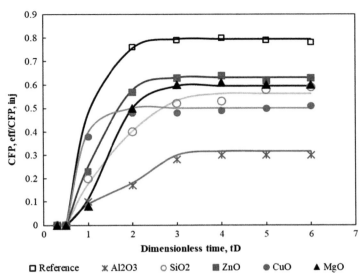

Figure 4.5 Comparison of fines effluent concentration history obtained from analytical models (solid line, Yuan et al., 2017b; 2018b) and laboratory experimental results (discrete points, Arab and Pourafshary 2013) for types of nanoparticles utilizations to control fines migration.

Within the time range from 1.0 to 1.13, due to the nanoparticle effects, there is no fines production at the outlet ($C_{FP,eff} = 0$), i.e., the rock grains with the effects of nanoparticles have retained all newly injected fine particles. After $t_{D1} = 3.3$, the maximum retention capacity of rock grains (with respect to fines) is reached; as a result, the newly injected fines cannot be attached onto the rock grains anymore. From that point, the injection–condition state ($C_{FP,inj} = 0.02 m^3/m^3$) spreads over the whole 1-D permeable medium, and the effluent concentration of fines also increases gradually to also reach 0.02 m^3/m^3 (i.e., the injection condition). The optimal usage of nanoparticles should be the amounts that have been injected before $t_{D3} = 3.3$, which is about 0.001 pore volume in total (i.e., a very small relative quantity). It is worth mentioning that the shadowed envelop ABCD in Fig. 4.6A represents the cumulative reduction quantity of fines production attributed to nanoparticle effects.

Scenario II demonstrates that before the breakthrough of injected fines at $t_{D1} = 1.0$, there are no fines produced at the outlet. Even after the breakthrough of injected fines, due to the positive effects of nanoparticle adsorption, there is an extended production period with very small amounts of fines (close to zero) produced at the outlet, as shown in

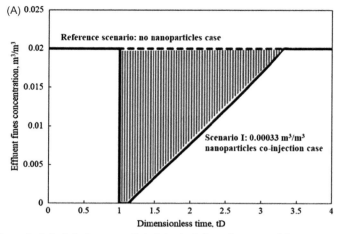

Scenario I: Co-inject nanoparticle into the injected stream of fines suspension

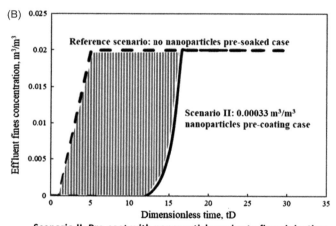

Scenario II: Pre-coat with nanoparticles prior to fines injection

Figure 4.6 Effluent history of fine particles concentration at the outlet for different scenarios of nanoparticles utilization (after Yuan et al., 2016).
(Scenario I: dashed line- without nanoparticle injection; solid line- with nanoparticle injection concentration 0.00033; t_{D1}, adsorption front breakthrough time; t_{D2}, time when injected fine particles start to produce; t_{D3}, the breakthrough time of injection point; Scenario II: dashed line: no nanoparticles used to precoat the porous medium; solid line: with 0.00033 concentration of nanoparticles used to precoat medium).

Fig. 4.6B. At $t_{cr}(x_D = 1)$, the effluent concentration of fines rapidly increases to equal the injected condition ($C_{FP,inj} = 0.02\text{m}^3/m^3$).

To evaluate the efficiency of nanoparticles to prevent fines migration, Mitigation Index (MI) was defined by Yuan and Moghanloo (2016) using Eq. (4.9). MI measures the reduction percentage of cumulative fines production using nanoparticles until the moment when the fines-effluent concentration increases to become the injected condition. By comparing the calculated MI of the two different approaches described, Yuan et al. (2018a,b,c) established that approach II (i.e., the pretreatment of the reservoirs or fracture packs with nanoparticles) performs better than approach I (i.e., to coinject nanofluids into fluid stream as additives) to fixate the injected fines and prevent fines moving further. This performance difference can be explained by the decrease of interaction efficiency between nanoparticles and fines, although both of them are mobile along with nanoparticles-fines reactions.

$$
MI = \begin{cases}
1 - \dfrac{C_{FP,ini}\phi(1 - S_{or})AL \times t_{D1} + \int_{t_{D2}}^{t_{D3}} C_{FP,eff}\phi(1 - S_{or})ALdt}{C_{FP,inil}\phi(1 - S_{or})AL \times t_{D1} + C_{FP,inj}\phi(1 - S_{or})AL \times (t_{D3} - t_{D1})} & \text{Scenario I} \\[4mm]
\dfrac{\int_{1}^{t_{\sigma}(x_D=1)} C_{FP,eff}\phi(1 - S_{or})ALdt}{C_{FP,inj}\phi(1 - S_{or})AL \times (t_{\sigma}(x_D = 1) - 1)} & \text{Scenario II}
\end{cases}
$$

$$(4.9)$$

For the case involving constant injection rates and constant injected nanoparticles concentrations, the pressure differentials along the 1-D permeable medium increase with the accumulation of fines attachment and straining. The severe permeability impairment can be indicated by the increase of injection pressure drop using Eq. (4.1) (Yuan et al., 2017b). Yuan et al. (2017a,b) developed analytical solutions to explain the increase

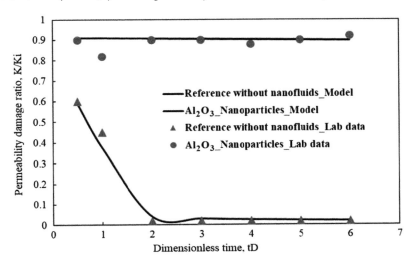

Figure 4.7 Permeability changes obtained from analytical models (solid line) and laboratory experimental results (discrete points) for both cases (case with nanoparticles effects; and reference case without nanoparticles (after Yuan et al., 2017b)).

of pressure drop caused by permeability damage as shown in Fig. 4.7. Because of the precoated nanoparticles onto rock grains prior to the invasion of fines, the impairment of Berea sandstone core permeability can be completely remedied from the damage case without the further injection of nanoparticles.

$$\Delta p(t_D) = \int_0^1 \frac{UL\mu(1 + \beta_a \sigma_{FP} + \beta_s S_{FP})}{k_0} dx_D \qquad (4.10)$$

4.5 USING NANOPARTICLES TO CONTROL FINES SUSPENSION IN OIL AND WATER-SATURATED POROUS SYSTEMS

Fines migration in two–phase-fluid (oil and water) flow occurs in many types of processes in the petroleum industry. As the injection of low–quality water with solids and liquid particles (produced water or waste water) continues for the purpose of waterflooding, both the newly invaded fine particles and the induced formation fines by injected fluids result in significant formation damage and impair well performance, i.e., well injectivity and oil recovery. Plan and design of those projects related to fines migration in two-phase flow and evaluation of various mechanisms by which nanoparticles control fines migration need reliable physical-based mathematical models.

4.5.1 Nanofluid coinjection to reduce fines migration in two mobile fluids

It is desirable to introduce nanoparticles as an additive continuously into the injection fluid stream to control injected fines suspension during various types of waterflooding. The coinjected nanoparticles are preferentially adsorbed onto the surfaces of mobile fines, which help fines more prone to be attached to become attached to the pore surfaces. As discussed by Yuan, 2017a; Yuan and Moghanloo, 2018c, the mass–balance equation of flowing nanoparticles considering their adsorption on mobile fines, and the mass-balance equation of flowing fines considering their attachment onto pore surfaces, and water component flowing through 1D permeable media can be written as:

Water component:

$$\frac{\partial f_w}{\partial x_D} + \frac{\partial S_w}{\partial t_D} = 0 \qquad (4.11a)$$

Fine particles component:

$$\frac{\partial(C_{FP}f_w)}{\partial x_D} + \frac{\partial(S_w C_{FP})}{\partial t_D} + \frac{1}{\phi}\frac{\partial \sigma_a}{\partial t_D} = 0, \quad \begin{cases} \sigma_a = \sigma_{cr}; \ \sigma_{cr} < \sigma_{cr,\max} \\ \sigma_a = 0; \ \sigma_{cr} = \sigma_{cr,\max} \end{cases} \qquad (4.11b)$$

Nanoparticles component:

$$\frac{\partial(C_{NP}f_w)}{\partial x_D} + \frac{\partial(S_w C_{NP})}{\partial t_D} + \frac{1}{\phi}\frac{\partial \hat{C}_{NP}}{\partial t_D} = 0, \quad \begin{cases} \hat{C}_{NP} = \dfrac{\hat{C}_{NP,\max}K_{NP}C_{NP}}{1 + K_{NP}C_{NP}}; \ \hat{C}_{NP} < \hat{C}_{NP,\max} \\ \hat{C}_{NP} = 0; \ \hat{C}_{NP} < \hat{C}_{NP,\max} \end{cases}$$

$$(4.11c)$$

The relative permeability of the wetting water-phase behaves according to a function of the retained fine particles concentration onto pore surfaces expressed as Eq. (4.12). The viscosity of flowing water is a function of fines concentration, which is modeled using the Flory-Huggins equation (Pope and Nelson, 1978):

$$\frac{k_{rw}(S_w, \sigma_a)}{\mu_w(C_{FP})} = \frac{k_{rw}\left(S_w, \sigma_{a,initial}\right)}{(1 + \beta\sigma_{FP})\mu_w(1 + aC_{FP})} \qquad (4.12)$$

where, β, a are constant coefficents.

Yuan (2017a) and Yuan and Moghanloo (2018c) introduced the splitting method (Borazjani et al., 2016, Borazjani and Bedrikovetsky, 2016) to reduce the above governing system (Eq. (4.9)) from the 3×3 system to two sub-systems, including, a nanoparticles and fines retention-kinetics auxiliary subsystem (Eq. (4.13)) and a lifting system for unknown water saturation (Eq. (4.14)). The procedure to solve the problem includes: (1) transformation of Eq. (4.11) using a stream-function and splitting technique; (2) analytical solutions of both the auxiliary system and lifting system using method of characteristics (MOC) (Appendix A); and (3) inversion of solutions by transforming the coordinates:

$$\frac{\partial C_{FP}}{\partial x_D} + \frac{1}{\phi}\frac{\partial \sigma_a}{\partial \varphi} = 0 \qquad (4.13a)$$

$$\frac{\partial C_{NP}}{\partial x_D} + \frac{1}{\phi}\frac{\partial \hat{C}_{NP}}{\partial \varphi} = 0 \qquad (4.13b)$$

$$\frac{\partial G}{\partial x_D} + \frac{\partial F}{\partial \varphi} = 0; \; G = \frac{1}{f_w}; \; F = -\frac{S_w}{f_w} \qquad (4.14)$$

In the analytical solutions using MOC, the dynamic relations between suspended nanoparticles and suspended fines concentration can be characterized along two different characteristic directions, i.e., a slow path and a fast path, as shown in Fig. 4.8. With the definition of initial and injection (boundary) conditions, both slow path and fast path, where the nanoparticles and fines concentrations vary, can be uniquely determined according to the Rankine–Hugoniot condition and concept of discontinuity (Lake, 1989; Hankins, 2004). After the suspended and retained fine particles concentrations are obtained from the above analysis, the characteristic form of the lifting equation (Eq. (4.14)) can be expressed as follows:

$$\left(\frac{d\varphi}{dx_D}\right)_2 = \sigma_2 = \frac{\partial F}{\partial G} = \frac{f_w(S_w, \sigma_a, C_{FP}) - S_w \frac{\partial f_w(S_w, \sigma_a, C_{FP})}{\partial S_w}}{\frac{\partial f_w(S_w, \sigma_a, C_{FP})}{\partial S_w}} \qquad (4.15a)$$

$$\frac{dS_w}{dx_D} = \frac{\frac{\partial f_w}{\partial \sigma_a}\left(S_w \frac{\partial \sigma_a}{\partial \varphi} - \frac{\partial \sigma_a}{\partial x_D}\right) + \frac{\partial f_w}{\partial C_{FP}}\left(S_w \frac{\partial C_{FP}}{\partial \varphi} - \frac{\partial C_{FP}}{\partial x_D}\right)}{\frac{\partial f_w}{\partial S_w}} \qquad (4.15b)$$

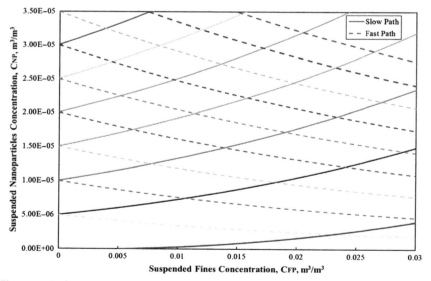

Figure 4.8 Composition diagram representing the relation between suspended nanoparticles and fines concentrations along the slow path and fast path in new coordinate of stream-function and distance (Yuan 2017a).

In Fig. 4.9A, the coordinates of the stream-function and distance, the corresponding characteristic lines of nanoparticles, and fines concentrations that vary from the injection condition to the initial condition, are obtained. This is followed by an inverse transformation of the stream-function back to time as the independent variable using Eq. (4.16). Along the characteristic lines in Fig. 4.8A, the constant values of component concentrations (nanoparticles, fines, and water) lead to the inversed transformation as straight-line images of t_D in Fig. 4.8B.

$$t_D = \int_{(0,0)}^{(x_D, \varphi)} \frac{1}{f_w(x_D, \varphi)} \, d\varphi + \frac{S_w(x_D, \varphi)}{f_w(x_D, \varphi)} \, dx_D \qquad (4.16)$$

Displayed in different coordinates, Fig. 4.9 indicates that the characteristic velocities (slope of characteristic lines) are different. Using the distance-time diagram of Fig. 4.8B, the intersection points of horizontal lines ($t_D = const.$) with the characteristic lines determine the profile of component concentrations through a 1-D porous medium. Accompanying the discontinuities of particles-concentration waves, c-shock1, and c-shock2, are the discontinuities of the water saturation wave, where water saturation decreases from 0.76 upstream to 0.696 downstream of c-shock1, and increases from 0.66 upstream to 0.693 downstream of c-shock2. In the initial-condition region with constant nanoparticles and fines concentration, analogos to the classical Buckley-Leverett problem, the Sw-shock appears because of the discontinuity of fraction flow curve, where water saturation decreases directly from 0.66 to the connate-water saturation 0.20. Nanoparticles coinjection can help enhance the attachment of fine particles onto pore surfaces (preferentially in the higher-Sw regions), which results in more damage to the water-phase relative permeability. Hence, as a positive consequence, nanoparticles utilization leads to the formulation of an "oil-bank" in the region between c-shock1 and c-shock2, as shown in Fig. 4.10.

4.5.2 Nanofluid preflush to control fines migration in a radial flow system saturated with two immiscible fluids

It is also recommended to precoat porous medium using nanoparticles prior to waterflooding to prevent the induced fines migration by injected fluids. In this section, fines migration affecting two-phase (oil and water) radial flow is discussed incorporating multiple fines capture phenomena, i.e., fines attachment and straining. The performance of nanofluid

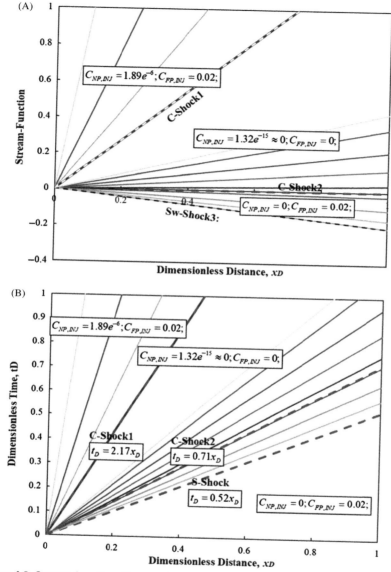

Figure 4.9 Stream function-distance (upper diagram) and distance-time relationships (lower diagram) (Yuan 2017a).
(Injection condition: $C_{NP} = 1.80E$-6, $C_{FP} = 0.02$; Constant-state condition: $C_{NP} = 1.32E$-15, $C_{FP} = 0$; Initial condition: $C_{NP} = 0$, $C_{FP} = 0.02$; Water-saturation wave: $S_{wj} = 0.80$, $Sw_{f1}^{+} = 0.76$, $Sw_{f1}^{-} = 0.696$, $Sw_{f2}^{+} = 0.66$, $Sw_{f2}^{+} = 0.693$, $Sw_{f3} = 0.66$, $S_{wi} = S_{wc} = 0.80$).

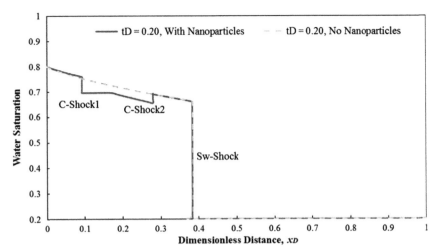

Figure 4.10 Profile of water saturation along 1-D porous medium (Dashed line: without nanoparticles; Solid line: with nanoparticles coinjection. The difference between two lines represents the "oil-bank" attributed to nanoparticles effects (Yuan, 2017a)).

pretreatment affecting both fines migration and waterflooding efficiency is also characterized. By implementing the splitting method and stream-function transformation already described, 3×3 governing equations can be reduced to a 2×2 auxiliary subsystem containing only particles components (nanoparticles and fines) and a lifting equation involving only phase saturation, as shown in Eq. (4.17).

Water component:

$$\frac{\partial G}{\partial x_D} + \frac{\partial F}{\partial \varphi} = 0; G = \frac{1}{f_w}; F = -\frac{S_w}{f_w} \tag{4.17a}$$

Fine particles component:

$$\frac{\partial C_{FP}}{\partial x_D} + \frac{\partial S_a}{\partial \varphi} + \frac{\partial S_s}{\partial \varphi} = 0 \tag{4.17b}$$

Kinetics equations for fine particles attachment and straining:

$$\frac{\partial S_a}{\partial \varphi} = \begin{cases} \dfrac{\Lambda_a C_{FP}}{2\sqrt{x_D}}; \sigma_a < \sigma_{cr}(x_D, C_{NP}) \\ \sigma_{cr}(x_D, C_{NP}) \end{cases} \tag{4.17c}$$

$$\frac{\partial S_s}{\partial \varphi} = \frac{\Lambda_s C_{FP}}{2\sqrt{x_D}} \tag{4.17d}$$

The formation damage caused by both attached and strained fines is incorporated into the retardation term in the relative permeability of the water phase. Hence, the fractional flow function can be updated as:

$$f_w(S_w, S_s, S_a) = \left(1 + \frac{k_{ro}\mu_w(1 + \phi\beta_s S_s + \phi\beta_a S_a)}{k_{rw}\mu_o}\right)^{-1} \tag{4.17e}$$

The existence of nanoparticles helps control fines migration by increasing the maximum (critical) retention concentration of fine particles onto rock grains, as shown in Eq. (4.17f):

$$\sigma_a(x_D, C_{NP}) = \left[1 - \left(\frac{\mu r_{FP}^2 \frac{q}{2\pi r}}{2\phi r_P \gamma \left(F_{ei} + \frac{128\pi r_{FP} n_\infty k_B T}{\kappa} e^{-\kappa h}\left[\frac{K_{NP}C_{NP}}{1 + K_{NP}C_{NP}}(\varsigma_{GS} - \varsigma_{NP})\varsigma_{FP}\right]\right)}\right)^2\right]\phi \tag{4.17f}$$

Here, introduce the following dimensionless variables for simplification, as shown in Eq. (4.17g):

$$x_D = \left(\frac{r}{r_e}\right)^2; \; t_D = \frac{qt}{\phi\pi r_e^2}; \; \Lambda_a = r_e\lambda_a; \; \Lambda_s = r_e\lambda_s; \; S_a = \frac{\sigma_a}{\phi}; \; S_s = \frac{\sigma_s}{\phi}; \; P = \frac{4k_0\pi}{q}p \tag{4.17g}$$

The initial conditions of both attached and suspended fines are summarized in Fig. 4.11. According to the assumptions of the maximum retention-concentration model (Bedrikovetsky et al., 2011), the release of the initial attached fines occurs instantly in conditions above the maximum limits. As shown in Fig. 4.11A, around the well vicinity, all the initially attached fines are released instantly because of the high flowing velocity. Within the range of medium flow velocities, the concentration of released fines equates to the difference between the initial attached fines concentration and the maximum fines retention concentration, depending on flow velocity at that location. In remote areas from the wellbore, there are no fines to be released because of sufficiently slow fluid velocities. In Fig. 4.11B, the adsorption of nanoparticles can enhance the attachment of fines effectively, i.e., no fines are released within the range of nanoparticles treatment. Even for the injection of new fines, within the nanofluid treatment range, the rock grains can have sufficient remaining capacity to capture those introduced fine particles, until the maximum fines retention concentration with nanoparticles effects is reached, which typically occurs after a sufficiently long-time interval.

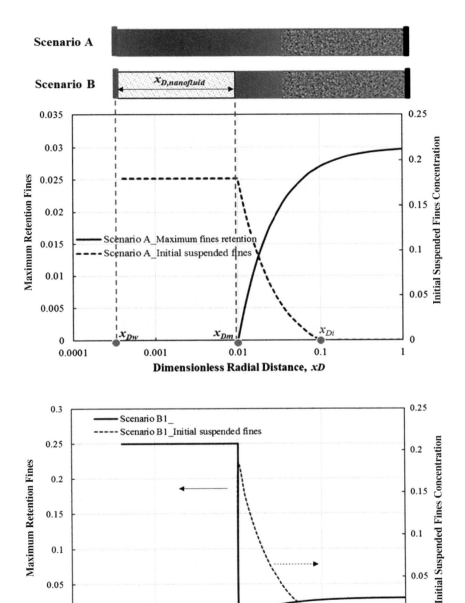

Figure 4.11 Maximum fines retention concentration profile in a radial flow system without and with nanoparticles pretreatment radius of 0.01 (revised after Yuan and Moghanloo, 2018c).

In view of the variation of released fines concentration depending on formation fluid velocities at different reservoir locations, it is necessary to optimize the radius of nanofluid pretreatment to maximize the efficiency of the nanoparticles treatment. Yuan (2017a) and Yuan and Moghanloo (2018c) presented the comparison between the case with 0.01 nanofluid pretreatment radius (Fig, 4.11B) and the reference case without nanofluid utilization (Fig. 4.11A).

In Fig. 4.12, the time-distance diagrams of three different cases are presented, including waterflooding without fines migration effects, water-flooding with fines migration effects, and waterflooding with nanofluids to control fines migration in the near-wellbore region. The time-distance diagram for the waterflooding case without fines migration is presented in Fig. 4.12A. The profiles of water saturation in an asymmetric system at different moments in time can be achieved by finding the intersection points of different horizontal lines (t_D = $const.$) with the characteristic lines.

In Fig. 4.12B, the end points of series of characteristic lines (propagation path of water saturation wave) reflect the front-saturation along with the waterflooding front shock. The values of front-saturation keep changing at different locations, which is attributed to the dependency of fines migration on the changing flowing velocities at different reservoir locations in a radial flow system. The line connecting those points represents the trajectory of the classical Buckley-Leverett waterflooding front. The dashed line (gray) in Fig. 4.12B indicates the trajectory of the erosion front (upstream, no fines attachment occurs; and downstream, fines attachment occurs). The trajectory of the erosion front becomes a vertical line, which indicates the erosion front is stationary at location of 0.1. In other words, the propagation of the erosion front does not affect the movement of water saturation waves.

Fig. 4.12C demonstrates the trajectories of the water–saturation waves, and the movements of both erosion front and saturation-shock, for the case with nanofluid treatment to control fines migration. In contrast to Fig. 4.12B, the trajectory of the erosion front passes through the set of saturation waves. In other words, the erosion front is not always ahead of the water saturation waves, which can affect the propagation of water-saturation waves.

Fig. 4.13 summarizes the evolution of the water saturation profiles in the radial flow system at different times for the above three different cases. First, the effects of fines migration (fines attachment, fines straining, and fines suspension) can slow down the movement of injected water. The

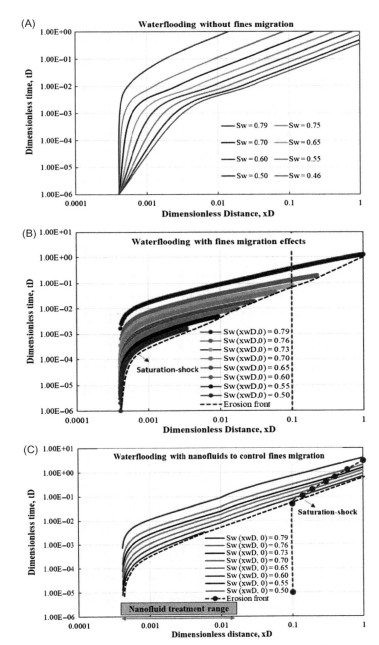

Figure 4.12 Time-distance diagrams indicating the propagation of water-saturation waves (revised after Yuan and Moghanloo, 2018c).

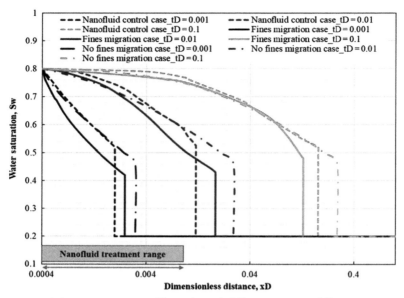

Figure 4.13 Water-saturation profile in the radial flow system at different time.

waterflooding front-saturation also varies at different locations within the test core (reservoir) because of velocity–dependent fines migration effects. With the attenuation of fines migration effects where slower velocity occurs far away from the inlets in the radial flow system, the difference of water-flooding front-saturation caused by fines migration decreases. However, the differences of front movement velocities continue increasing as injected water moves toward to the outlet. Second, within the region affected by the nanofluid treatment, the usage of nanofluids can slow down the move-ment of injected fluid further through enhancing the attachment of mobile fines. However, in the remote regions of the core (reservoir) where slower fluid velocities prevail, nanofluid treatment accelerates the advance of the waterflooding front to surpass that of case without nanofluid control. This phenomenon can be attributed to the attenuation of fines attachment and straining effects caused by the decreases of total mobile fines quantities in the formation fluids after the injected fluids have already passed through the nanofluid-treated region. Consequently, the attenuation of fines attach-ment and straining effects then impairs the improved mobility–control by the damage of water-phase relative permeability.

Fig. 4.14 presents the comparison of inlet–outlet pressure drop among the above three cases. The effects of fines migration and associated

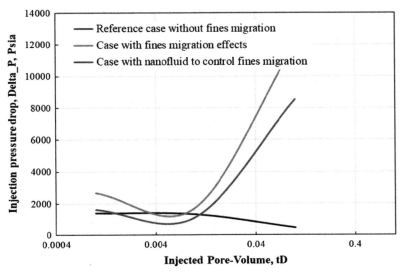

Figure 4.14 Changes of inlet-outlet pressure drop in radial flow system at different time (Yuan and Moghanloo, 2018c).

formation damage can be characterized by the increase of injection pressure loss as waterflooding continues. The existence of nanofluid to control fines migration can help mitigate the formation damage caused by fines migration. The mitigation of injection pressure loss would bring benefits to the management of the water injection surface facilities, leading to more cost-effective and profitable projects.

4.6 COMBINED NANOFLUIDS WITH LOW-SALINITY WATERFLOODING

During low-salinity waterflooding, the problems of fines migration induced by chemical environments of low-salinity fluids have aroused great debate. On the one hand, fines migration may carry small amounts of residual oil through the detachment of oil-coated particles from rock grains, i.e., improve the displacement efficiency (Aksulu et al., 2012; Bernard, 1967). In addition, the reduction of local water-phase effective permeability in water-swept areas caused by the blockage of mobile fines can also provide mobility-control to enhance the sweep efficiency (Lemon, 2011; Zeinijahromi et al., 2011, 2012, 2013).On the other, fines

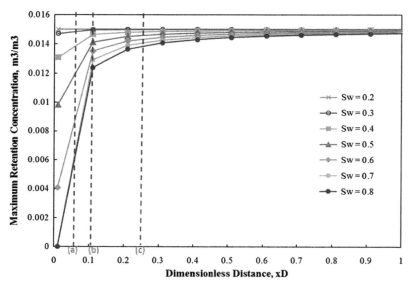

Figure 4.15 Variation of maximum retention concentration of fine particles in a radial flow system (Yuan and Moghanloo, 2018a).

migration and their size exclusion effects can also result in severe damage to reservoir permeability, which leads to the decline of well injectivity (or productivity in case of production wells). Therefore, a better understanding of avoiding versus encouraging fines migration in reservoirs is required. Fig. 4.15 presents the variation of maximum retention concentration of fine particles with the increase of distance away from an injection well. Within the near-wellbore region (less than about $x_D = 0.2$), the large flowing velocity exaggerates the problem of fines detachment and straining, and the blockage of the detached fine particles into pore-throats leads to significant loss of injection pressure. However, in the regions remote from the wellbore with slow flowing rates, even with changes to water saturation, very small reduction of fines retention concentration leads to negligible problems of fines detachment and associated formation damage.

Therefore, evoking the debate between the pros and cons of fines migration, it is desirable to control fines migration to take advantages of its positive effects far from the wellbore but minimize its weaknesses of inducing formation damage near the wellbore. This section develops the mathematical foundation for designing a nanofluid–slug pre-flush (treatment radiuses = 0.05, 0.10, and 0.25) to enhance well injectivity while

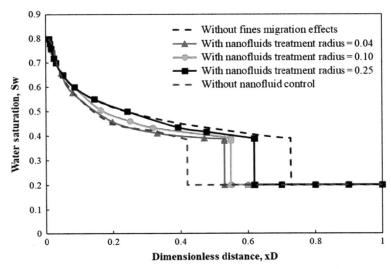

Figure 4.16 Comparison of water saturation profiles for cases with different nanofluid treatment radius prior to low-salinity waterflooding (Yuan et al., 2018b).

keeping the mobility-control assisted by fines migration to improve the performance of low-salinity waterflooding (both in terms of EOR and well injectivity) (Yuan and Moghanloo, 2018a; Yuan et al., 2018b).

Fig. 4.16 presents the profiles of water saturation in a radial flow system for cases with three different nanofluid treatment radiuses, at the same injected pore-volume (0.2). Nanofluid treatment can accelerate the movement of injected water by reducing the fines migration/straining effects. Without the effects of fines straining to improve mobility control, the propagation of waterflooding approximates the case of conventional waterflooding without fines migration, leading to the early breakthrough of injected water as a negative result. However, the increase of pressure drop can be controlled by nanofluids utilization, which mitigates the damage of fines migration as the low-salinity waterflooding continues. In Fig. 4.17, with the extension of the nanofluids treatment radius, the mitigation performance of injection pressure loss could be enhanced, but the trend of pressure loss mitigation slows down. Thus, the problem of fines migration weakens with the increase of distances away from the injection well, it is therefore not necessary to apply excessively large radiuses of nanofluids treatments. Therefore, by weighting the balance between the maintenance of well injectivity (minimizing pressure drop) and enhanced oil recovery, the optimal radius of nanofluid treatment can be determined as 0.10, approximately (Yuan et al., 2018b).

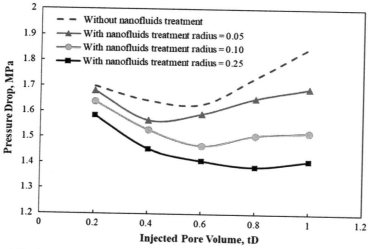

Figure 4.17 Comparison of inlet-outlet pressure drop for low-salinity waterflooding with different nanofluid treatment radius (Yuan et al., 2018b).

4.7 CONCLUSIONS

This chapter describes both theoretical and experimental works conducted to address the issues of nanofluid utilization to control fines migration in porous media. The key conclusions based on the results of this work are as follows:

- The important parameters pertinent to transport and capture of nanoparticles are characterized, i.e., the maximum adsorption concentrations, reversible or detachment adsorption concentrations, nanoparticles adsorption and straining rates, and formation damage coefficients.
- The mutual reactions among fines, nanoparticles, and rock grains are established to explain the reasons why nanoparticles can help mitigate fines migration and formation damage. MI is compared to reveal and explain the better controlling performance of nanoparticles pretreatment compared to coinjection of fines and nanoparticles.
- The combined applications of the splitting method and stream function with Method of Characteristic provide exact solutions of nanoparticles utilization to control fines migration in two-phase water and oil flow. The effects of fines migration and nanoparticles effects on the performance of waterflooding are quantified using changes of pressure drop, oil bank, front-saturation, and the breakthrough time of injected fluids.

- The maximum retention concentration of fine particles onto rock grains is dependent upon both water quality and flow velocity during low–salinity waterflooding. Nanofluid treatment prior to low salinity water injection can help control fines migration adjacent to injection wells, and significantly maintain long-term well injectivity, but with the negative consequence of accelerated breakthrough of injected water.

NOMENCLATURE

C_{NP}	Volumetric concentration of nanoparticles with respect to pore volume, m^3/m^3
C_{FP}	Volumetric concentration of nanoparticles with respect to pore volume, m^3/m^3
$C_{NP, inject}$	Injected nanoparticles concentration with respect to pore volume, m^3/m^3
$C_{FP, inj}$	Injected fine particles concentration with respect to pore volume, m^3/m^3
$C_{FP, ini}$	Initial fine particles concentration with respect to pore volume, m^3/m^3
$C_{FP, eff}$	Effluent fine particles concentration with respect to pore volume, m^3/m^3
$\hat{C}_{NP}$	Concentration of adsorbed nanoparticles with respect to bulk volume, m^3/m^3
S_{NP}	Concentration of straining nanoparticles with respect to pore volume, m^3/m^3
σ_{NP}	Concentration of retained nanoparticles with respect to bulk volume, m^3/m^3
S_a, S_{FP}	Concentration of straining nanoparticles with respect to pore volume, m^3/m^3
σ_{NP}, σ_a	Concentration of retained nanoparticles with respect to bulk volume, m^3/m^3
λ_s	Particles straining filtration coefficient
λ_{ad}	Nanoparticles adsorption filtration coefficient
λ_a	Fine particles attachment filtration coefficient
β_a	Formation damage coefficient related to particles adsorption
β_s	Formation damage coefficient related to particles straining
L	Length of core plug
A	Cross-section area of core plug
$\sigma_{NP,max1}$	Maximum total retention concentration of nanoparticles, m^3/m^3
$\sigma_{NP,max2}$	Maximum irreversible retention concentration of nanoparticles, m^3/m^3
q_{inj}	Fluid injection rate, mL/min
F_e	Electrostatic forces, N
F_{ei}	Electrostatic forces at the initial condition, N
K_{NP}	Langmuir adsorption constant of nanoparticle
T	Absolute temperature of reservoir, K

h	The surface-to-surface separation length, m
y	The ratio between the drag and electrostatic force
r_{FP}	Radium of fine particles, m
r_p	Pore radius, m
l	Characteristic wave length of interaction, $l = 100$ nm
n	Pore concentration, number/m^3
n_∞	Bulk number density of ions, 6.022×10^{25} number/m^3
$\varsigma_{FP}, \varsigma_{NP}, \varsigma_{GS}$	Zeta potentials for fine particles, nanoparticles and grain surfaces, mV
ϕ	Porosity of sand pack
ρ	Fluid density, kg/m^3
ρ_w	Water fluid density, kg/m^3
ρ_{NP}	Nanoparticles fluid density, kg/m^3
μ_w	Water viscosity, Pa.s
μ_o	Oil viscosity, Pa.s
r_e	Outer radius of radial flow system, 0.5 m
k_0	Permeability of core plug, mD
k_B	The Boltzmann's constant, 1.381×10^{-21} J/K
k_{rw}	Relative permeability of water phase
k_{ro}	Relative permeability of oil phase
Δp	Pressure drop, MPa
p	Flowing pressure at different location of flowing system, MPa
U	Fluid flowing velocity, m/s
q	Injection rate per formation height, m^2/s
n_∞	Bulk number density of ions, 6.022×10^{25} number/m^3
f_w	Fractional flow function
S_w	Water saturation, decimal
S_{or}	Residual oil saturation, decimal
t_{cr}	The breakthrough time of injected fines
φ	Stream function defined in this chapter
t_D	Dimensionless time or injected pore volume
x_D	Dimensionless distance

REFERENCES

Abousleiman, Y.N., Tran, M.H., and Hoang, B.O., et al., 2009. Geomechanics field and laboratory characterization of woodford shale: the next gas play. Paper SPE-110120-MS presented at *SPE Annual Technical Conference and Exhibition*, 11−14 November Anaheim, California, USA.

Achinta, B., Belhaj, H., 2016. Application of nanotechnology by means of nanoparticles and nano-dispersions in oil recovery- a comprehensive review. J. Nat. Gas. Sci. Eng. 34, 1284−1309.

Adkins, S., Gohil, D., Dickson, J., et al., 2007. Water-in-carbon dioxide emulsions stabilized with hydrophobic silica particles. Phys. Chem. Chem. Phys. 9, 6333−6343.

Ahmadi, M., Pourafshari, P., and Ayatollahi, S., 2011. Zeta potential investigation and mathematical modeling of nanoparticles deposited on the rock surface to reduce fine migration, Paper SPE-142633-MS presented at *SPE Middle East Oil and Gas Show and Conference*, 25−28 September, Manama, Bahrain.

Aksulu, H., Hamso, D., Strand, S., et al., 2012. The evaluation of low salinity enhanced oil recovery effects in sandstone: effects of temperature and pH gradient. Energy Fuels 26, 3497−3503.

Aminzadeh, B., DiCarlo, D.A., and Chung, D.H., et al., 2012. Effect of nanoparticles on flow alteration during CO_2 injection. Paper SPE-160052-MS presented at the *SPE Annual Technical Conference and Exhibition*, 8—10 October San Antonio, Texas, USA.

Arab, D., Pourafshary, P., 2013. Nanoparticles-assisted surface charge modification of the porous medium to treat colloidal particles migration induced by low salinity water flooding. Colloids Surf A: Physicochemical Eng. Asp. 436, 803—814.

Assef, Y., Arab, D., Pourafshary, P., 2014. Application of nanofluid to control fines migration to improve the performance of low salinity water flooding and alkaline flooding. J. Pet. Sci. Eng. 124, 331—340.

Bedrikovetsky, P., et al., 2011. Modified particle detachment model for colloidal transport in porous media. Transp Porous Media 86 (2), 353—383.

Berlin, J.M., Yu, J., Lu, W., et al., 2011. Engineered nanoparticles for hydrocarbon detection in oil-field rocks. Energy Environ. Sci. 4, 505—509.

Bernard, G.G. 1967. Effect of floodwater salinity on recovery of oil from cores containing clays, Paper SPE-1725-MS presented at *SPE California Regional Meeting*. Los Angeles, 26—27 October, Los Angeles, California.

Borazjani, S., Bedrikovetsky, P., 2016. Exact solutions for two-phase colloidal-suspension transport in porous media. Appl. Math. Model. 44, 296—320.

Borazjani, S., Behr, A., Genolet, L., et al., 2016. Effects of fines migration on low salinity waterflooding: analytical modelling. Transp Porous Media 116 (01), 213—249.

Contreras, O., Hareland, G., and Husein, M., et al., 2014. Application of in-house prepared nanoparticles as filtration control additive to reduce formation damage. Paper SPE-168116-MS presented at *SPE International Symposium and Exhibition on Formation Damage* Control, 26—28 February, Lafayette, Louisiana, USA.

Crews, J.B., and Gomaa, A.M., 2012. Nanoparticle-assisted surfactant micellar fluids: an alternative to crosslinked polymer systems. Paper SPE-157055-MS presented at *SPE International Oilfield Nanotechnology Conference*, 12—14 June, Noordwyk, The Netherlands.

Gonzenbach, U., Studart, A., Tervoort, E., et al., 2007. Tailoring the microstructure of particle-stabilized wet foams. Langmuir 23, 1025—1032.

Gruedes, R.G., Al-Abduwani, F., and Bedrikovetsky. 2006. Injectivity decline under multiple particles capture mechanisms. Paper SPE-98623-MS presented at *SPE International Symposium and Exhibition on Formation Damage Control*, 15—17 February, Lafayette, Louisiana, USA.

Gurluk, M.R., Nasr-El-Din, H.A., and Crews, J., 2013. Enhancing the performance of viscoelastic surfactant fluids using nanoparticles. Paper SPE-164900-MS presented at *EAGE Annual Conference and Exhibition*, 10—13 June, London, UK.

Hankins, N.P., 2004. Application of coherence theory to a reservoir enhanced oil recovery simulator. J. Pet. Sci. Eng. 42 (1), 29—55.

Hibbeler, J., Garcia, T., and Chavez, N. 2003. An integrated long-term solution for migratory fines damage. SPE 81017 presented at the *SPE Latin American and Caribbean Petroleum Engineering Conference* held in Port-of-Spain, Trinidad, West Indies, 27—30 April.

Huang, T., McElfresh, P.M., and Gabrysch, A.D. 2002. High temperature acidization to prevent fines migration. SPE 73745 presented at the *SPE International Symposium and Exhibition on Formation Damage Control* held in Lafayette, Louisiana, 20—21 February.

Huang, T., Crews, J., and Willingham, J.R., 2008. Using nanoparticles technology to control fine migration, Paper SPE-115384-MS presented at *SPE Annual Technical Conference and Exhibition*, 21—24 September, Denver, Colorado, USA.

Huang, T., Evans, B.A., and Crews, J.B., et al., 2010. Field case study on formation fines control with nanoparticles in offshore applications. Paper SPE-13508-MS presented at *SPE Annual Technical Conference and Exhibition*, 19—22 September, Florence, Italy.

Huang T., Han, J., and Agrawal, G., et al., 2015. Coupling nanoparticles with water flooding to increase water sweep efficiency for high fines-containing reservoir — lab and reservoir simulation results. Paper SPE-174802-MS presented at *SPE Annual Technical Conference and Exhibition*, 28—30, September, Houston, Texas, USA.

Jahagirdar, S.R., 2008. Oil-microbe detection tool using nano optical fibers. Paper SPE-113357-MS presented at the *SPE Western Regional and Pacific Section AAPG Joint Meeting*, 29 March—4 April Bakersfield, California, USA.

Jaramillo, O.J.; Romero, R.; Ortega, A.; Milne, A. and Lastre, M. 2010. Matrix acid systems for formations with high clay content. Paper SPE 126719 presented at the *2010 SPE International Symposium and Exhibition on Formation Damage Control* held in Lafayette, Louisiana, USA, 10—12 February.

Ju, B., Fan, T., 2009. Experimental study and mathematical model of nanoparticles transport in porous media. Powder Technol 192 (2), 195—202.

Ju, B., Fan, T., Ma, M., 2006. Enhanced oil recovery by flooding with hydrophilic nanoparticles. China Particuol. 4 (1), 41—46.

Kapusta, S., Balzano, L., and Riele, P., 2011. Nanotechnology application in oil and gas exploration and production. Paper SPE-15152-MS presented at *International Petroleum Technology Conference*, 15—17 November, Bangkok, Thailand.

Lake, L.W., 1989. Enhanced Oil Recovery. Prentice Hall Inc, United States: Old Tappan, NJ.

Lemon, P., 2011. Effects of injected-water salinity on waterflood sweep efficiency through induced fines migration. J. Can. Pet. Technol. 50, 9—10.

Li, S., Hendraningrat, L. and Torsæter, O. 2013. Improved oil recovery by hydrophilic silica nanoparticles suspension 2-phase flow experimental studies. Paper IPTC-16707 presented at *International Petroleum Technology Conference*, 26—28 March, Beijing, China.

Li, S., Jiang, M., and Torsæter, O. 2014. An experimental investigation of EOR mechanisms for nanoparticles fluid in glass micromodel. Paper SCA2014-022 presented at the *International Symposium of the Society of Core Analysts*, 8—11 September 2014, Avignon, France.

Li, S., Jiang, M., and Torsæter, O. 2015. An experimental investigation of nanoparticles adsorption behavior during transport in berea sandstone. Paper SCA2015-029 presented at International Symposium of the Society of Core Analysts held in St. John's Newfoundland and Labrador, Canada, 16—21 August 2015.

Massoudieh, A., Ginn, T.R., 2010. Colloid-facilitated Contaminant Transport in Unsaturated Porous Media, (Chapter 8)., Modelling of Pollutants in Complex Environmental System, Vol. II. ILM Publications, Glensdale.

McDonald, M.J., 2012. A novel potassium silicate for use in drilling fluids targeting unconventional hydrocarbons. Paper SPE-162180-MS presented at *SPE Canadian Unconventional Resources Conference*, 30 October—1 November, Calgary, Alberta, Canada.

Moghadam, T.F., Azizian, S., 2014. Effect of ZnO nanoparticles on the interfacial behavior of anionic surfactant at liquid/liquid interfaces. Colloids Surf. A 457, 333—339.

Moghanloo, R.G., 2012. Modeling the Fluid Flow of Carbon Dioxide through Permeable Media. The University of Texas at Austin, Austin, Texas.

Noh, M., and Lake, L.W. 2004. Implications of coupling fractional flow and geochemistry for CO_2 injection in aquifers. Paper SPE-89341-MS presented *at SPE/DOE Symposium on Improved Oil Recovery.* 17—21 April, Tulsa, Oklahoma.

Ogolo, N.A., Olafuyi, O.A., and Onyekonwu, M.O., 2012. Enhanced oil recovery using nanoparticles. Paper SPE-160847-MS presented at *SPE Saudi Arabia Section Technical Symposium and Exhibition*, 8—11 April, Al-Khobar, Saudi Arabia.

Pang, X., Boul, P.J., Jimenez, W., 2014. Nanosilicas as accelerators in oil well cementing at low temperatures. SPE Drilling Complet 29 (01), 98−105.

Pope, G.A., Nelson, R.C., 1978. A chemical flooding compositional simulator. SPE J 18 (05), 339−354.

Prigiobbe, V., Worthen, J.A., Johnston, P.K., et al., 2016. Transport of nanoparticles-stabilized CO_2-foam in porous media. Transp Porous Media 111, 265−285.

Santra, A.K., Boul, P., and Pang, X., 2012. Influence of nanomaterials in oil well cement hydration and mechanical properties. Paper SPE-156937-MS presented at *SPE International Oilfield Nanotechnology Conference and Exhibition*, 12−14 June, Noordwyk, The Netherlands.

Sharma, M.M., 1987. Transport of particulate suspensions in porous media: model formulation. AIChE J 33, 1636−1643.

Sharma, M.M., Chenevert, M.E., and Guo, Q., et al., 2012. A new family of nanoparticles based drilling fluids. Paper SPE-160045-MS presented *at SPE Annual Technical Conference and Exhibition*, 8−10 October San Antonio, Texas, USA.

Song, Y.Q., and Marcus, C., 2007. Hyperpolarized silicon nanoparticles: reinventing oil exploration? *Schlumberger Presentation*.

Vafai, K., 2005. Handbook of Porous Media, Second Edition CRC Press, Taylor & Francis Group, New York.

Van Zanten, R., Lawrence, B., and Henzler, S.J. 2010. Using surfactant nanotechnology to engineer displacement packages for cementing operations. Paper SPE-127885-MS presented at *IADC/SPE Drilling Conference and Exhibition*, USA, 2−4 February New Orleans, Louisiana.

Yuan, B., Moghanloo, R.G., 2016. Analytical solution of nanoparticles utilization to reduce fines migration in porous medium. SPE J 21 (06), 2317−2332.

Yuan, B., Moghanloo, R.G., and Purachet. P., 2015. Applying method of characteristics to study utilization of nanoparticles to reduce fines migration in deepwater reservoirs. Paper SPE-174192-MS presented at *SPE European Formation Damage Conference and Exhibition*, 3−5 June, Budapest, Hungary.

Yuan, B., 2017a. Modeling Nanofluids Utilization to Control Fines Migration. The University of Oklahoma, Norman, Oklahoma.

Yuan, B., Moghanloo, R.G., 2017b. Analytical modeling improved well performance by nanofluid pre-flush. Fuel 202, 380−394.

Yuan, B., Wang, W., Moghanloo, R.G., et al., 2017c. Permeability reduction of berea cores owing to nanoparticles adsorption onto the pore surfaces: mechanistic modeling and lab experiments. Energy Fuels 31 (01), 795−804.

Yuan, B., Moghanloo, R.G., Zheng, D., 2017d. A novel integrated production analysis workflow for evaluation, optimization and predication in shale plays. Int. J. Coal Geol. 180, 18−28.

Yuan, B., Moghanloo, R.G., 2018a. Nanofluid treatment, an effective approach to improve performance of low salinity water flooding. J. Petrol. Sci. Eng. in press. Available from: https://doi.org/10.1016/j.petrol.2017.11.032.

Yuan, B., Moghanloo, R.G., Wang, W., 2018b. Using nanofluids to control fines migration for oil recovery: nanofluids co-injection or nanofluids pre-flush? − a comprehensive answer. Fuel 215, 474−483.

Yuan, B., Moghanloo, R.G., 2018c. Nanofluid pre-coating, an effective method to reduce fines migration in radial system saturated with mobile immiscible fluids. SPE J. SPE 189464-PA . Available from: https://doi.org/10.2118/189464-PA.

Zeinijahromi, A., Lemon, P., Bedrikovetsky, P., 2011. Effects of induced fines migration on water cut during waterflooding. J Pet Sci Eng 78 (3−4), 609−617.

Zeinijahromi, A., Vaz, A., Bedrikovetsky, P., 2012. Well impairment by fines migration in gas fields. J Pet Sci Eng 88−89, 125−135.

Zeinijahromi, A., Nguyen, T.K.P., Bedrikovetsky, P., 2013. Mathematical model for fines-
    migration-assisted waterflooding with induced formation damage. SPE J 18 (03),
    518–533.
Zhang, J., Li, L., and Wang, S. 2015. Novel micro and nanoparticle-based drilling fluids:
    pioneering approach to overcome the borehole instability problem in shale formation.
    Paper SPE-176991-MS presented at the *SPE Asia Pacific Unconventional Resources
    Conference and Exhibition*, 9–11 November, Brisbane, Australia.
Zhang, T., Murphy, M., and Yu, H., et al., 2013. Investigation of nanoparticles adsorption
    during transport in porous media. Paper SPE-166346-MS presented at the *SPE
    Annual Technical Conference and Exhibition*. USA, 30 September–2 October 2013,
    New Orleans, Louisiana.

APPENDIX A: METHOD OF CHARACTERISTICS TO SOLVE SYSTEM OF QUASILINEAR FIRST-ORDER PARTIAL DIFFERENTIAL EQUATIONS (PDES)

Consider a general system of quasilinear first-order partial differential equations (PDEs) for two dependent variables (Rhee et al., 2001), C_1 and C_2, with two independent variables, x_D and t_D:

$$L_1 = A_1 \frac{\partial C_1}{\partial x_D} + B_1 \frac{\partial C_1}{\partial t_D} + C_1 \frac{\partial C_2}{\partial x_D} + D_1 \frac{\partial C_2}{\partial t_D} + E_1 = 0$$

$$L_2 = A_2 \frac{\partial C_1}{\partial x_D} + B_2 \frac{\partial C_1}{\partial t_D} + C_2 \frac{\partial C_2}{\partial x_D} + D_2 \frac{\partial C_2}{\partial t_D} + E_2 = 0$$

$$(A.1)$$

where $A_1, B_1, C_1, D_1, E_1, A_2, B_2, C_2, D_2, E_2$ are given continuous functions of C_1, C_2, x_D, and t_D.

The above PDEs are homogeneous if $E_1, E_2 = 0$. They are also called reducible although they are homogeneous and the coefficients are only functions of C_1 and C_2. For different cases of multiphase multicomponent flow with adsorption and chemical reactions, the functions of E_1, E_2 depend on the assumption of equilibrium or nonequilibrium reaction theory.

To ensure the characteristic lines of both variables C_1 and C_2, along the same direction, a linear combination of the above two equations is defined, as $L = \lambda_1 L_1 + \lambda_2 L_2$ to yield:

$$L = (\lambda_1 A_1 + \lambda_2 A_2) \frac{\partial C_1}{\partial x_D} + (\lambda_1 B_1 + \lambda_2 B_2) \frac{\partial C_1}{\partial t_D} + (\lambda_1 C_1 + \lambda_2 C_2) \frac{\partial C_2}{\partial x_D}$$

$$+ (\lambda_1 D_1 + \lambda_2 D_2) \frac{\partial C_2}{\partial t_D} + (\lambda_1 E_1 + \lambda_2 E_2) = 0$$

$$(A.2)$$

where L_1 and L_2 are the functions of derivatives of two variables; λ_1 and λ_2 are the line combination coefficients. If the directed derivatives of C_1 and C_2 will be collinear, it is necessary that:

$$\frac{dt_D}{dx_D} = \frac{\lambda_1 B_1 + \lambda_2 B_2}{\lambda_1 A_1 + \lambda_2 A_2} = \frac{\lambda_1 D_1 + \lambda_2 D_2}{\lambda_1 C_1 + \lambda_2 C_2} = \sigma \tag{A.3}$$

In matrix form, the system of Eq. (A.3) can be written as:

$$\begin{vmatrix} A_1\sigma - B_1 & A_2\sigma - B_2 \\ C_1\sigma - D_1 & C_2\sigma - D_2 \end{vmatrix} \begin{vmatrix} \lambda_1 \\ \lambda_2 \end{vmatrix} = 0 \tag{A.4}$$

If there are nontrivial values of λ_1 and λ_2, it becomes an eigenvalue problem; the determinate of coefficient matrix should be zero. Therefore, two characteristic directions can be obtained as:

$$\sigma^{\pm} = \frac{(A_1 D_2 - A_2 D_1 + B_1 C_2 - B_2 C_1) \pm \sqrt{(A_1 D_2 - A_2 D_1 + B_1 C_2 - B_2 C_1)^2 - 4(A_1 C_2 - A_2 C_1)(B_1 D_2 - B_2 D_1)}}{2(A_1 C_2 - A_2 C_1)} \tag{A.5a}$$

If the discriminant is a positive number,

$$(A_1 D_2 - A_2 D_1 + B_1 C_2 - B_2 C_1)^2 - 4(A_1 C_2 - A_2 C_1)(B_1 D_2 - B_2 D_1) > 0 \tag{A.5b}$$

This quadratic equation has two real roots and there are two families of characteristics C_+ and C_- presented in form of $\alpha(x_D, t_D) = const$ and $\beta(x_D, t_D) = const$, which are called as characteristic parameters. When the discriminant is zero or negative, the system is parabolic and elliptic, accordingly.

Since there are two characteristic directions for hyperbolic system of equations, there will be two different family of characteristics generated by two distinct values. The subscript $\pm$ in Eq. (A.5) indicates the corresponding terms to two roots of Eq. (A.4). The characteristic lines form a curvilinear map that serves as the possible solution route; the unique solution for different sets of initial and boundary conditions can be obtained along these characteristics lines.

Total derivatives of (C_1, C_2) in $x_D - \varphi$ domain will be only functions of ζ along each characteristic, $\alpha(x_D, t_D)$ and $\beta(x_D, t_D)$,

$$C_{1,\xi} = \frac{\partial C_1}{\partial x_D}\frac{\partial x_D}{\partial \xi} + \frac{\partial C_1}{\partial t_D}\frac{\partial \varphi}{\partial \xi} = C_{1,x_D} x_{D,\xi} + C_{1,t_D} t_{D,\xi}$$

$$C_{2,\xi} = \frac{\partial C_2}{\partial x_D}\frac{\partial x_D}{\partial \xi} + \frac{\partial C_2}{\partial t_D}\frac{\partial \varphi}{\partial \xi} = C_{1,x_D} x_{D,\xi} + C_{2,t_D} t_{D,\xi} \tag{A.6}$$

Replacing the coefficients of C_{1,t_D} and C_{2,t_D} in Eq. (A.2) by the numerators in Eq. (A.3) and substitution of total derivatives from Eq. (A.6):

$$(\lambda_1 A_1 + \lambda_2 A_2)C_{1,\xi} + (\lambda_1 C_1 + \lambda_2 C_2)C_{2,\xi} + (\lambda_1 E_1 + \lambda_2 E_2)x_{D,\xi} = 0$$

$$(A.7)$$

Alternatively, replace the coefficients of C_{1,x_D} and C_{1,x_D} in Eq. (A.2) by the numerators in Eq. (A.3) and substitution of total derivatives from Eq. (A.6):

$$(\lambda_1 B_1 + \lambda_2 B_2)C_{1,\xi} + (\lambda_1 D_1 + \lambda_2 D_2)C_{2,\xi} + (\lambda_1 E_1 + \lambda_2 E_2)t_{D,\xi} = 0 \quad (A.8)$$

For system of hyperbolic partial differential equations, two roots to the system of Eqs. (A.7-A.8) and two roots obtained from the system of Eq. (A.4) form the four-coupled ordinary differential equations which may be integrated simultaneously for (C_1, C_2) and (x_D, t_D) from an initial curve. The system of four above equations (Eq. A.4, A.7 and A.8) are summarized as,

$$\begin{cases} \left(A_1 C_{1,\xi} + C_1 C_{2,\xi} + E_1 x_{D,\xi}\right)\lambda_1 + \left(A_2 C_{1,\xi} + C_2 C_{2,\xi} + E_2 x_{D,\xi}\right)\lambda_2 = 0 \\ \left(B_1 C_{1,\xi} + D_1 C_{2,\xi} + E_1 t_{D,\xi}\right)\lambda_1 + \left(B_2 C_{1,\xi} + D_2 C_{2,\xi} + E_2 t_{D,\xi}\right)\lambda_2 = 0 \\ (A_1 \sigma_\pm - B_1)\lambda_1 + (A_2 \sigma_\pm - B_2)\lambda_2 = 0 \\ (C_1 \sigma_\pm - D_1)\lambda_1 + (C_2 \sigma_\pm - D_2)\lambda_2 = 0 \end{cases}$$

$$(A.9)$$

For nontrial value of two unknowns λ_1 and λ_2, we can choose any two equations from the system Eq. (A.9), and set the determinant of the coefficient matrix be zero. The below system Eq. (A.10) is the determinant of coefficient matrix of Eq. (A.9):

$$\begin{cases} (A_2 B_1 - A_1 B_2)C_{1,\xi} + [(A_2 C_1 - A_1 C_2)\sigma_\pm + (B_1 C_2 - B_2 C_1)]C_{2,\xi} \\ \quad + [(A_2 E_1 - A_1 E_2)\sigma_\pm + (B_1 E_2 - B_2 E_1)]x_{D,\xi} = 0 \\ [(A_1 C_2 - A_2 C_1)\sigma_\pm + (A_2 D_1 - A_1 D_2)]C_{1,\xi} + (C_2 D_1 - C_1 D_2)C_{2,\xi} \\ \quad + [(C_2 E_1 - C_1 E_2)\sigma_\pm + (D_1 E_2 - D_2 E_1)]x_{D,\xi} = 0 \\ [(A_2 B_1 - A_1 B_2)\sigma_\pm]C_{1,\xi} + [(A_2 D_1 - A_1 D_2)\sigma_\pm + (B_1 D_2 - B_2 D_1)]C_{2,\xi} \\ \quad + [(A_2 E_1 - A_1 E_2)\sigma_\pm + (B_1 E_2 - B_2 E_1)]t_{D,\xi} = 0 \\ [(B_1 C_2 - B_2 C_1)\sigma_\pm + (B_2 D_1 - B_1 D_2)]C_{1,\xi} + (C_2 D_1 - C_1 D_2)C_{2,\xi} \\ \quad + [(C_2 E_1 - C_1 E_2)\sigma_\pm + (D_1 E_2 - D_2 E_1)]t_{D,\xi} = 0 \end{cases}$$

$$(A.10)$$

As for the system of Eq. (A.11), it is a coupled four-equation system with four unknowns, $C_{1,\xi} \ C_{2,\xi} \ t_{D,\xi} \ x_{D,\xi}$. To obtain nonzero solutions, the determinant of coefficient matrix must be zero for system Eq. (A.10), in matrix form,

$$
\begin{pmatrix}
(A_2B_1 - A_1B_2) & [(A_2C_1 - A_1C_2)\sigma_\pm + (B_1C_2 - B_2C_1)] & [(A_2E_1 - A_1E_2)\sigma_\pm + (B_1E_2 - B_2E_1)] & 0 \\
[(A_1C_2 - A_2C_1)\sigma_\pm + (A_2D_1 - A_1D_2)] & (C_2D_1 - C_1D_2) & [(C_2E_1 - C_1E_2)\sigma_\pm + (D_1E_2 - D_2E_1)] & 0 \\
[(A_2B_1 - A_1B_2)\sigma_\pm] & [(A_2D_1 - A_1D_2)\sigma_\pm + (B_1D_2 - B_2D_1)] & 0 & [(A_2E_1 - A_1E_2)\sigma_\pm + (B_1E_2 - B_2E_1)] \\
[(B_1C_2 - B_2C_1)\sigma_\pm + (B_2D_1 - B_1D_2)] & (C_2D_1 - C_1D_2) & 0 & [(C_2E_1 - C_1E_2)\sigma_\pm + (D_1E_2 - D_2E_1)]
\end{pmatrix}
\begin{pmatrix}
C_{NP,\xi} \\ C_{FP,\xi} \\ x_{D,\xi} \\ t_{D,\xi}
\end{pmatrix}
= 0
$$

(A.11)

Combining the zero determinant of coefficient matrix in Eq. (A.4), and applying matrices transformation, the simplest form of coefficient matric for Eq. (A.11) becomes as follows:

$$
\begin{pmatrix}
(A_2B_1 - A_1B_2) & [(A_2C_1 - A_1C_2)\sigma_\pm + (B_1C_2 - B_2C_1)] & [(A_2E_1 - A_1E_2)\sigma_\pm + (B_1E_2 - B_2E_1)] & 0 \\
0 & 0 & 0 & 0 \\
0 & 0 & 1 & -\sigma_\pm \\
0 & 0 & 0 & 0
\end{pmatrix}
\begin{pmatrix}
C_{NP,\xi} \\ C_{FP,\xi} \\ x_{D,\xi} \\ t_{D,\xi}
\end{pmatrix}
= 0
$$

(A.12)

Therefore, the composition diagram is determined along the fast path and slow path, respectively, and obtain the composition path $C_{FP}(C_{NP})$:

$$
(A_2B_1 - A_1B_2)C_{1,\xi} + [(A_2C_1 - A_1C_2)\sigma_\pm + (B_1C_2 - B_2C_1)]C_{2,\xi}
$$
$$
+ [(A_2E_1 - A_1E_2)\sigma_\pm + (B_1E_2 - B_2E_1)]x_{D,\xi} = 0
$$

(A.13)

The existences of nonhomogeneous terms (E_1, E_2) make the derivation of $C_{FP}(C_{NP})$ difficult. In fact, for cases of two-phase multicomponent flow with different assumptions of chemical reactions or adsorptions, the nonhomogeneous terms are different.

CHAPTER FIVE

Formation Damage by Inorganic Deposition

Xingru Wu
The University of Oklahoma, Norman, OK, United States

Contents

5.1 INTRODUCTION

In nature, both formation water and surface water contain many anions and cations. The chemistry of brine in a reservoir is determined by past migration history and local conditions, and generally stable at reservoir conditions. The water used for injection is either sea water, fresh water, or produced water reinjection. Table 5.1 shows the major chemical species and their concentrations of brines from different fields and sea water. For both sea water and formation water, the salinity of the brine is higher than that of fresh water; however, both of them feature different ion species and compositions. For example, formation water usually has

Formation Damage during Improved Oil Recovery.
DOI: https://doi.org/10.1016/B978-0-12-813782-6.00005-1
© 2018 Elsevier Inc.
All rights reserved.

Table 5.1 Reported ion compositions in some sea waters and formation brines

Ion	Sea Water (mg/L)		Formation Water (mg/L)			
	Gulf of Mexico	North Sea (Mitchell et al., 1980)	Forties (Mitchell et al., 1980)	Miller (Adair and Smith, 1994)	Prudhoe Bay (Li et al., 1996)	A GoM Field
Sodium	12,300	11,000	30,200	28,780	7682	13,282
Calcium	475	403	3110	1060	247	302
Magnesium	1490	1320	480	115	35	87.3
Potassium	320	340	430	1830	90	75.4
Barium	—	—	250	1050	60	28
Strontium	—	—	660	110	21	43
Sulfate	2800	2480	0	2	140	26.7
Bicarbonate	162	135	360	2090	1978	748
Chloride	21,800	19,800	53,000	47,680	11,400	21,588

high concentrations of barium and strontium, but low concentrations of sulfate; whereas, sea water is just the opposite. Other than the shown species, both formation brine and sea water have other organic and inorganic ions, and some of them are also critical for scale management.

Formation damage, as the result of inorganic solids deposition, can occur mainly in two scenarios. The first one is self-scaling of the formation water, which occurs as a consequence of changing pressure, temperature, or composition of formation water because of reservoir depletion. For example, the formation of calcium carbonate scale can occur if CO_2 comes out of solution as the pressure goes down. The second scenario is induced scaling that is often related with waterflooding. In this process, an external water source is introduced to mix with the formation water. The external water source can be sea water for offshore reservoirs, produced water, or others. When the injected water is not compatible with the formation water, supersaturation and then precipitation can occur (Fig. 5.1).

Inorganic deposition can occur anywhere given the right conditions. Fig. 5.2 illustrates a water injection project in an offshore environment, and seven possible locations where scale can occur are identified. Scale can precipitates in a reservoir, which is usually not a major concern when it precipitates in at depth and remote from wellbores. However, once it precipitates near the wellbore, many field operations have shown that the precipitated scale can significantly reduce the well productivity (Tyler et al., 1985). Scales can also precipitate on the well completion tubulars

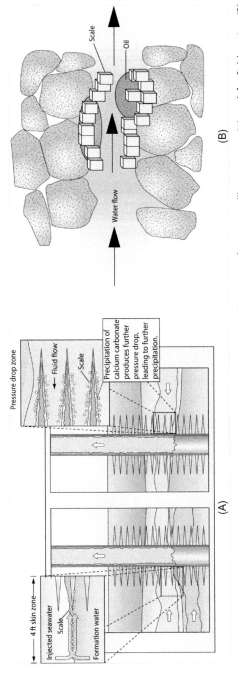

Figure 5.1 Diagrams for formation damage caused by inorganic solids deposition for near wellbore region (A) and far-field region (B). *Modified (Crabtree et al., 1999).*

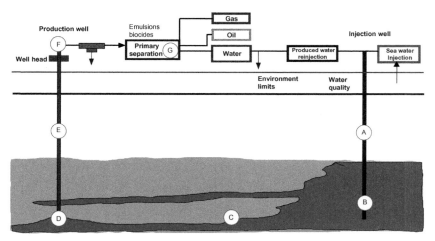

Figure 5.2 Offshore production and injection system with possible scaling locations. (A) Injection wellbore, (B) near injection wellbore region, (C) formation, (D) near production wellbore region, (E) production wellbore, (F) wellhead, and (G) separator.

and within the wellhead and Christmas trees. If scaling occurs along the wellbore, it will have a significant impact on well performance, integrity, and operability, and may eventually even lead to the cessation of production. Cases such as perforation scale off, gravel pack or screen blockages, tubing blockage, safety valve, or choke failure have been reported in literature due to inorganic deposition in reservoir and near wellbore location.

Options for scaling prevention and mitigation are available in practice. Chemical methods such as injecting scaling inhibitors and mechanical methods can be used if necessary. However, the operator should focus on preventing scaling deposition rather than remediation efforts, which are usually more time consuming and costly. Remediation and mitigation actions should be used as the last resort to overcome scale issues.

5.2 TYPES OF SCALES IN FORMATION DAMAGE

Many types of solid particles can precipitate when brine conditions change. Here, we classify them into three categories: Carbonate scales, Sulfates scales, and others.

5.2.1 Carbonate scales

Carbonate scales include precipitates containing CO_3^{2-}, which are mainly barium carbonate ($BaCO_3$), strontium carbonate ($SrCO_3$), calcium carbonate ($CaCO_3$), and iron carbonate ($FeCO_3$). Of the carbonate scales, calcium carbonate scale is probably the most commonly encountered scale in oil and gas production. Many of them can precipitate in different mineralogical forms. For example, calcium carbonate ($CaCO_3$) can exist as calcite, aragonite, vaterite, or witherite. The equilibrium of carbonate scales can be summarized by considering the two stoichiometry equilibria, with calcium used as an example:

$$C_a^{++} + 2(HCO_3) \leftrightarrow C_a(HCO_3)_2$$
$$C_a(HCO_3)_2 \rightarrow C_a(CO_3)\downarrow + CO_2\uparrow + H_2O$$

From the above equilibria, we can see that the presence of CO_2 or increasing CO_2 partial pressure leads to the increase of $CaCO_3$ solubility in water. When CO_2 comes out of solution as the pressure decreases, $CaCO_3$ solubility in water decreases. As $CaCO_3$ solubility in water decreases with the increase of temperature, changing temperature can also move the equilibrium to the right in the equation and increase the risk of carbonate scale forming. The solubility of $CaCO_3$ also increases as the level of dissolved salts increases (Plummer and Busenberg, 1982).

Many carbonate scales are acid soluble, which generally makes it easier to remediate than sulfate scales. However, damage in downhole frac-packs is very difficult to remove without very good treatment diversion or the use of coiled tubing. Therefore, a proactive approach to downhole carbonate scale prevention is essential.

5.2.2 Sulfate scales

The concentration of sulfate in formation water is usually low, but it is high in sea water as shown in Table 5.1. When sulfate-rich water is injected to flood the barium/strontium-rich formation water, the tendency of forming sulfate scales increases. The sulfate scales include barium sulfate ($BaSO_4$), strontium sulfate ($SrSO_4$) like celestite, and calcium sulfate ($CaSO_4$). calcium sulfate has three different crystalized forms known as gypsum ($CaSO_4 \cdot 2H_2O$), hemihydrate ($CaSO_4 \cdot \frac{1}{2}H_2O$) and anhydrate ($CaSO_4$). Gypsum is the most common calcium sulfate scale type at low temperatures. Anhydrate is the stable form of calcium sulfate scale at temperatures above $100°C$. Hemihydrate scale formation occurs at temperatures between $100°C$ and $120°C$ in brines with high ionic strength and in nonturbulent environments. Among these calcium sulfate scales, gypsum is more likely to form in the oil and gas reservoir conditions. Given right conditions, gypsum and anhydrate can transform from one to the other (Raju and Atkinson, 1990). As an example, the reaction of forming barium sulfate is as follows:

$$Ba^{++} + SO_4^{2-} \rightarrow BaSO_4 \downarrow$$

Fig. 5.3 shows the solubility of different scales at different conditions. Given that barium sulfate has a very low solubility in brine and even in mineral acids such as hydrochloric acid, the near wellbore formation damage caused by sulfate scale precipitation cannot be effectively remediated without a well workover. Field experience indicates that well productivity can be severely impacted by sulfate scale in a matter of days or weeks if left untreated in sea waterflooded fields. The solubility of barium sulfate slightly increases with temperature and the presence of mineral salts such as NaCl.

5.2.3 Other inorganic solids

Other inorganic solids can also precipitate during scaling process. For example, usually the formation brines contain a low concentration of iron, but some iron scales such as iron oxides and, hydroxides, and sulfides can occur as a result of corrosion when dissolved gases such as CO_2 and H_2S contribute to tubular corrosion, which leads to the formation of various iron compounds. Some organic scales such as calcium napthenates may also be generated in some particular conditions. Some other inorganic solids include lead sulfide, zinc sulfide, and calcium fluoride.

Some naturally occurring radioactive material (NORM) may also precipitate in reservoirs or wells. For example, Low Specific Activity (LSA) (^{226}Ra) can be coprecipitated with $BaSO_4$ and $CaSO_4$ scale, and it emits alpha, beta, and gamma radiations. Therefore, if this type of scale are produced in an open environment, special handling permits may be required from regulatory authorities to dispose of it, and it needs to be properly packed and shipped to registered recipient for proper disposal.

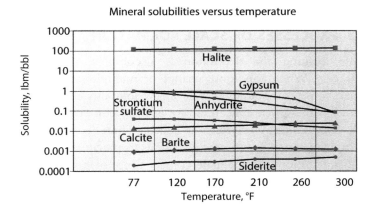

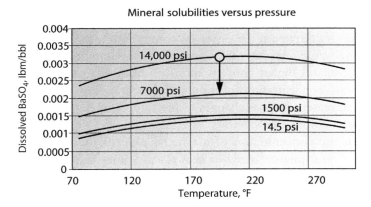

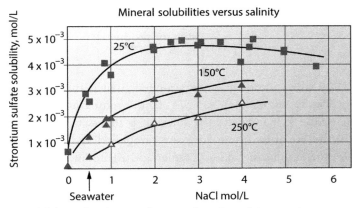

Figure 5.3 Solubility of various scales as a function of temperature, pressure, and salinity. *modified based on Crabtree et al. (1999).*

5.3 PROCESSES OF SCALE FORMATION

Scale forming and deposition is a complex crystallization and kinetic process. When conditions change or multiple streams of different waters mix, some compounds that have low solubility in the new conditions would form. When concentrations of these compounds exceed their solubility, they could precipitate as solids. Three conditions must be reached simultaneously for these compounds to precipitate: (1) supersaturation, (2) nucleation; (3) adequate contact time for crystal growth. The supersaturation state is thermodynamically determined, although the nucleation and crystal growth are affected by kinetic factors.

5.3.1 Solubility and supersaturation

A solution is saturated if the dissolved compounds reach their equilibrium with their solute. As the name suggests, supersaturation occurs when a solution contains higher concentrations of dissolved compounds than the equilibrium concentration of solute. Supersaturation can occur for a number of reasons such as changes in pressure, temperature, pH values, or mixing of incompatible waters.

When the concentration of a mineral exceeds the equilibrium concentration, precipitation starts to take place. Scaling tendency is used as a measure of the thermodynamic driving force for the scale formation. The scaling tendency is measured by the scaling index or the saturation ratio (Oddo and Tomson, 1994, Langelier, 1936) that is defined as the ratio of the ion activity product to the solubility product as follows:

$$SI \equiv \log\left\{\frac{\text{activity product}}{K_{eq}}\right\} \tag{5.1}$$

If $SI < 0$, the solution is undersaturated, which means that the solution can dissolve more solute. If $SI = 0$, the saturation is at equlirbium; and when $SI > 0$, it indicates the solution is supersaturated (or oversaturated). In the literature, there are many simplified and empirical correlations to estimate the scale index (Frenier and Ziauddin, 2008).

For example, calcite formation is based on the following reaction:

$$Ca^{2+} + CO_3^{2-} \rightarrow CaCO_3$$

Table 5.2 Coefficients of equation to calculate K value at equilibrium for calcite, aragonite, and vaterite

Species	A	B	C	D
Calcite	− 171.9065	− 0.07799	2839.319	71.595
Aragonite	− 171.9773	− 0.07799	2903.293	71.595
Vaterite	− 172.1295	− 0.07799	3074.688	71.595

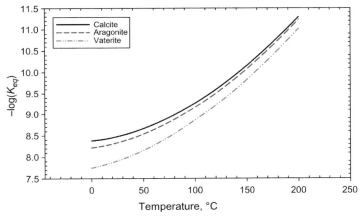

Figure 5.4 Calculate solubility product at the equilibrium conditions for calcite, aragonite, and vaterite.

The saturation index for calcite is therefore expressed as:

$$SI_{Calcite} = \log \left[\frac{\left(a_{C_a^{2+}} \right) \left(a_{CO_3^{2-}} \right)}{K_{eq}} \right] \tag{5.2}$$

where $\left(a_{C_a^{2+}} \right)$ and $\left(a_{CO_3^{2-}} \right)$ are the activity of C_a^{2+} and CO_3^{2-}, respectively. K_{eq} represents the solubility product of the ions, which is a function of pressure, temperature, and ion strength. For some selected carbonate scales Plummer and Busenberg (1982) gave the empirical equation (Eq. (5.3)) to calculate the solubility product K_{eq} at equilibrium, and for calcite, aragonite, and vaterite, their corresponding coefficients are given in Table 5.2.

$$\log\left(K_{eq} \right) = A + BT + \frac{C}{T} + D \log(T) \tag{5.3}$$

The calculated results, based on Eq. (5.3), are plotted in Fig. 5.4 for different temperatures, which indicate that the solubility of calcium carbonate declines with the increase of temperature.

Table 5.3 Coefficients for strontium sulfate, gypsum, anhydrite, and barium sulfate

Species	A	B	$C \times 10^{-6}$	I_h	I_g
Strontium Sulfate	641.541	− 1.90146	− 42.7605	−2,51,748	4102.24
Gypsum	763.714	− 2.04731	− 43.2002	−2,82,176	4837.58
Anhydrite	689.581	− 1.94455	− 45.0378	−2,87,889	4432.9
Barium Sulfate	594.534	− 1.91171	− 40.0731	−2,00,488	3840.12

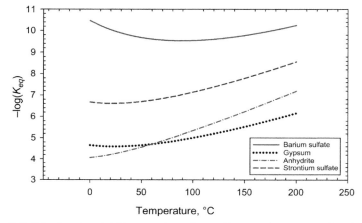

Figure 5.5 Calculated solubility product for sulfate scales using the Raju and Atkinson correlations.

Raju and Atkinson (1988), Raju and Atkinson (1989), Raju and Atkinson (1990) gave an empirical correlation to estimate the solubility product value at equilibria for strontium sulfate, gypsum, anhydrite, and barium sulfate as follows, and the related coefficients are given in Table 5.3.

$$\ln\left(K_{eq}\right) = \frac{A\ln(T)}{R} + \frac{BT}{2R} + \frac{C}{2RT^2} - \frac{I_h}{RT} - \frac{I_g}{R} \qquad (5.4)$$

Similarly, the calculated solubility products for these sulfate scales are plotted in Fig. 5.5, and they indicate that the solubility of sulfate scales may not be a monotonic relationship with changing temperature.

The activity of a species ion β is defined as follows:

$$\left(a_\beta\right) = \gamma_\beta[\beta] \qquad (5.5)$$

where γ_β is the activity coefficient of the species, and $[\beta]$ is the molality of species β (mol/kg of water). The activity coefficient is affected by pressure, temperature, and ion strength. Many models are available to estimate the activity coefficient of a particular species (Yuan and Todd, 1991), and the Pitzer theory is widely accepted for modeling ion-interaction and implemented in software package (He et al., 1997, Pitzer, 1973, Kan and Tomson, 2012).

Mineral precipitation or dissolution rate can be modeled as follows (Bethke, 2007):

$$r_\beta = A_\beta K_\beta \left(1 - \frac{Q_\beta}{K_{eq.\beta}} \right) \quad (5.6)$$

where r_β is the reaction rate, K_β is the rate constant (mole/m^{2}s), A_β is the reactive surface area of mineral β (m^2/m^3 of bulk volume of mineral), $K_{eq.\beta}$ is the chemical equilibrium constant, and Q_β is the activity product for mineral β which can be determined using Eq. (5.7):

$$Q_\beta = \prod_{k=1}^{n_{aq}} a_k^{v_{k\beta}} \quad (5.7)$$

where a_β is the activity of component k and $v_{k\beta}$ is the stoichiometry coefficients. The rate constant K_β for species β can be estimated using Eq. (5.8):

$$K_\beta = k_{0\beta} \exp\left[-\frac{E_\beta}{R} \left(\frac{1}{T} - \frac{1}{T_0} \right) \right] \quad (5.8)$$

where $k_{0\beta}$ is the reaction rate for reaction β at a reference temperature T_0, and it can be determined experimentally. E_β is the activation of reaction β.

For the purpose of illutration, we assume the reservoir pressure is 7000 psi, and temperature is 300°F, and at the wellhead, the pressure and temperature are 14.7 psi and 80°F, respectively. Mixing Miller field formation water samples with North Sea water using the composition shown in Table 5.1, the scale index and concentrations of different minerals can be calculated as shown in Fig. 5.6.

5.3.2 Dynamics of scale formation

In a supersaturated solution, both cations and anions are in constant motion and move in and out of the influence scopes of other ions or molecules. Ions with opposite electrically charges are attracted to form clusters. The cluster is not a stable stage until it grows large enough to

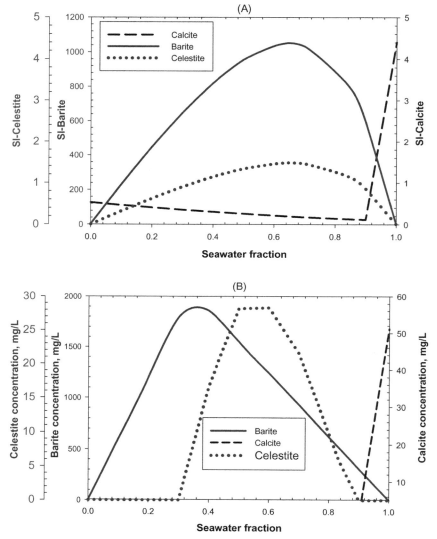

Figure 5.6 Scale index and concentration of North Sea Water mixing with Miller water at different pressures and temperatures. (A) Scale indices of calcite, barite, and celestite at reservoir condition; (B) concentrations of calcite, barite, and celestite at the reservoir conditions; (C) scale indices of calcite, barite, and celestite at wellhead condition; (D) concentration of calcite, barite, and celestites at wellhead condition.

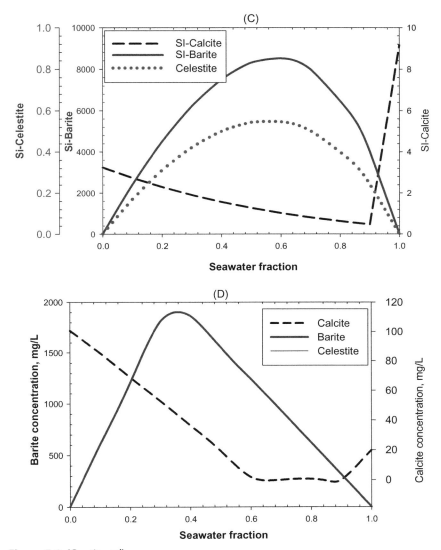

Figure 5.6 (Continued)

become a crystallite. The process of forming a cluster is known as nucleation, which is the initial formation of a precipitate. Nucleation has two mechanisms: homogeneous nucleation that does not require the presence of a foreign substance, and it is not a likely mechanism in the field; heterogeneous nucleation requires the presence of a foreign substance to initiate the nucleation, and the foreign substance can be solid particles in

Homogeneous nucleation   Heterogeneous nucleation

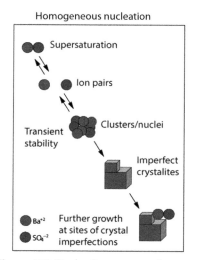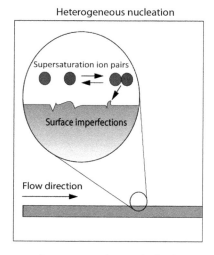

Figure 5.7 Nucleation process from supersaturation to crystal growth for homogeneous and heterogeneous nucleation. *Modified based on Crabtree et al. (1999).*

formation or imperfections on metal surfaces of tubing. For scale deposition in porous media resulting in formation damage, pore size distribution can also affect the nucleation process, because of its stochastic nature. The crystal growth is more limited in small pores because the probability of collision is less in smaller pores than larger pores. Owing to the complexity of scale deposition in porous media, the scaling prediction from a laboratory may result in significant errors if they are directly used to predict scaling in heterogeneous porous media (Fig. 5.7).

For scale to form from a supersaturated solution once nucleation has occurred, there must be sufficient contact time between the supersaturated solution and the nucleating sites on the reaction surface. The contact time required varies depending on the degree of supersaturation, type of mineral, temperature, pressure, agitation, and others. In the contacting duration, these factors can exacerbate the risk of scaling and scale precipitation by influencing the fate of the crystal nuclei. Given the nature of the kinetic process of scale formulation and precipitation, crystal growth remains understood only in qualitatively terms. The rate of precipitation from a scale-forming solution varies according to the crystalline form that the solid adopts. Also, scale deposits change with time as the continuous process of dissolution and precipitation takes place at the surface.

5.3.3 Formation damage from scale deposition

Field operations have observed scale-dropout issues, both in the reservoir and near wellbore, and recognized that they can have a significant impact on well performance and operations (Paulo et al., 2001, Sorbie and Mackay, 2000). However, there is no accepted model in practice to quantify the relationship between the amount of deposition and changes on reservoir properties. Some efforts to establish such a relationship are based on many assumptions of idealized models. Assuming the reservoir rock can be represented by a bundle of tubes of different radii, the modified Carman-Kozeny model can be used to understand the relationship (Hajirezaie et al., 2017).

Eq. (5.9) can be used to calculate the change in porosity resulting from mineral dissolution/precipitation:

$$\hat{\phi} = \phi^* - \sum_{\beta=1}^{n_m} \left(\frac{N_\beta}{\rho_\beta} - \frac{N_\beta^0}{\rho_\beta} \right) \tag{5.9}$$

where, $\hat{\phi}$ is the reference porosity when dissolution/precipitation is included, ϕ^* is the reference porosity without mineral dissolution/precipitation and ρ_β is the molar density of mineral β.

If there are n cylindrical tubes in the medium, the corresponding porosity is calculated based on the following equation:

$$\phi_1 = \frac{\sum_n \pi r^2 l'}{BV} \tag{5.10}$$

Here, BV stands for bulk volume, r denotes the pore radius, and l' is the length of tubes. Assuming that scale formation has occurred as shown in Fig. 5.8 with the same thickness of scale in each pore, the corresponding porosity is calculated based on the new radius of the tubes r' using Eq. (5.11):

$$\phi_2 = \frac{\sum_n \pi r'^2 l'}{BV} \tag{5.11}$$

Defining,

$$\frac{r'}{r} = X$$

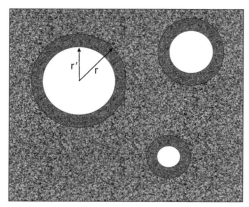

Figure 5.8 Porous medium with cylindrical tubes after scale formation.

The corresponding porosity ratio as follows:

$$\frac{\phi_1}{\phi_2} = X^2 \tag{5.12}$$

Assuming two cases for the Carman–Kozeny equation: before scale formation shown by K_1; and, after scale formation shown by K_2, then:

$$K_1 = \frac{1}{72\tau}\frac{\phi_1^3 D_s^2}{\left(1-\phi_1\right)^2} \tag{5.13}$$

$$K_2 = \frac{1}{72\tau}\frac{\phi_2^3 D_s^2}{\left(1-\phi_2\right)^2} \tag{5.14}$$

where τ is tortuosity, D_s is the representative diameter of rock, and ϕ_1 and ϕ_2 are porosities before and after scale deposition. Assuming that the characteristic diameter and tortuosity remain the same, the ratio of permeability contrast gives:

$$\frac{K_2}{K_1} = \frac{\dfrac{\phi_2^3}{\left(1-\phi_2\right)^2}}{\dfrac{\phi_1^3}{\left(1-\phi_1\right)^2}} \tag{5.15}$$

Substituting ϕ_2 with $\phi_1 x^2$ to get Eq. (5.16):

$$\frac{K_2}{K_1} = X^6\left(\frac{1-\phi_1}{1-\phi_1 X^2}\right)^2 \tag{5.16}$$

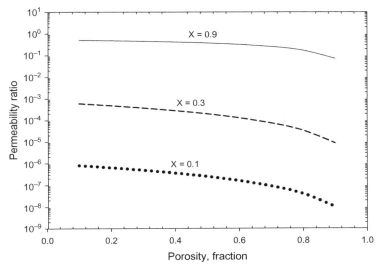

Figure 5.9 Impact of pore radius reduction on permeability and porosity as the result of scale deposition using the modified Carman-Kozeny model. Here, X is the radius ratio of reduced radius to the original radius.

Fig. 5.9 shows the severity of permeability reduction as a result of scale deposition for variable formations. The figure shows that the permeability ratio decreases significantly when the thickness of precipitation increases.

5.3.4 Scale inhibitors

Scale inhibitors are most often used as a prevention technique to reduce the scaling risks in near wellbore location and wellbore (Shaughnessy and Kline, 1983). From the process of scaling formation, scale inhibitors can work in one or more of three ways. Fig. 5.10 shows a summary diagram for main mechanisms of scale inhibitors.

Some scale inhibitors can interfere with the nucleation process by diffusing in the bulk liquid to reach the ion clusters either in the liquid or on a solid substrate. These inhibitor ions should be of a sufficiently large size to disrupt the scaling ion clustering and prevent further growth of existing clusters to the critical size where crystallites would form. A good nucleation inhibitor ion needs to be of a critical size but still be able to diffuse in the water at an acceptable rate.

Some inhibitors do not prevent the nucleation to cluster, but retard the crystal growth by diffusing over the crystal surface and active sites

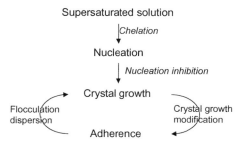

Figure 5.10 Summary of scale inhibitor mechanisms in different stage of scale-forming process.

through a strong affinity. This type of inhibitor has to be sufficiently small to be diffused, but sufficiently large in order not to be absorbed into the growing crystals. Different from the first kind inhibitor, this type of inhibitor promotes the formation of small crystals to reduce the supersaturation of the solution and lower the risk of scaling in following parts of flow conduit. If the inhibitor is absorbed into the crystal, it can be considered advantageous in that the resulting scale may be soft, friable, and easily removed.

Some inhibitors can modify the crystal surface. The transport theory and models of scale inhibitor can be found in the literature (Sorbie and Gdanski, 2005, Kahrwad et al., 2009). Molecules that are neither nucleation nor crystal growth modifiers are able to absorb strongly to crystal surfaces and prevent attachment of crystals.

Selecting the right inhibitor is critical in scale management and timing of inhibition is also important. Fig. 5.11 is an illustrative diagram for scale index change with time and when scaling inhibitors should be deployed. Even though the diagram is made for sulfate scales, other scales also qualitatively follow similar trends.

5.4 MANAGEMENT OF SCALING IN DEVELOPMENT AND PRODUCTION

The management of scaling occurrence during the development and production of an oil & gas field is done mainly to reduce the risk of scaling formation in reservoir and production facilities, especially in the near-wellbore formation and wellbore. It is possible to have a severe

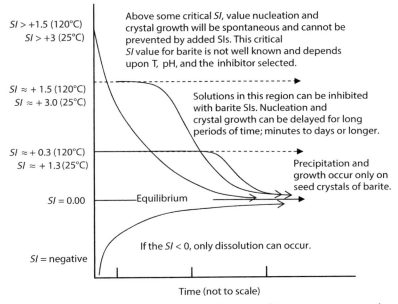

Figure 5.11 Illustrative schematic of the relationship of barite saturation index and kinetics. *Excepted from Kan and Tomson (2012).*

scaling issue even with low injected water percentages in the producing streams. As an example, Fig. 5.12 shows the amount of scale precipitation for a typical sea water injection. From the scaling tendency analysis, we can see that when the sea water amount accounts for less than 5% or more than 95%, the risk of scaling is relative low; although the sea water amount ranges from 20% to 80% in the water stream, we can expect high risk or amount of scaling precipitation.

5.4.1 Water sampling and analysis

The management of scaling issues starts with obtaining representative samples. It is very important to sample water properly before performing a water analysis or the analysis result could be misleading. Reservoir fluid samples are normally reconditioned to the sampling site conditions in laboratory.

To obtain the knowledge of the nature of brine in the formation of interest, representative brine samples should be taken with the right tools and proper guidelines in sampling, sample surfacing, transportation, and analysis. In practice, there are two ways to take brines samples: downhole

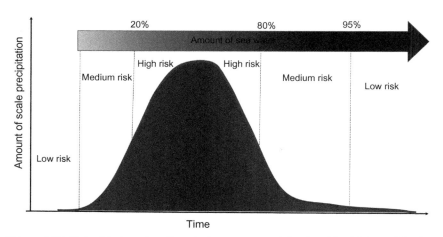

Figure 5.12 Risk of induced scaling during sea water injection. The highest risk after water breakthrough is around 20–80% of sea water in the effluent water stream.

sampling tools and surface sampling tools. The decision regarding which method to use to obtain water samples is a balance of the usage of the sample and operational cost and risks associated with sampling process.

Taking downhole samples is usually more challenging and costly operation wise. If the downhole sample is needed, it is preferred to use single-phase sampler with pressurization as water in the reservoir has associated gas with it that would be lost if the pressure is not maintained in the sampling and surfacing processes. In particular, the loss of CO_2 can affect the water chemistry. Single-phase sampling tools and technologies are available with oil service companies. Downhole brine samples are liable to be contaminated with other well fluids such as drilling muds. If oil-based muds have been used, these can affect the measured gas–water ratio because the gas from the saturated mud repartitions into the water phase. If a water-based mud is used, the chemistry of brine near the wellbore might have been altered during drilling, and a cleaning process has applied to obtain representative samples. Furthermore, taking downhole samples usually ends up with relative small volume that is constrained by the sampler volume and the operational cost of taking them.

Most brine samples are taken at atmospheric pressure. To determine the chemistry of water samples taken at atmospheric conditions, immediate analyses should be conducted on the brine for pH and bicarbonate before these values change. The solubility of carbonate and iron compounds is highly dependent on pH. For surface taken brine samples, it is also recommended to measure dissolved gas content, suspended solids, and bacterial population at the sample location. Sampling fluid should fill the sample container, leaving no ullage space to preserve any dissolved gas remaining at the sample location in solution. During the sample transportation to the lab, temperature can alter the solubility of salts and gasses or even lead to salt precipitation that may lead to inaccurate brine ion characterization in laboratory. Therefore, it is also a common practice in operation to take a series of brine samples in several sample containers. The first sample container only has the brine. In the second, the brine is sampled into a known volume of HCl (usually 5% of the total sample volume) to preserve the calcium ions in solution. In the third, the brine is sampled into a 10% Ethylenediaminetetraacetic acid (EDTA) solution to preserves the barium and strontium ions in solution by chelating them. In the fourth, the brine is sampled into an accurately known volume (approximately 5 vol%) of a reducing acid such as erythorbic or acetic present at a 5% concentration to preserve the irons in solution.

At the laboratory the brine can be flashed to atmospheric pressure, and the relative volumes of water and flashed gas are measured. Concentrations of key scaling ions such as $Ca^{++}, Ba^{++}, Sr^{++}, HCO_3^-$, and SO_4^{-2} should be analyzed along with other anions and cations, such as $Cl^{-1}, Na^+, K^+, Mg^{++}, Fe^{2+}$ and Fe^{3+}. Additionally, the determination of volatile fatty acids is important and best run at the same time as the bicarbonate determination. In the laboratory, bicarbonate is determined by a titration-based method of a fresh brine sample; however, this method actually titrates part of the fatty acid present as well, which leads to an overestimate of the bicarbonate present.

5.4.2 Options for scale prevention and remediation

The main challenge for the management of scales in the oil field is the control of downhole scaling in the near wellbore reservoir and the production wellbore. The two main strategies available for downhole scale control are prevention and remediation. There are other options to remove the precipitation of solids once they are deposited on wellbore, but they are out scope of this chapter that is dedicated to formation damage. For reservoir inside or near wellbore formation damages, there are three main types of preventative strategies that are discussed in the following sections.

5.4.2.1 *Removal of the scaling ions before mixing occurs*

This option generally refers to the removal of sulfate from sea water as injection water. One option is to reduce sulfate concentration through Sulfate Reduction Unit (SRU). Removing sulfate from the sea water not only lowers the risk of sulfate scaling issues, but also reduces the reservoir souring potential. On the operating cost side, this option can reduce or eliminate scale inhibitor squeeze treatments. Furthermore, this option can render the production free of LSA scales, which are usually coprecipitants of sulfate scales. Other technologies such as Reverse Osmosis (RO) can not only reduce sulfate concentration to a low level, but also remove other ionic species. For particular formations, injecting low–salinity water can also improve oil recovery. On the other hand, the technologies of reducing ionic concentration can require high initial capital expenditure and long-term operational expenditure because of the sulfate removal unit commitment.

5.4.2.2 *Passive scale treatment in well completion*

In well-completion stage, engineers can take preemptive actions to treat near wellbore condition during the completion process or prior to the start of well (Tyler et al., 1985). In wells completed over short reservoir intervals, it is possible to place the treatment effectively by bullheading the treatment from the surface. For many wells with frac-pack completion, fracture proppant impregnated with scale inhibitor or solid–scale inhibitor mixed with proppant are used. Alternatively, liquid-scale inhibitor is deployed with the frac fluid during fracturing operations. With this technology, large amounts of scale inhibitor, impregnated into proppant material, can be placed into a hydraulic fracture. However, the technical challenge of well interventions should not be underestimated.

5.4.2.3 *Periodic squeeze treatments*

In this method chemical inhibitor is pumped into the near well formation of a producing well to provide scale control for a limited period after which the process is repeated. Most of the conventional technology for the treatment of downhole scaling is essentially one form or another of batch treatment with scale inhibitor. Squeeze treatments being by far the most common method to control downhole scale problems (Mackay et al., 1998). With chemical squeeze inhibition, large volumes of scale inhibitor are injected into the formation and around the wellbore. An amount of the scale inhibitor is retained in the formation due to adsorption or precipitation on the rock surface or other mechanisms. After the well is placed back on production, the inhibitor slowly dissolves or desorbs from the rock and enters the produced water and sea water over time.

As the squeezed chemicals flow back into the well with the produced fluids, the inhibitors inhibit the formation of scale from the sand face to the topside facilities. Ideally, the inhibitor will return over a period of weeks or months at a low concentration just above the minimum inhibitor concentration (MIC) that is required to prevent scaling (Graham et al., 2005). The MIC is linked to the severity of the scale problem depending on the ratio of sea water and produced water and the concentration of scaling ions in the formation water. A high MIC could lead to more frequent treatments. However, the limitations of the methods have become apparent in more challenging environments such as deepwater or remote wells because these wells usually require intervention-free and totally reliable scale-prevention strategies are demanded. Scale inhibitor treatment frequencies are typically 2−3 per year per well depending on the volume of produced water treated. Squeeze lifetimes are sometimes referred to in total barrels of produced water treated. The squeeze lifetime is strongly influenced by the severity of the scaling environment, inhibitor chemistry, the MIC, and the formation properties. The general industry experience is that the first squeeze treatments performed on a well give the shortest lives leading to a higher intervention frequency during early sea water breakthrough.

Fig. 5.13 shows a typical scale inhibitor squeeze treatment. In the preflush stage, typically of about 50 barrels of flush fluid mixture is injected. The preflush fluids usually contain some biocide and scale inhibitors up to 200 ppm. The purpose of a preflush is to maximize the fluid contact with rock to reduce near wellbore damage. After the preflush, the main inhibitor treatment will be started with a volume of 250 to 500 barrels based on the treatment design. Following the main treatment, about 500 barrels or more fluid will be injected to ensure the accurate placement of chemicals. A typical treatment is usually less than 30 ft from the wellbore, and the reservoir gross thickness significantly affects the actual volume needed for each stage of injection during the squeeze treatment.

Some potential drawbacks of the squeeze technology should be taken into consideration before it is used. Firstly, squeeze lifetime is finite for a particular volume of water production, which indicates that wells need to be treated periodically to maintain scale control. Secondly, squeeze

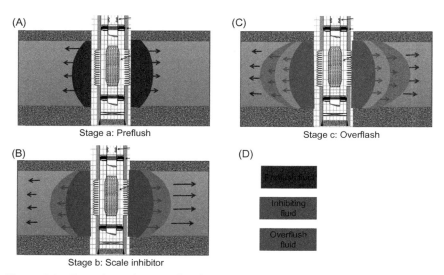

Figure 5.13 Procedure of a typical scale-squeeze operation.

treatments may result in reduced productivity as it partially blocks the flow paths in near wellbore regions. For subsea wells, squeeze treatments can either be bullheaded from the processing facility, which are susceptible to plugging by particles and debris, and have the potential to collect flowline debris such as wax, scale, and corrosion products. Maintaining the integrity of the squeeze package over long distances of flowline is also an issue. Additionally, in the offshore environment, some hydrate inhibitors can exacerbate the severity of scaling (Kan and Tomson, 2012).

5.5 SUMMARY

Inorganic deposition in the formation can significantly reduce the well productivity, and proactive actions should be taken to prevent the scaling of a formation in the near wellbore region or within the production facilities. Even through the formation brine in the reservoir condition is stable, solids can precipitate when pressure drops or mixed with the injected fluids. If the deposition occurs in the deep reservoir or far away from the production/injection well, the matrix can end up with a reduced permeability, but its impact on production performance is not dramatic. When the precipitation occurs at the near wellbore region, the productivity can be significantly compromised, even to the cessation of production. Unless sulfate-reduced brine or low-salinity water can be injected, chemical injection through squeeze treatments is often the technique employed to reduce the risk of scaling. The mechanisms of chemical inhibitors are closed related to the scaling crystallization process from the supersaturation condition. Cautions should be taken when selecting scale inhibitors as they should be compatible with other fluids and not cause other flow assurance problems or cause induced formation damaged.

REFERENCES

Adair, P., Smith, P., 1994. The Miller Field: Appraisal and Development Formation Damage Experiences. SPE Formation Damage Control Symposium. Society of Petroleum Engineers, Lafayette, Louisiana, USA.

Bethke, C.M., 2007. Geochemical and Biogeochemical Reaction Modeling. Cambridge University Press.

Crabtree, M., Eslinger, D., Fletcher, P., Miller, M., Johnson, A., King, G., 1999. Fighting scale—removal and prevention. Oilfield Rev. 11, 30−45.

Frenier, W.W., Ziauddin, M., 2008. Formation, Removal, and Inhibition of Inorganic Scale in the Oilfield Environment. Society of Petroleum Engineers,, Richardson, TX.

Graham, A.L., Boak, L.S., Neville, A., Sorbie, K., 2005. How Minimum Inhibitor Concentration (MIC) and Sub-MIC Concentrations Affect Bulk Precipitation and Surface Scaling Rates. SPE International Symposium on Oilfield Chemistry. Society of Petroleum Engineers, The Woodlands, Texas.

Hajirezaie, S., Wu, X., Peters, C.A., 2017. Scale formation in porous media and its impact on reservoir performance during water flooding. J. Nat. Gas. Sci. Eng. 39, 188−202.

He, S., Kan, A., Tomson, M., Oddo, J., 1997. A New Interactive Software for Scale Prediction, Control, and Management. SPE Annual Technical Conference and Exhibition. Society of Petroleum Engineers, San Antonio, Texas.

Kahrwad, M., Sorbie, K.S., Boak, L.S., 2009. Coupled adsorption/precipitation of scale inhibitors: experimental results and modeling. SPE Prod Ope 24, 481−491.

Kan, A., Tomson, M., 2012. Scale prediction for oil and gas production. SPE J. 17, 362−378.

Langelier, W.F., 1936. The analytical control of anti-corrosion water treatment. J. Am. Water Works Assoc. 28, 1500−1521.

Li, Y.-H., Crane, S.D., Scott, E.M., Braden, J.C., McLelland, W.G., 1997. Waterflood geochemical modeling and Prudhoe bay zone 4 case study. SPE J. 2, 58−69.

Mackay, E., Matharu, A., Sorbie, K., Jordan, M., Tomlins, R., 1998. Modeling of scale inhibitor treatments in horizontal wells: application to the alba field. SPE Prod. Facil. 15, 107−114.

Mitchell, R., Grist, D., Boyle, M., 1980. Chemical treatments associated with North Sea projects. J. Pet. Technol. 32, 904−912.

Oddo, J., Tomson, M., 1994. Why scale forms in the oil field and methods to predict it. SPE Prod Facil 9, 47−54.

Paulo, J., Mackay, E., Menzies, N., Poynton, N., 2001. Implications of Brine Mixing in the Reservoir for Scale Management in the Alba Field. International Symposium on Oilfield Scale. Society of Petroleum Engineers, Aberdeen, United Kingdom.

Pitzer, K.S., 1973. Thermodynamics of electrolytes. I. Theoretical basis and general equations. J. Phys. Chem. 77, 268−277.

Plummer, L.N., Busenberg, E., 1982. The solubilities of calcite, aragonite and vaterite in CO_2-H_2O solutions between 0 and 90°C, and an evaluation of the aqueous model for the system $CaCO_3$-CO_2-H_2O. Geochim. Cosmochim. Acta. 46, 1011−1040.

Raju, K.U., Atkinson, G., 1988. Thermodynamics of "scale" mineral solubilities. 1. Barium sulfate (s) in water and aqueous sodium chloride. J. Chem. Eng. Data. 33, 490−495.

Raju, K.U., Atkinson, G., 1989. Thermodynamics of "scale" mineral solubilities. 2. Strontium sulfate (s) in aqueous sodium chloride. J. Chem. Eng. Data. 34, 361−364.

Raju, K.U., Atkinson, G., 1990. The thermodynamics of "scale" mineral solubilities. 3. Calcium sulfate in aqueous sodium chloride. J. Chem. Eng. Data. 35, 361−367.

Shaughnessy, C., Kline, W., 1983. EDTA removes formation damage at Prudhoe Bay. J. Pet. Technol. 35, 1,783−1,791.

Sorbie, K., Mackay, E., 2000. Mixing of injected, connate and aquifer brines in waterflooding and its relevance to oilfield scaling. J. Pet. Sci. Eng. 27, 85−106.

Sorbie, K.S., Gdanski, R.D., 2005. A Complete Theory of Scale Inhibitor Transport, Adsorption/Desorption and Precipitation in Squeeze Treatments. SPE International Symposium on Oilfield Scale. Society of Petroleum Engineers, Aberdeen, UK.

Tyler, T., Metzger, R., Twyford, L., 1985. Analysis and treatment of formation damage at Prudhoe Bay, Alaska. J. Pet. Technol. 37, 1,010−1,018.

Yuan, M.D., Todd, A., 1991. Prediction of sulfate scaling tendency in oilfield operations (includes associated papers 23469 and 23470). SPE Prod. Eng. 6, 63−73.

CHAPTER SIX

Formation Damage by Organic Deposition

Rouzbeh G. Moghanloo, Davud Davudov and Emmanuel Akita

The University of Oklahoma, Norman, OK, United States

Contents

Formation Damage during Improved Oil Recovery.
DOI: https://doi.org/10.1016/B978-0-12-813782-6.00006-3

© 2018 Elsevier Inc.
All rights reserved.

6.1 INTRODUCTION, DEFINITION, EXISTENCE STATE OF ASPHALTENE IN CRUDE OIL, MOLECULAR STRUCTURE OF ASPHALTENE, MONITORING, AND REMEDIATION

6.1.1 Introduction

The issue of asphaltene deposition has plagued the oil and gas industry for decades since J.B. Boussinggault first identified the problem in 1837 and referred to this group of compounds as "asphaltenes" (Creek, 2005). It is therefore extremely important to understand the issue of asphaltene deposition and the factors affecting it, due to the huge costs associated with remediation (Gonzales et al., 2016). Crude oil has several fractions, and asphaltenes essentially tend to be its heaviest, polarizable fractions. They are known as the "cholesterol of petroleum" (Boek et al., 2010) due to their ability to precipitate as solids and subsequently deposit with changing pressure, temperature, and oil composition. This may affect both surface properties and crude oil rheology. In both upstream and downstream operations asphaltene deposition may cause severe problems. Asphaltenes precipitate and deposit on the pipeline surfaces, the bottom of distillation columns and heat exchangers as well, affecting efficiency and creating added economic costs to remediate (Davudov et al., 2017). Deposition can lead to fouling and/or damage during oil production or refining. During crude oil production, they can deposit in reservoir rock pores, thus leading to possible blocking of flow, particularly in the near well bore region. The asphaltene problem is the most challenging one in petroleum production, processing and transportation facilities. Asphaltene precipitation refers to the process when asphaltenes become a separate phase from the crude oil. They remain suspended in the liquid phase where the quantity and the size of the asphaltenes are relatively small. The precipitated asphaltenes clump together (aggregation) and form larger particles, also called flocs. The asphaltene aggregates are initially suspended in the crude oil. Subsequently, the flocs may attach to and accumulate on various surfaces, a process referred to as asphaltene deposition. (Seifried, 2016; Yuan et al., 2016).

Nonetheless, the properties of asphaltenes still remain to be properly understood. In the literature review section, a discussion of asphaltenes is provided, detailing its definition, it's existence state in crude oil, molecular structure, and composition. Issues and remediations are also addressed together with its formation in respect of its precipitation, aggregation, and deposition. Current models are described and some relevant research is briefly expounded on. Finally, some field examples are provided as practical scenarios of asphaltene problems facing the industry.

6.1.2 Definition

Asphaltenes are known to be susceptible to changing pressure, temperature, and oil composition leading to their precipitation and subsequent deposition. J.B. Boussinggault, the man to first describe asphaltenes in 1837 defined asphaltenes as "alcohol insoluble, turpentine-soluble solid found in the residue of a distillation" (Creek, 2005; Buckley, 2012). Speight and Long (1996) define asphaltenes in terms of their solubility, thus, as being soluble in benzene or toluene, and insoluble in pentane or heptane (Speight and Long, 1996).

Crude oil has several components, namely, resins, saturates, and aromatics. Another component of crude oil, the malteneses consist of the portion in which the asphaltenes are dispersed (Seifried, 2016). Seifried (2016) states that the fraction of crude containing asphaltene typically varies from 0% to nearly 20%. It is composed mainly of carbon and hydrogen elements and contains about 80.42%−88.65% carbon, 7.44%∼11.10% hydrogen, plus 0.3%−4.9% oxygen, 0.3%−10.3% sulfur, and 0.6%−3.3% nitrogen (Koots and Speight, 1975; Speight, 1996). Speight explains that there isn't a good grasp of the nature of the asphaltene molecules. He however notes that spectroscopic studies give credence to the idea that asphaltene contains aromatic ring systems with elements such as nitrogen, oxygen, and sulfur, scattered throughout its molecular structure (Speight, 1996). Its molecular structure depends on temperature and pressure, making its molecular weight difficult to measure (Speight, 1996). That notwithstanding, Mullins et al. (2012) estimated it to be about 750 g/mol using mass spectral methods, with most of the population being 500 and 1000 g/mol. Yarranton et al. (2013) on the other hand approximated the value to 850 g/mol, using vapor pressure osmometry (VPO). Another technique, the small angle X-ray scattering (SAXS) method gives values 10 times larger, and Yarranton explained that the discrepancy between SAXS and VPO is yet to be solved. Speight stated that a technique that prevents aggregation will result in accurate asphaltene molecular weight (Speight, 1996; Speight, 1999).

6.1.3 Molecular structure of asphaltene

Asphaltenes exhibit complex structures which comprise one or more poly-aromatic units, sulfur, and nitrogen linked by alkyl chains (Gray et al., 2011). The "Island" and "Archipelago" structure are the two ways recognized as

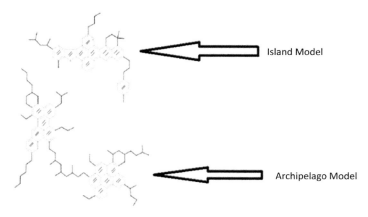

Figure 6.1 Asphaltene structure.

the variation for the chemical structure of asphaltene, as shown in the structure proposed in Fig. 6.1. Sabbah et al. (2011), states that the "Island" structure is the more dominant one, but ongoing debates still question the dominance. The "Island" structure has one single polycyclic aromatic hydrocarbons (PAH) core with cycloalkanes and external branched alkyl substituents, whereas the more recently proposed "Archipelago" structure comprises smaller condensed aromatic groups (Yarranton et al., 2013).

Pfeiffer and Saal (1940) described the phase separation of highly aromatic components with polar structures promoting dispersion. Mullins (2010, 2011) later described the self-formation of aromatic clusters from nanoaggregates by π—π-stacking of the aromatic rings in crude oil or toluene, referred to as the "Yen-Mullins-Model" in the literature. Yarranton et al. (2013) described the self-association concept by stating that asphaltenes in toluene comprise a mixture of associating and nonassociating species. Though techniques like VPO and elemental analysis, among others, it was concluded that approximately 90% of asphaltenes tend to self-associate with the nonassociated asphaltenes being smaller and more aromatic than the bulk asphaltenes (Seifried, 2016).

6.1.4 Monitoring and remediation

Remediation of asphaltene deposits involves using strong chemical solvents and exposing said deposits to turbulent flow. These chemicals have adverse effects on health, safety, and environment, along with production deferrals because of operation downtime. (Gonzales et al., 2016). A dynamic model provided by Dabir et al. (2016) enables the determination of the time at which there will be significant loss of productivity and permeability in the near wellbore region. This essentially means optimal time for remedial actions.

6.1.5 Experimental techniques to determine asphaltene-onset-pressure and wax-appearance-temperature

Fluid samples are obtained for asphaltene-onset-pressure (AOP) and wax-appearance-temperature (WAT) measurements. The samples are kept above reservoir pressure at all times to prevent reversibility of asphaltene precipitation, because they aggregate within an observed range of the bubble point value. When pressure drops from reservoir pressure during production, the density change causes a corresponding solubility change, which tends to trigger asphaltene precipitation, with the most precipitation happening close to bubble point pressure.

The AOP test is performed in a high pressure visual cell. The solid-deposition–system (SDS) technique includes a fixed wavelength laser-light source and a detector. The transmitted power of the near infrared light-scattering signal is recorded during the depressurization process. Before the onset, the signal transmitted is inversely proportional to the fluid density. During depressurization, the signal power increases. If particles appear, then the signal power decreases (Gonzales et al., 2016).

Asphaltene inhibitors (AI) are used to reduce the AOP and thus mitigate the asphaltene-deposition tendency. In situ Asphaltene tests are run, where state-of-the-art high-pressure/high temperature imaging technology is routinely used in the laboratory to detect asphaltene particles very clearly upon pressure depletion (Leontaritis et al., 2017).

6.2 ASPHALTENE FORMATION MECHANISMS REVIEW: PRECIPITATION, AGGREGATION AND DEPOSITION MECHANISM, SOLUBILITY

It is important to understand the differences between precipitation, aggregation and deposition in order to properly explain the behavior of asphaltene (Seifried, 2016). Experiments such as Electron Microscope, Electrical Effect, Ultracentrifuge, and Reversibility have been conducted to help determine existence states (Preckshot et al., 1943, Katz and Beu, 1945, Dykstra et al., 1944, Ray et al., 1957, Witherspoon and Munir,

1958, Hirschberg et al., 1982). This important concept provides better insight into its precipitation and deposition. The complex compatibility between maletenes and asphaltenes also needs to carefully considered, as they are intimately related to the stability of crude oil, and asphaltene onset, precipitation, and deposition (Bearsley et al., 2004).

Asphaltene precipitation refers to the process when asphaltenes become a separate phase from the crude oil. They remain suspended in the liquid phase where the quantity and the size of the asphaltenes are relatively small. Thereafter, the precipitated asphaltenes clump together (aggregation) and form larger particles, also called flocs. The asphaltene aggregates are initially suspended in the crude oil (Alian et al., 2011; Moghanloo et al., 2015). Furthermore, the process of deposition follows, the process whereby flocs attach to and accumulate on various surfaces.

6.2.1 Asphaltene precipitation

The factors which affect asphaltene precipitation are pressure, temperature, and crude oil composition

- Pressure:

    The influence of this factor comes into play at values above the bubble point. The Hassi Messaoud Algerian Field exemplifies this. In that field, asphaltene deposits were found on the production tubing because the wellhead pressure was near the crude oil saturation pressure, where the precipitation reaches as maximum (Seifried, 2016). If pressures remain above the bubble point, asphaltene solubility appears to decrease with decreasing pressure, reaching its minimum at the bubble point pressure. The solubility however increases with decreasing pressure below the bubble point Burke et al. (1990).

- Temperature:

    Temperature has been identified as a factor for precipitation, however, it's effects have not been well understood. The change in asphaltene solubility with a change in temperature depends on the reservoir temperature (Leontaritis, 1996), where the solubility increases with temperature as long as it remains below the reservoir temperature. When above the reservoir temperature however, the solubility decreases with an increase in temperature.

- Crude oil composition:

    Research shows that asphaltene solubility decreases as the amount of gas in solution increases Burke et al. (1990).

6.2.2 Solubility parameter

A (Hildebrand) solubility parameter is a parameter to estimate the situation where precipitation changes the solvent properties of the crude oil (Buckley, 1999). Defined as the square-root of the cohesive energy density (c) of the solvent, the Hildebrand solubility parameter describes the degree of solubility for nonpolar or slightly polar substances without hydrogen bonding. It also describes the amount of energy required to evaporate one-unit volume of a liquid (Hildebrand and Scott, 1964). Liquids with a similar solubility parameter are more likely to be miscible. This parameter is estimated from several techniques, namely, from dilution experiments, refractive index, and for supercritical fluids, by ideas based on the principles of thermodynamics.

6.2.3 Asphaltene precipitation models

The thermodynamic and the colloidal theory are the basis for the precipitation models widely described in literature. For the thermodynamic theory asphaltenes are considered as being part of a nonideal mixture which precipitate at solubility values below certain levels. Two main solubility theory approaches typically used are the regular solution theory and the equation of state (EoS) model. The colloidal theory on the other hand, describes asphaltenes as colloidal particles surrounded by adsorbed resins. The Flory–Huggins solution theory is the solubility model used to describe asphaltene precipitation. The chemical potential μ of the asphaltenes in the crude oil is given by Eq. (6.1) (Hirschberg et al., 1984; Andersen and Speight, 1999; Wang and Buckley, 2001):

$$\frac{\mu_a - \mu_a^{\theta}}{RT} = ln\varphi_a + \left(1 - \frac{V_{m,a}}{V_{m,s}}\right)\varphi_s + \frac{V_{m,a}}{RT}(\delta_a - \delta_s)^2\varphi_s^2, \qquad (6.1)$$

Flory–Huggins interaction parameter, χ

$$\chi = \frac{V_{m,a}}{RT}(\delta_a - \delta_s)^2 \qquad (6.2)$$

where μ = chemical potential; μ^{θ} = chemical potential at the reference state of pure asphaltene phase, φ = volume fraction; φ = solubility parameter; V_m = molar volume. The subscripts a and s denote asphaltenes and solvent containing all nonasphaltene components respectively.

6.2.3.1 Coloidal model

In this model, the surrounding adsorbed resins have lower molecular weight. The resins stabilize the asphaltenes by adsorbing on to their surface and their partitioning between the surface and surroundings determines the asphaltene solubility (Seifried, 2016).

6.2.3.2 Thermodynamic model

In this approach, resins are not taken into account and precipitation can be reversed. Reducing the asphaltene solubility can lead to phase separation. At thermodynamic equilibrium, each component's chemical potential value becomes equal in both phases (Wang and Buckley, 2001).

EoS model and the activity coefficients model (ACM) are used for the thermodynamic model approach. EoS models are used to relate pressure to temperature to predict phase behavior of hydrocarbons and asphaltenes. Whereas EoS assumes that all asphaltenes in the crude oil have the same size and molecular weight, ACM assume that precipitated asphaltenes do not affect the vapor liquid equilibrium (VLE), and are used to predict the activity coefficient γ in a mixture.

Furthermore, an example based on group-contribution is the predictive model UNIFAC-FV (Oishi and Prausnitz, 1978), which was specifically derived for polymer solvent mixtures. These models have limitations in the region of critical temperature and pressure. Tabibi et al. (2004) use the concept of perturbation theory to modify the Soave-Redlich-Kwong (SRK) EoS. This essentially was an asphaltene phase behavior predicting model, accurate for heavy oils.

6.2.4 Asphaltene deposition

Asphaltene deposition is the process whereby there is attachment of asphaltene aggregates onto a surface. Depending on the interaction between the surface and asphaltenes they may adsorb on the surface (Alian et al., 2011). Asphaltene deposition studies are mainly performed using microfluidics (more specifically capillary flow), Taylor-Couette (TC) cells or core flooding experiments through a porous media. Microfluidic studies provide researchers the chance to study phenomena such as colloidal dynamics at very low Reynolds numbers, where the small dimension of the capillary or the microfluidic device offers the suitable length scale (Saha and Mitra, 2012). Furthermore, the Taylor Couette device has the advantage of studying deposition at reservoir temperature and pressure conditions. Finally, core flood experiments use surface chemical techniques such as scanning electron microscope (SEM) with energy dispersive X-ray spectroscopy (SEM-EDS) and X-ray computed microtomography to study the crude oil volumes at the pore scale in order to quantify residual oil and wettability alteration (Seifried, 2016).

6.3 ISSUES WITH ASPHALTENE DEPOSITION

6.3.1 Asphaltene issues during oil production

Precipitation, aggregation, and deposition of asphaltenes are a huge problem in the field, especially with the injection of CO_2 as a means of enhanced oil recovery, since they lead to precipitation when crude oil is mixed with CO_2.

Crude oil exists in hydrocarbon reservoirs which are system consisting of reservoir rock and reservoir fluids. Porosity and permeability are the most important pore structure characteristics used in the discussion of hydrocarbon-reservoir-rock systems. For reservoir fluids, the compositions may vary from region to region, and typically overlay an aquifer, and are capped by a gas zone (gas cap). Due to multiphase flow behavior under varying pressure and temperature conditions, hydrocarbon systems are typically complex (Ahmed, 2006)

6.3.2 Formation damage and field experience

The precipitation of asphaltenes during all phases of crude oil recovery can lead to a decrease in production efficiency. Leontaritis et al. (1994) stated that generally four forms of formation damage induced by asphaltene deposition can be defined: (1) The physical blockage or permeability reduction, (2) wettability alteration, (3) a crude oil viscosity decrease, or (4) the formation of a water-in oil emulsion. Mechanism (1), however, seems to be the dominant one causing formation damage. The first step of asphaltene deposition is the adsorption of the asphaltene molecules on the surface of the rock, followed by hydrodynamic retention and/or trapping of the particles at the pore throat, eventually leading to a reduction of the effective hydrocarbon mobility (Nghiem et al., 1998; Al-Maamari and Buckley, 2003). Zekri et al. (2007) investigated how different flow rates in dynamic flow experiments affected formation damage due to asphaltene deposition. They stated that a decrease in the flow rate may reduce the formation damage for crude oils with a relatively high content of asphaltenes. According to Kokal and Sayegh (1995), some of the factors which influence adsorption of asphaltenes on the mineral surface are: (1) The morphology and chemistry of the mineral surface, (2) the amount of asphaltenes and resins present in the crude oil, (3) the composition and pH of the brines in the reservoir rock, and (4) pressure and temperature in the reservoir rock.

Leontaritis and Mansoori (1988) reviewed several experiences of the oil industry dealing with asphaltene issues during production in the Mata-Acema and Boscan fields, Venezuela. The reservoir in the Mata-Acema field is made out of sandstone where the producing fluid contains mainly C7 + fractions with an asphaltene content of 0.4%−9.8%. The asphaltene issues reported from this field were severe, whereas no asphaltene problems were reported from the Boscan field. The Boscan field is a sandstone and produces mainly heavy oils with an asphaltene content of 17.2%. Alian et al. (2011) and Behbahani et al. (2012) attributed the differences to the lower solubility of asphaltenes in lighter crude oils in the Mata-Acema field.

The asphaltene content from the stock tank oil produced from the Hassi Messaoud field sandstone reservoir (Algeria) was measured to be 0.062%. Despite the very low asphaltene content, deposition problems were reported from the start of the production onwards. Wells lost up to 25% of the wellhead pressure in only 2 weeks, causing significant loss in production. Furthermore, a solvent treatment and circulation of oil were used in the Ventura Avenue field (California) to avoid or reduce deposition, however, this did not lead to a great improvement in oil recovery.

6.3.3 Asphaltene deposition during CO_2 flooding

CO_2 can also induce asphaltene precipitation. Negahban et al. (2003) studied crude oil stability after the introduction of CO_2 and hydrocarbons. At reservoir conditions, the crude oil was understood to be stable in terms of asphaltene precipitation. However, whereas the addition of CO_2 did not show any such effect, the introduction of hydrocarbons seemed to have induced asphaltene precipitation. Gholoum et al. (2003) investigated the effects of C1−C7 and CO_2 on the asphaltene precipitation onset point on Kuwaiti reservoir fluids and stated that in their work, CO_2 is the most effective precipitant in terms of showing the lowest asphaltene onset pressure. The most important factors which determine asphaltene precipitation during CO_2 flooding are the CO_2 concentration and the pore topography (Srivastava and Huang, 1997).

It has been determined that in the Berea sandstone from west Virginia, deposition changes the rock wettability to oil wet, regardless of the initial rock wettability. Huang (1992) believed that, depending on the asphaltene content, the wettability of Berea sandstone would change from water wet to oil wet, resulting in lower crude oil recovery efficiency. Gonzalez et al. (2007) concluded that, depending on the range of pressure and temperature, CO_2 can either act as an inhibitor or enhancer for asphaltene precipitation.

It can be concluded that the main parameters determining the asphaltene precipitation and deposition are temperature, pressure, and oil composition, hence any change in those factors may affect the crude oil stability and trigger asphaltene precipitation. CO_2 injection may cause a change in oil composition (the solubility of asphaltenes in oil) and therefore acts as a precipitant for asphaltenes. That notwithstanding, there is still need for research on asphaltene deposition under dynamic conditions in porous media during CO_2 flooding and on modeling asphaltene solubility in CO_2 and oil mixtures (Seifried, 2016).

6.4 DEPOSITION IN POROUS MEDIA

As earlier stated, studies of asphaltene deposition are conducted by means of microfluidic experiments, TC cells or core flooding experiments through a porous media.

6.4.1 Microfluidic experiments

The use of microfluidic devices over the past 15 years has opened new possibilities in studying fluid mechanics and discovering new physical phenomena. As the dimension of the equipment shrinks, the bulk properties of the fluid become less important and interfacial properties and fluid–solid interactions dominate.

6.4.1.1 Asphaltene depostion in capillary flow

Seifried discusses Wang et al.'s (2004) asphaltene deposition study in a stainless steel capillary for two different crude oils. They reported a higher deposition rate with higher molecular weight precipitants (n-pentadecane) than with lower molecular weight components (n-heptane). Also, they didn't find any influence of the flow rate on the deposition rate. Wang et al. (2004) used a stainless steel capillary, which is a closer representation of process tubing, while Boek et al. (2008) investigated asphaltene deposition in a glass capillary, since a glass is a good representation of a sandstone reservoir rock. Boek et al. (2010) later studied the deposition in capillary flow of asphaltenes precipitated from the whole crude oil. They observed a deposition-erosion/entrainment cycle for higher flow rates, as well as higher pressure fluctuations when compared to experiments at lower flow rates. Broseta et al. (2000) investigated the onset point and deposition rate with respect to pressure, temperature, and varying fluid compositions. Broseta et al. (2000), Wang et al. (2004) and Lawal et al. (2012) put forward a Homogenous Deposition Model to determine the effective hydrodynamic thickness of a layer of deposited asphaltenes, assuming a homogeneous layer of deposits. The experimental pressure drop data were used to estimate the deposit layer of the asphaltenes by the change of the inner diameter of the capillary. The model is based on the assumption of a homogeneous deposition with a uniform thickness of the asphaltene deposit layer (Seifried, 2016).

6.4.2 Taylor Couette device studies

By varying the flow rate, the shear rate can be changed and the deposit mass growth can be measured over time (Buckley, 2012). Rahmani et al. (2003) studied asphaltene floc size distributions in a Couette device for an Athabasca bitumen (Alberta, Canada). This distribution was studied with respect to the contribution of varying shear rate to aggregation in Couette Flow. According to Rahmani et al. (2003), the frequency of collision and sticking probability control particle aggregation rate. They state that, on the other hand, fragmentation is not yet well understood and for that purpose they established a population balance.

6.4.3 (Imaging) Core flood experiments

X-ray computed tomography (CT) is a technique used to observe single and multiphase fluid flow in porous medium and to characterize the reservoir rock (heterogeneity or homogeneity, deposits and fractures). Knackstedt et al. (2011) imaged and studied the pore scale crude oil brine populations to quantify the residual oil phase and to estimate wettability alteration. For that purpose, several imaging techniques were applied, such as surface chemical techniques (SEM and SEM-EDS) and X-ray microCT. Using microCT, 3D images were obtained with a voxel size down to about 1.5 mm resolution (Seifried, 2016).

6.4.4 Porous media studies

Asphaltene deposition problems encountered deep down in rock reservoirs are extremely problematic, and very challenging to tackle, as opposed to production tubing deposition problems, which is typically the focus of many asphaltene studies. Minssieux (1997) studied various core samples with different rock characteristics in core-flooding experiments, with regards to porous media. He concluded that porous sample plugging only seemed to occur after enough oil had flowed through the sample, and that damage at earlier times was only observed in samples with a lower initial permeability (Seifried, 2016).

Behbahani et al. (2013) developed a new model based on multilayer adsorption kinetics and Pak et al. (2011) studied asphaltene deposition in a core holder containing a sandstone, in which asphaltene deposition was assessed through pressure drop measurements across the vessel. The amount of deposited asphaltenes within the core was evaluated through the difference of asphaltene content in the inlet and outlet stream. This indirect method was applied to study asphaltene deposition caused by three different processes: Recycled gas injection (most damage), CO_2 injection and natural depletion (least damage) (Pak et al., 2011). They have also observed that porosity reduction occurred mostly in the core inlet. A combined theoretical and experimental study was carried out by Mendoza De La Cruz et al. (2009), where the main mechanisms identified as significantly causing permeability reduction were: (1) The asphaltene deposits are significantly smaller than the size of pores and throats, leading to a gradual reduction of pore throat radii and (2) the deposition of aggregates are larger than the size of a given throat, hence leading to a reduction in the fluid flow area. However, this model fails to describe the deposition process under dynamic conditions because it doesn't account for entrainment of particles.

The mechanisms through which asphaltene damage can occur are surface deposition, entrainment, and pore throat plugging. As asphaltene deposits accumulate on the pore surface, the pore throat for the oil phase decreases, with some deposits being swept away and entrained by the liquid phase when increasing shear rate overcomes the critical value. When asphaltene deposits accumulate in front of a pore, they can plug them (Seifried, 2016).

Wang modified Civan's model for near wellbore asphaltene deposition, assuming negligible capillary pressure and one dimensional horizontal flow. As shown in Eq. (6.3) below, the overall deposition process is modeled by momentum and mass balance, with concepts from precipitation, deposition models, and porosity and permeability reduction equations (Wang, 2000, pp 53—58). The first term is the rate of asphaltene deposition on the pore surface, with coefficient α. The second is the entrainment of asphaltene deposits in the fluid, with the last term being the pore throat plugging rate. Rate of net deposition is therefore:

$$\frac{\partial E_A}{\partial t} = \alpha C_A \phi - \beta E_A \left(v_L - v_{\text{cr},L} \right) + \gamma u_L C_A \qquad (6.3)$$

where, E_A is volume fraction of asphaltenes deposition; v_L is interstitial velocity $(= u_L/\phi)$; $v_{\text{cr},L}$ is critical interstitial velocity; u_L is superficial velocity; α is surface deposition coefficient; β is entrainment rate coefficient; γ is pore throat plugging coefficient.

According to Boek et al. (2008) and Wang (2000), the surface deposition rate is directly proportional to the concentration of suspended particles concentration in the flowing fluid. For the second term it's been identified that entrainment becomes dominant once the critical interstitial velocity is exceeded. The magnitude of the difference between the average and critical velocities influences the transitional location of the deposited particles (Dabir et al., 2016). The value of the entrainment coefficient is set as β, when $v_L > v_{cr,L}$ and otherwise to zero. Finally, the last term describes the plugging contribution to the net deposition. Here, asphaltene precipitates in suspension, where the plugging rate is directly proportional to the superficial velocity. Wang (2000) defined the pore plugging coefficient, γ:

$$\gamma = \gamma_i (1 + \sigma E_a), \text{ if } 0 > R > R_c$$
$$\gamma = 0, \text{ otherwise} \qquad (6.4)$$

where, σ is the deposition constant; R refers to the ratio of particle size to pore throat size, R_c refers to the critical ratio of particle size to pore throat size. Basically, according to Eq. (6.4) pore throat plugging occurs at conditions where the critical pore throat diameter is greater than the average pore throat diameter.

Surface deposition, entrainment, and pore throat plugging are the three mechanisms contributing to asphaltene damage in porous media. There is however a postulated forth parameter, the Deep Bed Filtration (DBF) model by Wang and Civan (2001) and Wang (2000). DBF refers to the retention process of particles which occurs when the particulate suspension flow through a porous medium causes a separation between the solid phase and the liquid phase of the suspension. This is based on several mechanisms: Contact between the particles and the retention site, fixing of particle sites and finally the break off of formerly retained particles (Fallah et al., 2012).

6.5 PERMEABILITY DAMAGE MODELS

Several theoretical and empirical models have been developed to estimate permeability as a function of porosity (Pape et al., 2000; Civan, 2001; Zheng et al., 2016). It is well known that effective porosity can decrease owing to pore volume shrinkage and leading to permeability reduction. However, permeability can be also altered because of hydraulic conductivity/connectivity loss (coordination number reduction) owing to the pore plugging mechanism. In the extreme case where significant pore blockage occurs, the porosity may not even significantly decrease. As it will be discussed later in detail, when the rock has a large fraction of pores with the diameter comparable to the size of particles, pore throats can be easily plugged and blocked; this will lead to severe permeability reduction even when the large fraction of pore space yet remains intact. Thus, it is crucial to study asphaltene deposition in porous media via new permeability models which consider both porosity reduction and pore connectivity loss, especially for reservoirs with small size pores that are comparable to the particle size.

One of the fundamental permeability models is the Kozeny—Carmen (KC) equation that considers porous medium as a bundle of cylindrical tubes. The rock permeability is then related to porosity (φ), tortuosity (τ), and hydraulic pore radius (r_h) as follows:

$$k = \frac{r_h^2}{c}\frac{\phi}{\tau},\qquad(6.5)$$

When effective hydraulic pore radius, r_h is substituted by the surface area per unit of grain volume (S_{gv}), Eq. (6.5) can be rewritten as:

$$k = \frac{1}{c}\frac{\phi}{\tau}\frac{1}{S_{gv}}\left(\frac{\phi}{1-\phi}\right)^2,\qquad(6.6)$$

6.5.1 Permeability reduction: effect of surface deposition and pore plugging

Asphaltene deposition onto a rock surface may reduce porosity and permeability. Wang and Civan (2001) proposed that the instantaneous local porosity due to deposition can be calculated from the difference between the initial porosity, ϕ_i, and the fraction of asphaltene deposits, ε:

$$\phi = \phi_i - \varepsilon\qquad(6.7)$$

Further assuming that $1 - \phi \approx 1$ and considering constant tortuosity, Eq. (6.6) can be modified to estimate permeability change as a function of porosity (Civan, 2001; Kord et al., 2012):

$$\frac{k}{k_i} = \left(\frac{\phi}{\phi_i}\right)^3,\qquad(6.8)$$

However, connectivity loss (pore connectivity) has not been considered in Eq. (6.8) and permeability reduction is only attributed to pore volume reduction.

6.5.2 Plugging and nonplugging parallel pathways model

Gruesbeck and Collins (1982) proposed a permeability model based on porous media consisting of plugging and nonplugging parallel pathways. Pores with relatively large diameter are considered "non-plugging" pathways where permeability reduction is only occurring due to surface deposition of particles and thus pore volume shrinkage. The flow paths that are tortuous and have diameter relatively close to particle size are considered "plugging pathways." In the plugging pathways, pore throats will be blocked through retainment of particles that will potentially lead to severe conductivity and permeability reduction even when a large fraction of nonplugging pathways still remain intact. Note that the distinction between plugging and nonplugging pathways is subjective depending upon the size of particles relative to the average size of a pore.

Based on empirical correlations, the permeability of the plugging pathways can be written as follows (Civan, 1994; Gruesbeck and Collins, 1982):

$$\frac{k_p}{k_{pi}} = \exp\left(-\alpha\left(\phi_{p_i} - \phi_p\right)^{n_p}\right), \tag{6.9}$$

For nonplugging pores, the permeability ratio is expressed as:

$$\frac{k_{np}}{k_{npi}} = \left(\frac{\phi_{np}}{\phi_{np_i}}\right)^{n_{np}}, \tag{6.10}$$

where α, n_p, and n_{np} are constant fitting parameters.

Thus, total permeability reduction in porous media can be expressed as:

$$\frac{k}{k_i} = f_p \exp\left(-\alpha\left(\phi_{p_i} - \phi_p\right)^{n_p}\right) + f_{np}\left(\frac{\phi_{np}}{\phi_{np_i}}\right)^{n_{np}}, \tag{6.11}$$

where f_p and f_{np} represent the fractions of plugging and nonplugging initial permeability values, respectively, and where their sum is one. Note that Eq. (6.10) is identical to Eq. (6.8) when $n_{np} = 3$. In addition, examination of Eq. (6.9) reveals that permeability reduction follows an exponential trend due to pore throat blockage. Fig. 6.2 illustrates permeability reduction as a function of volumetric fraction of total deposited amount per unit pore volume, ε, using Eqs. (6.9) and (6.10). A significant permeability reduction can rapidly occur (steep reduction) when plugging pathways are considerable in a sample (Eq. (6.9)); on the contrary, Eq. (6.10) predicts much smoother permeability reduction when deposition mainly occurs along nonplugging pathways and the pore plugging mechanism is negligible. Thus, Eq. (6.8), which is usually used to model permeability reduction as a function of asphaltene deposition, should be modified to account for deposition onto plugging pathways/pore throats. In practice, when the ratio of asphaltene particle size to pore throat size is larger than a critical value, pore blockage will have a significant effect and Eq. (6.8) or Eq.(6.10) will not be sufficient to estimate permeability reduction.

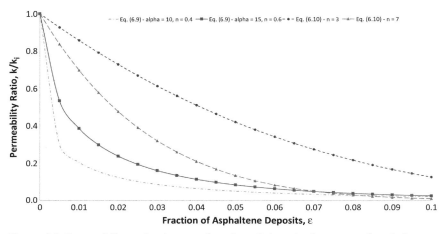

Figure 6.2 Permeability reduction as a function of deposited amount of asphaltene.

6.5.3 Power—law permeability reduction

Civan (2001) suggested that KC equation cannot properly address the gate or valve effect to predict permeability when pore throats are blocked and isolated pores are realized in the system. Therefore. he modified the equation by including an interconnectivity parameter, Γ:

$$k = \Gamma\phi\left(\frac{\phi}{1-\phi}\right)^{2\beta}, \tag{6.12}$$

Γ is a measure of the pore space connectivity, and it represents the valve effect for the pore throats controlling the pore connectivity to other pore spaces in an interconnected network (Civan, 2011).

Based on the permeability model described in Eq. (6.12), the combined effects of both surface deposition and pore plugging on permeability reduction can be analyzed. If β is considered to be unity, then Eq. (6.12) can be written as:

$$\underbrace{\frac{k}{k_i}}_{\text{Total Permeability Reduction}} = \underbrace{\frac{\frac{\phi^3}{(1-\phi)^2}}{\frac{\phi_i^3}{(1-\phi_i)^2}}}_{\text{Surface Deposition}} * \underbrace{\frac{\Gamma}{\Gamma_i}}_{\text{Pore Plugging}} \tag{6.13}$$

where first and second terms on the right side of Eq. (6.13) indicate permeability reduction due to surface deposition and pore plugging, respectively. If pore blockage is insignificant, then Eq. (6.13) will be reduced to:

$$\frac{k}{k_i} = \frac{\frac{\phi^3}{(1-\phi)^2}}{\frac{\phi_i^3}{(1-\phi_i)^2}} \tag{6.14}$$

Note that Eq. (6.14) is identical to both Eqs. (6.8) and (6.10) as it relates permeability reduction in porous media only due to pore surface deposition (pore volume reduction). In the case where pore plugging is not negligible, then Eq. (6.13) can be rewritten to estimate connectivity loss as follows:

$$\frac{\Gamma}{\Gamma_i} = \frac{k}{k_i} \Big/ \frac{\frac{\phi^3}{(1-\phi)^2}}{\frac{\phi_i^3}{(1-\phi_i)^2}} \tag{6.15}$$

6.5.4 Case studies: evaluation of surface deposition and pore plugging effects

To further illustrate the impact of surface deposition and pore plugging effects on permeability reduction, several experimental data sets obtained from limestone, sandstone and carbonate (dolomite) core samples are evaluated here. Using the experimental data sets, both initial porosity as well as the damaged/reduced porosity are calculated by subtracting total volume of deposited asphaltene from initial pore volume of the core sample. As expected, the porosity reduction and permeability damage are a function pore volume injected (PVI); the larger injected volume, the more reduction in both permeability and porosity is realized. A simple explanation is that more particle deposition is experienced when the system is more extensively exposed to particulate flow under pro-deposition conditions.

Next, permeability reduction as a function of total injected pore volume is calculated for the same experimental data sets. Thus, permeability reduction due to pore blockage and surface deposition can be evaluated using Eq. (6.13). Since porosity reduction as a function of injected pore volume is known (from the previous step), permeability reduction due to surface deposition (along nonplugging pathways) can be estimated using Eq. (6.14) and the ratio between predicted and actual permeability values

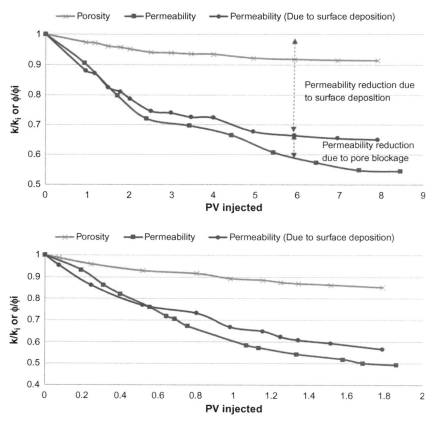

Figure 6.3 Impact of surface deposition and pore blockage on permeability reduction for limestone rock samples. *Experimental data is obtained from Behbahani et al. 2013 and Mousavi Dehghani et al., 2007.*

measured in the data set (Eq. (6.15)) can be attributed to the pore connectivity loss.

Fig. 6.3 illustrates porosity and permeability ratios obtained from experimental data and the predicted permeability ratios solely due to surface deposition (Eq. (6.14)) for limestone samples. As observed in both samples, permeability has been reduced by 50% and predicted values (Eq. (6.14)) and measured data are consistent.

Connectivity loss for the same samples can be calculated based on Eq. (6.15) as illustrated in Fig. 6.4. Results show that connectivity loss for both limestone samples are in the range of 12%−15%, from which it can be easily concluded that for limestone samples, major damage mechanism is surface deposition and pore blockage effect is small.

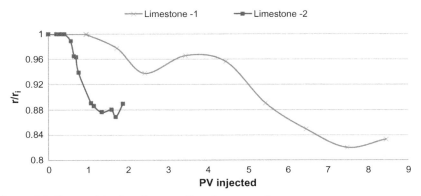

Figure 6.4 Connectivity loss for pure limestone samples.

On the other hand, when the experimental data obtained from sandstone samples are evaluated, results indicate that total permeability reduction is in the range of 30%—45% (Fig. 6.5), where connectivity loss can be up to 25% (Fig. 6.6). Thus, it is can be concluded that for the sandstone samples used in this study, the surface deposition mechanism and pore blockage mechanism have comparable contributions on the permeability reduction. Moreover, for one of the sandstone samples (sample #1), connectivity loss is recovered after initial decline (Fig. 6.6). Increased injection pressure (possibly to maintain a constant injection rate) yields sufficient drag force to remove previously deposited particles from the pore surface (Kord et al., 2012).

When experimental data related to carbonate (dolomite) samples are examined, our results show that the permeability reduction in carbonate samples (mainly dolomite) is sharper than pure limestone and sandstone rock samples, where porosity damage was small. Thus, it can be inferred that the plugging mechanism is very significant in the carbonate core samples we studied here (Fig. 6.7). Based on Eq. (6.15), the contribution of the pore throat plugging mechanism on permeability reduction is in the range of 70%—90% (Fig. 6.8). For these low permeability formations, pore size diameter is relatively small, and thus pore plugging becomes dominant. The pore plugging mechanism leads to steeper decline in permeability leaving the large fraction of pore space intact.

The results of this chapter are consistent with Nasri and Dabri (2014) study as well, where they have used network modeling to analyze asphaltene deposition in carbonates, and they have concluded that the main reason of absolute permeability reduction in carbonate samples is plugging

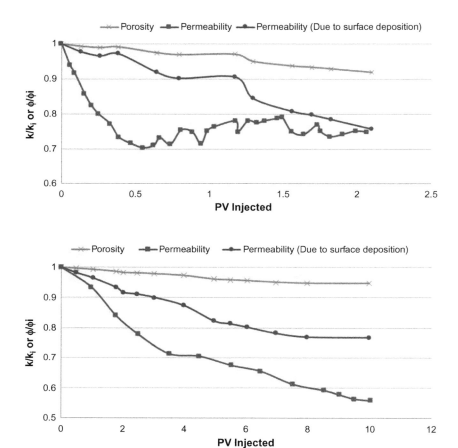

Figure 6.5 Impact of surface deposition and pore blockage on permeability reduction for sandstone rock samples. *Experimental data is obtained from Behbahani et al., 2013 and Mousavi Dehghani et al., 2007.*

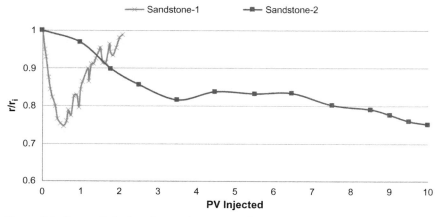

Figure 6.6 Connectivity loss for sandstone samples.

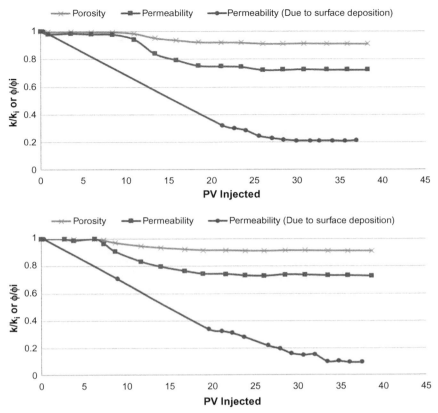

Figure 6.7 Impact of surface deposition and pore blockage on permeability reduction for carbonate rock samples. *Experimental data is obtained from Shedid, 2001.*

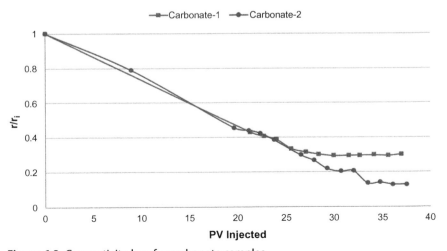

Figure 6.8 Connectivity loss for carbonate samples.

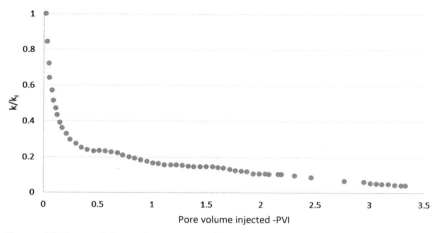

Figure 6.9 Permeability reduction as a function of PV injected for carbonate rock samples. *Experimental data is obtained from Kord et al., 2012.*

of the pore throats. They have shown a comparison between the throat size distribution before and after asphaltene deposition, where the number of throats was reduced and shifted to smaller sizes. Moreover, Shen and Sheng (2017) have reported that asphaltene deposition may have a tremendous impact on permeability in Eagle Ford shale samples as well, where pore size distribution is in micro-scale. Based on their results 83% of the total permeability reduction was due to pore plugging and 17 % reduction was due to adsorption mechanism.

6.5.4.1 Particle to pore size ratio

Experimental data adopted from Kord et al. (2012) illustrate permeability reduction as a function of PVI (Fig. 6.9). As can be seen, even after 0.5 pore volume injection, permeability reduction is close to 80%. Kord et al. (2012) suggested that this first sharp decline in permeability at an early time is due to surface deposition. However, if pore size and asphaltene particle size distributions (Fig. 6.10) are compared, it can be observed that particle to pore size ratio is close to one, which indicates that this significant reduction is because of pore blockage which is common for carbonate rock samples, consistent with our previous results.

Thus, it can be concluded that the relative difference between the size of deposited asphaltene particles and the pore size of a reservoir rock is a suitable measure and represents a parameter which is dominant in controlling asphaltene deposition in porous media. As illustrated in Fig. 6.11, in Case I the particle size is larger than the deposited particle size, thus

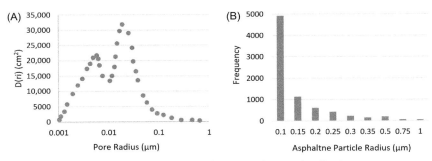

Figure 6.10 (A) Pore size distribution and (B) particle size distribution.

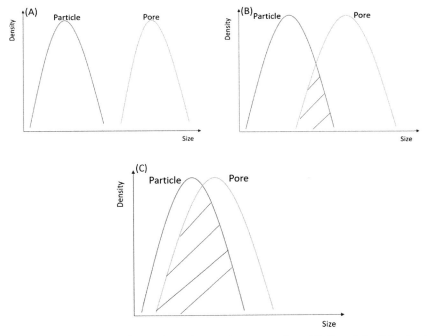

Figure 6.11 Schematic of particle and pore size comparison (A) Case I: Particle size is larger than deposited particle size, thus surface deposition is the dominant factor on permeability reduction; (B) Case II: Pore throat size and particle size are comparable close; both surface deposition and pore throat plugging contributes to permeability reduction and (C) Case III: Permeability reduction is steeper and sharper which is mainly because of pore throat plugging.

surface deposition is the dominant factor on permeability reduction; pore throat plugging barely contributes to permeability loss. This typically happens in limestone. Case II represents transition phase; at this phase pore throat size and particle size are comparable close; both surface deposition

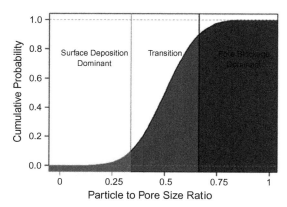

Figure 6.12 Generic cumulative density distribution for particle to pore size ratio.

and pore throat plugging contribute to permeability reduction. This typically happens in sandstone. Finally, Case III illustrates the phase at which permeability reduction is steeper and sharper which is mainly due to pore throat plugging. This typically happens in carbonate.

All three cases can be summarized as shown in Fig. 6.12, the cumulative distribution function (CDF) of particle to pore size ratio. The first region is the region dominated by surface deposition (Case I). As the particle size increases, one encounters the transition region (Case II). Further increase in particle size and/or reduction in pore throat size results in the pore throat plugging becoming the dominant controlling parameter on permeability reduction (Case III). There are no effective remedial actions for the pore throat dominant region. The only preventive method is not allowing precipitation to occur at the first place (Gonzales et al., 2016).

6.5.5 Effect of flow rate—particle entrainment

To evaluate the impact of fluid velocity on permeability reduction, several experimental data sets for sandstone and carbonate samples obtained from the literature are evaluated. As shown in Fig. 6.13, results show that an increase of flow rate has a significant effect on sandstone samples; the higher the flow rate, the less permeability reduction is realized. This can be explained through the entrainment of particles: Larger flow rates involve greater drag forces and remove previously deposited particles from the pore surface.

On the other hand, an increase in flow rate has no effect on permeability reduction in carbonate samples; the main explanation for

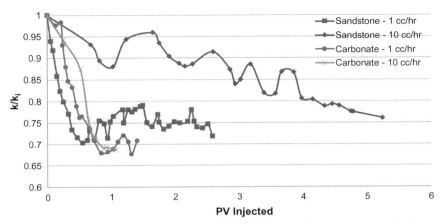

Figure 6.13 Impact of flow rate on permeability reduction for sandstone and carbonate rock samples. *Experimental data obtained from Behbahani et al., 2013.*

insensitivity of carbonate samples with change in flow rates has to do with the pore blockage mechanism that is the governing mechanism and particle entrainment has no effect on it.

6.6 CONCLUSIONS

In this chapter, asphaltene deposition in the near wellbore region is evaluated. The permeability reduction model was modified based on the historical porosity dependence and the added interconnectivity loss due to pore blockage. The major contributions are:

- In terms of permeability reduction, the dominance of the controlling mechanism of surface deposition or pore blockage is subject to the asphaltene particle to rock pore size ratio.
- Based on the evaluation of published experimental data, permeability reduction in limestone samples was observed to be due to surface deposition, where the effect of pore blockage is negligible. This is because of the large ratio of pore throat to the particle size.
- In sandstone samples, both the surface deposition and pore throat plugging contribute fairly equally to permeability reduction.
- In dolomite samples, the pore blockage is dominant. This results in an almost instantaneous sharp decrease in sample permeability.

REFERENCES

Ahmed, T., 2006. Reservoir Engineering Handbook. Gulf Professional Publishing, Houston, US.

Al-Maamari, R.S.H., Buckley, J.S., 2003. Asphaltene precipitation and alteration of wetting: the potential for wettability changes during oil production. SPE Reservoir Eval. Eng. 6 (4), 210–214.

Alian, S.S., Omar, A.A., Altáee, A.F., Hani, I., 2011. Study of asphaltene precipitation induced formation damage during CO_2 injection for a Malaysian light oil. J World Academy of Science, Engineering and Technology 78.

Andersen, S.I., Speight, J.G., 1999. Thermodynamic models for asphaltene solubility and precipitation. J. Pet. Sci. Eng. 22 (1), 53–66.

Bearsley, S., Forbes, A., Haverkamp, R.G., 2004. Direct observation of the asphaltene structure in paving-grade bitumen using confocal laser-scanning microscopy. J. Microsc. 215 (2), 149–155.

Behbahani, J., Ghotbi, T.C., Taghikhani, V., Shahrabadi, A., 2012. Investigation on asphaltene deposition mechanisms during CO_2 flooding processes in porous media: a novel experimental study and a modified model based on multilayer theory for asphaltene adsorption. Energy Fuels 26 (8), 5080–5091.

Behbahani, T.J., Ghotbi, C., Taghikhani, V., Shahrabadi, A., 2013. Experimental study and mathematical modeling of asphaltene deposition mechanism in core samples. Oil & Gas Science and Technology–Revue d'IFP Energies nouvelles 70, 1051–1074.

Boek, E.S., Ladva, H.K., Crawshaw, J.P., Padding, J.T., 2008. Deposition of Colloidal asphaltene in capillary flow: experiments and mesoscopic simulation. Energy Fuels 22 (2), 805–813.

Boek, E.S., Wilson, A.D., Padding, J.T., Headen, T.F., Crawshaw, J.P., 2010. Multiscale simulation and experimental studies of asphaltene aggregation and deposition in capillary flow. Energy Fuels 24 (4), 2361–2368.

Broseta, D., Robin, M., Savvidis, T., Féjean, C., Durandeau, M., Zhou, H., 2000. Detection of asphaltene deposition by capillary flow measurements. SPE/DOE Improved Oil Recovery Symposium. Society of Petroleum Engineers.

Buckley, J.S., 1999. Predicting the onset of asphaltene precipitation from refractive index measurements. Energy Fuels 13 (2), 328–332.

Buckley, J.S., 2012. Asphaltene deposition. Energy Fuels 26 (7), 4086–4090.

Burke, N.E., Hobbs, R.E., Kashou, S.F., 1990. Measurement and modeling of asphaltene precipitation (includes associated paper 23831). J. Pet. Technol 42 (11), 1–440.

Civan, F., 1994. Predictability of formation damage: an assessment study and generalized models. Final Report, U.S. DOE Contract No. DE-AC22-90-BC14658, April.

Civan, F., 2001. Scale effect on porosity and permeability: kinetics, model and correlation. AIChE J. 47 (2), 271–287.

Civan, F., 2011. Porous Media Transport Phenomena. Wiley, p. 15.

Creek, J.L., 2005. Freedom of action in the state of asphaltenes: escape from conventional wisdom. Energy Fuels 19 (4), 1212–1224.

Dabir, S., Dabir, B., Moghanloo, R.G., 2016. A new approach to study deposition of heavy organic compounds in porous media. Fuel 185, 273–280.

Davudov, D., Moghanloo, R.G., Flom, J., 2017. Scaling analysis and its implicationfor asphaltene deposition in a Wellbore. Soc. Petrol. Eng. J. Available from: https://doi.org/10.2118/187950-PA.

Dykstra, H., Beu, K., Katz, D.L., 1944. Precipitation of asphalt from crude oil by flow through silica. Oil Gas J. 30, 79–104.

Fallah, H., Ahmadi, A., Karaee, M.A., Rabbani, H., 2012. External cake build up at surface of porous medium. Open. J. Fluid. Dyn. 2 (4), 145.

Gholoum, E.F., Oskui, G.P., Salman, M., 2003. Investigation of asphaltene precipitation onset conditions for Kuwaiti reservoirs. Middle East Oil Show. Society of Petroleum Engineers.

Gonzales, D., Gonzales, F., Pietrobon, M., Haghshenas, M., Shurn, M., Mees, A., et al., 2016. Strategies to Monitor and Mitigate Asphaltene Issues in the Production System of Gulf of Mexico Deepwater Subsea Development SPE, BP Offshore Technology Conference, Houston, 2-5 May, 2016. J. Pet. Technol., 92—94.

Gonzalez, D.L., Vargas, F.M., Hirasaki, G.J., Chapman, W.G., 2007. Modeling study of CO_2 induced asphaltene precipitation. Energy Fuels 22 (2), 757—762.

Gray, M.R., Tykwinski, R.R., Stryker, J.M., Tan, X., 2011. Supramolecular assembly model for aggregation of petroleum asphaltenes. Energy Fuels 25 (7), 3125—3134.

Gruesbeck, C., Collins, R.E., 1982. Entrainment and deposition of fine particles in porous media. Soc. Petrol. Eng. J. 22 (6), 847—856.

Hildebrand, J.H., Scott, R.L., 1964. The Solubility of Nonelectrolytes. Dover Publication, New York, NY.

Hirschberg, A., De Jong, L.N.J., Schipper, B.A., Meyers, J.G., 1982. Influence of Temperature and Pressure on Asphaltene Flocculation, paper SPE 11202 presented at the SPE 57th Annual Technical Conference and Exhibition held in New Orleans, LA, September 26-29.

Hirschberg, A., DeJong, L.N.J., Schipper, B.A., Meijer, J.G., 1984. Influence of temperature and pressure on asphaltene flocculation. Soc. Petrol. Eng. J. 24 (3), 283—293.

Huang, E.T.S., 1992. The effect of oil composition and asphaltene content on CO2 displacement, SPE 24131. In: Eighth Symposium on Enhanced Oil Recovery of the SPE/DUE held in Tulsa, OK, April.

Katz, D.L., Beu, K.E., 1945. Nature of asphaltic substances. Ind. Eng. Chem. 37 (2), 195—200.

Knacktedt, M.A., Pinczewski, W.V., Fogden, A., Senden, T. (2011). Improved characterization of EOR process in 3D. Characterizing mineralogy wettability and residual fluid phases at the pore scale. In: SPE Enhanced Oil Recovery Conference. Society of Petroleum Engineers.

Kokal, S.L., Sayegh, S.G., 1995. Asphaltenes: the cholesterol of petroleum. Middle East Oil Show. Society of Petroleum Engineers.

Koots, J.A., Speight, J.G., 1975. Relation of petroleum resins to asphaltenes. Fuel 54, 79—184.

Kord, S., Miri, R., Ayatollahi, S., Escrochi, M., 2012. Asphaltene deposition in carbonate rocks: experimental investigation and numerical simulation. Energy Fuels 26 (10), 6186—6199.

Lawal, K.A., Crawshaw, J.P., Boek, E.S., Vesovic, V., 2012. Experimental investigation of asphaltene deposition in capillary flow. Energy & Fuels 26 (4), 2145—2153.

Leontaritis, K.J., 1996. The asphaltene and wax deposition envelopes. Fuel Sci. Technol. Int. 14 (1), 13—39.

Leontaritis, K.J., Mansoori, G.A., 1988. Asphaltene deposition: a survey of field experiences and research approaches. J. Pet. Sci. Eng. 1 (3), 229—239.

Leontaritis, K.J., Amaefule, J.O., Charles, R.E., 1994. A systematic approach for the prevention and treatment of formation damage caused by asphaltene deposition. SPE Prod. Facil. 9 (3), 157—164.

Leontaritis, K.J., Geroulis, E., Wilson, A., 2017. Laboratory Testing and Prediction of Asphaltene Depostion in Production Wells. SPE, Asphwax, Offshore Technology Conference, Houston, 1—4 May, 2016. J. Pet. Technol. pp. 95.

Mendoza De La Cruz, J.L., Argüelles-Vivas, F.J., Matías-Pérez, V., Durán-Valencia, C., López Ramírez, S., 2009. Asphaltene-induced precipitation and deposition during pressure depletion on a porous medium: an experimental investigation and modeling approach. Energy Fuels 23 (11), 5611−5625.

Minssieux, L., 1997. Core damage from crude asphaltene deposition. International Symposium on Oilfield Chemistry. Society of Petroleum Engineer.

Moghanloo, R.G., Yuan, B., Ingrahama, N., Krampf, E., Arrowooda, J., Dadmohammadi, Y., 2015. Applying macroscopic material balance to evaluate dynamic drainage volume and performance prediction of Shale oil/gas wells. J. Nat. Gas Sci. Eng 27 (Part 2), 466−478.

Mousavi Dehghani, S.A., Vafaie Sefti, M., Mirzayi, B., Fasih, M., 2007. Experimental investigation on asphaltene deposition in porous media during miscible gas injection. Iranian Journal of Chemistry and Chemical Engineering (IJCCE) 26 (4), 39−48.

Mullins, O.C., 2010. The modified Yen model. Energy Fuels 24 (4), 2179−2207.

Mullins, O.C., 2011. The asphaltenes. Annu. Rev. Anal. Chem. 4, 393−418.

Mullins, O.C., Sabbah, H., Eyssautier, J., Pomerantz, A.E., Barré, L., Andrews, A.B., et al., 2012. Advances in asphaltene science and the Yen Mullins model. Energy Fuels 26 (7), 3986−4003.

Nasri, Z., Dabir, B., 2014. Network modeling of asphaltene deposition during two-phase flow in carbonate. J. Petrol. Sci. Eng. 116, 124−135.

Negahban, S., Joshi, N., Jamaluddin, A.K.M., Nighswander, J., 2003. A systematic approach for experimental study of asphaltene deposition for an Abu Dhabi reservoir under WAG development plan. International Symposium on Oilfield Chemistry. Society of Petroleum Engineers.

Nghiem, L.X., Coombe, D.A., Ali, S.M., 1998. Compositional simulation of asphaltene deposition and plugging. SPE Annual Technical Conference and Exhibition. Society of Petroleum Engineers.

Oishi, T., Prausnitz, J.M., 1978. Estimation of solvent activities in polymer solutions using a group-contribution method. Ind. Eng. Chem. Process Des. Dev. 17 (3), 333−339.

Pak, T., Kharrat, R., Bagheri, M., Khalili, M., Hematfar, V., 2011. Experimental study of asphaltene deposition during different production mechanisms. Pet. Sci. Technol. 29 (17), 1853−1863.

Pape, H., Clauser, C., Iffland, J., 2000. Variation of permeability with porosity in sandstone diagenesis interpreted with a fractal pore space model. Pure Appl. Geophys. 157, 603−619.

Pfeiffer, J.P., Saal, R.N.J., 1940. Asphaltic bitumen as colloid system. J. Phys. Chem. 44 (2), 139−149.

Preckshot, G.W., DeLisle, N.G., Cottrell, C.E., Katz, D.L., 1943. Asphaltic substances in crude oils. Trans. AIME 151 (01), 188−205.

Rahmani, N.H.G., Masliyah, J.H., Dabros, T., 2003. Characterization of asphaltenes aggregation and fragmentation in a shear field. AIChE J. 49 (7), 1645−1655.

Ray, B.R., Witherspoon, P.A., Grim, R.E., 1957. A study of the colloidal characteristics of petroleum using the ultracentrifuge. J. Phys. Chem. 61, 1296−1302.

Sabbah, H., Morrow, A.L., Pomerantz, A.E., Zare, R.N., 2011. Evidence for island structures as the dominant architecture of asphaltenes. Energy Fuels 25 (4), 1597−1604.

Saha, A.A., Mitra, S.K., 2012. Microfluidics and Nanofluidics Handbook: Chemistry, Physics, and Life Science Principles.

Seifried, C.M., 2016. Asphaltene Precipitation and Depostion from Crude Oil with CO2 and Hydrocarbons: Experimental Investigation and Numerical Simulation. Imperial College London, Department of Chemical Engineering. Sept. 2016.

Shedid, S.A., 2001. Influences of asphaltene deposition on rock/fluid properties of low permeability carbonate reservoirs. SPE Middle East Oil Show. Society of Petroleum Engineers.

Shen, Z., Sheng, J.J., 2017. Investigation of asphaltene deposition mechanisms during CO_2 huff-n-puff injection in Eagle Ford shale. Petrol. Sci. Technol. 35 (20), 1960–1966. Available from: https://doi.org/10.1080/10916466.2017.1374403.

Speight, J.G., 1996. Asphaltenes in crude oil and bitumen: structure and dispersion. In: Schramm, L.L. (Ed.), Suspensions: Fundamentals and Applications in the Petroleum Industry. American Chemical Society, Washington, DC, pp. 377–401.

Speight, J.G., Long, R.B., 1996. The concept of asphaltenes revisited. Fuel Sci. Technol. Int. 14 (1&2), 1–12.

Speight, J.G., 1999. The chemical and physical structure of petroleum: effects on recovery operations. J. Petrol. Sci. Eng. 22, 3–15.

Srivastava, R.K., Huang, S.S., 1997. Asphaltene deposition during CO2 flooding: laboratory assessment. SPE Production Operations Symposium. Society of Petroleum Engineers.

Tabibi, M., Nikookar, M., Ganbarnezhad, R., Pazuki, G.R., Hosienbeigi, H.R., 2004. Phase behavior assessment of deposition compound (Asphaltene) in heavy oil. Soc. Pet. Eng . Available from: https://doi.org/10.2118/88753-MS.

Wang, J.X., Buckley, J.S., 2001. A two-component solubility model of the onset of asphaltene flocculation in crude oils. Energy Fuels 15 (5), 1004–1012.

Wang, J., Buckley, J.S., Creek, J.L., 2004. Asphaltene deposition on metallic surfaces. J. Dispersion Sci. Technol. 25 (3), 287–298.

Wang, S., 2000. Simulation of asphaltene deposition in petroleum reservoirs during primary oil recovery. PhD dissertation, University of Oklahoma, Norman, Oklahoma,

Wang, S., Civan, F., 2001. Productivity decline of vertical and horizontal wells by asphaltene deposition in petroleum reservoirs. SPE International Symposium on Oilfield Chemistry. Society of Petroleum Engineers.

Witherspoon, P.A., Munir, Z.A., 1958. "Sine and Shape of Asphaltic Particles in Petroleum", paper SPE 1168-G presented at the Fall Meeting of the Los Angeles Basin Section held in Los Angeles, October 16-17.

Yarranton, H.W., Ortiz, D.P., Barrera, D.M., Baydak, E.N., Barre, L., Frot, D., et al., 2013. On the size distribution of self-associated asphaltenes. Energy Fuels 27 (9), 5083–5106.

Yuan, B., Moghanloo, R.G., Zheng, Da., 2016. Analytical evaluation of nanoparticles application to reduce fines migration in porous media. SPE J. 21 (06), 2317–2332.

Zekri, A.Y., Shedid, S.A., Almehaideb, R.A., 2007. An experimental investigation of interactions between supercritical CO_2, asphaltenic crude oil, and reservoir brine in carbonate cores. International Symposium on Oilfield Chemistry. Society of Petroleum Engineers.

Zheng, D., Yuan, B., Moghanloo, R.G., et al., 2016. Analytical modeling dynamic drainage volume for transient flow towards multi-stage fractured wells in composite shale reservoirs. J. Pet. Sci. Eng. 149, 756–764.

FURTHER READING

Civan, F., 1995. Modeling and simulation of formation damage by organic deposition. Presented at the First International Symposium in Colloid Chemistry in Oil Production: Asphaltenes and Wax Deposition, ISCOP'95, Rio de Janeiro- RJ- Brazil, November 26-29, pp. 102–107.

Civan, F., Wang, S., 2005. Modelling formation damage by asphaltene deposition during primary oil recovery. ASME, J. Energy Resour. Technol. 127, 310–317.

Leontaritis, K.J., Mansoori, G.A., 1987. Asphaltene flocculation during oil production and processing: a thermodynamic colloidal model. In SPE International Symposium on Oilfield Chemistry. Society of Petroleum Engineers.

CHAPTER SEVEN

Formation Damage During Chemical Flooding

Caili Dai
China University of Petroleum (East China), Qindao, China

Contents

7.1 INTRODUCTION

Tertiary oil recovery aims at exploiting more crude oil, which cannot be recovered during the secondary oil recovery, including remaining oil and residual oil (Herbeck et al., 1976). Remaining oil refers to the crude oil, which cannot be swept due to the formation heterogeneity. While residual oil refers to the crude oil residue in the reservoir pores, which cannot be displaced even after water flooding. Oil recovery can be enhanced through two aspects. One aspect is to increase the sweep efficiency of the injection fluids in the oil reservoir. The commonly used

Formation Damage during Improved Oil Recovery.
DOI: https://doi.org/10.1016/B978-0-12-813782-6.00007-5
© 2018 Elsevier Inc.
All rights reserved.
275

method is to reduce the effects of oil reservoir heterogeneity through improving the mobility-control of the displacing phase, which can be mainly achieved by raising the viscosity of displacing phase to reduce the remaining oil. Another aspect is to increase the oil displacement efficiency, which is achieved by changing the wettability of the rock surface and reducing the oil–water interfacial tension (IFT), thus to decrease the residual oil saturation (Wagner and Leach, 1959). The technologies employed in tertiary oil recovery are divided into four types generally, including chemical flooding, thermal recovery, gas flooding, and microbial flooding. Among them, chemical flooding mainly refers to the method to enhance oil recovery by injecting chemical agents, which can increase sweep efficiency or displacement efficiency, including polymer flooding, surfactant flooding, alkaline flooding, foam flooding, and combination flooding (Nelson and Pope, 1978). Combination flooding refers to the binary combination flooding (surfactant/polymer flooding, alkaline/polymer flooding) and ternary combination flooding (alkaline/surfactant/polymer flooding, etc.). Nowadays polymer flooding, surfactant/polymer binary combination flooding and alkaline/surfactant/polymer ternary combination flooding have been frequently applied in oilfields (Deng et al., 2002). All these flooding technologies play an important role in enhancing the recovery in oil and gas fields, but at the same time can generate certain damage in the reservoir formation. In this chapter, the various damage mechanisms generated by the three aforementioned oil displacement methods and their impacts are introduced and evaluated.

7.2 FORMATION DAMAGE BY POLYMER FLOODING

Polymer flooding refers to the oil displacement method using polymer solutions as the oil displacing agents. During the application of polymer flooding, the formation damage is generated mainly due to the following reasons (Vela et al., 1976).

- The adsorption and retention of polymers in the porous medium;
- The incompatibility of polymer solutions with formation fluids and minerals;
- Polymers facilitate the migration of loose sand particles.

7.2.1 The retention of polymers in the porous medium

The retention of polymers in the porous medium is likely to occur for many reasons. Appropriate retention of polymers can reduce the water-phase permeability, which is beneficial to decrease the mobility ratio of displacing fluids to water, and then enhance the sweep efficiency. However, the excessive retention of polymers is likely to cause formation damage to the porous medium, as results, decreasing permeability and affecting oil displacement efficiency. The retention of polymers in the formation porous medium can be divided into three types: Adsorption retention, mechanical trapping, and hydraulic retention (Cools et al., 1994).

7.2.1.1 Adsorption retention

Adsorption retention refers to the phenomenon that polymer molecules accumulate on the surface of rock pore structure by physical interactions, such as Van der Waals force and hydrogen bonding with the hydroxyl groups present at the rock surfaces. This type of adsorption is usually called as physical adsorption, but not chemical adsorption, as no chemical bonds are formed during the adsorption process. For well dispersed polymers, adsorption retention is the main mechanism of retention of polymer. When evaluating the polymer flooding in Dos Cuadras offshore oilfield in 1990, Dovan et al. found that the blockage caused by the mutual entanglement of polymers after adsorption retention is the main reason for the difficulties in the further injection of polymers after long-term injection (Dovan et al., 1990). In 2001, Tang et al. studied the adsorption behavior of polymers on the sandstone reservoir in the Daqing oilfield (China) (Tang et al., 2001). The results indicated that adsorption retention occurs easily when polymers flow in the sandstone pores. The adsorption capacity of polymers on clay minerals is usually larger than that for the rock matrix. In 2011, Guo et al. studied the aggregation morphology of the polymer after injection into the formation using a high-magnification scanning electron microscope (Guo, 2011). It can be observed that after polymer flooding, the polyacrylamide (PAM) is mainly retained in the formation by means of flocculent aggregation and adsorption onto the sandstone surfaces.

7.2.1.1.1 Mechanism of polymer adsorption retention

- Electrostatic interaction

     Partially hydrolyzed polyacrylamide (HPAM) is a commonly used chemical agent in polymer flooding. HPAM is the hydrolysis product of PAM after the treatment of alkali. After dissolved in water, the sodium carboxylate in HPAM can ionize to generate negatively charged $-COO^-$ group, which is in ionization balance with $-COONa$. Electric charges exist not only on the polymer surfaces, but also on the mineral surfaces. For clay minerals with layer structures, they are comprised by silica tetrahedron and alumina octahedron as a certain alternating order, with interactions including Van der Waals force, hydrogen bond and electrostatic attractions between adjacent crystal layers. Before polymer injection, ions with similar radii, but different charges, would replace each other, rendering the minerals mostly negatively charged. For example, a negative charge is generated after the replacement of Si^{4+} by Al^{3+}. This kind of charge is caused by a material's internal structure, which is independent of the solution's concentration. Besides, the hydroxylated ions generated from the rupture of end face Si-O and Al-O bonds can render the side face (or end face) of mineral particles negatively charged. This kind of end-face charge is related to the solution concentration or pH. Under acidized conditions, the end face of a clay mineral chip is positively charged but the surface is negatively charged. However, under alkaline conditions, both the end face and the surface of clay mineral chips are negatively charged. Therefore, the charge distribution of clay mineral particles is uneven. For clay particles, both the end face and internal face are likely to bear charges at the point of zero-charge, and thus, they possess the unique property of zero-point adsorption. For other minerals, due to the dissociation of surface hydroxyl groups in different pH conditions, the mineral surfaces would bear charges (usually they would be positive charges under low pH condition). While HPAM would bear negative charges when dissolved in water. The electrostatic attraction between positive and negative charges is the main reason for the electrostatic adsorption of polymers on the mineral surface (Fuoss and Strauss, 1948).

- Hydrogen bonding

Besides the adsorption driven by electrostatic attraction, there is also another type of adsorption caused by the hydrogen bonding. Under the long-term water invasion conditions, hydroxylation reactions would occur on the surfaces of reservoir rocks to generate hydroxyl groups. The hydroxyl groups on the rock surfaces can bond with the carboxylate radical and the amide groups through the

hydrogen bonds. The large amounts of groups, which can form hydrogen bonds on the polymer molecular chains, facilitate the adsorption retention of polymers (Owens and Wendt, 1969).
* Dispersion force

   Dispersion force exists among all molecules or atoms. For electrons, in the moment of high-speed rotation, the central position of the electron does not overlap with the center-of-gravity of the mass of the molecule. The momentary mismatching of centers of gravity between positive charge and negative charge leads to the generation of a transient dipole, by which two molecules can be attracted together. The force generated from the transient dipole is called dispersion force (Yang et al., 2004). Dispersion force is one kind of Van der Waals force. The higher the molecular weight, the larger the dispersion force. As the molecular weight of polymer applied in polymer flooding is usually quite high, ranging between 10 million to 40 million g/mol, the dispersion force is a very important function mechanism in polymer adsorption.

7.2.1.1.2 Factors influencing adsorption retention

The adsorption retention of polymers onto rock grains is related to many factors. The first is the intrinsic properties of the polymers, such as type, molecular weight, degree of hydrolysis, and concentration (Shaw and Stright, 1977). The second kind of factor is the properties of water used for preparing the polymer solution, such as the degree of mineralization, ion type, etc. The third kind of factor is the in situ reservoir condition, such as the mineral composition, surface property of rock particles, and reservoir temperature. In general, the higher degree of mineralization of the water solution is beneficial to the polymer adsorption. The surfaces of carbonate rock (usually positively charged) are more prone to adsorb polymer molecules than those of sandstone. Elevated temperature is not favorable to the polymer adsorption, as the adsorption capacity is raised with increasing polymer concentration.
* Effect of molecular weight, degree of hydrolysis, and polymer concentration

   The molecular weight and degree of hydrolysis have relatively significant effects on the polymer adsorption. When the molecular weight and degree of hydrolysis of HPAM increases, the volume of each molecule coil expands as the density decreases. Thus, the adsorptions amount and retention amount of HPAM in the various porous

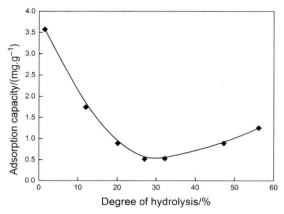

Figure 7.1 Effect of degree of hydrolysis on the adsorption capacity. *Zhang X., Zhao F. and Liu H., Study on static adsorption of polymer, Contemporary Chem. Industry 46 (3), 2017, 396–399.*

media would be reduced. When polydispersity is low, the polymer coil density is mainly determined by the numbers of carboxyl groups in the molecular chains. Thus, the adsorption loss of polymer is mainly determined by the degree of hydrolysis, followed by the molecular weight. In 2017, by studying the static adsorption rule of polymers in Xing 13block of the Daqing oilfield, Zhang et al. found that the adsorption of polymers on natural sandstone cores decreases with the increase of the degree of hydrolysis, and reaches the lowest adsorption capacity when the degree of hydrolysis is 28%, as shown in Fig. 7.1. However, when the degree of hydrolysis is further increased, the adsorption capacity of a polymer is gradually enhanced (Zhang et al., 2017). As shown in Fig. 7.2, with the increase of polymer concentration and molecular weight, the static adsorption of polymer increase gradually (Li et al., 2016). This can be attributed to the intermolecular entanglement of polymer chains facilitated by the further increase in polymer concentration. As results, the reduced amounts of adsorption sites result in the decrease of adsorption capacity at higher polymer concentration.

- Effect of electrolyte

After the dissolution of HPAM into water, the repulsion among anions leads to the expansion of the molecular coil and the mutual entanglement of molecular chains. However, the existence of monovalent and divalent cations can restrain the dissociation of carboxyl, generating an

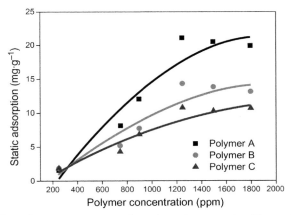

Figure 7.2 Effect of concentration on the adsorption capacity. The molecular weight (M_w, g/mol) of three polymers are 25×10^6 (A), 20×10^6 (B), and 15×10^6 (C). *Li Q., Pu W., Wei B., Jin F. and Li K., Static adsorption and dynamic retention of an anti salinity polymer in low permeability sandstone core, J. Appl. Polym. Sci. 134 (8), 2016. Copyright © 2016 Wiley Periodicals, Inc.*

ion screening effect to reduce the repulsion in molecular chains. Thus the volume of the molecule coil shrinks, and the mutual entanglement of polymer chains reduces, as a result, the polymer adsorption capacity is enhanced (Dovan et al., 1990). When multivalent metal salts exist, the adsorption capacity increases with the increasing ion dissociation degree and the valence state of the cations. Ion exchange occurs between the sodium ions dissociated from the polymer molecule and the divalent ions in clay mineral. The divalent ions compress the polymer coil structures, resulting in an enhanced polymer adsorption capacity. In 2010, Dai et al. studied the effects of various factors on the adsorption capacity of polymers through laboratory tests (Dai et al., 2010). As shown in Fig. 7.3, the adsorption capacity of polymers is enhanced by the increased degree of mineralization of water (though relatively slowly), which can be attributed to the following reasons.

- Effect of intermolecular reactions

    With the increase of mineralization degree of water, the intermolecular reactions among water molecules are strengthened, reducing the solubility of polymers which is controlled by the formation of hydrogen bonds with water, and thus, leading to the enhanced adsorption capacity of the polymer.

- Effect of the cations concentration

    With the increase of mineralization degree of water, the concentration of cations in the solution increases, reducing the electrostatic charges

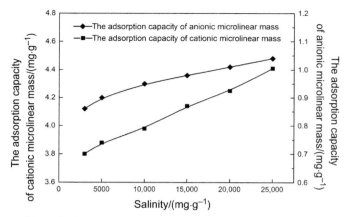

Figure 7.3 Effect of salinity on the adsorption capacity of polymers. *Dai C., Zhao J., Jiang H., Wang X., Lv X., Zhao G., et al., Alternative injection of anionic and cationic polymers for deep profile control in low-permeability sandbody reservoir, Acta Petrolei Sinica 31 (3), 2010, 440–444.*

on the surfaces of the formation and HPAM molecules. The electrostatic repulsions between the insitu rocks and HPAM molecules are weakened, resulting in the increased adsorption capacity of HPAM onto the rock grain surfaces. On the other hand, the increased number of cations in the solution would inhibit the dissociation of carboxyl groups in HPAM, resulting in the decrease of effective diameter of anionic polymer molecular coils and the increase of polymer adsorption capacity.

- Effect of wettability

    The adsorption of HPAM onto the in situ rocks is largely dependent on the wettability of the rock surfaces in oil reservoirs. The adsorption of HPAM onto water-wet rocks is much larger than that onto the oil-wet rocks. As HPAM is commonly water-soluble, HPAM has greater affinity interactions with water than oil.

7.2.1.2 Mechanical trapping

Mechanical trapping is a kind of filtration of the porous medium to the polymers. When a polymer passes through the porous medium, its hydrodynamic radius may surpass the pore throat diameter of the porous medium. Thus, parts of the polymers are trapped in the porous medium, reducing the core permeability and porosity. At the core entrances, the mechanical trapping of polymers performs the most serious, while the trapping phenomenon alleviates with the increasing distance to the core entrances (Garti and Zour, 1997). For polymers with ultrahigh molecular weight, mechanical trapping is more serious. The micro-gel formation near the polymer flooding injection well is caused by the mechanical trapping of large amounts of polymers in the near-wellbore area.

7.2.1.3 Hydraulic retention

The retention of polymer caused by the change of flow direction or velocity increase is called hydraulic retention. While the small pores or particle angles are blocked by polymers due to the mechanical trapping, the flowing streamline would be forced to change, resulting in the further retention of polymers. The detained polymers would not be flushed out of the porous medium due to the increased velocity; instead the retention phenomenon would be aggravated. Hydraulic retention would be more serious while the molecular weight of the polymer increases and the core permeability decreases. Hydraulic retention can result in the significant increase of the retention amounts of polymers inside the cores, and also the arise of the injection pressure (Bhardwaj et al., 2007). The hydraulic retention process of polymers is usually reversible. When the injection velocity or pressure is reduced, the molecular conformation of polymer will change, as a result, the polymer can be released from the retention sites back to the flowing stream of water, meanwhile generating an "additional residual effect". If this effect can be applied in subsequent water injection, the mobility control performance can be further improved. In 2004, Zhou proved that the reservoir is susceptible to velocity-dependent damage caused by the polymer solution through lab experiments, and calculated the distance-in-effect around the polymer injection well at which velocity-dependent damage would occur (Zhou, 2004; Yuan and Moghanloo, 2017). In 2009, Fu et al. found that the polymer adsorption and retention capacity increases with the injection rate, and that the polymer adsorption and retention capacity in low-permeability cores is much more significant than that in high-permeability cores (Fu and Xiong, 2009). As shown in Fig. 7.4, for the cores with higher permeability, with the increase of injection rates, parts of the retained molecules are prone to be separated from their original retentions sites back to the flowing stream. As results, the retention of polymer decreases. However, for the core with low permeability, with the increase of injection rates, the polymer molecules can enter the smaller pore-throats by tensile deformation, and hence resulting in the more serious problems of retention of polymer in cores.

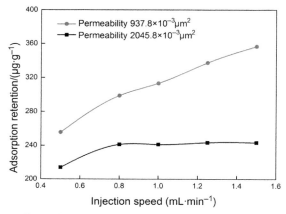

Figure 7.4 The effect of injection speed (mL/min) on the adsorption retention for samples with different permeability. *Fu M. and Xiong F., Study on the plugging mechanism of polymer flooding in Henan oilfield, Drilling Production Technol. 32 (4), 2009, 77—79.*

7.2.2 Incompatibility of polymer solution with formation

7.2.2.1 Incompatibility of polymer solution with formation fluids

Formation fluids mainly consist of oil, gas and water, among which formation oil and water are the main fluids that would lead to the blockage of polymer due to their incompatibility with polymer solution during the polymer flooding. In 2004, Zheng et al. studied the core wafer in the polymer injection well using X-ray diffraction imaging technology (Zheng et al., 2004). It was found that polymers formed micelle-like plugs with an inner colloidal nucleus, and over 50% of the matter constituting the colloidal nucleus is inorganic.

7.2.2.1.1 Incompatibility of polymer solution with formation water

- The clay particles as well as inorganic salt precipitates of calcium, magnesium, and iron, which enter the oil reservoirs together with the polymer solution, would block the reservoirs by themselves. It is considered that inorganic particles play major contribution to the blockage of polymers (Yan et al., 2000; Yuan et al., 2018). In the solution, the cations of inorganic salts with stronger electrophilicity will replace water molecules with priority, and form ion pairs with the carboxyl groups on the PAM molecular chains, which can screen the negative charges on the polymer chains to certain extent. In this way the electrostatic repulsion among molecules is reduced, resulting in the conformation change of molecules from an extended state to a coiled state. Therefore the effective volume of molecule is reduced, resulting in the contraction of molecular coils and the generation of precipitation.

The ferric ions with a high valent state (Fe^{3+}) in the formation tend to have cross-linking reactions with polymers to form hydrogels that block the formation (Fletcher et al., 1992). Taking into account experiments using ferric sulfide, as an example, when the concentration of Fe^{3+} is lower than 1 mg/L, polymer precipitates may be produced; when the concentration of Fe^{3+} is larger than 1 mg/L, apparent plugging would be generated, resulting in an increase of the injection pressure. This is attributed to that Fe^{3+} as one of the nucleating materials for inducing polymer molecules to form hydrogels, can accelerate the formation rate and yield of polymer gel.

- The injection of polymer solutions into the formation will also lead to a temperature drop inside formation. When ions required in generating precipitates are contained in the formation water, inorganic precipitate such as $BaSO_4$ can be generated that block the formation.
- To prevent the oxidative degradation of polymers during polymer flooding, deoxidants such as Na_2SO_3 and $NaHSO_3$ are usually added. However, these deoxidants can react with some ions in the formation water to generate inorganic precipitates after entering the formation. For example, when Na_2SO_3 is applied as deoxidant, it can generate SO_4^{2-} after removing oxygen. If there is Ba^{2+} in the formation water, $BaSO_4$ precipitate can be produced.

7.2.2.1.2 Incompatibility of polymer solution with crude oil

When a polymer solution is injected into the formation, the formation temperature will be decreased. When the crude oil temperature reduces below the solidifying point of paraffin, paraffin will precipitate out from the crude oil and deposit in the pore channel, resulting in the generation of organic deposition and consequently the plugging of formation.

7.2.2.1.3 Incompatibility of polymer solution with formation rocks

The adsorption of PAM onto the oil film onto the surfaces of oil-wet rocks or sand particles is almost zero, however, its adsorption is extremely large in super-hydrophilic (completely water-wet) conditions. However, the wettability of the rock is usually heterogeneous, which means that the smaller pores and gelation sites may be water-wet, but the inner surfaces of larges pores may be oil-wet, which leads to the nonuniform distribution of the injected polymers adsorption. The adsorption of polymers in the smaller pores or gelation sites will generate resistance to the flowing capacity of fluids. In 2002, Lu et al. studied the relationship between the pore-throat diameter of the formation and the radius of gyration of polymer coils (Xiang et al., 2002). It was found that the relative molecular weight of polymers applicable to the Daqing oilfield Pu I_{1-4} layer should be no larger than 8.2×10^6 g/mol. However, if the relative molecular weight of polymers applied is larger than the practical requirement, as a result, formation damage would be induced.

7.3 FORMATION DAMAGE BY SURFACTANT/POLYMER BINARY COMBINATION FLOODING

The surfactant/polymer binary combination flooding is a kind of polymer flooding with low IFT by adding surfactant into the polymer solution. The binary combination method can exert the synergistic effect of both surfactant and polymer to enhance oil recovery. When the concentration of surfactant is higher than the critical micelle concentration (or even higher), micro-emulsion/polymer miscible flooding can be formed, leading to much higher oil-displacement efficiency.

In the surfactant/polymer binary combination flooding, the addition of surfactant can effectively reduce the IFT and increase the displacement efficiency. Under certain conditions, surfactants can form chelation structures with polymers, which is also beneficial to increase the viscosity of the compound system. The emulsification effect of surfactant can also increase the viscosity of the displaced phase. The addition of polymers can effectively increase the viscosity of the displacing phase and improve the mobility control performance, and hence increase the swept efficiency of the surfactant washing out the residual oil. In addition, polymers can also act as sacrificial agents to react with the divalent ions in the formation to reduce the consumption of surfactants and enhance the stability of the emulsions. Besides, the combination of surfactant and polymer bears advantages attributed to their synergistic effect. During the injection of a binary combination flooding slug, with the presence of polymers, the surfactants will not form precipitates or be deactivated due to the effect of metal ions. Besides, the ultralow IFT generated by the surfactant also guarantees that the viscoelasticity of polymers can be fully realized. (Cui et al., 2011)

Surfactants which can be applied in oil displacement generally should meet the following requirements. First, the surfactants should have wide raw materials sources and be cost-effective and easily soluble, and display excellent compatibility with the formation fluids; and have good stability after injection into the formation at small dosages; and help achieve enough low IFTs; and bear the tolerance to certain temperature and salt concentration; and have low adsorption retention onto the reservoir rock surface. Nowaday, the commonly used surfactants in tertiary oil recovery include anionic surfactant, nonionic surfactant, nonionic—anionic surfactant, etc., among which the mostly widely applied are anionic surfactants such as petroleum sulfonate, carboxylate and lignosulfonate and OP type and Tween type nonionic surfactants.

Besides, the damage generated by polymers, the formation damage generated by surfactant/polymer flooding during the tertiary chemical-combination oil recovery can also include the precipitation of surfactants, emulsification adsorption and phase separation.

7.3.1 Precipitation of the surfactants

For anionic surfactants, when in contact with the cationic medium in the formation, incompatibility reactions are likely to occur, which will not only deprive the surfactants of the capability to reduce the IFT, and also result in the loss of surfactants, and form precipitates that damage the formation. (Stellner and Scamehorn, 1986) For example, petroleum sulfonate can react with the multivalent cations in both formation water and clays to form sulfonate precipitates (such as calcium sulfonate and magnesium sulfonate), which will not only cause the loss of surfactants, but also block pore-throat system, and hence bring damage on the performance of the compound oil displacement system. During oil displacement, once precipitation occurs, on one hand, the precipitates would block pores leading to the decrease of system permeability, and thus degrade the displacement performance. On the other hand, for clay minerals, the water sensitivity of clays is significantly improved due to the replacement of Ca^{2+} by Na^+. In 1978, Somasundaran et al. studied the precipitation effect of sodium dodecyl benzene sulfonate (SDBS) in a $CaCl_2$ solution (Somasundaran and Hanna, 1977). In 1985, Yang et al. studied the precipitation loss of petroleum sulfonate in the oil displacement process in Yumen oilfield, and discussed the impact of ionic strength, degree of mineralization, temperature and pH on the precipitation loss of petroleum sulfonate (Ananthapadmanabhan and Somasundaran, 1985). The results indicated that appropriate amount of monovalent cations in the formation water could reduce the precipitation loss of petroleum sulfonate.

In this work, three distinct factors affecting the precipitation generation of surfactants are identified as follows.

- Effect of degree of mineralization on the precipitation generation

    The degree of mineralization is one of the most important factors affecting the precipitation of surfactants (Trujillo, 1983). Generally speaking, the contents of Ca^{2+} and Mg^{2+} would increase with the increase of the mineralization degree of the formation water. Then, the precipitation–dissolution equilibrium of sulfonate would move toward the trend of precipitation generation, resulting in the increase of the precipitations of calcium sulfonate and magnesium sulfonate, and thus the loss of petroleum sulfonate (Martin et al., 1985). Therefore, for oil reservoirs with high-salinity formation water, especially that with higher divalent ions (Ca^{2+} and Mg^{2+}), formation pre-flushing should be preferably conducted before the introduction of surfactant/polymer flooding.

- Effect of temperature on the precipitation generation

    The formation temperature also affects the precipitation loss of surfactants (Trujillo, 1983). Generally, the precipitation amount and the loss of petroleum sulfonate decreases with an increase of temperature (Amante et al., 1991). The reason why the enhanced temperature leads to the reduced precipitation loss of petroleum sulfonate is that the solubility of calcium (magnesium) sulfonate increases with the system temperature, thus the precipitation amount will be reduced, which is beneficial to its oil displacement work.

- Effect of pH on the precipitation loss

    The pH value can also affect the precipitation loss of the surfactants (Fig. 7.5) (Liao et al., 2004). In general, the precipitation loss of petroleum sulfonate decreases with the pH increase. This is attributed to that when the pH value of the system increases, the concentration of OH^- in the solution would increase, as a result, more amounts of $Mg(OH)_2$ can be generated. As a consequence, the concentrations of free Ca^{2+} and Mg^{2+} are reduced, resulting in the mitigation of precipitation loss of petroleum sulfonate (Martin et al., 1985). Therefore, the oil displacement performance can be improved by properly increasing the pH value of surfactant/polymer flooding system.

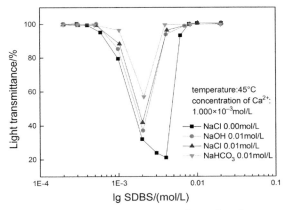

Figure 7.5 The effect of pH on the precipitation loss of surfactant sodium dodecyl benzene sulphonate (SDBS). *Liao L., Zhao S., Zhang L., Chen T. and Yu J., Effect of different additives on the precipitaitng behavior of anionic surfactant and multivalent cation, Petroleum Exploration and Development 31 (S1), 2004, 13—16.*

7.3.2 Emulsification

During surfactant/polymer combination flooding, another source of formation damage is the blockage of pore-throat system caused by the water-oil emulsion. The emulsion is usually generated due to the emulsification of the injected fluid with the in situ crude oil in reservoirs. Due to the presence of surfactants in the compound displacement system, the emulsion could be stabilized. In the oil-wet reservoirs, the water-in-oil emulsion is usually formed with higher viscosity and stronger stability, rendering more severe damage in the formation. While in the water-wet reservoirs, the oil-in-water emulsion can be formed with higher viscosity, which can result in more damage in the formation (Garti and Aserin, 1996).

Both water-in-oil and oil-in-water emulsions could enhance the flow resistance of the fluids through increasing the viscosity of flowing fluids. Besides, the existence and movement of the oil-water interfaces in the emulsion can also bring the momentary fluctuation of pressure in the irregular channels, which can facilitate the transport efficiency of formation rock particles through the porous system. Some studies also show that with the presence of surfactants, the particles fixed onto rock surfaces due to the effects of wettability and interfacial tension can be released to generate the problems of large amounts of fines migration (Li et al., 1999). Meanwhile, while large amounts of clays (such as kaolin and montmorillonite) are contained in the formation, adsorption of surfactants will occur (adsorption loss), which will dramatically weaken the performance of oil displacement using surfactant.

7.3.3 Phase separation

In surfactant/polymer combination flooding, when surfactant and polymer are injected simultaneously into the formation, phase separation would occur under certain degrees of mineralization to generate a highly viscous, surfactant-enriched phase, which would not only deactivate the primarily prepared displacement system, but also increase the flow resistance with formation damage (Nelson and Pope, 1978).

7.4 FORMATION DAMAGE BY TERNARY COMBINATION FLOODING

The methodology of ternary combination flooding was developed and tested in the early 1980s. It is a multicomponent compound oil recovery technology employing the mixture of alkaline, surfactant and polymer, usually abbreviated as ASP. ASP compound has various advantages to enhance oil recovery. Attributed to the function of polymers, the viscosity of displacement fluids can be enhanced to decrease the oil–water mobility ratio to enhance the sweep efficiency. In addition to the effects of surfactants and alkali, the oil/water IFT can be effectively reduced as below as 10^{-3} mN/m to enhance the displacement efficiency. Therefore, ASP has both advantages of enhancing sweep efficiency and improving displacement efficiency for improving oil recovery (Deng et al., 2002). The combination of three kinds of displacement agents is also beneficial to broaden the range of surfactant concentration and salt concentration which keep the IFT enough low, in other words, it can have stronger adaptability into types of oil reservoirs, though emulsion may occur during the process as negative effects of formation damage.

7.4.1 Roles of each component in ASP system
7.4.1.1 Roles of alkali in ASP systems
- The alkane carboxylic soap and naphthenic soap generated from the reaction between alkali and the organic acid inside the oil can be

adsorbed onto the oil–water interfaces to reduce the IFT, weaken the capillary force retardation, as a result, more additional amounts of the trapped oil can be activated and flow together with the displacement fluids.

- Alkali can have synergistic effects with the added surfactants in system, which also enhance the interfacial activity and broaden the effective range of surfactant.
- Alkali can react with the organic polar matter in colloid, asphaltene, paraffin and porphyrin existing in the oil–water interface. This may result in the rupture of the rigid film in the oil–water interface, facilitating the dissolution of organic polar matter and enhancing the transforming and flowing capability of oil drops in the pore throats.
- Under the effects of alkali, the moving oil droplets can self-emulsify into oil-in-water emulsions with different sizes, and then be carried by the flowing water.
- Alkali can have ion exchanges with the minerals on the rock surface, which can facilitate compositional changes of the minerals at the rock surface, improve the surface electrical properties of rock mineral particles, and reduce the adsorption and retention loss of surfactants and polymers onto the rock surfaces (Shutang and Qiang, 2010).

7.4.1.2 Role of surfactant in ASP system

- Surfactant can reduce the oil-water IFT, changing more amounts of residual oil into mobile oil.
- Surfactants can change the wettability of the rock surface. The adsorption of rock surface to oil is weakened, resulting in enhanced oil displacement efficiency.
- Surfactant micelles have dissolving effects on the crude oil, and can emulsify crude oil to enhance its mobility. A sufficient amount of micelles is required to achieve miscible displacement.
- The existence of surfactant is beneficial to the conducting of saponification reactions. Their synergistic effect can lead to the further reduction of IFT.
- Surfactants can perform compensation effects when the ionic strength and the concentration of divalent cations are high, which can broaden the interfacial activity range of the system, and also broaden the salt content or pH value range where self-emulsification may occur (Sheng, 2014).

7.4.1.3 Role of polymer in ASP system

- Polymer can increase the viscosity of the displacement fluids and reduce the mobility of water, thus significantly reducing the water-oil mobility ratio and retarding the fingering phenomenon. In this way, the effects of microscopic and macroscopic heterogeneity on swept efficiency can be mitigated, increasing the area swept by the displacement fluid.
- Polymer can improve the distribution ratio of displacement fluids in the vertical direction of the oil reservoir, adjust the water injection profile and increase the water adsorbing capacity of the low permeability layer and the upper part of positive rhythm settled layer. In this way the phenomenon that water flees into high permeability layer is improved, enhancing the sweep efficiency of displacement fluids.
- Polymer can protect surfactant from excessive reactions with high valent cations such as Ca^{2+} and Mg^{2+}, to prevent the surfactants from losing their surface activity.

7.4.2 Formation damage by ASP flooding

Despite ASP flooding bears many advantages such as apparently reduced IFT, preferable mobility control, reduced adsorption of chemical agents for enhanced oil recovery (EOR), ASP can introduce very severe formation damage with scaling problems. After entering the formation, ASP mixed fluids destroys the original equilibrium state of the formation fluids and rocks, but generates a new complex environment consisting of a ternary displacement fluid, formation water, formation rock, and crude oil containing dissolved gas. ASP systems can not only mix with the in situ fluids in the formation for physic-chemical reaction, but also react with the rock minerals. ASP systems can directly or indirectly lead to the corrosion and dissolution of minerals in reservoirs, the ion exchange of minerals with formation water, the oxidation of minerals and hydrocarbons, and the changes of temperature and pH of the formation. Under certain temperature, pressure, and pH conditions, after the scaling ions reach oversaturation, chemical precipitation will be generated with the formation of scale (Karazincir et al., 2011). The chemical agents applied in ASP play different roles to contribute to the corrosion and scale formation. Among them, alkaline is the main factor causing the scaling associated with ASP flooding. A strong alkali has stronger corrosion ability than a weak alkali, thus a strong alkali is much easier to cause scale formation.

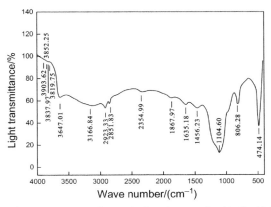

Figure 7.6 Infrared spectroscopy of a typical scale sample. *Liu D., Li J., Li T., Zheng Y., Zhang X. and Wang W., Scaling Characteristics of Silicon and Anti-scaling Measures in ASP Flooding with Alkali, Acta Petrolei Sinica 28 (5), 2007, 139–141.*

Liu et al. (2007) found that the scale produced in an oil well is not made up of a single component, but a mixture of silicate scale, carbonate scale and organic matter, as illustrated in Fig. 7.6. The major component of the scale sample is amorphous SiO_2, along with small amounts of calcium carbonate and ferric compounds.

7.4.2.1 Scale formation due to alkaline

Alkaline can involve alkali corrosion reactions with the minerals in rocks (such as kaolinite, feldspar, and montmorillonite), and the generated soluble matters can enter the formation fluid. Besides, alkali can also have physiochemical reactions with the formation fluids. The alkali in ASP flooding can have the following reactions with both formation water and minerals.

(1) Reactions of alkali with formation water

$$Ca^{2+} + 2OH^- \rightleftharpoons Ca(OH)_2 \downarrow$$

$$Mg^{2+} + 2OH^- \rightleftharpoons Mg(OH)_2 \downarrow$$

$$HCO_3^- + OH^- \rightleftharpoons H_2O + CO_3^{2-}$$

$$Ca^{2+} + CO_3^{2-} \rightleftharpoons CaCO_3 \downarrow$$

$$Mg^{2+} + CO_3^{2-} \rightleftharpoons MgCO_3 \downarrow$$

(2) Reactions of alkali with subsurface minerals

The chemical formula of feldspar is $K[AlSi_3O_8]$.

$$K[AlSi_3O_8] + Ca[AlSi_3O_8] + 6OH^- \rightleftharpoons 2Al(OH)_3\downarrow + K^+ + Ca^{2+} + 6SiO_3^{2-}$$

The major component of quartz is SiO_2.

$$SiO_2 + 2OH^- \rightarrow SiO_3^{2-} + H_2O$$

The clay minerals in the formation, such as illite, chlorite, kaolinite, and montmorillonite, contain different kinds of cations; however, their reactions with alkali are almost the same as the reactions of feldspar with alkali.

(3) The inorganic salt precipitations probably produced from the reaction of alkali with subsurface fluids and rock minerals are shown as follows (Sui, 2006)

Carbonate scale:

$$Ca^{2+} + CO_3^{2-} \rightleftharpoons CaCO_3\downarrow$$
$$Mg^{2+} + CO_3^{2-} \rightleftharpoons MgCO_3\downarrow$$

Hydroxide scale:

$$Ca^{2+} + 2OH^- \rightleftharpoons Ca(OH)_2\downarrow$$
$$Mg^{2+} + 2OH^- \rightleftharpoons Mg(OH)_2\downarrow$$

Silicic acid scale:

$$SiO_4^{4-} + 4H^+ \rightleftharpoons H_4SiO_4\downarrow$$
$$SiO_3^{2-} + 2H^+ \rightleftharpoons H_2SiO_3\downarrow$$

Silicate scale:

$$Ca^{2+} + SiO_3^{2-} \rightleftharpoons CaSiO_3\downarrow$$
$$Mg^{2+} + SiO_3^{2-} \rightleftharpoons MgSiO_3\downarrow$$

The various ions dissolved in the formation fluid would precipitate mainly through three distinct processes to deposit scale on the inner surface of oil reservoir pore-throats. First, with the removal of produced fluid from the reservoir, the changes of reservoir conditions such as temperature, pressure, Ca^{2+}, Mg^{2+}, Ba^{2+}, Si^{4+}, and their corresponding acid-ion compositions would result in the decreased solubility of these salts in the produced fluids, and thus the generation of crystallization precipitation and forming crystalline scale due to oversaturation of the produced water solution. Second, when water solutions with different ion compositions mix with each other, precipitates will be separated from the formation fluids in the reservoir due to oversaturation of the solution, forming crystalline scale or particle scale. Third, the local ambient temperature changes caused by the friction inside the fluids, and the friction between the fluids and the internal wall of pore-throats, would lead to the change of dissolution capacity of some materials [such as $Ca(OH)_2$] in the produced fluid, resulting in the generation of crystalline scale and corrosion scale.

Since 1980, the United States has conducted alkaline flooding to one of the blocks in Long Beach field reservoir. After 1 year, the fluid yield of the production wells near the alkali injection well has decreased a lot. It was found that the formation far away from the wellbore was partially blocked, due to the alkaline scale and silica–aluminate scale formed by the reaction of alkali with the reservoir fluid and rock minerals. When initiating the North Ward-Estes oilfield alkaline flooding pilot tests, Raimondi et al. (1977) found that the amount of gypsum scale generated in the production well increased. In 2016, Cheng et al. studied the effects of temperature, pressure, pH, and ion strength on the deposition of $CaCO_3$ scale and silica scale, and established a prediction equation for the deposition of calcium/silicon mixed sediment (Cheng et al., 2016). The scaling quantification prediction method for the calcium/silicon mixed scale was established in the Daqing oilfield for ASP flooding, demonstrating an accuracy rate of over 90%.

7.4.2.2 Scale formation due to surfactant
When ASP mixture is injected into the reservoir system, surfactant can help reduce the oil–water IFT and decrease the scale deposition velocity; although it demonstrates an extremely strong adsorption preference with calcite. During the injection-production process, only a small amount of surfactants are degraded due to environmental effects, while most of surfactants are adsorbed by the inorganic salt precipitates generated in the system to form scale.

7.4.2.3 Scale formation due to polymer
In the ASP injection system, the addition of polymer increases the viscosity of the system; meanwhile the polymer can have synergistic effects with surfactant to retard the deposition of inorganic salt scale. Polymer and its degradation product can wrap and adsorb the inorganic scale, turning into

organic scale. Calcium can also react with the carboxylic groups generated from the hydrolysis of PAM to form calcium carboxylate precipitation. Therefore, the addition of polymer is one of the reasons for scaling occurring in association with ASP flooding (Wang and Cheng, 2003).

7.4.3 Main factors affecting the scaling during ASP flooding

During ASP flooding, the underground environment of an oilfield reservoir becomes very complicated. There are various factors affecting its scaling, such as, formation mineral composition, solution component, colloidal substance, underground temperature, and pressure, pH, flow velocity, and flow state of the fluids (Wang et al., 2004; Li et al., 2009).

7.4.3.1 Mineral composition

The mineral compositions in different reservoirs are quite different, the ionic composition and concentration of injection fluid are also not constant throughout different types of reservoirs. Thus the chemical compositions of generated scales would be also different. For instance, ions such as Ca^{2+} and Mg^{2+} in the formation rocks can be released into the oil-reservoir water as impurity elements, during the kaolinization of potassium feldspar, plagioclase, and microcline. Under the long-term scouring of injection fluid, minerals such as dolomite, siderite, calcite, and barite can slowly be dissolved with the increase of the concentration of Ca^{2+} and Mg^{2+} in the produced water, which providing an environment with abundant materials for the generation of scale. What's more, many factors such as the heterogeneity in the distribution of formation rock mineral composition, mineral particle size, and pore-throat size, rock surface roughness and adsorption capacity, fluid flowing state and velocity, reservoir temperature, pressure, and pH will have effects on the formation and deposit distribution of scale. In 2003, Wang et al. (LU et al., 2003) studied the mineral and chemical composition of the scale associated with ASP flooding in an oil production well of the Daqing oilfield using X-Ray Diffraction (XRD), Scanning Electron Microscope (SEM), and Energy Dispersive X-Ray spectroscopy (EDX) techniques. The results showed that the major mineral components in the scale samples are silica, calcite, quartz, and clay particles. In 2003, Wang et al. found that ASP flooding can cause different degrees of corrosion to the primary reservoir minerals such as kaolinite and potassium feldspar, and also change the mineral composition (Wang and Cheng, 2003).

7.4.3.2 Ionic concentration

Ionic concentration is one of the major factors directly affecting the scale formation. While all the other conditions are roughly the same, the higher ionic concentration would facilitate the scale formation. For those precipitates commonly found in oil fields, such as $MgCO_3$ and $CaCO_3$, their solubility are 3.162×10^{-3} mol/L and 6.928×10^{-5} mol/L at the temperature of 25 °C, respectively. The hardness of their saturated mixture solution is 323.128 mg/L. Therefore, the critical hardness of the solution with $MgCO_3$ and $CaCO_3$ as scale forming matter is 323.128 mg/L, indicating that when the concentration of Ca^{2+} and Mg^{2+} surpasses 323.128 mg/L (25 °C) and the concentration of CO_3^{2+} exceeds this value, precipitates of $MgCO_3$ and $CaCO_3$ will definitely be generated. Hongwei Zhu (Zhu, 2012) investigated the scaling law of the ASP flooding oil well in the Daqing oilfield and found that the increase of ionic concentration accelerated the scaling rates after the fracturing treatment of the oilfield.

7.4.3.3 Colloidal matters

During ASP flooding, the solution contains colloidal material, bacteria, and organisms that can facilitate the mutual bonding of precipitates such as carbonate, silicate and sulfate into blocks, and the adherence of those precipitates onto rock surfaces. At the earlier stage of the adsorption of scale particles onto the surfaces, the colloidal matters in the solution can change the surface properties of scale particles, and then providing superior conditions to enhance its adsorption and crystal growth. In addition, the adhesive colloidal matter roles as the beam in the concrete, making the scale particles adhere to the equipment surface more firmly. Therefore, in the ASP flooding system, the formation of hard scale is largely affected by the presence of colloidal matters, and the accumulation of hard scale is also affected by the content of colloidal matter.

7.4.3.4 Temperature and pressure

From the total depth location of a well to the ground surface, the borehole temperature and pressure change with the depth of the well. Changes in temperature and pressure accelerate or restrain the scale formation reactions by first affecting the solubility of the scale forming matter, and then changing the thermodynamic conditions of the scale forming reaction. For example, when temperature increases, the solubility of $CaCO_3$ is decreased. The increase in temperature also facilitates the decomposition of Ca

$(HCO_3)_2$ into $CaCO_3$. Therefore, a large amount of $CaCO_3$ are precipitated out in higher temperature zones, resulting in severe scaling problems. Since it is not easy to control the strata temperature, the scale formation can only be reduced by changing the solubility of the water ions. Pressure also affects the solubility of scale. Take $CaCO_3$ scale as an example, under certain temperature, the solubility of $CaCO_3$ increases with the pressure. The solubility of $CaCO_3$ changes little with pressure in high temperature conditions; however it changes obviously with pressure in low temperature conditions. What's more, the thermodynamic conditions of various physical and chemical reactions change with the pressure, resulting in the destruction of the chemical equilibrium.

7.4.3.5 pH value

As the forming environments of different kinds of precipitates are different, the pH of solution directly affects the types and amounts of the precipitate. Carbonate and hydroxide scales form easily in the basic environment, while the silicate scales tend to form in acidic condition. As carbonate and hydroxide are common in the stratum environment, higher pH would facilitate the generation of scale. According to the literature research, the underground and surface water contain different extents of carbonic acid, while the pH determines the concentration proportion of three forms of carbonic acids (carbonic acid, carbonate, and bicarbonate) in the final equilibrium shown in Fig. 7.7 (Chen, 2016).

This proportion ratio is completely correlated with the pH of water. At higher pH, the generation of $CaCO_3$ precipitates is easier, with a larger precipitation quantity, while at lower pH, $CaCO_3$ precipitates are hard to form. Dang et al. (2013) studied the relationship between the barium scaling and the pH of the ASP flooding in Daqing oilfield. The results showed that with the pH increasing, the scaling speed of $BaCO_3$ was accelerated. When the pH reached 9.5, the scaling amount of $BaCO_3$ would be significantly improved.

7.4.3.6 Flow velocity and flow state

The main effect of the flowing velocity of downhole fluid on the scale formation is attributed to the changes of shear force of the moving fluids onto the deposited scale particles. Although the increase of flowing velocity can improve the scale precipitation rate, however, the shear force applied to the scale unstably bonded with the formation surfaces is also

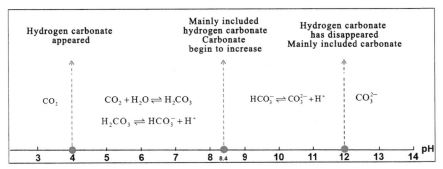

Figure 7.7 Variations of forms of carbonic acids in solution with changes in pH. *Chen Y., Regularities and Prediction of Corrosion and Scaling in Formation during the Procedure of ASP Flooding, 2016, Northeast Petroleum University, Daqing, China.*

significantly increased to enhance the detachment of scale particles from surfaces. Therefore, assuming all the other conditions fixed, the scaling rate decreases with the increasing flowing velocity (Nawrath et al., 2006). The flow states of fluids include laminar flow and turbulent flow. In the turbulent flow state, the serious turbulence will cause more collisions of crystal nuclei, resulting in the severe crystal precipitation, which will also accelerate the formation speeds of crystal nuclei. Therefore, the scale particles are easily formed in the high turbulent flow region. Additionally, the abrupt changes of flowing velocity and flowing direction can also aggravate the scale formation (Wang et al., 2008).

7.4.4 Scaling mitigation and prevention techniques

To solve the scaling problem during the ASP flooding, diverse antiscaling techniques have been designed on the basis of the composition, formation processes and various influencing factors of the scale, which mainly include chemical techniques and physical techniques to prevent and mitigate scaling (Li et al., 2009).

7.4.4.1 Chemical techniques to prevent scaling

The chemical antiscaling technique use is kind of approach to introduce the chemical antiscaling agents to prevent the scale formation in its growth period. The advantages of this technique include lower cost, obvious antiscaling effect, simple operation, etc., making it one of the mostly most widely used techniques in the oilfields worldwide. The mechanisms of antiscaling agents include coordinating solubility effect, low-dosage effect, lattice distortion effect, cohesion, and dispersion effect and electrostatic repulsion. At present, the most widely used antiscaling agents include organic phosphonic acid type, polymer type, chelating agent type, etc. Among them, the organic phosphonic acid type antiscaling agents are used widely in the oilfield due to their lower cost, good chemical stability, strong heat resistance and excellent antiscaling properties. The polymer type antiscaling agents include natural polymers such as lignin, tannin, starch and chitosan, and synthetic polymers include such as lignin, tannin, starch, and chitosan acrylic acid copolymer and maleic acid copolymer, and novel eco-friendly polymers such as polyaspartic acid and polyepoxysuccinic acid. The chelating agents mainly consist of organic phosphonate, hydroxyl carbonic acid, aminocarboxyl chelating agent, macrocyclic polyether, polyamine, etc. The mechanism of this type of antiscaling agent to prevent scale formation is that the alternative formation of coordination compounds between the chelating agent and scale-forming metal ions through coordination bonds, which are more stable and water-soluble than the scale instead, and can be discharged together with the fluid, thus achieving the antiscaling objective. This kind of antiscaling agent is mainly used to remove acid-resistant scales such as silicate and sulfate. Liying Xu (Xu et al., 2001) obtained oxy-starch(OS) from the modification of natural macromolecule starch through chemical oxidation, and the results showed that the anti-scaling rate reached 100% in systems whose calcium concentration were 50−300 mg/L.

7.4.4.2 Physical techniques to prevent scaling

Physical antiscaling techniques can prevent the deposition of inorganic salts on the pipeline surface through physical or mechanical methods or smearing fluid-repellent organic coatings on the inner surfaces of the oil pipelines. This strategy possesses various advantages including no dosing, no pollution, energy saving, and environmental friendly outcomes; however the issue of poor stability which means short duration of efficacy needs to be addressed. The physical antiscaling technique mainly takes effect in the stage of scale particles adhering with the equipment surfaces. Types of commonly applied physical antiscaling techniques include supersonic treatment, magnetic antiscaling technique, high-frequency pulse antiscaling technique and antiscaling coating onto the metal surfaces of the facilities.

Besides the above mentioned antiscaling techniques, the scale formation can also be destroyed by changing the properties of displacement
fluid or adjusting the facility shapes and surface properties (Wang et al.,
2013). Some of controlling measures are concludes as follows:

- Conducting ionic analysis on the injection water and stratum water to
  confirm the injection water resource having good compatibility with
  formation fluids and minerals;
- Reducing the injection rates and inducing a pressure drop in the oil
  well by using W/O emulsions generated by the downhole choke;
- Reducing the friction between the produced liquid and the surfaces of
  facilities by using small interference screw pump and long plunger
  short pump barrel pumping unit to achieve the purpose of scale
  prevention.

7.5 SUMMARY AND CONCLUSIONS

Nowadays polymer flooding, surfactant/polymer binary combination flooding, and alkaline/surfactant /polymer ternary combination
flooding are frequently applied in oilfields. However, all these kinds of
chemical flooding technologies can generate certain damage in the reservoir formation. Among them, during the application of polymer flooding, the formation damage can be generated mainly due to the following
reasons: (1) The adsorption and retention of polymers in the porous
medium; (2) the incompatibility of polymer solutions with formation
fluids and minerals; (3) polymers facilitate the migration of loose sand
particles. In addition, the formation damage issues generated by surfac-
tant/polymer flooding during the tertiary oil recovery also include the
precipitation of surfactants, emulsification, and phase separation. And the
main mechanism of formation damage caused by ternary chemicals-
combination flooding is scaling problem, as a result of interactions among
different chemical agents. To prevent or mitigate the scaling problems
during the ASP flooding, diverse antiscaling techniques have been
designed on the basis of the composition, formation processes and various
influencing factors of the scaling. In this work, both the chemical and
physical antiscaling techniques are summarized. The chemical antiscaling
technique is kind of approach to introduce the chemical antiscaling agents

to prevent the scale formation in the stage of scaling growth. Physical antiscaling techniques can prevent the deposition of inorganic salts through physical or mechanical methods, or smearing fluid-repellent organic coatings.

REFERENCES

Amante, J.C., Scamehorn, J.F., Harwell, J.H., 1991. Precipitation of mixtures of anionic and cationic surfactants: II. Effect of surfactant structure, temperature, and pH. J. Colloid Interface Sci. 144 (1), 243−253.

Ananthapadmanabhan, K., Somasundaran, P., 1985. Surface precipitation of inorganics and surfactants and its role in adsorption and flotation. Colloids Surf. 13, 151−167.

Bhardwaj, A., Shainberg, I., Goldstein, D., Warrington, D., Levy, G.J., 2007. Water retention and hydraulic conductivity of cross-linked polyacrylamides in sandy soils. Soil Sci. Soc. Am. J. 71 (2), 406−412.

Chen, Y., 2016. Regularities and Prediction of Corrosion and Scaling in Formation during the Procedure of ASP Flooding. Northeast Petroleum University Daqing, China.

Cheng, J., Wang, Q., Wang, J., Li, C., Shi, W., 2016. Ca/Si scale sedimentary model in the strong-base ASP flooding block and scaling prediction. Acta Petrolei Sinica 37 (5), 653−659.

Cools, P.J., Van Herk, A.M., German, A.L., Staal, W., 1994. Critical retention behavior of homopolymers. J. Liq. Chromatogr. Relat. Technol. 17 (14−15), 3133−3143.

Cui, Z.-G., Song, H.-X., Yu, J.-J., Jiang, J.-Z., Wang, F., 2011. Synthesis of N-(3-Oxapropanoxyl) dodecanamide and its application in surfactant-polymer flooding. J. Surfactants Deterg. 14 (3), 317−324.

Dai, C., Zhao, J., Jiang, H., Wang, X., Lv, X., Zhao, G., et al., 2010. Alternative injection of anionic and cationic polymers for deep profile control in low-permeability sandbody reservoir. Acta Petrolei Sinica 31 (3), 440−444.

Dang, Q.G., Bian, J.P., Cheng, H.L., 2013. The prediction of scaling time in producing wells of ASP flooding in the fourth plant of Daqing oilfield. Adv. Mater. Res. 734−737, 1313−1316.

Deng, S., Bai, R., Chen, J.P., Yu, G., Jiang, Z., Zhou, F., 2002. Effects of alkaline/surfactant/polymer on stability of oil droplets in produced water from ASP flooding. Colloids & Surfaces A Physicochemical & Engineering Aspects 211 (2−3), 275−284.

Dovan, H., Hutchins, R., Terzian, G., 1990. Dos Cuadras Offshore Polymer Flood. SPE California Regional Meeting, Society of Petroleum Engineers.

Fletcher, A., Lamb, S., Clifford, P., 1992. Formation damage from polymer solutions: factors governing injectivity. Soc. Pet. Eng. Reservoir Eng. 7 (2), 237−246.

Fu, M., Xiong, F., 2009. Study on the plugging mechanism of polymer flooding in Henan oilfield. Drilling Production Technol. 32 (4), 77−79.

Fuoss, R.M., Strauss, U., 1948. Electrostatic interaction of polyelectrolytes and simple electrolytes. J. Polymer Sci. Part A: Polymer Chem. 3 (4), 602−603.

Garti, N., Aserin, A., 1996. Double emulsions stabilized by macromolecular surfactants. Adv. Colloid Interface Sci. 65, 37−69.

Garti, N., Zour, H., 1997. The effect of surfactants on the crystallization and polymorphic transformation of glutamic acid. J. Cryst. Growth 172 (3), 486−498.

Green D.W. and Willhite G.P., Enhanced Oil recovery, 1998, Society of Petroleum Engineers, TX, United States.

Guo, L., 2011. Study of polymer's adsorption and retention law and performance change. Chem. Eng. Oil Gas 40 (6), 587−589.

Karazincir, O., Thach, S., Wei, W., Prukop, G., Malik, T., Dwarakanath, V., et al., 2011. Scale formation prevention during ASP flooding. SPE International Symposium on Oilfield Chemistry, Society of Petroleum Engineers.

Lu, Y.-J., Fang, B., Jiang, T.-Q., Fang, D.-Y., Shu, Y.-H., 2003. Formation and rheologic behavior of viscoelastic micelle fracturing fluids. Oilfield Chem. 20 (4), 78−80.

Li, J., Li, T., Yan, J., Zuo, X., Zheng, Y., Yang, F., 2009. Silicon containing scale forming characteristics and how scaling impacts sucker rod pump in ASP flooding. Asia Pacific Oil and Gas Conference & Exhibition, Society of Petroleum Engineers.

Li, Z., Zeng, H., Jiao, L., 1999. The nature of colloid and interface of miscible phase system in alkaline-surfactant combination flooding. J. Dispersion Sci. Technol. 20 (4), 1143−1162.

Li, Q., Pu, W., Wei, B., Jin, F., Li, K., 2016. Static adsorption and dynamic retention of an anti salinity polymer in low permeability sandstone core. J. Appl. Polym. Sci. 134 (8), 44487.

Liao, L., Zhao, S., Zhang, L., Chen, T., Yu, J., 2004. Effect of different additives on the precipitaitng behavior of anionic surfactant and multivalent cation. Petroleum Exploration and Development 31 (S1), 13−16.

Liu, D., Li, J., Li, T., Zheng, Y., Zhang, X., Wang, W., 2007. Scaling Characteristics of Silicon and Anti-scaling Measures in ASP Flooding with Alkali. Acta Petrolei Sinica 28 (5), 139−141.

Martin, M.M., Rockholm, D.C., Martin, J.S., 1985. Effects of surfactants, pH, and certain cations on precipitation of proteins by tannins. J.Chem. Ecol. 11 (4), 485−494.

Nawrath, S.J., Khan, M.M.K., Welsh, M.C., 2006. An experimental study of scale growth rate and flow velocity of a super-saturated caustic−aluminate solution. Int. J. Miner. Process. 80 (2−4), 116−125.

Nelson, R., Pope, G., 1978. Phase relationships in chemical flooding. Soc. Pet. Eng. J. 18 (5), 325−338.

Owens, D.K., Wendt, R., 1969. Estimation of the surface free energy of polymers. J. Appl. Polym. Sci. 13 (8), 1741−1747.

Raimondi, P., Gallagher, B.J., Ehrlich, R., Messmer, J.H., Bennettal, G.S., 1977. Alkaline waterflooding: design and implementation of a field pilot. J. Petrol. Technol 29 (10), 1359−1368.

Shaw, R., Stright Jr, D., 1977. Performance of theTaber south polymer flood. J. Can. Petrol. Technol. 16 (01), 35−40.

Sheng, J.J., 2014. A comprehensive review of alkaline−surfactant−polymer (ASP) flooding. Asia Pac. J. Chem. Eng. 9 (4), 471−489.

Shutang, G., Qiang, G., 2010. Recent Progress and Evaluation of ASP Flooding for EOR in Daqing Oil Field. SPE EOR Conference at Oil & Gas West Asia, Society of Petroleum Engineers.

Somasundaran, P., Hanna, H.S., 1977. Physico−Chemical Aspects of Adsorption at Solid/Liquid Interfaces: I. Basic Principles - Improved Oil Recovery by Surfactant and Polymer Flooding. 205−251.

Stellner, K.L., Scamehorn, J.F., 1986. Surfactant precipitation in aqueous solutions containing mixtures of anionic and nonionic surfactants. J. Am. Oil Chem. Soc. 63 (4), 566−574.

Sui, X., 2006. Study on Mechanism and Domain Factors of Scale Formation in Alkaline-surfactant-polymer Flooding. Northeast Petroleum University, Daqing, China.

Tang, H., Meng, Y., Yang, X., 2001. A study of adsorption consumption of polyacrylamide on reservoir minerals. Oilfield Chem. 18 (4), 343−346.

Trujillo, E.M., 1983. Static and dynamic interfacial tensions between crude oils and caustic solutions. SPE Repr. Ser. (US) spe−10917 (4), 645−656.

Vela, S., Peaceman, D., Sandvik, E.I., 1976. Evaluation of polymer flooding in a layered reservoir with crossflow, retention, and degradation. Soc. Pet. Eng. J. 16 (02), 82—96.

Wagner, O., Leach, R., 1959. Improving oil displacement efficiency by wettability adjustment. Transactions of the AIME 216 (01), 65—72.

Wang, Y.-P., Cheng, J.-C., 2003. The scaling characteristics and adaptability of mechanical recovery during ASP flooding. J.-Daqing Pet. Inst. 27 (2), 20—22.

Wang, B., Li, C., Zhu, W., Xu, Q., Liu, Y., 2008. Application of de-fouling technology in pipelines. Corrosion & Protection in Petrochemical Industry 25 (1), 28—30.

Wang, Y., Liu, J., Liu, B., Liu, Y., Wang, H., Chen, G., 2004. Why Does Scale Form in ASP Flood? How to Prevent from It?--A Case Study of the Technology and Application of Scaling Mechanism and Inhibition in ASP Flood Pilot Area of N-1DX Block in Daqing. SPE International Symposium on Oilfield Scale, Society of Petroleum Engineers.

Wang, Z., Pang, R., Le, X., Peng, Z., Hu, Z., Wang, X., 2013. Survey on injection—production status and optimized surface process of ASP flooding in industrial pilot area. J. Pet. Sci. Eng. 111, 178—183.

Xiang, L.U., Chen, H., Shan, Y., 2002. The origines of blockage creation in polymer injection wells at west of district N2 of Northern Saertu in Daqing. Oilfield Chemistry 3, 257—259.

Xu, L., He, Y., Xi, H., 2001. Research on oxidized starch used as water treatment agent. Water Purif. Techno. 20 (2), 27—29.

Yan, H., Blanford, C.F., Holland, B.T., Smyrl, W.H., Stein, A., 2000. General synthesis of periodic macroporous solids by templated salt precipitation and chemical conversion. Chem. Mater. 12 (4), 1134—1141.

Yang, B.D., Yoon, K.H., Chung, K.W., 2004. Dispersion effect of nanoparticles on the conjugated polymer—inorganic nanocomposites. Mater. Chem. Phy. 83 (2), 334—339.

Yuan, B., Moghanloo, R.G., 2017. Analytical modeling improved well performance by nanofluid pre-flush. Fuel 202, 380—394.

Yuan, B., Moghanloo, R., Wang, W., 2018. Using nanofluids to control fines migration for oil recovery: nanofluids co-injection or nanofluids pre-flush? A comprehensive answer. Fuel 215, 474—483.

Zhang, X., Zhao, F., Liu, H., 2017. Study on static adsorption of polymer. Contemporary Chem. Industry 46 (3), 396—399.

Zheng, J.D., Zhang, Y.Z., Ren, H., Liu, Y., 2004. Blockage mechanism and blockage reducer for polymer-injection well. Petroleum Exploration and Development 6, 108—111.

Zhou, W.-F., 2004. Analysis of the polymer injection wells plugging. J. Daqing Pet. Inst. 28 (2), 40—42.

Zhu, H., 2012. The rule of fouling of the three-element compound oil-driven well in Lamadian oilfield. Oil-Gas Field Surf. Eng. 31 (12), 18—19.

CHAPTER EIGHT

Formation Damage Problems Associated With CO_2 Flooding

Xuebing Fu[1], Andrew Finley[2] and Steven Carpenter[3]

[1]University of Wyoming, Laramie, WY, United States
[2]Goolsby, Finley & Associates, LLC, Casper, WY, United States
[3]University of Wyoming, Casper, WY, United States

Contents

8.1 INTRODUCTION

8.1.1 Review of CO_2 flooding sites

CO_2 flooding has been used as a commercial enhanced oil recovery (EOR) process since the 1970s. After secondary water flooding, many light and medium oil reservoirs are suitable for miscible or even immiscible CO_2 flooding. Two main injection strategies exist for CO_2 injection: Continuous gas injection or water alternating gas (WAG) injection. Significant amounts of oil can be recovered CO_2 flooding post primary and secondary production.

Formation Damage during Improved Oil Recovery.
DOI: https://doi.org/10.1016/B978-0-12-813782-6.00008-7

© 2018 Elsevier Inc.
All rights reserved.

The first two large-scale projects consisted of the SACROC flood in Scurry County, TX (Texas, United States), implemented in January 1972, and the North Crossett flood in Crane and Upton Counties, TX initiated in April 1972. These two projects are still active until today (Melzer, 2012). The carbon dioxide for these two projects came from CO_2 separated from produced natural gas processed and sold in the south region of the Permian Basin. Later, however, development of large natural sources of CO_2 in Colorado (McElmo Dome/Doe Canyon and Sheep Mountain) and New Mexico (Bravo Dome) along with construction of high-volume CO_2 pipelines enabled CO_2 EOR to achieve its first burst of growth in the Permian basin starting in the 1980s. Subsequent development of natural CO_2 supplies at Jackson Dome, Mississippi, and the capture of vented CO_2 at the massive LaBarge natural gas processing plant in western Wyoming provided the foundations for the second round of CO_2 EOR growth at the turn of the century in the Gulf Coast and the Rocky Mountains (Kuuskraa and Wallace, 2014).

Fig. 8.1 provides the state-by-state locations for the 136 active CO_2 EOR projects in 2014 (Kuuskraa and Wallace, 2014). Much of the activity was in West Texas with 77 projects, followed by Mississippi (19 projects), and Wyoming (14 projects). The number of projects must have included pilots, as there are only five commercial scale CO_2 EOR

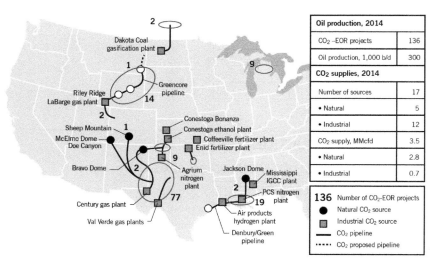

Figure 8.1 CO_2 EOR operations and CO_2 (Kuuskraa and Wallace, 2014). *Source: Advanced Resources International Inc. based on OGJ EOR/heavy Oil Survey 2014 and other sources.*

projects in Wyoming currently—to take Wyoming as an example. Fig. 8.1 also shows the location of CO_2 supply sources in year 2014. Much of the CO_2 was from natural CO_2 fields, but industrial sources also provided CO_2 in growing numbers to the EOR industry. Over 3000 miles of CO_2 pipelines link CO_2 supply areas with oil fields in the United States.

Fig. 8.2 demonstrates the historical oil production from CO_2 EOR operations. CO_2 EOR projects are usually long-lived projects. Although fluctuations of oil prices have an effect of temporarily decreasing the pace of project starts, CO_2 EOR production has been increasing quite steadily over the past three decades. The oil price crash of 1986 resulted in a drop of oil prices into single digits (\$/bbl) in many regions. The economics of flooding reservoir to enhance oil recovery was crippled at that time; capital for new projects was not available. However, due to the long-term nature of CO_2 flooding, the EOR projects survived the crash with fairly minor long-term effects (Melzer, 2012). The next oil price crash in 1998 also merely plateaued the growth temporarily—growth curve resumed in 2002. Besides low oil prices, slow growth of CO_2 EOR production is due primarily to limits on accessible and affordable supplies of CO_2. Fig. 8.2 also demonstrates the percentage of production from different regions—Permian

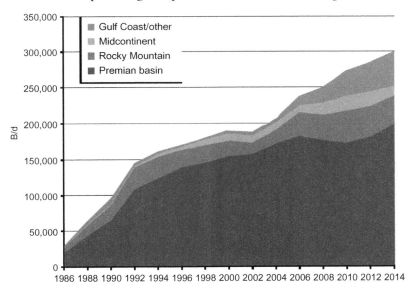

Figure 8.2 Historical CO_2 EOR production in the United States (Kuuskraa and Wallace, 2014). *Source: Advanced Resources International Inc. adjustment to OGJ EOR/ Heavy Oil Survey 2014.*

basin has remained to be the primary production region since the beginning of CO_2 EOR.

An outlook for regional oil production from CO_2 EOR projects predicted that oil production would increase from 300,000 b/d in 2014 to 638,000 b/d in 2020; correspondingly the CO_2 injection would increase from 3.5 bcf/d (68 mmton/y) in 2014 to 6.5 bcf (126 mmton/y) (Kuuskraa and Wallace, 2014). These estimates were based on both existing and planned/potential CO_2 EOR floods, and a detailed list of these projects including their key parameters were also included the study cited.

Uptake of CO_2 EOR has been relatively slow outside of the United States. A comparison of the number of CO_2 EOR projects from worldwide, United States and the Permian Basin is shown in Fig. 8.3 (Melzer, 2012). Some notable CO_2 EOR fields outside the United States include Weyburn oil field in Canada, Bati Raman heavy oil field in Turkey, and certain heavy oil fields in Trinidad. More recently, interest in CO_2 EOR has emerged in oil fields in Abu Dhabi, Brazil, China, Malaysia, the North Sea, and other areas in the world.

Geology is a critical element in reservoir development and exploitation using CO_2 EOR. All reservoir lithology, including siliciclastic, carbonate, and others, are suitable for CO_2 EOR. The oil recovery is influenced by geologic features such as rock and fluid characteristics, porosity, permeability, and structural or stratigraphic features such as faults and other barriers to oil or gas movement (Moghanloo et al., 2015;

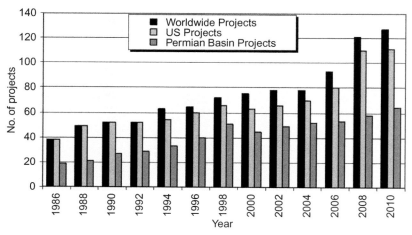

Figure 8.3 Active worldwide, United States and Permian basin CO_2 EOR project counts: 1986–2010 (Melzer, 2012).

Yuan et al., 2015, 2016, and 2017). The vast majority of the CO_2 EOR projects are miscible CO_2 processes—roughly 90% of the total CO_2 EOR oil production came from miscible CO_2 in 2014 and the rest came from immiscible CO_2 flooding (Koottungal, 2014).

A miscible CO_2 tertiary flood can recover 8%—25% of the original oil in place (OOIP) as demonstrated by field-scale pilot tests (Cao and Gu, 2013). In general, the injected CO_2 cannot achieve the first-contact miscibility (FCM) with the reservoir crude oil, but it can gradually develop the so-called multicontact miscibility (MCM) under the actual reservoir conditions (Green and Willhite, 1998). The minimum miscibility pressure (MMP) of a given crude oil—CO_2 system is defined as the minimum operating pressure at which CO_2 can reach the dynamic or MCM with the crude oil. Miscible CO_2 flooding can be achieved if the operating pressure in the reservoir is above the MMP, and usually a minimum reservoir depth of 2500 ft is required to reach the MMP (Taber et al., 1997). The main mechanisms contributing to oil recovery by CO_2 flooding process are oil-viscosity reduction, oil-swelling effect, and changes in the interfacial properties of the crude oil—CO_2 system, which results in high sweep efficiency.

8.1.2 Formation damage in CO_2 EOR fields

The application of CO_2 to enhance oil recovery can induce both inorganic and organic depositions. Likely due to higher complexity or prevalence of organic depositions, almost all reported field issues are related to precipitation and deposition of asphaltene. Asphaltene deposition can occur during the production and processing of oils, possibly causing permeability reduction, wellbore, and downhole facilities plugging, seizure of downhole valves and hindrance in wireline operations, upsets of flow lines, separators, and downstream equipment. These issues can result in production delays and costly clean up procedures that negatively impact the economics of a project.

The first cause of asphaltene deposition is reservoir depressurization, if the oil has a tendency for asphaltene flocculation at reduced pressures. Many reservoirs fall into this category. For example, the Ventura field in California had encountered significant problems during the wells' early production stages since 1944 (Tuttle, 1983). Deposition of asphaltene were partly responsible for the plugging of well bores and tubing. Both crude oil and various solvents were tried at the wells to dilute the produced oil and reduce asphaltene precipitation, but with limited success. Interestingly, the problems diminished after the bottom-hole pressure decreased further and fell below the bubble point pressure of the crude oil. For example, one well had plugged 15 times before the bubble point was reached; as the reservoir pressure reduced below the bubble point, the plugging problems were reduced significantly. Many wells were re-drilled during the early history of this field because of the asphaltene deposition problem. However, the wells have produced trouble-free from asphaltene since the early 1970s.

The second cause of asphaltene deposition is solvent injection during EOR processes, including CO_2 injection. Asphaltene dropout during CO_2 EOR is usually more difficult to quantify compared to that in primary or secondary production, as an additional component, CO_2, is introduced into the system. The dropout can occur anywhere in the reservoir, although typically the problem occurs at the production wellbore after solvent breakthrough (Leontaritis et al., 1994). A few asphaltene deposition issues in CO_2 EOR fields have been reported.

Asphaltene deposition is more likely to be a problem during CO_2 EOR if this were a problem encountered during primary production at a field. For example, at the Ventura field mentioned above, asphaltene deposition problem also occurred during CO_2 EOR (Tuttle, 1983; Leontaritis et al., 1988). Fields with asphaltene deposition issues during primary or secondary production are strong candidates for CO_2-induced deposition evaluations before considerations for CO_2 EOR.

Asphaltene deposition issues in CO_2 EOR have also been reported at fields without prior issues during primary production at a field. In the Little Creek CO_2 flooding EOR pilot plant in Mississippi, asphaltene deposition occurred in the well tubing (Tuttle, 1983; Leontaritis et al., 1994). No such problem had been observed during the primary and secondary production phases of this field. The deposition was thought to have been caused by CO_2-induced deposition of the asphaltenes and rigid films at the oil/water interface due to pH reduction. Well stimulation and acidizing fluids had a damaging effect in wells that produce asphaltenic crude oils since the acid can further induce asphaltene deposition. Attempts to remove the deposited asphaltenes were often unsuccessful and in some cases permanently damaged the well. In cases where asphaltenes had already precipitated, mechanical cleaning techniques were found to be the most effective.

Another example of asphaltene deposition in CO_2 EOR was found at a tertiary CO_2 food pilot in the Midale Unit in southeastern

Saskatchewan, Canada (Novosad and Costain, 1990). The Midale reservoir is an extensively fractured carbonate formation. Shortly after the CO_2 breakthrough in production wells, a considerable amount of solid deposit consisting of asphaltene, wax, and trapped oil formed in the wellbore equipment and downhole facilities. Again no such problems occurred during the field primary and secondary production. Another unreported case is known by the authors at the Big Sand Draw CO_2 flooding project in Wyoming (Flaten, 2017), where no asphaltene deposition problem was reported during primary production (natural bottom aquifer drive). During CO_2 flooding, severe asphaltene deposition problems were encountered, causing malfunctions of ESPs and wellbore plugging. Injecting asphaltene inhibitors downhole into producers was found effective in solving the problem. At the Big Sand Draw, scale was also a problem during primary production and CO_2 flooding. Scale inhibitors were used to effectively address the problem. Scale inhibitors are also used at the Salt Creek field, which is the largest producing CO_2 EOR field in Wyoming (Scherlin and Larson, 2017).

An oil that contains asphaltenes will not necessarily cause asphaltene problems during production. The Boscan crude in Venezuela contains more than 17 wt% asphaltenes, but it has not caused any asphaltene problems. In fact, it can mitigate the asphaltene flocculation propensity when mixed with other Venezuelan crudes (Lichaa, 1977). Conversely, some crude oils with low asphaltene contents have been associated with serious asphaltene deposition problems during oil recovery and processing. For example, the Hassi Messaoud oil (Algeria) which contains only 0.1 wt% asphaltenes, has caused serious asphaltene deposition problems (Haskett and Tartera, 1965). In general, asphaltene flocculation is dependent on the thermodynamics of the system—as will be discussed later in this chapter.

8.2 CO_2 FLOODING FORMATION DAMAGE MECHANISMS

8.2.1 Overview of formation damage induced by CO_2 injection

During CO_2 flooding, the injected CO_2 will interact with both rock and fluids in the reservoir. Fig. 8.4 provides an overview of the various effects of CO_2 on crude oil reservoirs (Okwen, 2006).

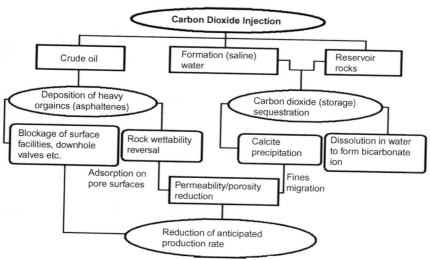

Figure 8.4 A sketch of the distribution and effects of injected CO_2 on reservoir fluids and rocks (Okwen, 2006).

As indicated in Fig. 8.4, CO_2 reacts with formation water containing calcium ions (Ca^{2+}) to precipitate calcite ($CaCO_3$), or it can react with water to form hydrogen ions (H^+) which have the potential to dissolve various minerals in reservoir rocks to produce high concentrations of ions especially calcium ions, which can precipitate again once the water chemistry is no longer favorable for dissolution. Calcite is a common precipitate in this process. Mineral dissolution and precipitation also lead to fines migration, which in turn causes pore blockages. On the other hand, CO_2 precipitates heavy organic components (asphaltenes are often assumed) in the crude oil, which leads to blockages of reservoir rocks and downhole/surface facilities. Also, asphaltene precipitation induces reservoir rock wettability reversal, which potentially reduces the oil relative permeability. Additionally, not listed in Fig. 8.4, tight emulsions may form during the process of CO_2 flooding, potentially causing formation damage. These mechanisms are discussed in greater details in the following sections.

8.2.2 Interactions between CO_2 and rock minerals

Interactions (diagenesis) between CO_2 and reservoir rock consist of the dissolution and precipitation (alteration) of minerals through formation water. The introduction of CO_2 into a system that has been at reservoir conditions for millions of years drives a series of reactions that vary based on component minerals, reservoir fluid composition, pH, pressure, temperature, etc. The transfer of ions between reservoir minerals and fluids create a continuously changing chemical environment which may drive reactions in different directions through time.

8.2.2.1 Dissolution

CO_2 injection into a reservoir results in the formation of carbonic acid, which lowers pH and increases both Eh (activity of electrons) and total dissolved solids. The magnitude of pH drop is primarily buffered by the presence of carbonate minerals. Multiple studies have illustrated the buffering potential of calcite through dissolution during injection [Fig. 8.5, Eqs. (8.1—8.3)]. Under certain conditions, if dolomite is present in addition to calcite, dolomite can dissolve preferentially [Fig. 8.6, Eqs. (8.4—8.5)] (Jin et al., 2016). Injection can also result in the dissolution of quartz [Fig. 8.7, Eq. (8.6)], feldspar, mica [Figs. 8.8—8.9, Eqs. (8.7—8.9)], barite, and clays and cements including kaolinite [Fig. 8.10, Eq. (8.10), illite (Fig. 8.8), smectite [Fig. 8.11, Eq. (8.11)], anhydrite [Fig. 8.12, Eq. (8.12)], and siderite (Marbler et al., 2013; Rathnaweera et al., 2016).

Figure 8.5 Field emission scanning electron microscope micrographs of calcite (Chopping and Kaszuba, 2012). Micrographs illustrate smooth, unreacted calcite (upper figure) and etched, dissolved calcite (lower figure) after reaction with injected CO_2.

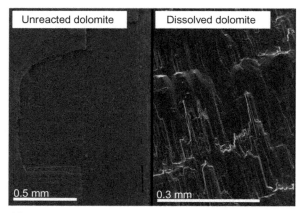

Figure 8.6 Field emission scanning electron microscope micrographs of dolomite (Chopping and Kaszuba, 2012) illustrating unreacted dolomite on left and etched/dissolving dolomite on right.

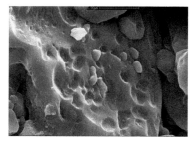

Figure 8.7 Micrograph illustrating pitting from quartz dissolution (Australian National University website, 2017).

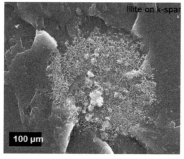

Figure 8.8 Illite precipitates on potassium feldspar (Rosenqvist et al., 2014).

Figure 8.9 Albite dissolution pits in potassium feldspar matrix after reaction with CO_2 (Rosenqvist et al., 2014).

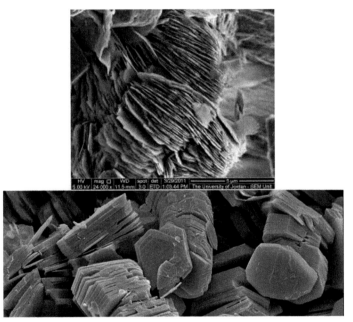

Figure 8.10 SEM examples of Kaolinite (Mahmoud, 2017; Corex website 2017). Kaolinite forms "booklets" which are susceptible to fines migration which can plug pores and decrease permeability.

Dissolution of minerals during injection can have multiple results. The liberation of ions through dissolution can cause precipitation and alteration of other minerals in the system. The nature of component minerals like dolomite and quartz can affect the mechanical integrity of the reservoir. In the case of quartz, corrosion can change the structure

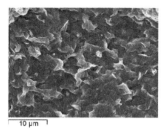

Figure 8.11 SEM micrograph of Montmorillonite (smectite) (Kelessidis and Maglione, 2008). Smectite is highly susceptible to swelling through reaction with water resulting in potentially significant permeability degradation.

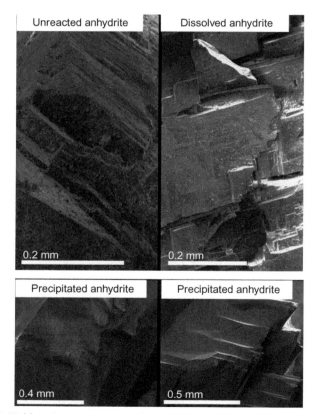

Figure 8.12 Field emission scanning electron microscope micrographs of unreacted anhydrite (upper left), dissolved (upper right) and precipitated anhydrite (lower left and right). (Chopping and Kaszuba, 2012).

of pores. Silicate mineral rims around quartz grains reduce the strength of grain to grain contacts, which can significantly change pore pressure (Marbler et al., 2013). If dolomite is present as framework grains and not as cement, dissolution can result in the reduction of mechanical integrity of the reservoir. Finally, injection and diagenesis can cause the liberation of fines which can result in pore throat plugging.

8.2.2.2 Precipitation

Changes in fluid/mineral chemistry caused by CO_2 injection and subsequent dissolution can lead to mineral precipitation. If CO_2 concentration is high enough to dry the formation, halite, and anhydrite precipitation can occur. If sulfur is present in the system, anhydrite and pyrite can precipitate (Kaszuba et al., 2011). Multiple studies indicate the potential formation of carbonate, quartz, and kaolinite from aluminosilicates (Kaszuba and Janecky, 2009; Gunter et al., 2000; Hitchon, 1996). However, Kaszuba and Janecky (2009) noted that during silicate dissolution, acidity enhances silica solubility and inhibits quartz precipitation.

Mineral precipitation during CO_2 injection can result in porosity and permeability reduction. The effects of porosity reduction are generally offset by dissolution but the effects of permeability reduction can be significant. The precipitation of clay minerals can result in the plugging of pore throats and/or decrease permeability through clay swelling because of pore fluid interactions [Fig. 8.13, Eqs. (8.7−8.11)]. The likelihood of carbonate precipitation increases with distance from the CO_2 injection site due to changes in pore fluid chemistry. Finally, desiccation of the formation during the injection process can result in halite precipitation, pore plugging, and permeability destruction (Fig. 8.14).

8.2.2.3 Geochemistry of mineral reactions and illustrations of impact at the mineral grain scale

The following equations and images (Figs. 8.5−8.14) illustrate examples of common mineral dissolution and precipitation reactions that can occur during CO_2 injection. The following interactions represent examples of the complex interaction between rocks and fluids during the injection process. As the arrows indicate, these reactions move in both directions depending on the geochemistry of the environment, which, as these reactions illustrate, can constantly change during CO_2 injection.

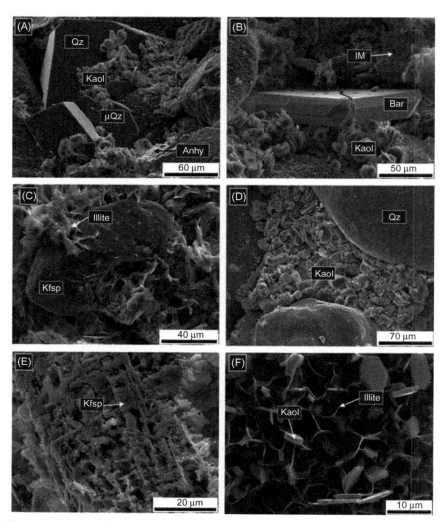

Figure 8.13 Secondary electron microscopy (SEM) images of detrital and authigenic minerals in Rotliegend sandstones. (A) Two quartz generations with early microcrystalline quartz and a later, partly euhedral formed quartz overgrowth coexisting with kaolinite (sample: Aw03-01). (B) Early diagenetic meshwork illite is followed by kaolinite and barite precipitation (sample: Aw03-0). (C) Illitization of a K-feldspar grain (sample: Cw02-01). (D) Arrangement of kaolinite "booklets" in a pore throat surrounded by detrital quartz grains with irregular distributed voids on the surface (sample: Aw02-12). (E) Partly dissolved K-feldspar grain (sample: Cw04-05). (F) Single kaolinite crystals growing on the surface of a detrital grain, partly covered by fibrous illite (sample: Bw01-01). Anhy = anhydrite, Bar = barite, IM = meshwork illite, Kaol = kaolinite, Kfsp = K-feldspar, Qz = quartz, μ Qz = microcrystalline quartz. These images illustrate examples of porosity and permeability reduction through mineral precipitation (Waldmann et al., 2014).

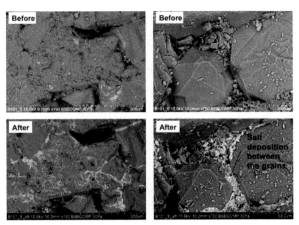

Figure 8.14 SEM images illustrating porosity and permeability destruction due to halite precipitation formed during desiccation of the formation during CO_2 injection (Agersborg et al., 2011).

Water and carbon dioxide react to form carbonic acid and result in a decline in pH:

$$H_2O + CO_2 \Leftrightarrow H_2CO_3 \Leftrightarrow H^+ + HCO_3^- \qquad (8.1)$$

Calcium carbonate dissolves in a strong acid to yield carbon dioxide:

$$CaCO_3 + 2H^+ \Leftrightarrow Ca^{2+} + CO_2 + H_2O \qquad (8.2)$$

Calcium carbonate dissolves in a weak acid to yield the bicarbonate ion:

$$CaCO_3 + H^+ \Leftrightarrow Ca^{2+} + HCO_3^- \qquad (8.3)$$

Dolomite at normal temperatures:

$$CaMg(CO_3)_2 \Leftrightarrow Ca^{2+} + Mg^{2+} + 2CO_3^{2-} \qquad (8.4)$$

Dolomite at higher temperatures:

$$CaMg(CO_3)_2 \Leftrightarrow CaCO_3 + Mg^{2+} + CO_3^{2-} \qquad (8.5)$$

Quartz dissolution at low pH:

$$SiO_2 + 4H^+ \Leftrightarrow SI^{4+} + 2H_2O \qquad (8.6)$$

Potassium Feldspar/Mica react with CO_2 in water:

$$3KAlSi_3O_8 + 2CO_2 + 2H_2O \Leftrightarrow Muscovite/Illite + 6SiO_2 + 2K^+ + 2HCO_3^- \qquad (8.7)$$

Or

$$KAlSi_3O_8 + Na^+ + CO_2 + H_2O \Leftrightarrow NaAlCO_3 + 3SiO_2 + K^+ \qquad (8.8)$$

Calcium/sodium (plagioclase) feldspar react with CO_2, water, and oxygen:

$$CaAl_2Si_2O_8 + H_2CO_3 + 1/2O_2 \Leftrightarrow Al_2Si_2O_5(OH)_4 + Ca^{2+} + CO_3^{2-}$$
$$(8.9)$$

Kaolinite dissolution at low pH:

$$Al_2Si_2O_5(OH)_4 + 6H^+ \Leftrightarrow 2Al^{3+} + 2SiO_2 + 5H_2O \qquad (8.10)$$

Smectite dissolution in water:

$$K_{0.19}Na_{0.51}Ca_{0.21}Mg_{0.08}\left(Al_{2.56}Fe_{0.42}Mg_{1.02}\right)(Si_{7.77}Al_{0.23})O_{20}(OH)_4 + 20H_2O$$
$$\Leftrightarrow 0.19Na^+ + 0.51K^+ + 0.21Ca^{2+} + 1.1Mg^{2+} + 0.42Fe^{3+}$$
$$+ 7.77H_4SiO_4 + 1.76OH^-$$
$$(8.11)$$

Anhydrite react with bicarbonate and magnesium ions in solution:

$$2CaSO_4 + HCO_3^- + Mg^{2+} \Leftrightarrow 3H^+ + 2SO_4^{2-} + CaCO_3 + CaMg(CO_3)_2$$
$$(8.12)$$

In summary, the injection of CO_2 into a reservoir changes in situ equilibrium conditions which can drive changes in mineral diagenesis. Although porosity and permeability can be enhanced during the injection process, formation damage can also result. The main types of potential formation damage during CO_2 injection include permeability reduction through the movement of fines, precipitation of clays which may react with pore fluids or precipitate in pore throats and decrease in porosity and permeability through mechanical compaction caused by corrosion of framework grains. The results of fluid/mineral interactions can cause mineral dissolution and precipitation processes which can both enhance and decrease reservoir permeability (and porosity). Injection can drive mineral reactions that change with time and distance from the injection wells, resulting in permeability enhancement in some areas of the reservoir and reservoir damage in other portions. As with all hydrocarbon producing operations, a good understanding of fluid and mineral properties is necessary to minimize the potential for reservoir damage during CO_2 injection and maximize the potential profitability of any EOR project.

8.2.3 Interactions between CO_2 and crude Oil

Formation damage can be caused by precipitation of different components within crude oil, most commonly paraffin wax and asphaltene. However, almost all previous studies regarding CO_2-induced crude oil precipitations have been focused on asphaltene precipitation, although wax is sometimes observed to co-exist with asphaltene in the precipitated solids (Novosad and Costain, 1990). One possible explanation is that paraffin wax has a good solubility in CO_2 at supercritical conditions (Crause and Nieuwoudt, 2000; Chartier et al., 1999), and most studies are conducted at temperatures and pressures above the critical point of CO_2 (31.1°C, 1071 psi).

Asphaltene is typically defined as the fraction of crude oil that is soluble in light aromatics such as benzene and toluene but insoluble in straight-chain solvents such as pentane or heptane. The definition of asphaltene also relates directly to saturate, aromatic, resin, and asphaltene (SARA) analysis, which is a common experimental method to determine the Saturates, Aromatics, Resins, and Asphaltenes fractions of heavy crudes (Speight, 2014). In terms of chemical composition, asphaltene are complex organic compounds consisting of aromatic and naphthenic rings containing nitrogen, sulfur, and oxygen atoms. Fig. 8.15 illustrates some possible idealized structures of asphaltene molecules (Groenzin and Mullins, 2000). Asphaltene structures can vary largely depending on the crude oil. Revelation of the structures of asphaltene has been a constant effort since the 1960s. A number of physical and chemical methods are available for

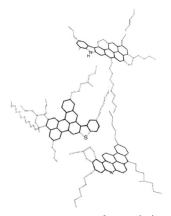

Figure 8.15 Idealized molecular structures for asphaltene consistent with overall molecular size, aromatic ring systems, and chemical speciation. The aromatic rings are shown with darker lines (Groenzin and Mullins, 2000).

construction of model structures for asphaltene: Physical methods include mass spectrometry, X-ray, infrared (IR), nuclear magnetic resonance (NMR), electron spin resonance (ESR), ultracentrifugation, electron microscopy, vapor pressure osmometry (VPO), gel permeation chromatography (GPC), etc., and chemical methods involve oxidation, hydrogenation, etc. Very recently atomic-force microscopy (AFM) was used in measurement of asphaltene structures directly (Schuler et al., 2015).

Although the exact nature of how asphaltene exist in petroleum fluids is still under investigation, one characteristic is that asphaltene tend to form aggregates in the crude oil solutions. These aggregates are micelles—they are colloidal suspensions in the hydrocarbon medium. Diameters of asphaltene molecules may be around 1−2 nm, while the diameter of asphaltene micelles may be approximately 6 nm (Groenzin and Mullins, 1999). An earlier study also reported micelle sizes within the range of 5−35 nm through ultrafiltration of petroleums (Neumann et al., 1981).

One commonly held view is that the colloids are stabilized by resins adsorbed on their surface (Pfeiffer and Saal, 1940; Leontaritis and Mansoori, 1987). Asphaltenes and resins combined constitute the polar fraction of a crude oil—the difference being resins are smaller molecules and dissolve in pentane or heptane. Fig. 8.16 schematically shows asphaltene-resin micelles that are suspended in the oil (Leontaritis, 1989). Because asphaltenes are colloids, they can flocculate and precipitate when the solution is destabilized. Asphaltene precipitation can be caused by pressure drop, shear (turbulent flow), acids, CO_2 injection and miscible gas injection, mixing of incompatible crude oils or other conditions or materials that alters the solubility of asphaltenes in oil. Flocculated asphaltene particles can often reach >100 nm in size (Leontaritis et al., 1994).

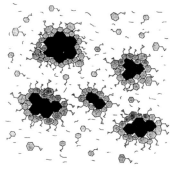

Figure 8.16 Asphaltene-resin micelles. Asphaltenes are shown in black and resins in gray (Leontaritis, 1989).

Asphaltene flocculation and *asphaltene precipitation* are often interchangeable in definition: Both refer to a stage when the suspended asphaltene colloids come out of the oil in the form of floc and create a separate solid phase. After asphaltene flocculation/precipitation, *asphaltene deposition* will occur. Asphaltene deposition refers to the settling of the precipitated asphaltene onto solid surfaces, such as the rock surface in a porous medium. Asphaltene deposition leads to permeability reduction and wettability change, both are important types of formation damage and have an effect on EOR performance.

8.2.3.1 The nature of CO_2-induced asphaltene precipitation

In a study conducted by Monger and Trujillo (Monger and Trujillo, 1991), 17 stock tank oils (STOs) were used to determine the amount of CO_2 induced precipitation in relation to the asphaltene amount in the original oils. These 17 STOs came from fields located in seven US states and Canada, and they ranged in gravity from 19.5° to 46.5° API. Table 8.1 lists the SARA analysis results of the 17 STOs. The asphaltene contents were measured both in-house and by a service company, showing good agreements in results. A high-pressure high-temperature variable-volume circulating cell (VVCC) was used in this study, which allowed oil to mix with increasing amount of CO_2 (injected into the system gradually during the experiment) and circulate through a 0.5 μm filter under constant temperature and pressure, so that the precipitated asphaltene could be captured by the filter. When the differential pressure across a filter reached 150 psi, the filter would be replaced by a new one. Four to eleven stainless-steel filters were required per run before the end of the experiments.

At the end of the experiments (when CO_2 reached 96 mol%), the precipitate was air dried and the total amounts of organic deposition were determined gravimetrically. Table 8.2 shows the amounts of CO_2-induced organic deposition measured for the 17 STOs examined in decreasing order of deposition extent. This same order was used in Table 8.1 to list STO compositions. Also listed in Table 8.2 were the pressure conditions for all runs and the MMP ranges for all the STO samples. It was found that the mass of CO_2-induced organic deposition was largely independent of the oil heaviness (properties that indicate the heaviness were listed in Monger and Trujillo (1991)), but was strongly correlated to the asphaltene content (correlation coefficient of 0.8). If results for one STO (Huntington Beach) were removed from the data base, the linear correlation coefficient would be $+0.94$ for CO_2 deposition versus either

Table 8.1 Stock tank oil (STO) compositions (Monger and Trujillo, 1991)

Sample	STO	C_{15+} Contents (wt%)[a]				STO Asphaltenes[b] (wt%)
		Saturate	Aromatic	Resin	Asphaltene	
1	Cedar Creek	34.2	38.6	12.9	14.2	13.87
2	Sample C	36	35.3	19.8	8.8	8.27
3	Brookhaven	57.3	28.1	12	2.6	1.8
4	Sample Et	61.6	29.5	7.2	1.7	0.89
5	Beechwood	62.5	25.9	10.9	0.7	1.32
6	Sample D	43.7	43.7	10	2.7	2.1
7	West Nancy	55.4	36.6	5.7	2.3	2.32
8	Huntington Beach	22.4	41	26.6	10	9.42
9	Sample A	58.1	31.6	10.1	0.3	0.17
10	Big Sinking	54	32	12.6	1.4	0.23
11	Sample B	39.8	41.5	17	1.7	1.29
12	Charenton	51.6	33.9	14.1	0.4	0.14
13	Matilda Gray	46.9	38.4	14.3	0.3	0.43
14	South Summerland	79.2	15.4	4.4	1	0.08
15	Timbalier Bay	48.1	31.4	18.7	1.8	0.49
16	Hilly Upland	75.3	15.6	8.8	0.3	0.06
17	Bath	70.4	19.7	8.6	1.3	0.21

[a]Measured by a service company.
[b]Measured in house with STOs filtered under ambient conditions. The in-house determinations of STO asphaltenes are in good agreement with the service company's C15 + asphaltene results.

determination of asphaltene content. Fig. 8.17(A) and (B) illustrates this relationship on log–log scales.

Although the correlation between the mass of CO_2-induced organic deposition and the asphaltene content was strong, the CO_2-induced organic deposition was not identical to asphaltene content (determined by pentane addition). For some oils, notably Cedar Creek, Sample C, and Huntington Beach (Sample numbers 1, 2, and 8, respectively), the extent of CO_2-induced organic deposition was substantially less than the asphaltene content. For other oils, notably Brookhaven and Sample A (Samples numbers 3 and 9, respectively), the extent of CO_2-induced organic deposition exceeded asphaltene content. Therefore, it was clear that the precipitation of asphaltenes by CO_2 was neither complete nor exclusive. Some asphaltenes could remain suspended, and other heavy organics could be precipitated.

Table 8.2 CO_2-induced organic deposition from Stock tank oil (STO) samples
(Monger and Trujillo, 1991)

Sample	STO	Deposition (wt%)	Run pressure (psi)	MMP (psi)
1	Cedar Creek	5.41	2810	2,330−2,970
2	Sample C	4.38	2040	1,820−1,880
3	Brookhaven	2.95	3150	1,790−2,160
4	Sample Et	1.56	2160	1,670−1,990
5	Beechwood	1.32	3300	2,840−3,250
6	Sample D	1.3	2480	2290−2,310
7	West Nancy	1.3	3130	2,130−2,580
8	Huntington Beach	1.14	4300	3,380−3,490
9	Sample A	1.02	2980	1,360−1,370
10	Big Sinking	0.983	3150	1,910−2,230
11	Sample B	0.967	1940	1,750−1,770
12	Charenton	0.9	3280	3,340−4,710
13	Matilda Gray	0.819	3840	3,700−5,290
14	South Summerland	0.793	3170	1,690−1,750
15	Timbalier Bay	0.733	3150	1,960−1,990
16	Hilly Upland	0.608	3150	1,930−2,130
17	Bath	0.576	3150	1,850−1,960

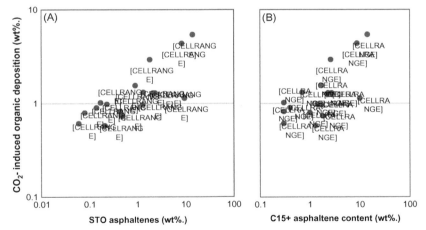

Figure 8.17 Relationship between CO_2-induced organic deposition and asphaltene
fraction for 17 STOs. CO_2 precipitates determined in VVCC experiments at conditions
above MMP: (A) asphaltenes measured in-house with cold-filter test and (B) by the
service company as C15 + pentane-insolubles. Data point labels are STO sample
numbers (Tables 8.1 and 8.2) (Monger and Trujillo, 1991).

The nature of CO_2-induced organic deposition was further tested by
compositional analysis. Table 8.3 lists precipitate compositions in terms of
saturate, aromatic, resin, and asphaltene contents for four of the STOs
(Samples 3−7). This analysis confirmed that both asphaltenes and other

Table 8.3 Precipitate compositions from stock tank oil (STO) samples (Monger and Trujillo, 1991)

Sample	Oil	Precipitating agent	Pressure (psi)	Temperature (°F)	Extractable contents (wt%)			
					Saturate	Aromatic	Resin	Asphaltene
5	Beechwood	CO_2	3300	118	60.6	25.1	13.4	0.9
7	West Nancy	CO_2	3130	116	47.9	31.3	11.5	9.3
3	Brookhaven	CO_2	3150	111	30.4	16.5	9.6	43.5
	Brookhaven reservoir	CO_2	3150	111	28	15.7	7.7	48.6
	Brookhaven flashed	CO_2	3150	111	23.9	14.9	8.6	52.6
4	Sample E	CO_2	2160	116	46	22.4	20.1	11.5
	Sample E	C_3	2530	134	19.1	12	19.8	49.1

fractions of heavy organics were precipitated upon mixing with CO_2. Compared to the original compositions listed in Table 8.1, all asphaltene portions of precipitates were higher than the corresponding STO asphaltene levels except in Sample 5. However, the asphaltene contents did not necessarily constitute the major portion of the precipitants. Resin was also preferentially precipitated to some extent. Saturates simply mirrored the compositions of the corresponding STOs. Although it was mentioned at the beginning of this section that CO_2 is a good solvent for paraffin and almost all CO_2-induced organic precipitation studies had been focusing on "asphaltene precipitation," the precipitates were not exclusively asphaltenes—indeed in some cases asphaltenes made up a relatively small component of the precipitates (Table 8.3).

Although the results listed above were derived from asphaltene precipitants collected by filters, an earlier published study conducted by Monger and Fu (1987) indicted that the extent of deposition using Berea (sandstone) cores were comparable to the ultimate deposition obtained using stainless-steel filters at the same conditions, based on three runs with Berea cores with a highly asphaltic oil (Brookhaven) in comparison with runs conducted with filters using the same VVCC equipment. This similarity between deposition captured by filters and sandstone cores suggests that organic deposition is not sensitive to pore topography. In the three runs with Berea cores, multiple Berea cores were also required to overcome the differential pressure build-up exceeding the limit of 150 psi during the deposition process, as well as to capture all the precipitants. Significant permeability reduction was associated with organic deposition as indicated by the pressure differential increases during the experiments. Fig. 8.18 shows the three solid-phase formation profiles using CO_2/Brookhaven-stock-tank-oil mixtures and native Berea core traps. Interestingly the build-up of differential pressure across the cores among the three Berea core tests was not reproducible, suggesting that variations of pore topography do have a significant effect on permeability reduction during organic deposition induced by CO_2 injection.

8.2.3.2 Onset conditions of CO_2-induced organic precipitation

Under stable reservoir conditions, saturates, aromatics, resins, and asphaltenes within the crude oil are in thermodynamic equilibrium. This equilibrium is disturbed by change in thermodynamic conditions: Pressure, temperature, and composition, organic precipitation can occur. Without the influence of injected gas, in a simple pressure depletion process, the

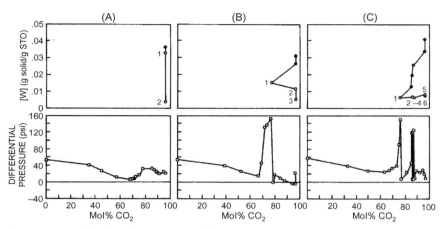

Figure 8.18 Solid-phase formation profiles generated at 111°F (317K) and 3150 CO_2/Brookhaven-stock-tank-oil mixtures and Berea core traps. Closed circles are cumulative measurements, and open circles with core numbers are individual measurements of organic deposition. A and B are for 1.5-in. long by 2-in. diameter cores, and C is for 0.4-in. long by 1-in. diameter cores (Monger and Fu, 1987).

amount of precipitation can be plotted against pressure at certain temperatures or against temperature at certain pressures. The effects of pressure and temperature on precipitation can be clearly observed in this way. If CO_2 is injected, then the precipitation versus CO_2 concentration can be plotted, usually against pressure at certain temperatures. Bahrami et al. (2015) studied an Iranian heavy oil (20.3°API) and systematically presented data on precipitation with pressure, temperature, and composition variations.

Fig. 8.19 demonstrates the asphaltene precipitation experimental results for depletion tests at three different temperatures. The original reservoir pressure is 30.82 MPa, and the bubble point pressures at 60°C, 80°C, and 96.1°C are 8.32, 9.10, and 9.88 MPa, respectively. From Fig. 8.19, maximum precipitation occurs near the bubble point pressure at each temperature, and above or below this pressure, the amount of precipitation reduces. The study used solubility theory to describe the asphaltene precipitation behavior—above the bubble point, pressure reduction causes more expansion in the light end components than in the heavy end components of crude oil and as a result, the density of oil and the asphaltene solubility parameter decreases, so more asphaltene precipitation is observed; below the bubble point, pressure reduction causes solution gas to be liberated from the liquid phase, hence the crude oil becomes richer in heavy end components and an increase in the solubility

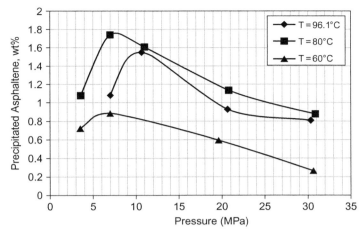

Figure 8.19 Experimental results of precipitated asphaltene, wt% versus pressure for an Iranian heavy oil (Bahrami et al., 2015).

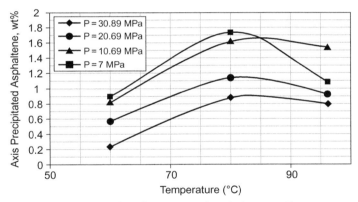

Figure 8.20 Experimental results of precipitated asphaltene, wt% versus temperature for an Iranian heavy oil (Bahrami et al., 2015).

parameter is observed. Results show that the solubility of asphaltene particles is very sensitive to pressure changes under saturation pressure at all different temperatures. Also among the three temperatures, 80°C resulted in higher precipitation than 60°C and 96.1°C, and this point is further revealed in Fig. 8.20.

Fig. 8.20 shows that the effect of temperature on asphaltene precipitation at four pressures using the same data that constructed Fig. 8.19. Based on the results, asphaltene precipitation increases from 60°C to 80°C and decreases from 80°C to 96.1°C. Since the stabilizing action of

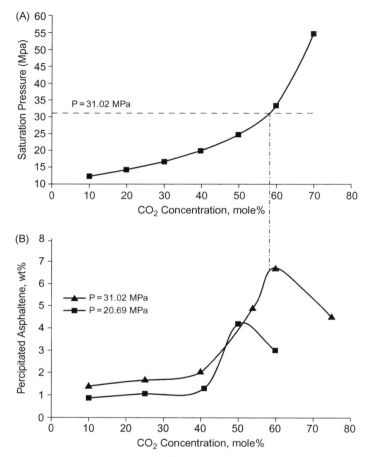

Figure 8.21 Experimental results of (A) saturation pressure of the mixture during CO_2 gas injection and (B) effect of CO_2 gas injection on asphaltene precipitation for an Iranian heavy oil (Bahrami et al., 2015).

the resins works through the mechanism of polar interactions, their effect becomes weaker as the temperature rises, that is, flocculation will occur as the temperature increases. However, at a specified temperature, solution entropy increases and asphaltene melts and redissolves in the oil as temperature increases. Around 80°C represents the threshold temperature where the precipitation trend reverses.

Fig. 8.21 shows CO_2 gas injection laboratory tests carried out at two pressures. By adding more CO_2 into the system, saturation pressure of the mixture increases, as shown in Fig. 8.21(A). The MMP is 26.2 MPa based on the slim tube experiment. Under a system pressure of 31.02 MPa, injected CO_2 gas will fully mix with the fluid until the CO_2 reaches about 60% (mole), when the saturation pressure reaches 31.02 MPa. During this process, the density of the oil decreases and the solubility of asphaltene in oil also decreases, resulting in increasing amounts of asphaltene precipitation. As CO_2 mol% increases beyond 60%, the saturation pressure becomes higher than the system pressure, as a consequence dissolved gas leaves the mixture and the asphaltene solubility increases, resulting in a decreased amount of asphaltene precipitation. The experiments performed at 20.69 MPa behave in a similar fashion, except that less CO_2 is dissolved throughout CO_2 injection. Therefore, less precipitation will occur at 20.69 MPa than at 31.02 MPa. Also the maximum asphaltene precipitation at 20.69 MPa occurs at about 50 mol% CO_2, at which state there is free gas in the system as 20.69 MPa is below the bubble point pressure at this composition, indicating maximum precipitation is not necessarily associated with bubble point pressure in CO_2/crude oil systems. Overall, CO_2 injection tests yielded much higher precipitation (maximum of 6.7 wt%) than that in natural depletion (maximum of 1.8 wt%), demonstrating the importance of monitoring asphaltene precipitation in a CO_2 flooding process. A visual study on CO_2-induced asphaltene deposition conducted by Zanganeh et al. (2012) also observed much higher amounts of deposition with increasing mole percent of injected CO_2 using high-pressure visual cells.

In a different study conducted by Nakhli et al. (2011), experiments were carried out upon two Iranian live oil samples: A heavy oil (19.7°API) with bubble point pressure of 7513 kPa at reservoir temperature of 96°C and a light oil (33.6°API) with bubble point pressure of 23,877 kPa at reservoir temperature of 113°C. Result of pressure depletion tests at these two temperatures is presented by Fig. 8.22. With pressure reduction, the amount of precipitation continues to increase until the bubble point pressure is reached; below the bubble point pressure, asphaltene precipitation decreases with decreasing pressures. This result is similar to that from the pressure depletion tests presented by Bahrami et al. (2015), and can be explained in the same way. Interestingly, temperature variation has impacts on the amount of asphaltene precipitation for two samples in opposite direction: When temperature goes up the amount of precipitated asphaltene increases in light oil, while it decreases for the heavy oil indicating the dependence of the temperature effect on

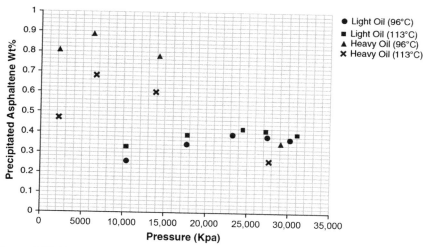

Figure 8.22 Experimental results of precipitated asphaltene, wt% versus pressure for a heavy oil and a light oil during pressure depletion at two temperatures (Nakhli et al., 2011).

oil composition. The behavior represents different threshold temperatures at which the precipitation trend reverses (as shown in Fig. 8.20) for the two different oils.

Furthermore, when CO_2 was mixed with the heavy oil (Nakhli et al., 2011), similar behavior as shown in Fig. 8.21 was observed. However, when CO_2 was mixed with the light oil, an unusual trend of precipitation versus CO_2 mole fraction was observed. CO_2 was injected into the light oil at 31,018 kPa and two temperatures of 96°C and 113°C. Fig. 8.23 shows the amount of precipitation due to CO_2 injection. Surprisingly, the amount of precipitation reduces with increasing CO_2 mole fraction up to a minimum point thereafter precipitation increases again—this observation was completely contrary to that observed for the heavy oil. It was concluded that CO_2 hindered the effect of pressure and temperature because the amount of precipitation was below the base line, which corresponded to the precipitation at reservoir pressure and temperature without gas injection. It was explained that the higher density of CO_2 compared to reservoir oil at 31,018 kPa resulted in the opposite effect of CO_2 on the solubility parameter of the mixture of oil/CO_2 in contrast to CO_2 injection in the heavy oil sample. This phenomenon could also be explained by the solubility parameter theory presented by Gonzalez et al. (2007). There is a crossover point temperature, below

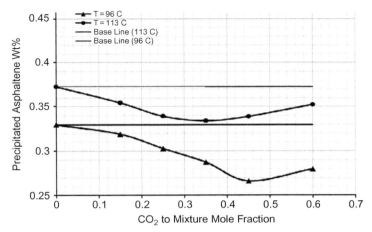

Figure 8.23 Experimental results of precipitated asphaltene, wt% versus CO_2 injection for a light oil (Nakhli et al., 2011).

which the solubility parameter of CO_2 is greater than the solubility parameter of the oil, thereby the addition of CO_2 under these conditions increases the solubility of asphaltene; above the crossover temperature, the addition of CO_2 reduces the solubility parameter of the oil and results in higher asphaltene precipitation. Based on this theory, the two temperatures used for obtaining the results in Fig. 8.23 should be lower than the crossover temperature.

In a number of additional studies (Srivastava and Huang, 1997; Huang et al., 2010), the onset conditions of CO_2 induced asphaltene precipitation were more clearly observed—a particular CO_2 mole fraction would correspond to the onset of asphaltene precipitation at a certain pressure and temperature. An example is presented by Srivastava and Huang (1997) in Fig. 8.24—CO_2 was injected into a crude oil from southeast Saskatchewan, Canada (Weyburn crude W2, 36°API) at 61°C and 16 MPa, and asphaltene flocculation with CO_2 concentration was plotted in this figure. No asphaltene flocculation was obtained at concentrations of less than 41 mol%, and a smooth increase in asphaltene flocculation with CO_2 concentration was observed above this critical value of 41 mol %, which is the onset concentration of CO_2 for this crude oil at the pressure and temperature. Another example is presented by Huang et al. (2010) in Fig. 8.25, where the onset concentrations of CO_2 (the left six points on the horizontal axis) for a crude oil of unspecified composition from China were experimentally measured at six different pressures. The

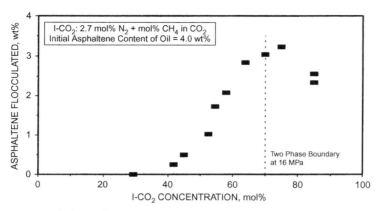

Figure 8.24 Asphaltene flocculation in Weyburn reservoir fluid W2 as a function of I-CO$_2$ concentration at 61°C and 16 MPa (Srivastava and Huang, 1997).

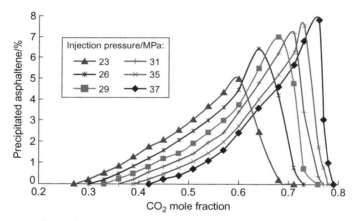

Figure 8.25 Relationship between precipitated asphaltene quantity and CO$_2$ mole fraction at different injection pressures for a Chinese crude of unspecified composition (Huang et al., 2010).

rest of the curves were constructed by model calculations with a method proposed by Nghiem et al. (1993) and equation of state proposed by Anderko (1992) after matching PVT experimental data of this crude. Also constructed by Huang et al. (2010) was Fig. 8.26, in which a phase envelope is shown on a pressure versus CO$_2$ concentration plot. The six gray (dark) points again represent the onset concentrations measured at the six different pressures, and the gray (light) dots represent measured bubble point pressures of crude/CO$_2$ mixtures at different CO$_2$ concentrations. The rest of the data points and curves were obtained by the same

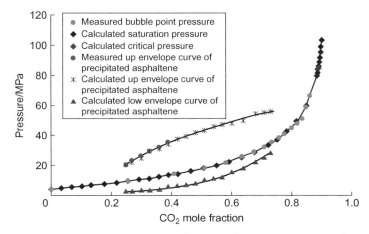

Figure 8.26 Pressure-composition phase diagram of experimental CO_2-oil system for a Chinese crude of unspecified composition (Huang et al., 2010).

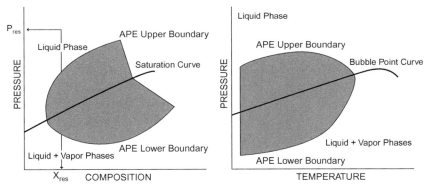

Figure 8.27 Pressure-composition and pressure-temperature APEs (Leontaritis, 1996; Anderko, 1992).

methodology in constructing Fig. 8.25. Based on the model, the amount of asphaltene precipitation reaches a maximum around the bubble point pressure at a certain CO_2 concentration.

Fig. 8.26 represents an asphaltene precipitation envelope (APE) (Leontaritis, 1996), or sometimes referred to as the asphaltene deposition envelope (ADE). There are two types of APEs: Pressure–composition APE (P-x APE) and pressure–temperature APE (P-T APE). Fig. 8.26 is a P-x APE. Model graphs of the two types of APEs are shown in Fig. 8.27 (Leontaritis, 1996; Anderko, 1992). In either graph, an upper

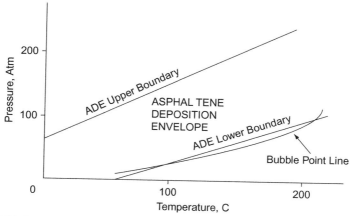

Figure 8.28 P-T ADE measured for Prinos oil (Leontaritis et al., 1994).

and a lower envelope curve above and below the bubble point pressure curve define the onset conditions of asphaltene precipitation. Within the APE, the amount of precipitated asphaltene increases as pressure decreases from the upper onset pressure to the saturation pressure of the oil, and the amount of precipitation reaches maximum at the bubble point pressure (or the saturation pressure) and decreases as pressure decreases below the bubble point pressure. The model graphs provide a general expectation of the asphaltene precipitation region in relationship to the bubble point pressure line. In reality, however, the bubble point line does not necessarily lie in the middle between the ADE upper and lower boundaries, or represent the conditions that result in maximum precipitation. For example, The P-T ADE of Prinos oil measured by Institut Francais du Petrole (IFP) is shown in Fig. 8.28 (Leontaritis et al., 1994). This figure shows that the reservoir fluid suffered asphaltene flocculation at some pressure above the bubble point. The bubble point pressure line is outside of the ADE lower boundary in some regions. As another example, the P-x ADE of Brookhaven crude measured by Monger and Fu is shown in Fig. 8.29 (Monger and Fu, 1987). This figure shows that the ADE is completely below the bubble point line.

The APEs can be measured by a number of different techniques, including: Gravimetric (Kabir and Jamaluddin, 1999), acoustic-resonance (Sivaraman et al., 1998), light scattering (Hammami and Raines, 1997), and filtration (Jamaluddin et al., 2002). Jamaluddin et al. conducted comprehensive comparison of measurements with the gravimetric,

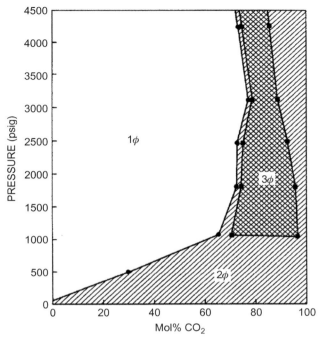

Figure 8.29 P-x ADE measured by a variable-volume circulating cell for mixtures of CO_2 and Brookhaven stock tank oil at 111°F (Monger and Fu, 1987).

acoustic–resonance, light–scattering, and filtration techniques on the same oil (Jamaluddin et al., 2002). These methods determine both the upper and lower APE pressure except for the acoustic–resonance technique. The acoustic–resonance technique normally defines only the upper onset pressure. In addition to APE pressures, the gravimetric and filtration techniques also provide the amount of precipitated asphaltene within the precipitation region, but they are more time consuming than the acoustic–resonance and light–scattering techniques. Additional techniques include electrical–conductance (Fotland et al., 1993; MacMillan et al., 1995) and viscometric (Escobedo and Mansoori, 1995) methods. The electrical–conductance technique can determine both precipitation onset and amounts of precipitate that are consistent with the gravimetric technique. The viscometric technique is advantageous in its applicability to heavy crude oil, which may pose difficulties for light–scattering techniques, and in the low–cost equipment. Two measurement techniques can be applied to the same oil to enhance the reliability of data interpretation.

Experimental measurements are very expensive for obtaining APEs, and usually limited data are available. In order to have a better understanding of asphaltene precipitation in a wide range of conditions, different models have been proposed and studied. Subramanian et al. (2016) did a comprehensive review on asphaltene precipitation models over the last 30 years. Prediction and modeling of asphaltene precipitation are based on either solubility theory or colloidal theory. The solubility approach assumes that the asphaltenes are dissolved in crude oil and the precipitation occurs if the solubility falls below a certain threshold level (Ting et al., 2003). The colloidal theory assumes that the asphaltenes exist as colloidal particles stabilized by resins adsorbed onto their surfaces (Leontaritis et al., 1988). The partitioning of resins between the colloidal surface and the surrounding medium controls the asphaltene solubility. If a sufficient amount of resins desorbs, then the asphaltenes will be destabilized and precipitate (Agrawala and Yarranton, 2001). In that review (Subramanian et al., 2016), the applications and limitations of different models in predicting the onset and the amount of asphaltene precipitation were also discussed. Asphaltene precipitation and deposition continue to be major challenges faced by the oil industry. The quality of asphaltene precipitation prediction is continuously improved with the developments in the fluid characterization technique and accounting for asphaltene polydispersity in modeling.

The APEs represent the thermodynamics of crude oil/CO_2 asphaltene precipitation systems. The kinetics of CO_2 induced asphaltene precipitation has also been studied. Idem and Ibrahim (2002) evaluated the kinetics of CO_2-induced asphaltene precipitation for three Saskatchewan crude oils under isothermal and isobaric reservoir conditions in a pressure volume temperature (PVT) cell. The results show that the rate of asphaltene precipitation depends on both the asphaltene and CO_2 contents of the oil. The orders of reaction for CO_2 and asphaltene (n and m, respectively) change with temperature, indicating the mechanism for CO_2-induced asphaltene precipitation is temperature dependent. The order of reaction with respect to CO_2 (n) is much larger than the order with respect to asphaltene (m) implying that CO_2-induced asphaltene precipitation is more sensitive to the CO_2 added than the asphaltene content of the oil. Also, the overall order of reaction (m + n) is very large, suggesting that CO_2-induced asphaltene precipitation is not an elementary process.

In asphaltene precipitation kinetics studies, another focus has been on the reversibility of the precipitations (Mullins et al., 2007; Hammami et al., 2000). Reversibility is a requirement for the application of equilibrium thermodynamics to predict the phase behavior of mixtures. Determining whether the asphaltene precipitation process is reversible or not will provide valuable insight into determining whether colloidal or solubility models are appropriate to model asphaltene phase behavior. From a thermodynamic standpoint, asphaltene flocculation was observed to be either reversible or irreversible (Lichaa, 1977). The colloidal models consider precipitations to be irreversible once the asphaltenes are destabilized, while solubility approaches use thermodynamic phase behavior models to describe the solution behavior of asphaltenes and treats precipitation as a reversible process. Therefore colloidal models are more suitable to investigate asphaltene precipitation that is irreversible and solubility approaches are more appropriate to model asphaltene precipitation that is reversible (Chaisoontornyotin et al., 2017).

8.2.4 Wettability alterations

Asphaltene deposition is considered to be a major cause of wettability reversal within an oil reservoir. Wettability is defined as "the tendency of one fluid to spread on or adhere to a solid surface in the presence of other immiscible fluids," and it is a key property that controls multiphase fluid flow in oil recovery processes. The degree of wettability alteration due to asphaltene deposition depends upon oil composition, brine chemistry, surface mineralogy, pressure, and temperature (Anderson, 1986). A number of laboratory experiments in literature have investigated the wettability alteration during CO$_2$ flooding processes.

Monger and Fu (1987) conducted experiments on CO$_2$/crude oil injection into Berea sandstone cores and examined the wettability change due organic deposition. They concluded that sufficient levels of CO$_2$-induced organic deposition typically made Berea cores less water wet and more oil-wet regardless of the initial core wettability. Native cores could be altered to a mixed-wettability condition, and artificially neutral cores became strongly oil wet. No relationship between deposition extent and the degree of wettability alteration attained was observed. The presence of brine reduced, but did not eliminate, CO$_2$-induced organic deposition in strongly water-wet cores. Clays appear to be important deposition sites, as clay stabilization (by KOH treatment) enhanced CO$_2$-induced organic deposition, and once deposition occurs, the sensitivity of the core to fresh water was significantly reduced.

Leontaritis et al. (1994) indicated that flocculated asphaltene possess a positive charge (Leontaritis et al., 1988; Nicksic and Jeffries-Harris, 1968; Lichaa and Herrera, 1975), and thus can behave as cationic surfactants, with the water-soluble portion of the particles being positively charged. The positive charge of asphaltenes facilitates their penetration of connate water to attach to the negatively charged water-wet mineral surfaces (e.g., sand and clays), so as to convert the surfaces from water-wet to oil-wet. As cationic surfactants, flocculated asphaltenes are normally expected to (Hirschberg et al., 1984) oil-wet the sand, shale, or clay; water-wet the limestone or dolomite up to pH 8; oil-wet the limestone or dolomite if pH = 9.5 or higher; break the oil-in-water emulsions; emulsify water in oil; disperse clays or fines in oil; and flocculate clays in water. The adsorption of flocculated asphaltenes onto rock surfaces often results in the irreversible alteration of mineral properties, especially those of clays, simply because some asphaltenes are no longer available to redisperse in the oil. There is evidence that, even after deflocculation, the flocculated asphaltene particles do not return to exactly the preflocculated asphaltene (Katz and Beu, 1945).

Huang and Holm (1988) investigated the effect of rock wettability on CO_2 flooding oil recovery. They created cores of different initial wettabilities by treating Berea sandstones with different wetting agents. Then they conducted CO_2 flooding using these cores and compared the amount of oil remaining in the cores after 1 PV of fluid injection. By comparing the results, they concluded that the amount of oil trapped in the water-wet cores was much higher than that trapped in either the mixed-wet or oil-wet cores. Oil trapping caused by the water-blocking effect is significant, especially at higher water saturations (Lin and Huang, 1990; Tiffin and Yellig, 1983; Shelton and Schneider, 1975). On the other hand, experiments for mixed-wet and oil-wet cores suggest that oil trapping is insignificant after large PV's of injection (Tiffin et al., 1991; Magruder et al., 1990). Therefore, wettability reversal in sandstone reservoirs due to asphaltene precipitation may provide a mechanism for improved oil recovery when the precipitation is less extensive, as indicated by Monger and Fu (1987), although the overall outcome of asphaltene precipitation is normally undesired and destructive to the CO_2 flooding process.

Along with wettability change, flocculated asphaltenes can also disperse fine particles in the oil and affect fines migration. If fines migration is expected to be a problem in a waterflooded reservoir, the flocculated asphaltenes will tend to stabilize the clays and make them hydrophobic and less sensitive to the injected water. Conversely, if oil is the primary flowing phase, the oil-wet clay particles and fines will tend to mobilize because of the increase in the hydro-dynamic drag forces (Tiffin et al., 1991). Therefore, reservoirs that are produced in the primary mode with asphaltene flocculation may be expected to suffer fines migration near the producers. If so, the fines are expected to become oil-wet as a result of asphaltene flocculation. Also, the presence of other solids with water in the produced fluid can worsen the consequences of asphaltene precipitation, generating a greater mass of solids and/or stable emulsions (Sarbar and Wingrove, 1997). The stabilized emulsions can be sufficiently strong to plug production. Fine migrations and stable emulsion formation facilitated by asphaltene precipitation provide additional mechanisms for formation damage besides porosity and permeability reduction caused directly by asphaltene precipitation.

8.2.5 Effect of asphaltene precipitation on oil recovery

Asphaltene deposits may heavily damage the reservoir and negatively influence production. On the other hand, formation damage can potentially have a positive impact on oil recovery. Monger and Fu (1987) suggested a few mechanisms on how CO_2-induced organic deposition may improve oil recovery. First, by analogy to polymer flooding, CO_2-induced organic deposition may lower the mobility of the displacing phase, which had been observed and lead to improved oil recovery. Secondly, changes in wettability can improve waterflood performance through the establishment of continuous flow paths for oil. Since a wetting phase can ultimately be completely displaced, this would permit the attainment of ultra-low residual oil saturations during CO_2 flooding. They suggested that when organic deposition is less extensive, smaller regions of the reservoir would become oil wet and oil recovery would be improved through the coalescence of residual oil blobs and the localized imbibition of bypassed oil into zones swept by CO_2. Srivaslava et al. (1993) suggested that permeability reduction can potentially plug high permeability areas of the reservoir and divert drive fluid to low permeability zones where oil is often trapped, which provides an additional EOR mechanism.

All these proposed EOR mechanisms are possible in theory, but in reality, the total effect of CO_2-induced asphaltene precipitation must be considered; frequently asphaltene formation damage leads to reduced oil recovery, aside from raising severe production concerns. Also, for any crude oil with a potential to precipitate asphaltenes upon CO_2 injection, an interesting and practical issue would be how operating conditions (especially pressure) would affect oil recovery. A number of laboratory and simulation studies have explored how asphaltene precipitation affects EOR and how operating conditions impact the oil recovery through coreflood experiments.

Hamouda et al. (2009) investigated miscible CO_2 flooding of oils containing asphaltene in carbonate chalk cores. Three types of oils were used in this study: (1) n–decane (reference), (2) model oil (0.35 wt % asphaltene dissolved in toluene and 0.005 M stearic acid dissolved in n–decane), and (3) crude oil (containing 10% asphaltene, 30° API gravity). Oil-saturated core samples were placed in a core holder under a net overburden pressure of 20 bar. Then CO_2 was injected into the core at a constant pressure (90, 120, or 140 bar) and temperature (50°C, 70°C, or 80°C)—all in a miscible condition. The results of the nine sets of oil recovery experiments are listed in Table 8.4. Oil recovery from n–decane is used as a reference oil and showed an increase of recovery from 81% to 89% at 50°C and 90 bar and 80°C and 140 bar, respectively. Both model and crude oils showed a reduction in the oil recovery from 78.92% to 70.4% and from 37.6% to 36.6%, respectively, at the same conditions as that for the reference oil. The insignificant change in the oil recovery among the three conditions for the crude oil was attributed to high asphaltene content by the author.

Although there was no comparison among certain key properties (such as viscosity) of the three oils, the significant total oil recovery differences among the three oils seemed to indicate severe negative effects of

Table 8.4 Oil recovery from three oils at three different temperature and pressure conditions (Hamouda et al., 2009)

Oil used	Temperature and pressure	Recovery (%) before BT	Total recovery (%)
n–decane	80°C and 140 bar	52.58	89
	70°C and 120 bar	47.7	86
	50°C and 90 bar	50.11	81
Model oil	80°C and 140 bar	41.73	70.4
	70°C and 120 bar	42.76	75.5
	50°C and 90 bar	55.4	78.92
Crude oil	80°C and 140 bar	2.74	36.6
	70°C and 120 bar	13.52	36.9
	50°C and 90 bar	6.11	37.6

asphaltene precipitation on oil recovery. From their results, before CO_2 breakthrough, oil production was due to a nearly piston-like displacement mechanism with a higher oil recovery compared to after breakthrough for all of the cases studied, except for the crude oil. The authors considered that the higher oil recovery in the case of n-decane and model oil may be attributed to miscible bank displacement accompanied with interfacial tension reduction; for the crude oil, lower recovery was attributed to early breakthrough as a result of channeling. In the crude oil case, lighter oil was produced after breakthrough (visually observed), which may indicate that the dominate mechanism for oil recovery after CO_2 breakthrough was solvent extraction.

Chukwudeme and Hamouda (2009) conducted reservoir simulations using Eclipse based on laboratory results obtained in the same rock and fluids system mentioned above (Hamouda et al., 2009), to investigate the effect of CO_2-oil recovery on a field scale. The relative permeability curves that were fed into reservoir simulations were obtained from Sendra Simulator, which relies on experimental parameters and results as an input. Three phase model (water, oil and gas) and miscible option were used for both asphaltenic and nonasphaltenic oils, and for the asphaltenic oil, asphaltene precipitation option was selected. The results showed that the oil recovery for nonasphaltenic oil was much higher than that for asphaltenic oil on a field scale. Both the residual oil saturation and the cross points of the relative permeability curves were highly unfavorable for oil flow in the asphaltenic oil case. Also very high reservoir pressure response from asphaltene precipitation was observed when the amount of precipitation was large. The low oil recovery for asphaltenic oil was attributed to fingering and oil trapping.

Cao and Gu (2013) investigated oil recovery and asphaltene precipitation in immiscible and miscible CO_2 flooding processes. A series of five CO_2 coreflood tests were conducted at immiscible, near-miscible, and miscible conditions using tight sandstone reservoir core plugs and a light crude oil containing 0.26 wt% asphatene. The oil recovery factor became higher at a higher pressure during the immiscible CO_2 flooding. Once the pressure exceeded the MMP, the oil recovery increased slightly and eventually reached an almost constant maximum value in the miscible CO_2 flooding. These observations are consistent with the expectations from a typical CO_2 flooding experiment using a nonasphaltenic oil, but inconsistent with what was observed by Hamouda et al. (2009). A possible explanation is that the typical trend of incremental oil recovery with increasing pressure until MMP can be maintained if the amount of asphaltene precipitation is small.

In the studies conducted by Hamouda et al. (2009), it was also found that the produced oil in all five tests resulted in lower asphaltene content than the original oil, and the asphaltene content of the produced oil decreased with increasing pore volumes of injection of CO_2, indicating more and more asphaltene deposition within the cores during CO_2 flooding. Additionally, the average asphaltene content of the produced oil was found to be higher in an immiscible condition compared to a miscible condition, indicating less asphaltene precipitation at lower pressures. On the other hand, the oil effective permeability reduction was also higher in an immiscible condition. The authors (Hamouda et al., 2009) speculated that both the asphaltene precipitation and wettability alteration jointly contributed to a larger oil effective permeability reduction in an immiscible CO_2 flooding case. Also in a miscible CO_2 flooding case, although more heavy hydrocarbons, including the precipitated asphaltenes, were left in the reservoir, they were more likely produced during the final original crude oil reinjection after CO_2 flooding during permeability measurements. As miscible flooding conditions resulted in higher oil recovery and lower produced asphaltene content in their experiments, the authors (Hamouda et al., 2009) concluded that the results explained why a near-miscible or miscible CO_2 flooding process is always pursued in CO_2 field applications if CO_2-induced asphaltene precipitation is not of major concern.

8.3 CO_2 FLOODING FORMATION DAMAGE MONITORING, PREVENTION, AND REMEDIATION

Asphaltene deposition in reservoirs, wells, and facilities severely impacts the oil production economics. It reduces flow as the deposit build-up in the production system, causing formation damage and wellbore plugging, and in some extreme cases causing the wellbore to be completely plugged. Removal or reduction of asphaltene deposits from oil wells can be highly expensive, and the loss in revenue due to reduced flow can be also significant. A remediation treatment with coiled tubing could cost approximately from USD $0.5 MM on land or in shallow water, to USD $3 MM or more for a deepwater well where entry into the wellbore is required (Creek, 2005). For the Gulf of Mexico oil fields, the economic impact associated with asphaltene deposition has been estimated in USD $70 M per well (wet tree) when well shut in for ring interventions is required (Melendez–Alvarez et al., 2016). If the deposition occurs in the surface controlled subsurface safety valve, the cost increases to USD $100 M per well (Melendez–Alvarez et al., 2016). In more severe situations of losing a well and replacing the lost well with a sidetrack, the cost could go up to $150 MM (Melendez–Alvarez et al., 2016).

In prevention or reduction of asphaltene deposition by injecting chemicals, chemical additive costs roughly between USD $330,000 and USD $390,000 per well per year for typical Gulf of Mexico production, and USD $31,000 to USD $46,000 per well per year for Middle Eastern fields, depending on the oil production rate of a well (Melendez–Alvarez et al., 2016). Moreover, given a production rate of 10,000 bbl/day at $60 USD/bbl, downtime loss reaches $600,000 per day. Seeing that asphaltene deposition significantly impacts the economics of a project, it should be taken into consideration when a CO_2 EOR project is designed or implemented.

8.3.1 Monitoring and identification

Monitoring and identification of asphaltene damage are prerequisites for successful prevention and remediation as well as performance evaluation of inhibitors/dispersants. In the laboratory, quite a few methods are well developed for the evaluation of asphaltene precipitation and deposition, such as the asphaltene dispersion test, solid detection system, microscopy, and filtration (Melendez–Alvarez et al., 2016). Either visual observation, microscopic, or spectroscopic methods are employed to detect the onset of precipitation. In the field, asphaltene damage in the near–wellbore regions impairs the formation's permeability, resulting in a production rate decline (Thawer et al., 1990). The extent of asphaltene damage in the near–wellbore region can be quantified by increased skin factors measured from well testing, given that it has been identified as the dominant damage mechanism.

Laboratory and field evaluation techniques are commonly used in combination for predicting and designing field applications. Shen and Sheng (2016) employed nano-size membranes to filtrate and analyze the asphaltene particles aggregated during CO_2 and CH_4 injection in shale cores. In comparison to pore sizes of Wolfcamp shale, Eagle Ford shale, and Mancos outcrop cores measured by mercury intrusion, most asphaltene particles have larger diameters, suggesting that shale matrix pores can be easily blocked by asphaltene aggregates during CO_2 or CH_4 injection. Recently, analytical centrifuge stability analysis for asphaltenes (ACSAA) was used to continuously monitor the asphaltene treatment effectiveness in a WAG CO_2 flooding field project in North America (Jennings et al., 2016). ACSAA uses a centrifuge than can spin multiple fluid samples while measuring their transmittance to near-IR light. Well and equipment failures have been identified as a consequence of asphaltene deposition in this field. By combining inhibitor injection and continuous ACSAA analysis, the crude oil instability index has been well correlated to inhibitor concentration, treatment effectiveness, and performance evaluation of different inhibitors with good precision. This field monitoring method also provides an effective means to detect the critical inhibitor concentration for stabilizing asphaltenes and is used to determine the optimal interval for inhibitor treatments.

8.3.2 Prevention and remediation

Since asphaltene starts to precipitate at certain pressure, temperature, and solvent mixing conditions, to keep the operating parameters outside the deposition envelope is a straightforward way to avoid asphaltene induced formation damages (Leontaritis et al., 1994). In the Hassi Messaoud field in Algeria, based on the observation that asphaltene deposition did not occur below the bubble point and previous deposits could be repeptized by the two phase stream, chokes were installed in five wells to lower the tubing pressure and productivity improvements were achieved in all these wells (Leontaritis et al., 1988). In the Ventura Avenue field in California, asphaltene deposition was drastic in the early production phase, but the asphaltene deposition problem was found to diminish at pressures lower than the bubble point. As the field depleted, wells had been producing damage free since the early 1970s (Tuttle, 1983). Sometimes maintenance of high pressures in production is viable in preventing asphaltene deposition. For example, according to the ADE, producers in the sour Prinos field in north Aegean Sea operated at high wellhead pressures to inhibit asphaltene deposition (Leontaritis et al., 1988). As another example, the Thunder Horse field in the deepwater Gulf of Mexico typically maintain higher pressures in the production system than the bubble point pressure to prevent severe asphaltene deposition.

During operation, as reservoir pressure depletes and water or gas injection starts, operation control parameters reside in a limited window, avoiding asphaltene flocculation may not always be possible. Various techniques, including injection of inhibitors and dispersants, chemical solvent washing, physical/mechanical treatments, and combinations of these methods, have been developed for preventing or removing asphaltene deposition from the near-wellbore region, wellbore, downhole assemblies, production tubing, and surface facilities.

8.3.2.1 *Chemical treatments*

Chemical treatment is a common technique for alleviating asphaltene damage. Chemical additives can be continuously injected into the crude oil stream at either the wellhead or the bottom-hole through a capillary tube. In addition, chemical additives can be squeezed into the near-wellbore region and mixed into complex bottom-hole assemblies beyond the reach of regular mechanical tools, for example the gas lift systems (Yen et al., 2001) and downhole pumps (Allenson and Walsh, 1997).

Chemical additives can be classified into two types: Inhibitors/dispersants and solvents.

8.3.2.1.1 Asphaltene inhibitors/dispersants

Asphaltene inhibitors and dispersants are chemical additives that prevent or interrupt the processes of asphaltene precipitation and deposition. Asphaltene inhibitors are believed to act as resins and maltenes that naturally exist in crude oil to stabilize asphaltene micelles in solution but with much stronger associations with asphaltenes (Allenson and Walsh, 1997). The performance of additives is a function of a series of variables, including temperature, pressure, asphaltene structure and properties, additive structure, and adsorption (Melendez-Alvarez et al., 2016). The delay in the onset of asphaltene precipitation is typically used as a measure to evaluate the effectiveness of the chemical additives (Tavakkoli et al., 2015).

Conventional asphaltene dispersion tests and solid detection systems are used to determine the asphaltene onset conditions for crude oil samples with or without dispersants. However, due to the insufficient resolution of optical or near-IR spectroscopy, direct observations can only capture asphaltene aggregates down to 1 μm, while in effect small asphaltene particle sizes are around 100 nm when precipitation starts. That is, these conventional methods are unable to differentiate the asphaltene precipitation and aggregation processes. Based on centrifugal analyses, Melendez-Alvarez et al. (2016) concluded that dispersants do not seem to delay the asphaltene onset as have been widely accepted, but they can slow down the subsequent aggregation. Meanwhile, to replace the onset concept, they proposed an index to quantify the dispersive performance instead of the precipitation prevention performance for dispersants.

González and Middea (1991) evaluated the peptizing performance of different oil soluble amphiphiles and found that nonyl phenol is a good candidate capable of preventing the asphaltene precipitation and reducing their adsorption onto quartz. By incorporating a few oxyethylene units (e.g., two or four) onto the nonyl phenol, asphaltene adsorption on quartz increases; grafting a large quantity of oxyethylene units on the other hand can effectively prevent the asphaltene adsorption. Nevertheless, these peptizing agents are unable to desorb the asphaltene that are already adsorbed onto the quartz surface. Chang and Fogler (1994) further varied the polar head and alkyl tail to investigate the performance of alkyl phenols in stabilizing the asphaltenes using Fourier transform IR spectroscopy and small-angle X-ray scattering techniques. Consistently, it was found that an addition of two oxyethylene units worsened asphaltene stability, while using a head group with higher polarity or acidity (i.e., $-SO_3H$) improved the amphiphile's effectiveness in stabilizing asphaltenes. Also, increasing the alkyl tail length from a nonyl to a dodecyl improved the chemical performance, as longer tails are more favorable for forming a stable steric layer around asphaltenes. In addition, disperse media made of lighter alkanes tend to reduce the association between asphaltenes and amphiphiles and weaken the effectiveness of amphiphiles.

Ibrahim and Idem (2004) fitted an empirical power law to the kinetics of n-pentane induced asphaltene precipitation from three crude oils and tested the performance of toluene, nonyl phenol, and dodecylbenzene sulfonic acid on precipitation inhibition. All these three additives drastically reduced the asphaltene precipitation rates in three crude oils but to different extents, demonstrating that the precipitation rate is a strong function of the asphaltene content and characteristics, such as content of heteroatoms, paraffin fraction, and aggregation propensity. Using toluene and a more polar solvent o-dichlorobenzene as the dispersion media, Barcenas et al. (2008) tested and simulated the inhibition performance of nonyl phenol and two oxazoline derivatives of polyalkyl or polyalkenyl N-hydroxyalkyl on Puerta Ceiba asphaltene. It was observed that inhibitors at low concentration of 0.1 g/L notably reduced the asphaltene aggregation number in toluene at 50°, but the inhibition effectiveness diminished at high concentration of 0.3 g/L. Based on vapor pressure osmomentry measurements and Monte Carlo modeling of inhibitor adsorption onto asphaltene structures, the authors suggested that at high concentrations the inhibitors may tend to self-associate in the bulk solvent rather than adsorb to the asphaltene surface, thus worsening the inhibition effectiveness. In the case of 6 g asphaltene in 1 L o-dichlorobenzene at 90°C, nonyl phenol even promoted the asphaltene agglomeration. Vapor pressure osmomentry measurements indicated that inhibitor significantly self-associates into complexes in this more polar solvent, possibly enhancing the asphaltene adsorption and agglomeration onto them. Therefore, there exists an optimum concentration that maximizes the inhibitor's performance, and high concentrations of inhibitors could be just counteractive. The existence of such an optimal inhibitor dosage was also predicted by a free energy based molecular thermodynamics model, which can rationalize the effect of asphaltene, inhibitor, and solvent characteristics on asphaltene aggregation, such as the number of active sites on asphaltene molecules, and the molecular structure and solubility of inhibitors (Rogel, 2010).

Besides these relatively low molecular weight (M) dispersants, polymeric additives, such as polyolefin amide alkeneamine with M $\sim$ 2000 and an alkylated phenol with M $\sim$ 4000, were also demonstrated to be capable of reducing asphaltene particle size and greatly delaying the occurrence of asphaltene sedimentation (Hashmi et al., 2010).

It is well-known that the performance of asphaltene inhibitors is specific to crude oils. Through measuring the heteroatom abundance, Smith et al. (2008) tried to explain the inhibitor specificity from the acid-base interactions between the inhibitors and the polar components in crude oils. Wang et al. (2009) analyzed the functional groups and measured the zeta potential of two kinds of asphaltenes, one of which was negatively charged with rich carboxyl and calcium, and the other was positively charged with pyrrolic and nickel/vanadium. By using dodecyl benzene sulfonic acid and dodecyl trimethyl ammonium bromide as inhibitors, it was demonstrated that cationic amphiphiles tended to work with negatively charged asphaltenes while anionic amphiphiles were compatible with positively charged asphaltenes. Therefore, for different oil reservoirs laboratory screening of proper asphaltene inhibitor/dispersant is critical for successful field applications.

Screening of asphaltene inhibitors generally involves oil sample characterization (SARA-saturate, aromatic, resin, and asphaltene) and different asphaltene precipitation tests in searching for effective additives. For example, Firoozinia et al. (2016) adopted microscopic observation, and viscosity and turbidity measurements at laboratory conditions and solid detection system and filtration measurements to evaluate the performance of four dispersants (nonyl phenol-formaldehyde modified by polyamines, rapeseed oil amide, polyisobutylene succinimide, and polyisobutylene succinic ester) on two crude oils. They found that nonyl phenol-formaldehyde modified by polyamines exhibited the best performance on inhibiting asphaltene precipitation. In addition, asphaltene phase behavior modeling provides a verification option for experimental results and can help extend the experimental conditions. By integrating these techniques for the Marrat reservoir in southern Kuwait, Kabir and Jamaluddin (1999) established the ADE and screened a combined formula consisting of deasphalted oil and dispersant, which was more cost-effective and environment-friendly than the widely used toluene. With SARA results and a solid detection system, Yen et al. (2001) selected an inhibitor that was compatible with the live crude, and successfully applied the inhibitor to a dual string well in Alaska, demonstrating that laboratory evaluation and field performance can be closely correlated.

8.3.2.1.2 Solvents
Aromatics, primarily xylene, and toluene, are the most commonly used solvents for dissolving asphaltene depositions. Xylene has been applied successfully in many cases in restoring the asphaltene damage. Some well-known limiting factors of using aromatic solvents include chemical consumption, failure to maintain production increase, ineffectiveness in water-wet reservoirs, and environmental and safety requirements. Several other solvent options have been investigated in the laboratory and the field in the past few decades, including aromatic distillates, nonaromatic naphthas, allyl chloride, water-soluble thiocarbamates, nonaromatic alkenes, diethyl ether, their blends, etc. (Samuelson, 1992). By adding aliphatic amines or dodecylbenzylsulfonic acid, solvency and stability of asphaltenes or asphaltene/paraffin sludge can be further enhanced in aromatic and nonaromatic solvents (Samuelson, 1992). As a biodegradable chemical, d–limonene, a natural terpene extracted from Brazilian oranges, was used successfully as a substitute for xylene and toluene (Rae and Lullo, 2001).

Cosolvents of xylene and alcohols have been tested and demonstrated better performance than xylene or mechanical scraping alone in removing asphaltene deposits. Alcohol does not improve the dissolving capacity of xylene, the advantage it brings is to help penetrate water barriers, such as water wet surfaces, emulsions, or sludge. Formation squeezing (5−50 gallons/ft pay zone) and soaking (24−48 hours) treatments using these cosolvents in 31 wells in Alaska, Alberta, Wyoming, etc. to remedy asphaltene damage showed that higher production rates were maintained for longer periods of time than straight xylene, attributing to the hydrophilic characteristics of alcohol (Trbovich and King, 1991). Deasphalted crude oil has also been widely used in field tests, as it contains original resins and maltenes that can help peptize and stabilize asphaltene particles within crude oil. Dispersant addition could further increase the asphaltene solubilizing capacity of deasphalted oils. Jamaluddin et al. (1996) and Kabir and Jamaluddin (1999) experimentally evaluated the asphaltene dissolution potential of deasphalted oil and additive mixtures, and found that addition of alkylbenzenesulphonate significantly improved the solubilizing power of deasphalted oils. Several mixture formulas that outperformed conventional solvents, that is, toluene, were screened. Additional benefits of using deasphalted oils for asphaltene damage remediation include higher efficiency in transporting produced sands to the surface and much less cost.

8.3.2.2 Mechanical and physical remediation

Conventional mechanical tools used for removing asphaltene deposition include rod and wireline scrapers. Scrapers are attached to the rod or wireline to cut the accumulated asphaltenes from the wall of the tubing. High velocity jetting was also integrated to help pulverize the asphaltene deposits. For example, Al-Ghazi and Lawson (2007) used a reciprocating tool combing a vibrating chisel and high velocity jet nozzle to restore an asphaltene blocked well in the Ghawar Field (Saudi Arabia), turning out to be a great success as compared to the ineffective water bullheading and xylene washing. Application of mechanical tools is limited to borehole systems of regular shapes.

Ultrasonic irradiation has also been tested as a potential method for removing asphaltene depositions. Experimental results indicated that ultrasonic irradiation is capable of disrupting asphaltene depositions (Gollapudi et al., 1994) and remedying the permeability damage caused by asphaltenes (Shedid, 2004). Thermal methods are also options for removing asphaltenes from the borehole, although the cost of maintenance is high (Gharbi et al., 2017). Additionally, laser had been investigated as an option for asphaltene cleaning and the treatment improved the permeability of damaged limestone cores (Zekri et al., 2001). Incapability of reversing the wettability or penetrating into the near-wellbore region is a common disadvantage shared by mechanical and physical remedial methods as compared to chemical methods.

8.3.3 Field case studies

In field applications, chemical inhibitors, dispersants, solvents, and mechanical or physical methods are often used in combination for synergetic and long-lasting benefits. Shahreyar (2000) used chemical solvents to soften or partially peptize the deposited asphaltenes, and then applied mechanical tools to scrape the deposits off the tubing surfaces. Newberry and Barker (2000) reported the design and implementation of several field cases based on SARA analysis and asphaltene stability tests. Chemical solvent and asphaltene dispersant blends were injected into the perforated intervals and then were soaked for 24 hours. As recommended by the authors (Newberry and Barker, 2000), retreatments should be implemented regularly to remove the ongoing asphaltene damage for maximizing economic benefits.

Brownlee and Zern (1997) reported case studies on removing formation damage in Saskatchewan heavy oil wells by combining a chemical solvent system and stimulation treatments. The oil-based chemical solvent system contained 15% asphaltene dispersant for all wells except well #6, which used 20%. The average oil production rates before and after the treatments were reported based on at least one month's production. Results of these nine wells treated in 1995 and 1996 are summarized in Table 8.5.

As denoted in the table, not all the nine wells were damaged solely by asphaltene deposition; some of the damage involved sand and clay production. Also, along with the injection and soaking (a minimum of 16 hours) of the chemical system containing asphaltene dispersant, some wells were reperforated and/or the pumps and oil tubings were upgraded. It is clear that all of the treatments led to production enhancement, which lasted for several months. The highest enhancement achieved was eight times as high as the original.

Table 8.5 Chemical stimulation effectiveness for nine heavy oil wells in Saskatchewan, Canada (Brownlee and Zern, 1997)

Well #	Pre-treatment prod. m^3/day	Two-day treatment, m^3 chemical/m perforations	Post-treatment prod. m^3/day	Major damage type	Year
1	2.2	2.0/8.9, twice	3.6	Sand, clay, and asphaltene	1995
2	1.3	2.7/6.7	10.1	Asphaltene, sand	1995
3	1.4	0.75/1.8	5	Undetermined	1995
4	7.3	0.85/2.5 + reperf.	10.7	Sand, asphaltene	1995
5	7.0	3.5/7.9 + reperf. + upgrade	20.7	Asphaltene	1995
6	6.4	2.3/7.0 + reperf.	9.5	Asphaltene, sand	1996
7	3.2	2.2/15.3 + reperf. + upgrade	3.5	Undetermined	1996
8	3.9	1.5/4.4 + upgrade	9.4	Asphaltene	1996
9	5.8	1.0/3.0 + upgrade	8.6	Asphaltene	1995

In the Boquerón field in eastern Venezuela, an electrostatically charged polymeric inhibitor that can adsorb onto the pore surface was used to treat two wells damaged by asphaltene (Villard et al., 2016). The inhibitor concentrations were 100 ppm and 300 ppm, and the designed penetration depths into the near-wellbore region were 5 ft and 3 ft, respectively. After 24 hours' soaking, both wells achieved an incremental production of 400 bbl/day and the improved rates were maintained for 180 days.

8.4 SUMMARY AND CONCLUSIONS

To ensure a high success rate in removing asphaltene damage from oil wells, a complete workflow consisting of detailed laboratory analyses and sound field design is needed. Laboratory tests should include SARA analysis, inhibitor/dispersant and solvents screening, concentration, and dosage optimization with crude oil samples. If possible, live crude oil samples should be tested under reservoir conditions. Injection slug design, soaking period, operating parameter control, and other associative measures should be considered in designing field pilots. As a supplemental verification, asphaltene phase modeling can be carried out to extrapolate experimental findings and well-based simulations to predict production enhancement.

REFERENCES

Agersborg, R., Johansen, T.A., Mavko, G., Vanorio, T., 2011. Modeling of elasticity effects of sandstone compaction using coated inclusions. Geophysics 76 (3), E69—E79.
Agrawala, M., Yarranton, H.W., 2001. An asphaltene association model analogous to linear polymerization. Ind. Eng. Chem. Res. 40 (21), 4664—4672.
Al-Ghazi, A.S., Lawson, J., 2007. Asphaltene cleanout using VibraBlaster tool. SPE Saudi Arabia Section Technical Symposium. Society of Petroleum Engineers, Dhahran, Saudi Arabia.
Allenson, S.J., Walsh, M.A., 1997. A novel way to treat asphaltene deposition problems found in oil production. International Symposium on Oilfield Chemistry. Society of Petroleum Engineers, Houston, TX.
Anderko, A., 1992. Modeling phase equilibria using an equation of state incorporating association. Fluid Ph. Equilibria 75, 89—103.
Anderson, W.G., 1986. Wettability literature survey-part 1: rock/oil/brine interactions and the effects of core handling on wettability. J. Pet. Tech. 38 (10), 1125—1144.
Australian National University, 2017. Research School of Chemistry, SEM of quartz dissolution, ANU website.
Bahrami, P., Kharrat, R., Mahdavi, S., Ahmadi, Y., James, L., 2015. Asphaltene laboratory assessment of a heavy onshore reservoir during pressure, temperature and composition variations to predict asphaltene onset pressure. Korean J. Chem. Eng. 32 (2), 316—322.
Barcenas, M., Orea, P., Buenrostro-González, E., Zamudio-Rivera, L.S., Duda, Y., 2008. Study of medium effect on asphaltene agglomeration inhibitor efficiency. Energy Fuels 22 (3), 1917—1922.
Brownlee, D., Zern, K., 1997. Chemical stimulation of heavy oil wells. Annual Technical Meeting. Petroleum Society of Canada, Calgary, Alberta.
Cao, M., Gu, Y., 2013. Oil recovery mechanisms and asphaltene precipitation phenomenon in immiscible and miscible CO_2 flooding processes. Fuel 109, 157—166.
Chaisoontornyotin, W., Bingham, A.W., Hoepfner, M.P., 2017. Reversibility of asphaltene precipitation using temperature-induced aggregation. Energy Fuels 31 (4), 3392—3398.
Chang, C.L., Fogler, H.S., 1994. Stabilization of asphaltenes in aliphatic solvents using alkylbenzene-derived amphiphiles. 1. Effect of the chemical structure of amphiphiles on asphaltene stabilization. Langmuir 10 (6), 1749—1757.
Chartier, T., Delhomme, E., Baumard, J.F., Marteau, P., Subra, P., Tufeu, R., 1999. Solubility, in supercritical carbon dioxide, of paraffin waxes used as binders for low-pressure injection molding. Ind. Eng. Chem. Res. 38 (5), 1904—1910.
Chopping, C., Kaszuba, J.P., 2012. Supercritical carbon dioxide-brine-rock reactions in the Madison Limestone of Southwest Wyoming: an experimental investigation of a sulfur-rich natural carbon dioxide reservoir. Chem. Geol. 322, 223—236.
Chukwudeme, E.A., Hamouda, A.A., 2009. Enhanced oil recovery (EOR) by miscible CO_2 and water flooding of asphaltenic and non-asphaltenic oils. Energies 2 (3), 714—737.
Corex (UK) Ltd., 2017. SEM of Kaolinite, Corex website.

Crause, J.C., Nieuwoudt, I., 2000. Fractionation of paraffin wax mixtures. Ind. Eng. Chem. Res. 39 (12), 4871–4876.

Creek, J.L., 2005. Freedom of action in the state of asphaltenes: escape from conventional wisdom. Energy Fuels 19 (4), 1212–1224.

Escobedo, J., Mansoori, G.A., 1995. Viscometric determination of the onset of asphaltene flocculation: a novel method. SPE Prod. Facil. 10 (2), 115–118.

Firoozinia, H., Abad, K.F., Varamesh, A., 2016. A comprehensive experimental evaluation of asphaltene dispersants for injection under reservoir conditions. Pet. Sci. 13 (2), 280–291.

Flaten, D., 2017. Personal Conversations Over Field Operations at Big Sand Draw, Wyoming.

Fotland, P., Anfindsen, H., Fadnes, F.H., 1993. Detection of asphaltene precipitation and amounts precipitated by measurement of electrical conductivity. Fluid Ph. Equilibria 82, 157–164.

Gharbi, K., Benyounes, K., Khodja, M., 2017. Removal and prevention of asphaltene deposition during oil production: a literature review. J. Pet. Sci. Eng. 158, 351–360.

Gollapudi, U.K., Bang, S.S., Islam, M.R., 1994. Ultrasonic treatment for removal of asphaltene deposits during petroleum production. SPE Formation Damage Control Symposium. Society of Petroleum Engineers, Lafayette, Louisiana.

González, G., Middea, A., 1991. Peptization of asphaltene by various oil soluble amphiphiles. Colloids Surf. 52, 207–217.

Gonzalez, D.L., Vargas, F.M., Hirasaki, G.J., Chapman, W.G., 2007. Modeling study of CO_2-induced asphaltene precipitation. Energy Fuels 22 (2), 757–762.

Green, D.W., Willhite, G.P., 1998. Enhanced Oil Recovery. Textbook Series, Vol. 6. SPE, Richardson, TX.

Groenzin, H., Mullins, O.C., 1999. Asphaltene molecular size and structure. J. Phys. Chem. A 103 (50), 11237–11245.

Groenzin, H., Mullins, O.C., 2000. Molecular size and structure of asphaltenes from various sources. Energy Fuels 14 (3), 677–684.

Gunter, W.D., Perkins, E.H., Hutcheon, I., 2000. Aquifer disposal of acid gases: modeling of water-rock reactions for trapping of acid wastes. Appl. Geochem. 15 (8), 1085–1095.

Hammami, A., Raines, M.A., 1997. Paraffin deposition from crude oils: comparison of laboratory results to field data. SPE Annual Technical Conference and Exhibition. Society of Petroleum Engineers, San Antonio, Texas.

Hammami, A., Phelps, C.H., Monger-McClure, T., Little, T.M., 2000. Asphaltene precipitation from live oils: an experimental investigation of onset conditions and reversibility. Energy Fuels 14 (1), 14–18.

Hamouda, A.A., Chukwudeme, E.A., Mirza, D., 2009. Investigating the effect of CO_2 flooding on asphaltenic oil recovery and reservoir wettability. Energy Fuels 23 (2), 1118–1127.

Hashmi, S.M., Quintiliano, L.A., Firoozabadi, A., 2010. Polymeric dispersants delay sedimentation in colloidal asphaltene suspensions. Langmuir 26 (11), 8021–8029.

Haskett, C.E., Tartera, M., 1965. A practical solution to the problem of asphaltene deposits-Hassi Messaoud Field, Algeria. J. Pet. Technol. 17 (04), 387–391.

Hirschberg, A., DeJong, L.N., Schipper, B.A., Meijer, J.G., 1984. Influence of temperature and pressure on asphaltene flocculation. Society of Petroleum Engineers Journal 24 (03), 283–293.

Hitchon,, B. (Ed.), 1996. Aquifer disposal of carbon dioxide: hydrodynamic and mineral trapping — proof of concept. Geoscience Publishing Ltd., Sherwood Park, Alberta, Canada.

Huang, E.T., Holm, L.W., 1988. Effect of WAG injection and rock wettability on oil recovery during CO_2 flooding. SPE Res. Eng. 3 (01), 119–129.

Huang, L., Shen, P., Jia, Y., Ye, J., Li, S., Bie, A., 2010. Prediction of asphaltene precipitation during CO_2 injection. Petroleum Exploration and Development 37 (3), 349–353.

Ibrahim, H.H., Idem, R.O., 2004. Interrelationships between asphaltene precipitation inhibitor effectiveness, asphaltenes characteristics, and precipitation behavior during n-heptane (light paraffin hydrocarbon)-induced asphaltene precipitation. Energy Fuels 18 (4), 1038–1048.

Idem, R.O., Ibrahim, H.H., 2002. Kinetics of CO_2-induced asphaltene precipitation from various Saskatchewan crude oils during CO_2 miscible flooding. J. Pet. Sci. Eng. 35 (3), 233–246.

Jamaluddin, A.K., Nazarko, T.W., Sills, S., Fuhr, B.J., 1996. Deasphalted oil-A natural asphaltene solvent. SPE Prod. Facil. 11 (3), 161–165.

Jamaluddin, A.K., Creek, J., Kabir, C.S., McFadden, J.D., D'cruz, D., Manakalathil, J., et al., 2002. Laboratory techniques to measure thermodynamic asphaltene instability. J. Can. Pet. Technol. 41 (7), 44–52.

Jennings, D.W., Chao, K.P., Cable, R., Oczkowski, W., Cote, D., 2016. Method for Field monitoring asphaltene treatment programs. Offshore Technology Conference. Society of Petroleum Engineer, Houston, Texas.

Jin, M., Ribeiro, A., Mackay, E., Guimarães, L., Bagudu, U., 2016. Geochemical modelling of formation damage risk during CO_2 injection in saline aquifers. J. Nat. Gas Sci. Eng. 35, 703–719.

Kabir, C.S., Jamaluddin, A.K., 1999. Asphaltene characterization and mitigation in south Kuwait's Marrat reservoir. Middle East Oil Show and Conference,. Society of Petroleum Engineers, Bahrain.

Kaszuba, J.P., Janecky, D.R., 2009. Geochemical impacts of sequestering carbon dioxide in brine formations. In: Sundquist, E., McPherson, B. (Eds.), Carbon Sequestration and Its Role in the Global Carbon Cycle, Geophysical Monograph 183. American Geophysical Union, Washington, DC, pp. 239–247.

Kaszuba, J.P., Navarre-Sitchler, A., Thyne, G., Chopping, C., Meuzelaar, T., 2011. Supercritical carbon dioxide and sulfur in the Madison Limestone: a natural analog in southwest Wyoming for geologic carbon-sulfur co-sequestration. Earth and Planet Sci. Lett. 309 (1), 131–140.

Katz, D.L., Beu, K.E., 1945. Nature of asphaltic substances. Ind. Eng. Chem. 37 (2), 195–200.

Kelessidis, V.C., Maglione, R., 2008. Yield stress of water–bentonite dispersions. Colloids Surf. A Physicochem. Eng. Asp. 318 (1), 217–226.

Koottungal, L., 2014. Survey: miscible CO_2 continues to eclipse steam in US EOR production 2014 worldwide EOR survey. Oil Gas J. 112 (4), 78–91.

Kuuskraa, V., Wallace, M., 2014. CO_2-EOR set for growth as new CO_2 supplies emerge. Oil Gas J. 112 (4), 66–77.

Leontaritis, K.J., 1989. Asphaltene deposition: a comprehensive description of problem manifestations and modeling approaches. SPE production operations symposium. Society of Petroleum Engineers, Oklahoma City, Oklahoma.

Leontaritis, K.J., 1996. The asphaltene and wax deposition envelopes. Fuel Sci. Technol. Int. 14 (1–2), 13–39.

Leontaritis, K.J., Mansoori, G.A., 1987. Asphaltene flocculation during oil production and processing: a thermodynamic collodial model. SPE International Symposium on Oilfield Chemistry. Society of Petroleum Engineers, San Antonio, Texas.

Leontaritis, K.J., Mansoori, G.A., Jiang, T.S., 1988. Asphaltene deposition in oil recovery: a survey of field experiences and research approaches. JPSE 1, 229–239.

Leontaritis, K.J., Amaefule, J.O., Charles, R.E., 1994. A systematic approach for the prevention and treatment of formation damage caused by asphaltene deposition. Soc. Pet. Eng. Prod. Facilities 9 (3), 157–164.

Lichaa, P.M., 1977. Asphaltene deposition problem in Venezuela crudes-usage of asphaltenes in emulsion stability. Oil Sands 609 (1).

Lichaa, P.M., Herrera, L., 1975. Electrical and other effects related to the formation and prevention of asphaltene deposition problem in Venezuelan crudes. SPE Oilfield Chemistry Symposium. Society of Petroleum Engineers, Dallas, Texas.

Lin, E.C., Huang, E.T., 1990. The effect of rock wettability on water blocking during miscible displacement. SPE Res. Eng. 5 (2), 205–212.

MacMillan, D.J., Tackett Jr., J.E., Jessee, M.A., Monger-McClure, T.G., 1995. A unified approach to asphaltene precipitation: laboratory measurement and modeling. J. Pet. Technol. 47 (9), 788–793.

Magruder, J.B., Stiles, L.H., Yelverton, T.D., 1990. Review of the means San Andres Unit CO_2 tertiary project. J. Pet. Technol. 42 (5), 638–644.

Mahmoud, W., 2017. Kaolinite Clay Sheets, ThermoFisher Scientific website.

Marbler, H., Erickson, K.P., Schmidt, M., Lempp, C., Pöllmann, H., 2013. Geomechanical and geochemical effects on sandstones caused by the reaction with supercritical CO_2: an experimental approach to in situ conditions in deep geological reservoirs. Environ. Earth Sci. 69, 1981–1998.

Melendez-Alvarez, A.A., Garcia-Bermudes, M., Tavakkoli, M., Doherty, R.H., Meng, S., Abdallah, D.S., et al., 2016. On the evaluation of the performance of asphaltene dispersants. Fuel 179, 210–220.

Melzer, L.S., 2012. Carbon Dioxide Enhanced Oil Recovery (CO_2 EOR): Factors Involved in Adding Carbon Capture, Utilization and Storage (CCUS) to Enhanced Oil Recovery. Melzer Consulting,, Midland Texas. Available from: http://neori.org/Melzer_CO2EOR_CCUS_Feb2012.pdf.

Moghanloo, R.G., Younas, D., Yuan, B., et al., 2015. Applying fractional flow theory to evaluate CO2 storage capacity of an Aquifer. J. Pet. Sci. Eng. 125, 154–161.

Monger, T.G., Fu, J.C., 1987. The nature of CO_2-induced organic deposition. SPE Annual Technical Conference and Exhibition. Society of Petroleum Engineers, Dallas, Texas.

Monger, T.G., Trujillo, D.E., 1991. Organic deposition during CO_2 and rich-gas flooding. SPE Res. Eng. 6 (1), 17–24.

Mullins, O.C., Sheu, E.Y., Hammami, A., Marshall, A.G., 2007. Asphaltenes, Heavy Oils, and Petroleomics. Springer Science & Business Media, New York.

Nakhli, H., Alizadeh, A., Moqadam, M.S., Afshari, S., Kharrat, R., Ghazanfari, M.H., 2011. Monitoring of asphaltene precipitation: experimental and modeling study. J. Pet. Sci. Eng. 78 (2), 384–395.

Neumann, H.J., Paczynska-Lahme, B., Severin, D., 1981. Composition and Properties of Petroleum. Halsted Press, New York.

Newberry, M.E., Barker, K.M., 2000. Organic formation damage control and remediation. SPE International Symposium on Formation Damage Control. Society of Petroleum Engineers, Lafayette, Louisiana.

Nghiem, L.X., Hassam, M.S., Nutakki, R., George, A.E., 1993. Efficient modelling of asphaltene precipitation. SPE Annual Technical Conference and Exhibition. Society of Petroleum Engineers, Houston, Texas.

Nicksic, S.W., Jeffries-Harris, M.J., 1968. Acid precipitation of crude oil asphaltenes-structural implications. J. Inst. Pet. 54, 532.

Novosad, Z., Costain, T.G., 1990. Experimental and modeling studies of asphaltene equilibria for a reservoir under CO_2 injection. SPE Annual Technical Conference and Exhibition. Society of Petroleum Engineers, New Orleans, Louisiana.

Okwen, R.T., 2006. Formation damage by CO_2 asphaltene precipitation. SPE International Symposium and Exhibition on Formation damage control. Society of Petroleum Engineers, Lafayette, Louisiana.

Pfeiffer, J.P., Saal, R.N., 1940. Asphaltic bitumen as colloid system. J. Phys. Chem. 44 (2), 139–149.

Rae, P., Lullo, G.D., 2001. Towards environmentally-friendly additives for well completion and stimulation operations. SPE Asia Pacific Oil and Gas Conference and Exhibition. Society of Petroleum Engineers, Jakarta, Indonesia.

Rathnaweera, T.D., Ranjith, P.G., Perera, M.S.A., 2016. Experimental investigation of geochemical and mineralogical effects of CO_2 sequestration on flow characteristics of reservoir rock in deep saline aquifers. Sci. Rep. 6, 19362.

Rogel, E., 2010. Effect of inhibitors on asphaltene aggregation: a theoretical framework. Energy Fuels 25 (2), 472–481.

Rosenqvist, J., Kilpatrick, A.D., Yardley, B.W.D., Rochelle, C.A., 2014. Feldspar dissolution at CO_2-saturatied conditions. Geophys. Res. Abs.

Samuelson, M.L., 1992. Alternatives to aromatics for solvency of organic deposits. SPE Formation Damage Control Symposium. Society of Petroleum Engineers, Lafayette, Louisiana.

Sarbar, M.A., Wingrove, M.D., 1997. Physical and chemical characterization of Saudi Arabian crude oil emulsions. SPE Annual Technical Conference and Exhibition. Society of Petroleum Engineers, San Antonio, Texas.

Scherlin, J., Larson, D., 2017. Personal Conversations Over Field Operations at Salt Creek, Wyoming.

Schuler, B., Meyer, G., Peña, D., Mullins, O.C., Gross, L., 2015. Unraveling the molecular structures of asphaltenes by atomic force microscopy. J. Am. Chem. Soc. 137 (31), 9870–9876.

Shahreyar, N., 2000. Review of Paraffin Control and Removal in Oil Wells Using Southwestern Petroleum Short Course Searchable Database [thesis]. Texas Tech University.

Shedid, S.A., 2004. An ultrasonic irradiation technique for treatment of asphaltene deposition. J. Pet. Sci. Eng. 42 (1), 57–70.

Shelton, J.L., Schneider, F.N., 1975. The effects of water injection on miscible flooding methods using hydrocarbons and carbon dioxide. Soc. Pet. Eng. J. 15 (03), 217–226.

Shen, Z., Sheng, J.J., 2016. Experimental study of asphaltene aggregation during CO_2 and CH_4 injection in Shale oil reservoirs. SPE Improved Oil Recovery Conference. Society of Petroleum Engineers, Tulsa, Oklahoma.

Sivaraman, A., Imer, D., Thomas, F.B., 1998. Defining SLE and VLE conditions of hydrocarbon fluids containing wax and asphaltenes using acoustic resonance technology. AIChE Spring National Meeting. American Institute of Chemical Engineers, New Orleans, Louisiana.

Smith, D.F., Klein, G.C., Yen, A.T., Squicciarini, M.P., Rodgers, R.P., Marshall, A.G., 2008. Crude oil polar chemical composition derived from FT-ICR mass spectrometry accounts for asphaltene inhibitor specificity. Energy Fuels 22 (5), 3112–3117.

Speight, J., 2014. The Chemistry and Technology of Petroleum, 5th ed. CRC Press.

Srivaslava, R.K., Huang, S.S., Dye, S.B., Mourits, F.M., 1993. Quantification of asphaltene flocculation during miscible co flooding in the weyburn reservoir. Technical Meeting/Petroleum Conference of The South Saskatchewan Section. Petroleum Society of Canada, Regina.

Srivastava, R.K., Huang, S.S., 1997. Asphaltene deposition during CO_2 flooding: a laboratory assessment. SPE Production Operations Symposium. Society of Petroleum Engineers,, Oklahoma City, Oklahoma.

Subramanian, S., Simon, S., Sjöblom, J., 2016. Asphaltene precipitation models: a review. J. Disper. Sci. Technol. 37 (7), 1027−1049.

Taber, J.J., Martin, F.D., Seright, R.S., 1997. EOR screening criteria revisited-Part 1: Introduction to screening criteria and enhanced recovery field projects. Soc. Pet. Eng. Res. Eng. 12 (3), 189−198.

Tavakkoli, M., Grimes, M.R., Liu, X., Garcia, C.K., Correa, S.C., Cox, Q.J., et al., 2015. Indirect method: a novel technique for experimental determination of asphaltene precipitation. Energy Fuels 29 (5), 2890−2900.

Thawer, R., Nicoll, D.C., Dick, G., 1990. Asphaltene deposition in production facilities. SPE Prod. Eng. 5 (4), 475−480.

Tiffin, D.L., Yellig, W.F., 1983. Effects of mobile water on multiple-contact miscible gas displacements. Soc. Pet. Eng. J. 23 (3), 447−455.

Tiffin, D.L., Sebastian, H.M., Bergman, D.F., 1991. Displacement mechanism and water shielding phenomena for a rich-gas/crude-oil system. SPE Res. Eng. 6 (02), 193−199.

Ting, D.P., Hirasaki, G.J., Chapman, W.G., 2003. Modeling of asphaltene phase behavior with the SAFT equation of state. Pet. Sci. Technol. 21 (3−4), 647−661.

Trbovich, M.G., King, G.E., 1991. Asphaltene deposit removal: long-lasting treatment with a co-solvent. SPE International Symposium on Oilfield Chemistry. Society of Petroleum Engineers, Anaheim, California.

Tuttle, R.N., 1983. High pour point and asphaltic crude oils and condensates. JPT 35 (6), 1192−1196.

Villard, Y., Fajardo, F., Milne, A., 2016. Enhanced oil recovery using innovative asphaltene inhibitors in East Venezuela. SPE International Conference and Exhibition on Formation Damage Control. Society of Petroleum Engineers, Lafayette, Louisiana.

Waldmann, S., Busch, A., van Ojik, K., Gaupp, R., 2014. Importance of mineral surface areas in Rotliegend sandstones for modeling CO_2-water-rock interactions. Chem. Geol. 378, 89−109.

Wang, J., Li, C., Zhang, L., Que, G., Li, Z., 2009. The properties of asphaltenes and their interaction with amphiphiles. Energy Fuels 23 (7), 3625−3631.

Yen, A., Yin, Y.R., Asomaning, S., 2001. Evaluating asphaltene inhibitors: laboratory tests and field studies. SPE International Symposium on Oilfield Chemistry. Society of Petroleum Engineers, Houston, Texas.

Yuan, B., Moghanloo, R.G., 2016. Analytical evaluation of nanoparticles application to reduce fines migration in porous media. Soc. Pet. Eng. J. 21 (06), 2317−2332.

Yuan, B., Su, Y., Moghanloo, R.G., 2015. A new analytical multi-linear solution for gas flow toward fractured horizontal well with different fracture intensity. J. Nat. Gas Sci. Eng. 23, 227−238.

Yuan, B., Moghanloo, G.R., Zheng, D., 2017. A novel integrated production analysis workflow for evaluation, pptimization and predication in shale plays. Int. J. Coal Geol. 180, 18−28.

Zanganeh, P., Ayatollahi, S., Alamdari, A., Zolghadr, A., Dashti, H., Kord, S., 2012. Asphaltene deposition during CO_2 injection and pressure depletion: A visual study. Energy Fuels 26 (2), 1412−1419.

Zekri, A.Y., Shedid, S.A., Alkashef, H., 2001. A novel technique for treating asphaltene deposition using laser technology. SPE Permian Basin Oil and Gas Recovery Conference. Society of Petroleum Engineers, Midland, TX.

CHAPTER NINE

Formation Damage by Thermal Methods Applied to Heavy Oil Reservoirs

Jiaming Zhang[1] and Zhangxin Chen[2]

[1]CNPC Economics & Technology Research Institute, Beijing, China
[2]University of Calgary, Calgary, AB, Canada

Contents

Only about 30% of the world's oil reserves are light crude oil, while about 70% are heavy crude oil (Prats, 1982; Giacchetta et al., 2015). Increasing the recovery of heavy oil can increase about 300 billion barrels of crude oil production (Doe, 2014; Morrow et al., 2014). Heavy oil reservoirs are mainly developed by means of thermally enhanced oil recovery (TEOR) (Prats, 1982; Hashemi-Kiasari et al., 2014; Chen, 2007; Chen et al., 2006). Thermal methods have produced billions of barrels of oil (Nwidee et al., 2016). Early TEOR dates back to 1865 (Prats, 1982), but the first significant industrial thermal method project was deployed in Woodson,

Formation Damage during Improved Oil Recovery.
DOI: https://doi.org/10.1016/B978-0-12-813782-6.00009-9
© 2018 Elsevier Inc.
All rights reserved.
361

Texas in 1931 (Stovall, 1934). Thermal methods have achieved success in Canada, the United States, Venezuela, Indonesia, China, and other countries (Prats, 1982; De haan and Schenk, 1969; Doe, 2014). In particular, from 1980 to 2002, the United States produced more than 4 billion barrels of oil through steam flooding (SF) (Doe, 2014; Hashemi et al., 2014).

TEOR is one of the most suitable methods for recovery of heavy oil and oil sands (Thomas, 2008; Zhao et al., 2013). First, TEOR introduces heat to heavy oil reservoirs and evaporates some heavy components. In addition, TEOR can bring a significant reduction in crude oil viscosity, thereby increasing the mobility ratio. There are other mechanisms such as rock and fluid expansion, compaction, and distillation (Prats, 1982; Zhang et al., 2015a). During the process, a shift in rock wettability occurs usually, which enhances the chance for better oil recovery (Kovscek, 2012).

Most of heavy oil reservoirs are relatively shallow and loose sandstone reservoirs. Generally, these reservoirs are prone to formation damage during oil production due to the characteristic of poorly consolidation, and even thermal recovery methods can further exacerbate the extent of formation damage, including sands and fines migration, clay swelling, organic precipitation, mineral dissolution and precipitation, wettability alteration, scales, emulsion, and rock compaction (Tague, 2000; Zhang et al., 2015b; Thomas, 2008; Yuan et al., 2015). Due to the relative high permeability of heavy oil reservoirs, to some extent, formation damage is easily overlooked, resulting in sub-optimal production of heavy oil (Okoye et al., 1990). However, it is still very favorable to prevent the occurrence of formation damage to improve more economic values of TEOR applied in heavy oil reservoirs.

Hence, this chapter will systematically summarize various thermal recovery methods and their associated EOR mechanisms, and analyze the reservoir damage they potentially cause.

9.1 MECHANISMS OF THERMALLY ENHANCED OIL RECOVERY METHODS

As illustrated in Fig. 9.1, there are two types of thermal EOR methods, in situ combustion (ISC) and steam injection (Nwidee et al., 2016). In ISC, air is injected into viscous oil reservoirs to generate heat

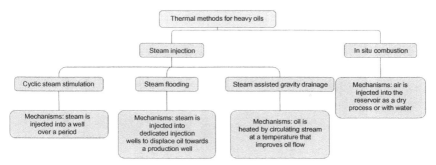

Figure 9.1 Methods of thermally enhanced oil recovery.

by burning a portion of the existing oil (Moore et al., 1995). The steam injection process is through injecting steam into shallow, thick and highly permeable heavy oil reservoirs (Kokal and Al-Kaabi, 2010; Kuuskraa et al., 2011) to enhance oil recovery. Steam injection methods include SF, cyclic steam stimulation (CSS; sometimes referred to as "huff and puff"), and steam assisted gravity drainage (SAGD).

ISC has been experimented in Romania, the United States, Canada, India, and China (Kovscek, 2012; Zhang et al., 2013). Steam injection projects have been implemented in California, Indonesia, Oman, Alberta (Canada), Venezuela, Brazil, China, Trinidad, Tobago and the former Soviet Union (Moritis, 2010; Kovscek, 2012). CSS has been widely used in California, Alberta and Venezuela. SAGD is mainly used in the province of Alberta, Canada, for recovery of bitumen. SAGD has also been practiced in Venezuela, although it has not been successful (Kovscek, 2012; Moritis, 2010). ISC is not as popular as steam injection because it is complex and difficult to control, but usually can achieve much higher oil recovery. Using the ISC process, as high as 95% oil displacement can be achieved (Hascakir, 2017).

9.1.1 Cyclic steam stimulation

The process of CSS, as shown in Fig. 9.2, specifically for a single well, can be divided into three stages (Owens and Suter, 1965; Lake and Walsh, 2008; Robertson et al., 1989):

1. Injection stage. In this stage, a sufficient amount of steam is injected into a reservoir through a well for about 1 month.
2. Soaking stage. In this stage, the well is shut-in for several days or weeks to heat the formation and the crude oil, thereby reducing the viscosity of the crude oil.

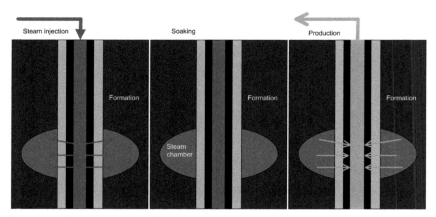

Figure 9.2 Cyclic steam stimulation.

3. Production stage. In this stage, oil production will rise rapidly initially and maintain this rate in a short time, and then production will begin to decline (Robertson et al., 1989; Alvarez and Han, 2013).

With the temperature of both the formation and saturated oil declining, the oil production rates decrease slowly. As the production rates decrease below an economic limit, the cycles can be repeated (Robertson et al., 1989; Alvarez and Han, 2013). As for cyclic steam injection, the steam injection, and oil production take place in the same well, and enhance oil recovery with the effect of reducing the viscosity of the crude oil. Especially in thick heavy oil reservoirs, CSS has a great potential for enhancing oil recovery and is simple to set up. However, compared with SF, the CSS recovery factor is relative lower by about 20% (Thomas, 2008). Because the CSS process involves significant heat loss (Chunze et al., 2008), CSS is not economic in a thin layer of a heavy oil reservoir (Zhao et al., 2013).

9.1.2 Steam flooding

Similar to water flooding, SF, as shown in Fig. 9.3, is one kind of flooding methods by continuously injecting steam to form a steam zone and displacing the crude oil through reducing the viscosity of crude oil (Ali and Meldau, 1979). Its performance highly depends on well patterns and geological conditions (Stokes and Doscher, 1974). The residual oil saturation can be decreased as low as 10% in swept areas (Thomas, 2008), usually with oil recovery factors about 50% (Farouq Ali, 1982).

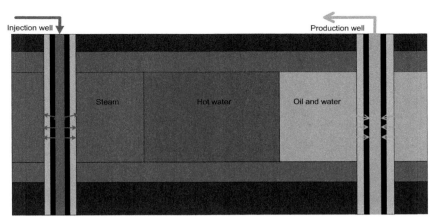

Figure 9.3 Steam flooding.

Steam is injected from an injection well, and the injected steam forms a steam chamber around the injection well (Bridgwater, 2003; Nabipour et al., 2007). The steam chamber is expanded toward a production well and displaces the reservoir fluid into a production well with reducing the viscosity of heavy oil (Shafiei et al., 2013; Lake and Walsh, 2008).

Compared with CSS, SF is more efficient. But SF is much more expensive than CSS. There is significant heat loss during the process of a large amount of steam injection, such as heating the well tubing, overburden, and underburden rocks (Ali, 2003). Early steam breakthrough at production wells after steam channeling can also reduce the performance of oil recovery (Baibakov et al., 2011). Avoiding sand plugging at the bottom of a well, preventing steam channeling, and improving steam displacement efficiency are still key challenges for the success of SF (Ali, 2003; Baibakov et al., 2011; Yanbin et al., 2012).

9.1.3 Steam assisted gravity drainage

SAGD was first developed by Butler by arranging a pair of parallel horizontal wells in reservoirs for oil recovery (Butler, 1985). The schematic diagram of SAGD is shown in Fig. 9.4. In the SAGD process, steam is injected from the upper horizontal well, and oil is produced from the lower horizontal well. The steam heats heavy oil to reduce oil viscosity and also provides a driving force for the flowing of crude oil to the production well (Hashemi-Kiasari et al., 2014; Gates et al., 2008). The steam rises to the top of a formation to form a steam chamber, which causes

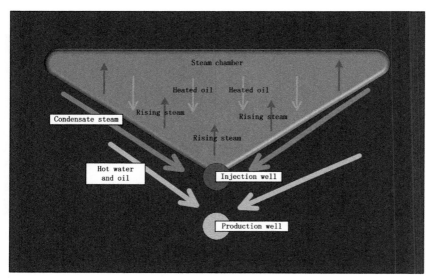

Figure 9.4 Steam assisted gravity drainage.

the viscosity of the crude oil to decrease. The oil drains down by gravity, and is collected by the lower horizontal well. The continuous injection of steam ensures that the steam chamber expands and spreads laterally in the reservoir. High vertical permeability is critical to ensure the successful implementation of SAGD. SAGD has made Alberta's heavy oil well developed in past years (Thomas, 2008).

The mobility of heavy oil and bitumen is very low, which leads to the formation of steam chambers rather than steam channels. Steam channels can be caused by the specific nature of steam fingering phenomenon during SAGD operation (Ito and Ipek, 2005). The efficiency of SAGD process is mainly controlled by fluid mobility. SAGD requires a lot of water to produce steam and much natural gas to generate sufficient heat to produce steam. The energy consumption of SAGD is therefore very high. If the following conditions are met, SAGD can be successfully implemented.

First, the steam injected from an upper horizontal well should have enough capacity to form a steam chamber. As a result, the temperature in the steam chamber significantly reduces the viscosity of heavy oil or bitumen, thereby improving the mobility of the crude oil (Gates et al., 2008). At the edge of the steam chamber, the steam can also release heat by condensation to reduce the viscosity of the crude oil.

Second, the heated crude oil should be drained from the steam chamber to a production well by gravity (Elliott and Kovscek, 1999). Steam injection and oil production are carried out simultaneously (Giacchetta et al., 2015). The oil production is determined by the expansion of a steam chamber (Giacchetta et al., 2015; Hashemi-Kiasari et al., 2014).

SAGD is very effective for bitumen and heavy oil recovery, and its efficiency increases with an increase in the target oil layer thickness (Shin and Polikar, 2006). SAGD oil production is considered uneconomical when the reservoir thickness is less than about 15 m (Edmunds and Chhina, 2001), due to the high steam-to-oil ratio (SOR) required.

The improper control of operation will result in the loss of steam and then impair oil recovery, as the process is susceptible to low mobility control and rapid gravity separation (Coskuner, 2009).

9.1.4 In situ combustion

ISC is also called fire flooding (Chu, 1977; Chieh, 1981; Zhang et al., 2013). In this method, air or oxygen is injected into a reservoir to generate heat by burning parts of the crude oil (about 10%) (Kokal and Al-Kaabi, 2010; Doe, 2014; Slider, 1983), as shown in Fig. 9.5. A high temperature of $450°C-600°C$ is reached in the smaller zone where the viscosity of crude oil can be reduced significantly. The thermal efficiency of ISC is higher than SF due to the relatively smaller heat loss to the overburden or underburden rock, and no surface or wellbore heat loss. In

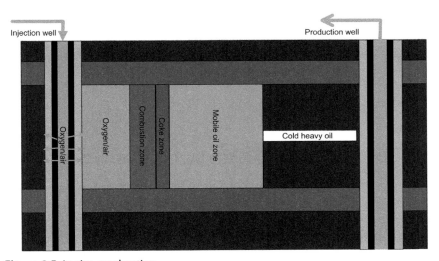

Figure 9.5 In situ combustion.

some cases, additives are injected with the air, primarily to increase the heat recovery.

ISC can be forward or reverse, which is mainly dependent on the combustion front. In forward ISC, ignition occurs near an injection well and the combustion front moves in the direction of the air flow. In reverse ISC, ignition occurs near a production well and the combustion front moves in the opposite direction to the air flow.

In practice, forward ISC is usually adopted. The combustion reactions between the injected air and the crude oil produce a lot of heat to decrease the oil viscosity and achieve oil recovery. As air is constantly injected, the combustion front will propagate toward production wells (Moore et al., 1995; Satter et al., 2008). Reservoir fluids can be displaced in the condition of high temperature (600°C–700°C) toward production wells. The heavy components are burned to produce large amounts of flue gas (Moore et al., 1995; Satter et al., 2008), and the lighter component will move downstream and mix with the crude oil as mobile oil (Moore et al., 1995).

ISC can effectively displace the oil in contact with the hot fluids within the combustion zone. The ISC can achieve as high as 95% oil displacement (Hascakir, 2017), with negligible control by reservoir permeability (Satter et al., 2008). In addition, by using air for ignition, ISC is usually much more economical than SF (Satter et al., 2008).

As negative sides, the ISC process is usually very complex and difficult to control (i.e., control a combustion front) (Green and Willhite, 1998). ISC requires amounts of heat energy to achieve enough high temperature, about 700°C and above. Hence, the condition of high temperature of ISC can often damage tubing and other equipment (Ali, 2003; Akkutlu and Yortsos, 2003).

9.2 FORMATION DAMAGE BY THERMAL METHODS

Generally, heavy oil reservoirs are prone to formation damage due to their poor consolidation, and thermal recovery techniques can further exacerbate the extent of formation damage (Civan, 2007b). There are commonly three major mechanisms of formation damage induced by thermal methods, as shown in Fig. 9.6, such as thermal mechanisms,

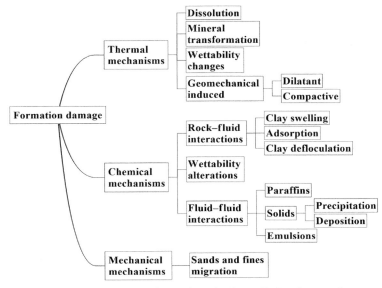

Figure 9.6 Formation damage by thermal methods applied to heavy oil reservoirs.

chemical mechanisms, and mechanical mechanisms. These mechanisms may not only exist alone, but also interact with each other to affect the performance of oil recovery systematically.

9.2.1 Sands and fines migration

Most of heavy oil reservoirs are shallow with weak cementation. In the production of heavy oil, sands plugging caused by sands production is one of the major formation damage issues (Gruesbeck and Collins, 1982; Yuan et al., 2016b, 2018). Any abrupt changes in pressure or flowing velocities in a porous system can cause the changes and re-distribution of the effective stress. As the effective stress exceeds a limit, the rock frame-work is damaged and then sands production is activated.

That little consolidation occurs is usually the result of clays and shales found in a formation (Neasham, 1977; Tague, 2000; Yuan et al., 2016a). Sands can be liberated by cement dissolution, as shown in Fig. 9.7. In San Joaquin Valley and other areas, a sandstone formation also contains a high proportion of feldspar. Feldspar is usually very unstable and can be easily crushed to generate many small particles or fines powder. The hydrodynamic forces exerted by the flowing fluids in production help remove the original consolidated clay and shale onto pore surfaces (Muecke, 1979).

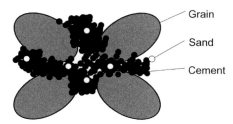

Figure 9.7 Liberation of sand by cement dissolution.

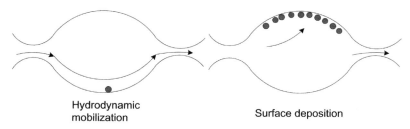

Figure 9.8 Particles on pore surfaces.

During a thermal recovery process, the excessive contact of sands with the high temperature steam will also promote the migration of sands and fines (Tague, 2000).

Due to hydrodynamic mobilization, colloidal expulsion, cement dissolution and surface deposition, sands, and fines can attach onto pore surfaces (Fig. 9.8). They plug pore throats and form internal or external cake formations, as shown in Fig. 9.9. Migrations of sands and fines can cause the extent of formation damage, and manifest itself in a porosity and permeability reduction (Muecke, 1979; Bennion et al., 1992).

9.2.2 Clay swelling and mineral transformations

Most of heavy oil reservoirs contain more or less clay. The problematic clay minerals include montmorillonite, illite, and kaolinite. Illite and kaolinite are migrating clays, and montmorillonite is swelling clay.

Due to an uneven charge distribution of the montmorillonite structure, fresh steam condensate can combine with it easily, resulting in clay structure expansion, as shown in Fig. 9.10. When fully hydrated, the size of the montmorillonite crystal can be increased from 9.2 Angstroms to 17 Angstroms (Bennion et al., 1992). Obviously, due to the presence of montmorillonite in the matrix of the formation rock, the sensitivity to the low salinity steam condensate will be a potential threat of formation

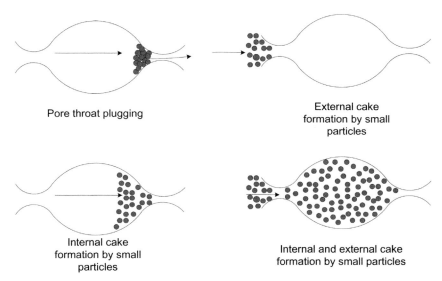

Pore throat plugging

External cake formation by small particles

Internal cake formation by small particles

Internal and external cake formation by small particles

Figure 9.9 Particles in pore throats and in the pore volume.

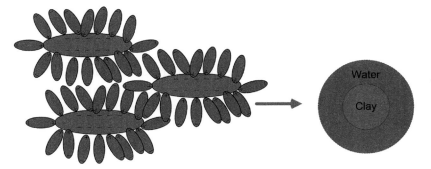

Water

Clay

Figure 9.10 Clay swelling.

damage (Zhang et al., 2015b). By adding additives into the steam, it is possible to avoid clay swelling caused by steam injection (Scheuerman and Bergersen, 1990).

Many heavy oil reservoirs also contain migrating clays, such as illite and kaolinite. These types of clay exhibit balanced charges and, therefore, are not readily swelled. The mechanism of formation damage caused by migrating clays is similar to that of sand and fines migration.

Although they exhibit the ability to resist flocculation or migration, high temperature operations may cause mineral transformation to promote the conversion of relatively inert kaolinite into water-sensitive montmorillonite (Gruesbeck and Collins, 1982; Hutcheon et al., 1989). Newly formed expandable clay is in contact with steam condensate and the condensate is combined with it to cause clay swelling and a permeability reduction (Abercrombie, 1988; Perry and Gillott, 1979). In general, high temperatures above 200°C can cause mineral transformation (Hutcheon et al., 1989; Bennion et al., 1995; Bennion, 2002). The contact of the steam condensate with newly converted water-sensitive clay results in significant swelling and expansion (Bennion, 2002; Reed, 1979). The permeability of a porous system is greatly reduced by such mechanisms.

9.2.3 Minerals dissolution and precipitation

Some minerals are exposed to high temperature steam and will undergo transformation and dissolution. This kind of damage usually occurs in CSS, which is caused by direct injection of high temperature steam into wells. Many studies have shown that sandstone reservoirs are more susceptible to dissolution, some of which will be precipitated in the form of silica at well perforations (Hollman and Tague, 1999; Reed, 1979). Mineral dissolution and precipitation can greatly reduce the permeability of near-well areas.

Fig. 9.11 reflects the phenomenon of mineral dissolution and precipitation. There are two forms of damage caused by mineral dissolution. First, when a reservoir or the saturated hot water cools causing the dissolved minerals to be precipitated, the sediment will move into the

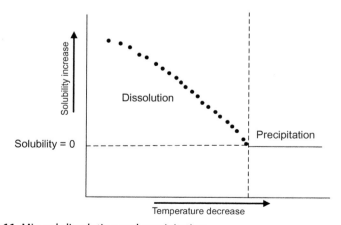

Figure 9.11 Mineral dissolution and precipitation.

reservoir, causing formation damage. Second, in general, debris of the soluble material contains encapsulated insoluble particles. When the soluble material is dissolved at high temperature, the insoluble particles are released. These particles migrate along with a fluid and may cause bridging and plug pore throats (Bennion et al., 1992).

In general, the solubility of the minerals increases with temperature. Precipitation may occur when a hot fluid moves further into the reservoir or into production wells (Bennion, 2002; Gupta and Civan, 1994). Increased temperature may cause the increase of carbonate and silica solubility. This may lead to partial reservoir dissolution. When a portion of the carbonaceous or silicate material is dissolved, particles are released and mobilized from their original immobilized positions. The particles can migrate into pore throats and lead to a decrease in formation permeability. When the saturated solution is further introduced into the formation, it slowly cools and the dissolution capacity decreases. This can lead to precipitation of calcium, magnesium, and silicate minerals. The size and location of precipitation may also lead to a decrease in reservoir permeability.

9.2.4 Wettability alteration and change

As temperature increases, formations generally become more water-wet (Bennion, 2002). Polar compounds and asphaltic compounds which may be physically adsorbed onto rock surfaces contributing to more oil-wet behavior may be desorbed at elevated temperature conditions causing the rock to behave in a more water-wet fashion. Temperature-dependent effects are of particular importance in thermal recovery processes associated with heavy oil recovery (Bennion et al., 1993; Punase et al., 2014). This is schematically illustrated in Fig. 9.12.

The medium is gradually altered to strongly oil-wet in the temperature range of $150°C-400°C$ and subsequently changed to water-wet as the temperature is increased (Escrochi et al., 2008). As formation temperature increases toward crude oil's bubble point, its molar volume increases and solubility of asphaltene decreases. This leads to asphaltene precipitation and alteration of wettability toward strongly oil-wet condition. At temperatures higher than the crude oil bubble point, precipitated asphaltene start to dissolve, mainly due to evaporation of crude oil saturates. This stimulates a wettability shift toward more water-wet condition (Escrochi et al., 2008; Kar et al., 2015).

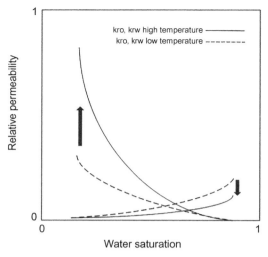

Figure 9.12 Illustrative example of influence of temperature on relative permeability/wettability.

The relative permeability to oil is increased and the water relative permeability is reduced, which are beneficial to oil production. In particular, for CSS, the mobility ratio increases dramatically, resulting in an increase in crude oil production. If water relative permeability is severely depressed by temperature, it may result in a reduction or restriction in injection of hot water or steam (Bennion et al., 1995; Bennion et al., 1992).

However, there are isolated circumstances of transitions to oil–wet behavior on the application to formations of superheated steam (Bennion, 2002). In SF projects, injected steam does not only change fluid properties, but also affects particles and porous media properties by rock surface processes such as rock wettability changes (Permadi et al., 2016). As shown in Fig. 9.13, it may be described in a simple case as follows: the pore in a sandstone formation is naturally water-wet, at high temperature condition, in which the residual oil is trapped in the center of the pore. The oil is encapsulated and inhibited from contacting with the rock matrix by surrounding liquid-phase hot water. When the transition to SF occurs, the encapsulating water is vaporized allowing the residual oil to contact with the rock surface directly. This allows adsorption to happen, which in many cases, even at elevated temperature, can cause the wettability state to be altered rapidly. This phenomenon will affect the irreducible water saturation, the efficiency of immiscible displacement, and thus the efficiency of oil recovery (Bennion et al., 1992; Naser et al., 2014).

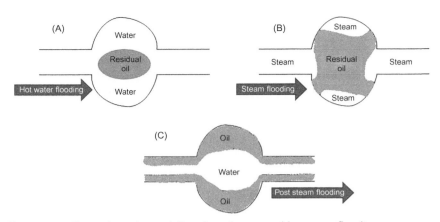

Figure 9.13 Illustration of wettability alteration caused by steam flooding.

In general, as temperature increases, a reservoir becomes water-wet. However, only when superheated steam is applied, the reservoir rock will change to oil–wet (Bennion, 2002).

9.2.5 Deposition

Deposition usually occurs in two forms, such as organic deposition and inorganic deposition. Heavy oil by nature is prone to organic deposition.

Asphaltene is an organic deposition. Most heavy crude oils do not naturally destabilize asphaltene with normal reductions in temperature or pressure but may be fairly susceptible to the formation of asphaltic sludge when contacted with some agents (Permadi et al., 2012). It can be induced by high flow rates, high temperature, a sharp drop in pressure, and contact with steam (Leontaritis et al., 1994; Kokal and Sayegh, 1995). Asphaltene needs to be identified by extensive laboratory testing. Asphaltene deposition occurs in downhole perforations and in near-wellbore locations within a reservoir formation. Asphaltene adsorption on rock surfaces will create an oil-wet sandstone. Therefore, it can be a wettability reversal agent.

Paraffins are more problematic in some situations with heavy oils and are generally controlled by reductions in temperature. In many situations, paraffin problems tend to be more a production issue rather than a downhole issue. Generally, at reservoir temperature conditions the temperature remains sufficiently high enough to inhibit the formation of crystalline waxes (Permadi et al., 2012).

Calcium carbonate scale and gypsum scale are common inorganic deposition forms. Iron-rich ion deposition and silica scale are less common and more difficult to deal with. Inorganic deposition clogs a formation to reduce fluid flow. There are many ways to remedy inorganic deposition, including simple jet washing and acid washing.

A pore surface deposition rate is given by (Civan, 2007a):

$$\frac{d\varepsilon_d}{dt} = k_d(\alpha + u)\sigma_p \varnothing^{2/3} \tag{9.1}$$

The initial condition is given by (Civan, 2007a):

$$\varepsilon_d = \varepsilon_{d0}, t = 0 \tag{9.2}$$

where ε_d is the volume fraction of the particles deposited at pore surfaces, σ_p is the particle volume fraction in the flowing pore fluid, $\varnothing^{2/3}$ is taken as a measure of the pore surface available, k_d is the deposition rate coefficient, and α is the stationary deposition factor accounting for no fluid flow conditions.

9.2.6 Emulsions and foams

Emulsions are a problem associated with many heavy oil operations where both oil and water are simultaneously being produced (Permadi et al., 2012). With the formation of emulsions in situ at elevated temperatures in porous media with respect to water and oil emulsions, as shown in Fig. 9.14, two different types of emulsions are possible: the water-in-oil emulsion, which tends to be the most problematic as it exhibits very high apparent viscosity in comparison to clean oil, and the less problematic oil-in-water emulsion. The water-in-oil emulsions are generated by a number of documented phenomena including turbulence, the presence of sand, silt or dispersed fines, paraffins, iron sulfide, asphaltenes and resins, a variety of organic acids, and cyclic and aromatic hydrocarbon compounds (Bennion et al., 1993).

Cyclic steam operations introduce a high level of energy into a reservoir, a primary element needed for the formation of stable emulsions in highly porous sands. Several factors associated with CSS can contribute to the creation of in situ emulsion (Castro, 2001). These components, an energy source, immiscible fluids, formation fines, and oil–wet surfaces, are all naturally present in most SF reservoirs (Juprasert and Davis, 1996; Krueger, 1988). Significant fluid velocities and shear forces during steam cycles combine these elements in the wellbore much like the action of a

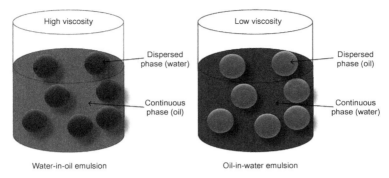

Figure 9.14 Water-oil emulsion types (Yahaya khan et al., 2014).

Waring blender in the laboratory. This leads to the creation of stable in situ emulsions which can partially or wholly block the production paths into the wellbore.

"Foamy oil" also falls into the category of a stabilized emulsion where the oil forms the external phase and small bubbles of trapped gas form the internal phase. Usually associated with high viscosity "heavy" oils, these fluids have been documented to have viscosities substantially higher than nonfoamy fluids (Bennion, 2002).

9.2.7 Dilatation and compaction

When temperature of an injected fluid is different than that of a formation, permeability may vary significantly and affect the ability of fluids to flow through porous media (Civan, 2006; Zheng et al., 2017). Permeability may be altered by expansion/contraction of grains and other constituents in the porous formation and by creation of micro fractures owing to thermal shock caused by suddenly induced thermal stress (Arias buitrago et al., 2016; Yuan et al., 2017). Thermal stress cracking and manufacturing of mobile and damaging fines have also been observed at high temperatures in some isolated studies (Bennion, 2002). In general, increases in temperature may result in an increase in compression on the grain-to-grain contacts caused by thermal expansion effects associated with the porous media (Permadi et al., 2012). Thermal expansion of minerals by temperature increase causes pore-throat contraction in porous rocks, which in turn increases tortuosity and the formation resistivity factor and hence decreases permeability (Civan, 2008).

The absolute permeability of the reservoir rock changes because of a change in pore channel sizes. A permeability change solely due to thermal changes in a mean particle diameter, grain matrix volume, and effective length of pore channels can be expressed as a ratio of the permeability at temperature T to that at a reference temperature, T_0 as follows (Gupta and Civan, 1994):

$$\left(\frac{k_T}{k_{T_0}}\right) = exp\left[-\left(\frac{3-\overline{\varnothing}}{\overline{\varnothing}}\right)\beta_{gv}(T-T_0)\right] \qquad (9.3)$$

where β_{gv} is the coefficient of volumetric thermal expansion for the rock matrix and $\overline{\varnothing}$ is the mean porosity over the temperature range T_0-T. This relationship indicates that the temperature sensitivity of permeability due to the thermal expansion effect is dependent on rock porosity. Therefore, the lower the porosity, the greater the temperature sensitivity of permeability.

However, absolute permeability measurements at elevated temperature are difficult to conduct in a reproducible fashion, because a considerable amount of permanent physical alteration in the rock character occurs from the start to the conclusion of the test sequence, therefore, eliminating the possibility for reproducibly repeating control permeability measurements (Bennion et al., 1995; Bennion, 2002).

9.3 DISCUSSION

There have been numerous classical topics about formation damage in a near-wellbore region of the rock matrix resulting from an application of TEOR including sands and fines migration, clay swelling, mineral transformation, minerals dissolution and precipitation, wettability alteration, scale, emulsions, rock dilatation, and compaction. However, formation damage is difficult to quantify in many cases. This is due to the inability of reservoir engineers to retrieve exact samples and conduct detailed measurements on the area of interest.

Future research topics on the formation damage by TEOR for heavy oil reservoirs are expected to remain those that have been researched historically. The fundamental processes causing formation damage identified by historical research and well observations are those of thermal mechanisms, chemical mechanisms, and mechanical mechanisms will be studied and quantified extensively through laboratory and field tests.

9.4 CONCLUSIONS

Heavy oil reservoirs are prone to formation damage during production, and thermal recovery can further exacerbate the damage. A review of various thermal recovery methods and their associated formation damage mechanisms in heavy oil reservoirs is presented. The main conclusions are:

- The thermal recovery methods involving CSS, SF, SAGD, and ISC are described physically.
- A porosity and permeability reduction can be caused by migrations of sands and fines. Due to hydrodynamic mobilization, colloidal expulsion, cement dissolution and surface deposition, sands, and fines can attach onto pore surfaces.
- The permeability of a porous system can be reduced by clay swelling and mineral transformations.
- Minerals dissolution and precipitation may lead to a decrease in reservoir permeability.
- As temperature increases, formations generally become more water-wet. However, there are isolated circumstances of transitions to oil-wet behavior in an application to formations of superheated steam.
- Deposition usually occurs in two forms, which are organic deposition and inorganic deposition. Fluid inflow can be reduced by depositions.
- Production paths into the wellbore can be blocked by emulsions and foams.
- Dilatation and compaction can decrease permeability during a thermal recovery process, due to thermal expansion effects associated with porous media.

REFERENCES

Abercrombie, H., 1988. Water-Rock Interaction During Diagenesis and Thermal Recovery. Ph.D. Thesis, University of Calgary, Cold Lake, Alta.

Akkutlu, I.Y., Yortsos, Y.C., 2003. The dynamics of in-situ combustion fronts in porous media. Combust. Flame 134 (3), 229–247.

Ali, S.F., 2003. Heavy oil—evermore mobile. J. Pet. Sci. Eng. 37 (1–2), 5–9.

Ali, S., Meldau, R., 1979. Current steamflood technology. J. Pet. Technol. 31 (10), 1332–1342.

Alvarez, J., Han, S., 2013. Current overview of cyclic steam injection process. J. Pet. Sci. Res. 2 (3), 116.

Arias buitrago, J., Alzate-Espinosa, G., Arbelaez-Londono, A., Morales, C., Chalaturnyk, R., Zambrano, G., 2016. Influence of confining stress in petrophysical properties changes during thermal recovery in silty sands Colombia. SPE Latin America and Caribbean Heavy and Extra Heavy Oil Conference. Society of Petroleum Engineers.

Baibakov, N.K., Garushev, A.R., Cieslewicz, W., 2011. Thermal Methods of Petroleum Production. Elsevier, Amsterdam.

Bennion, D.B., 2002. An overview of formation damage mechanisms causing a reduction in the productivity and injectivity of oil and gas producing formations. J. Can. Pet. Technol. 41.

Bennion, D.B., Thomas, F., Sheppard, D., 1992. Formation damage due to mineral alteration and wettability changes during hot water and steam injection in clay-bearing sandstone reservoirs. SPE Formation Damage Control Symposium. Society of Petroleum Engineers.

Bennion, D.B., Sarioglu, G., Chan, M., Hirata, T., Courtnage, D., Wansleeben, J., 1993. Steady-State Bitumen–Water Relative Permeability Measurements at Elevated Temperatures in Unconsolidated Porous Media. SPE International Thermal Operations Symposium. Society of Petroleum Engineers.

Bennion, D., Thomas, F., Bennion, D., Bietz, R., 1995. Mechanisms of formation damage and permeability impairment associated with the drilling, completion and production of low API gravity oil reservoirs. SPE International Heavy Oil Symposium. Society of Petroleum Engineers.

Bridgwater, A.V., 2003. Renewable fuels and chemicals by thermal processing of biomass. Chem. Eng. J. 91(2–3), 87–102.

Butler, R., 1985. A new approach to the modelling of steam-assisted gravity drainage. J. Can. Pet. Technol. 24, 42–51.

Castro, L.U., 2001. Demulsification Treatment and Removal of In-Situ Emulsion in Heavy-Oil Reservoirs. SPE Western Regional Meeting. Society of Petroleum Engineers.

Chen, Z., 2007. Reservoir Simulation: Mathematical Techniques in Oil Recovery. SIAM, Dallas, Texas.

Chen, Z., Huan, G., Ma, Y., 2006. Computational Methods for Multiphase Flows in Porous Media. SIAM,, Dallas, Texas.

Chieh, C., 1981. State-of-the-art review of fireflood field projects. Getty Oil Co.

Chu, C., 1977. A study of fireflood field projects (includes associated paper 6504). J. Pet. Technol. 29, 111–120.

Chunze, L., Linsong, C., Yang, L., 2008. Calculating models for heating radius of cyclic steam stimulation and formation parameters in horizontal well after soaking. Acta Pet. Sinic. 29 (1), 101–105.

Civan, F., 2006. Including Non-isothermal Effects in Models for Acidizing Hydraulically Fractured Wells in Carbonate Reservoirs. International Oil Conference and Exhibition in Mexico. Society of Petroleum Engineers.

Civan, F., 2007a. Formation damage mechanisms and their phenomenological modeling-an overview. European Formation Damage Conference. Society of Petroleum Engineers.

Civan, F., 2007b. Reservoir formation damage, Fundamentals, Modeling, Assessment and Mitigation, second ed. Gulf Professional Publishing, Amsterdam (Chapter 2).

Civan, F., 2008. Correlation of permeability loss by thermally-induced compaction due to grain expansion. Petrophysics 49 (4).

Coskuner, G., 2009. A new process combining cyclic steam stimulation and steam-assisted gravity drainage: hybrid SAGD. J. Can. Pet. Technol. 48 (1), 8–13.

De haan, H., Schenk, L., 1969. Performance analysis of a major steam drive project in the Tia Juana Field, Western Venezuela. J. Pet. Technol. 21 (1), 111–119.

Doe, U., 2014. Annual Energy Outlook 2014 with Projections to 2040. US Energy Information Administration, US Department of Energy,, Washington, DC.

Edmunds, N., Chhina, H., 2001. Economic optimum operating pressure for SAGD projects in Alberta. J. Can. Pet. Technol. 40 (12), 13—17.

Elliott, K., Kovscek, A., 1999. Simulation of early-time response of single-well steam assisted gravity drainage (SW-SAGD). SPE Western Regional Meeting. Society of Petroleum Engineers.

Escrochi, M., Nabipour, M., Ayatollahi, S.S., Mehranbod, N., 2008. Wettability alteration at elevated temperatures: the consequences of asphaltene precipitation. SPE International Symposium and Exhibition on Formation Damage Control. Society of Petroleum Engineers.

Farouqali, S., 1982. Steam injection theories: a unified approach. SPE Repr. Ser., United States, Conference: California Regional SPE Meeting, San Francisco, CA, USA.

Gates, I., Adams, J., Larter, S., 2008. The impact of oil viscosity heterogeneity on the production characteristics of tar sand and heavy oil reservoirs. Part II: Intelligent, geotailored recovery processes in compositionally graded reservoirs. J. Can. Pet. Technol. 47 (9).

Giacchetta, G., Leporini, M., Marchetti, B., 2015. Economic and environmental analysis of a Steam Assisted Gravity Drainage (SAGD) facility for oil recovery from Canadian oil sands. Appl. Energ 142, 1—9.

Green, D., Willhite, G., 1998. Enhanced Oil Recovery; SPE Textbook Series, Vol. 6. Society of Petroleum Engineers (SPE), Richardson, TX.

Gruesbeck, C., Collins, R., 1982. Entrainment and deposition of fine particles in porous media. Soc. Pet. Eng. J. 22, 847—856.

Gupta, A., Civan, F., 1994. Temperature sensitivity of formation damage in petroleum reservoirs. SPE Formation Damage Control Symposium. Society of Petroleum Engineers.

Hascakir, B., 2017. Introduction to Thermal Enhanced Oil Recovery (EOR) (Special Issue). Elsevier, Amsterdam.

Hashemi, R., Nassar, N.N., Almao, P.P., 2014. Nanoparticle technology for heavy oil in-situ upgrading and recovery enhancement: opportunities and challenges. Appl. Energ. 133, 374—387.

Hashemi-Kiasari, H., Hemmati-Sarapardeh, A., Mighani, S., Mohammadi, A.H., Sedaee-Sola, B., 2014. Effect of operational parameters on SAGD performance in a dip heterogeneous fractured reservoir. Fuel 122, 82—93.

Hollman, G., Tague, J., 1999. The Long-Term Effect of Cyclic Steam on Gravel-Packed Liners. SPE Western Regional Meeting. Society of Petroleum Engineers.

Hutcheon, I., Abercrombie, H., Shevalier, M., Nahnybida, C., 1989. A comparison of formation reactivity in quartz-rich and quartz-poor reservoirs during steam assisted recovery. Fourth UNITAR/UNDP International Conference on Heavy Crude and Tar Sands, pp. 747—757.

Ito, Y., Ipek, G., 2005. Steam fingering phenomenon during SAGD process. SPE International Thermal Operations and Heavy Oil Symposium. Society of Petroleum Engineers.

Juprasert, M.S., Davis, B.W., 1996. Stimulation by defoaming increases thermal oil production. SPE Western Regional Meeting. Society of Petroleum Engineers.

Kar, T., Yeoh, J.J., Ovalles, C., Rogel, E., Benson, I., Hascakir, B., 2015. The impact of asphaltene precipitation and clay migration on wettability alteration for Steam Assisted Gravity Drainage (SAGD) and expanding Solvent-SAGD (ES-SAGD). SPE Canada Heavy Oil Technical Conference. Society of Petroleum Engineers.

Kokal, S., Al-Kaabi, A., 2010. Enhanced oil recovery: challenges & opportunities. World Petroleum Council: Official Publication 64.

Kokal, S.L., Sayegh, S.G., 1995. Asphaltenes: the cholesterol of petroleum. Middle East Oil Show. Society of Petroleum Engineers.

Kovscek, A., 2012. Emerging challenges and potential futures for thermally enhanced oil recovery. J. Pet. Sci. Eng 98, 130–143.

Krueger, R.F., 1988. An overview of formation damage and well productivity in oilfield operations: an update. SPE California Regional Meeting. Society of Petroleum Engineers.

Kuuskraa, V.A., Van leeuwen, T., Wallace, M., Dipietro, P., 2011. Improving domestic energy security and lowering CO2 emissions with "next generation" CO2-enhanced oil recovery (CO2-EOR). National Energy Technology Laboratory, Pittsburgh, PA, USA.

Lake, L.W., Walsh, M.P., 2008. Enhanced Oil Recovery (EOR) Field Data Literature Search. Department of Petroleum and Geosystems Engineering University of Texas at Austin,, Austin, TX.

Leontaritis, K., Amaefule, J., Charles, R., 1994. A systematic approach for the prevention and treatment of formation damage caused by asphaltene deposition. SPE Prod. Facil. 9, 157–164.

Moore, R.G., Laureshen, C.J., Belgrave, J.D., Ursenbach, M.G., Mehta, S.R., 1995. In situ combustion in Canadian heavy oil reservoirs. Fuel 74, 1169–1175.

Moritis, G., 2010. CO$_2$, miscible, steam dominate EOR processes. Oil Gas J. 108 (14).

Morrow, A.W., Mukhametshina, A., Aleksandrov, D., Hascakir, B., 2014. Environmental impact of bitumen extraction with thermal recovery. SPE Heavy Oil Conference-Canada. Society of Petroleum Engineers.

Muecke, T.W., 1979. Formation fines and factors controlling their movement in porous media. J. Pet. Technol. 31, 144–150.

Nabipour, M., Escrochi, M., Ayatollahi, S., Boukadi, F., Wadhahi, M., Maamari, R., et al., 2007. Laboratory investigation of thermally-assisted gas–oil gravity drainage for secondary and tertiary oil recovery in fractured models. J. Pet. Sci. Eng 55, 74–82.

Naser, M.A., Permadi, A.K., Bae, W., Ryoo, W.S., Park, Y., Dang, S.T., et al., 2014. Steam-induced wettability alteration through contact angle measurement, a case study in X field, Indonesia. SPE International Heavy Oil Conference and Exhibition. Society of Petroleum Engineers.

Neasham, J.W., 1977. The morphology of dispersed clay in sandstone reservoirs and its effect on sandstone shaliness, pore space and fluid flow properties. SPE Annual Fall Technical Conference and Exhibition. Society of Petroleum Engineers.

Nwidee, L.N., Theophilus, S., Barifcani, A., Sarmadivaleh, M., Iglauer, S., 2016. EOR processes, opportunities and technological advancements. In: Romero-Zeron, L. (Ed.), Chemical Enhanced Oil Recovery (cEOR): A Practical Overview, pp. 1–50. InTech. doi: 10.5772/64828.

Okoye, C., Onuba, N., Ghalambor, A., Hayatdavoudi, A., 1990. Characterization of formation damage in heavy oil formation during steam injection. SPE Formation Damage Control Symposium. Society of Petroleum Engineers.

Owens, W., Suter, V., 1965. Steam stimulation—newest form of secondary petroleum recovery. Oil Gas J. 90 (2), 82–87.

Permadi, A.K., Naser, M.A., Mucharam, L., Rachmat, S., Kishita, A., 2012. Formation damage and permeability impairment associated with chemical and thermal treatments: future challenges in EOR applications. The Contribution of Geosciences to Human Security, Chapter 7. Logos Verlag Berlin GmbH, Berlin, Germany.

Permadi, A., Naser, M., Bae, W., Ryoo, W., Siregar, S., 2016. Investigating the effect of steam saturation properties on wettability through contact angle measurement using a novel experimental method. Int. J. Appl. Eng. Res. 11 (1), 177–184.

Perry, C., Gillott, J., 1979. The formation and behaviour of montmorillonite during the use of wet forward combustion in the Alberta oil sand deposits. Bull. Can. Petrol. Geol. 27 (3), 314–325.

Prats, M., 1982. Thermal Recovery. SPE Monograph Series Vol. 7. Society of Petroleum Engineers.

Punase, A., Zou, A., Elputranto, R., 2014. How do thermal recovery methods affect wettability alteration? J. Pet. Eng. Available from: http://dx.doi.org/10.1155/2014/538021.

Reed, M., 1979. Gravel pack and formation sandstone dissolution during steam injection. SPE, Chevron Oil Field Research Co.

Robertson, J., Chilingarian, G., Kumar, S., 1989. Surface Operations in Petroleum Production, II. Elsevier, Amsterdam.

Satter, A., Iqbal, G.M., Buchwalter, J.L., 2008. Practical Enhanced Reservoir Engineering: Assisted with Simulation Software. Pennwell Books,, Tulsa, USA.

Scheuerman, R.F., Bergersen, B.M., 1990. Injection-water salinity, formation pretreatment, and well-operations fluid-selection guidelines. J. Pet. Technol. 42, 836—845.

Shafiei, A., Dusseault, M.B., Zendehboudi, S., Chatzis, I., 2013. A new screening tool for evaluation of steamflooding performance in naturally fractured carbonate reservoirs. Fuel 108, 502—514.

Shin, H., Polikar, M., 2006. Experimental investigation of the fast-SAGD process. Canadian International Petroleum Conference. Petroleum Society of Canada.

Slider, H.C., 1983. Worldwide Practical Petroleum Reservoir Engineering Methods. PennWell Books.

Stokes, D., Doscher, T., 1974. Shell makes a success of steam flood at Yorba Linda. Oil Gas J. 72 (35), 71—78.

Stovall, S., 1934. Recovery of oil from depleted sands by means of dry steam. Oil Weekly 73, 17—24.

Tague, J.R., 2000. Overcoming formation damage in heavy oil fields: a comprehensive approach. SPE/AAPG Western Regional Meeting. Society of Petroleum Engineers.

Thomas, S., 2008. Enhanced oil recovery-an overview. Oil Gas Sci. Technol.-Revue de l'IFP 63 (1), 9—19.

Yahaya khan, M., Abdul karim, Z., Hagos, F.Y., Aziz, A.R.A., Tan, I.M., 2014. Current trends in water-in-diesel emulsion as a fuel. Sci. World J. 15.

Yanbin, C., Dongqing, L., Zhang, Z., Shantang, W., Quan, W., Daohong, X., 2012. Steam channeling control in the steam flooding of super heavy oil reservoirs, Shengli Oilfield. Pet. Explor. Dev. 39, 785—790.

Yuan, B., Su, Y., Moghanloo, R.G., Rui, Z., Wang, W., Shang, Y., 2015. A new analytical multi-linear solution for gas flow toward fractured horizontal wells with different fracture intensity. J. Nat. Gas Sci. Eng. 23, 227—238.

Yuan, B., Moghanloo, R.G., Shariff, E., 2016a. Integrated investigation of dynamic drainage volume and inflow performance relationship (transient IPR) to optimize multistage fractured horizontal wells in tight/shale formations. J. Energ. Res. Technol. 138, 052901.

Yuan, B., Moghanloo, R.G., Zheng, D., 2016b. Analytical evaluation of nanoparticle application to mitigate fines migration in porous media. SPE J. 21 (6), 2317—2332.

Yuan, B., Moghanloo, G.R., Zheng, D., 2017. A novel integrated production analysis workflow for evaluation, optimization and predication in shale plays. Int. J. Coal Geol. 180, 18—28.

Yuan, B., Moghanloo, R., Wang, W., 2018. Using Nanofluids to Control Fines Migration for Oil Recovery: Nanofluids Co-injection or Nanofluids Pre-flush? — A Comprehensive Answer. Fuel. 215, 474—483.

Zhang, J.M., Wu, X.D., Li, S.D., Zhang, J., Zhang, H.H., Qi, B.L., et al., 2013. Utilizing of in situ combustion process in Xinjiang Oil Field through analysis of produced fluids. Adv. Mater. Res. 772, 751—754.

Zhang, J., Wu, X., Chen, Z., Han, G., Wang, J., Ren, Z., et al., 2015a. Application and experimental study of cyclic foam stimulation. RSC Adv. 5, 76435–76441.

Zhang, J., Wu, X., Han, G., Zhang, K., Wang, J., Zhang, J., et al., 2015b. Damage by swelling clay and experimental study of cyclic foam stimulation. Acta Geol. Sin. 89 (S1), 215–216.

Zhao, D.W., Wang, J., Gates, I.D., 2013. Optimized solvent-aided steam-flooding strategy for recovery of thin heavy oil reservoirs. Fuel 112, 50–59.

Zheng, D., Yuan, B., Moghanloo, R.G., 2017. Analytical modeling dynamic drainage volume for transient flow towards multi-stage fractured wells in composite shale reservoirs. J. Pet. Sci. Eng. 149, 756–764.

CHAPTER TEN

A Special Focus on Formation Damage in Unconventional Reservoirs: Dynamic Production

Davud Davudov, Rouzbeh G. Moghanloo and Bin Yuan

The University of Oklahoma, Norman, OK, United States

Contents

10.1 INTRODUCTION

As shale plays maintain their role as one of the main energy resources in the United States, production sustainability remains the key decision–making parameter for investors. Shale resources have distinctive

Formation Damage during Improved Oil Recovery.
DOI: https://doi.org/10.1016/B978-0-12-813782-6.00010-5

© 2018 Elsevier Inc.
All rights reserved.

characteristics dissimilar to the conventional hydrocarbon reservoirs, such as nano-size pores and ultra-low permeability. Three main type of pores are observed in shale formations: Intergranular porosity, intercrystalline matrix porosity, and organic porosity in the kerogen (Passey et al., 2010; Loucks et al., 2012). Inorganic matrix is mainly composed of calcite, quartz, silicates, and clays which all are water-wet, whereas organic pores usually exist in mixed-wet conditions depending on the maturity level of the shale (Hu et al., 2015). Based on Loucks et al. (2012), pore diameter in these formations ranges from a few nanometers to a few micrometers causing very low permeability.

Moreover, shale gas formations have very complex gas storage mechanisms and fluid flow behavior as opposed to the conventional reservoirs. The gas storage in shale rocks occurs in the form of compressed free gas in organic, nonorganic matrix pores and natural fractures, adsorbed gas in organic matter, and dissolved gas in kerogen (Javadpour et al., 2007; Ghanbarnezhad Moghanloo et al., 2013). Moreover, the modeling of gas transport through the matrix is a complex task that includes evaluation of hydraulic connectivity and non-Darcy flow mechanisms (Yuan et al., 2016, 2017). Pore diameter in shale formations is in the order of nanometer causing violation of the Darcy's law. Depending on pore size and gas properties, non-Darcy flow regimes such as slip-flow, molecular diffusion, Knudsen diffusion of free gas molecules, and surface diffusion of adsorbed phase can affect the formation/matrix deliverability (Javadpour, 2009; Civan, 2010; Davudov et al., 2017a; Yuan et al., 2015a). The key parameter to classify flow regimes in nanoscale pores is Knudsen number, which can be expressed as the ratio of mean free path of the gas molecules to the pore radius (Roy et al., 2003).

Despite all of these complexities, production from tight formations has been achieved with great success owing to the recent technological advancements in horizontal drilling and hydraulic fracturing (Saldungaray and Palisch, 2012; Yuan et al., 2015b). Especially, hydraulic fracturing allows a large volume of a reservoir to be made accessible by creating high conductive pathways. Hence, drilling horizontal wells with multistage hydraulic fracturing has become the industry's standard practice for shale development.

10.1.1 Damage mechanisms in shale reservoir

Fluid production from shale reservoirs is mainly controlled through the combination of an induced fracture network (and potential interaction with natural fractures) and interconnected pores in the matrix

(Shaoul et al., 2011; Zheng et al., 2016). Although, multistage fracturing treatment has made shale resources economically viable, the success of that operation depends on the quality of the created fracture network and the matrix deliverability around the wellbore. The sustainability of production from shale formations is mainly influenced by the combination of formation properties, completion parameters, and the level of damage that takes place during fracturing and post-fracturing production. Usually, the observed sharp decline in production is strongly affected and controlled by the reduction of fracture conductivity and reservoir matrix deliverability due to damage mechanisms (Shaoul et al., 2011; Guo et al., 2011; Bahrami et al., 2012; Li et al., 2012).

Formation damage can be described as the reduction of the fluid deliverability owing to various processes; these processes often change the characteristics of the near wellbore region. In the context of shale resources, formation damage can be characterized as a combination of various phenomena that jeopardize flow deliverability of the stimulated reservoir volume (SRV). Formation damage can be also expressed as the hindrance of the fluid mobility toward the wellbore owing to the alteration of flow parameters and interactions of different fluid phases with each other, as well as with the boundaries of the flow paths.

Xu et al. (2010) classified the primary potential damage mechanisms as physical and chemical. Physical damage mechanisms are related to the interaction between the formation and in-situ fluids, resulting in permeability reduction of the formation (Bennion, 2002; You et al., 2015; Yuan and Moghanloo, 2016). Some of the physical damage mechanisms in shale reservoirs can be categorized as fines migration, particle invasion from the well into the reservoir, phase trapping damage, and stress damage. On the other hand, chemical damage mechanisms can result due to an incompatibility between reservoir fluid and fracturing fluid.

10.2 FRACTURE DAMAGE

A comprehensive hydraulically created fracture network with sustainable connectivity is required for economically viable production of shale resources. Fracture connectivity is the key parameter to control the performance of the fracture network. Permeability of a proppant-packed fracture can be estimated as a function of the median particle diameter of the proppant (Schubarth and Milton–Tayler, 2004) as expressed by Eq. (10.1):

$$K_f = 5.1 \times 10^{-6} \phi^5 d_{50}^2 \exp[-1.4(d_{90} - d_{50})], \qquad (10.1)$$

where, ϕ is porosity, d_{50} is the median particle size (mm), $d_{90} - d_{50}$ is difference between 90th and 50th percentile particle size (mm). In a dimensionless form fracture connectivity, (C_f) can be expressed [Eq. (10.2)] as a function of fracture permeability (K_f), fracture width (w), fracture half-length (x_f), and reservoir permeability (K):

$$C_f = \frac{K_f w}{K x_f}, \qquad (10.2)$$

For practical purposes, if C_f is greater than 300, then the fracture network is considered to have infinite conductivity where pressure drop across the fractures is considered to be negligible (Kelkar, 2008). However, in low permeability formations, hydraulic fractures cannot achieve infinite conductivity; because the conductivity is often significantly reduced owing to formation damage induced by various factors, such as proppant transport and placement, proppant embedment and crushing, fine migration and plugging, gelling damage, non-Darcy flow, and multiphase flow effect. Several studies have addressed these damage-related factors and their impacts on fracture connectivity loss and reduction of production rate.

Romero et al. (2002) studied the performance of fractured wells considering fracture face and choke skin factor, from which they concluded that proppant type and fracturing fluid are significant factors affecting long-term fracture performance. Similarly, Saldungaray and Palisch (2012) studied the influence of different proppant types on fracture conductivity. They suggested that the correct proppant selection can be a critical factor; because damage mechanisms can vary depending on proppant type and, in some cases, proppant-pack conductivity can be reduced by two orders of magnitude owing to the damage-related parameters. Vincent (2009) reported that fracture connectivity values measured in the lab based on the API standard are overestimated and that real-field conductivity is less than 10% of the lab-derived values. The major reason of this overestimation is because the API test does not consider real-field factors such as non-Darcy and multiphase flow, proppant embedment, fine migration and plugging, proppant diagenesis, or gelling damage from the gel residue of fracturing fluid.

When multiple formation-damage parameters are considered together, the overall conductivity can diminish by 90%, which will have significant effects on production rate (Palisch et al., 2007; Pattamasingh, 2016). Thus, it has been suggested that all of these factors (including cross effects) should be considered in the fracture conductivity models. Further understanding of the impact of each parameter on conductivity loss can improve prediction of well performance and fracture-stimulation design. It is worthwhile to mention that, although all formation-damage parameters can diminish fracture conductivity at some level, the contribution of each individual damage mechanism can vary depending on the circumstances (e.g., reservoir characteristics and fluid compositions) of each specific case.

10.2.1 Proppant transport and placement

Proppant transport and placement are important parameters controlling fracture geometry and conductivity over time. To properly place proppants inside created fractures (especially far from the perforations), fracturing fluid should maintain sufficient viscoelasticity to keep the proppant in suspension. Water-based conventional fracturing fluids can be classified as noncross-linked fluids and cross-linked fluids. Noncross-linked fluids such as slick water that does not have sufficiently high apparent viscosity to keep the proppant suspended will result in faster proppant settlement. However, cross-linked fluids have high apparent viscosity and particle settling is more efficiently controlled (i.e., delayed within the fracture network). On the other hand, cross-linked fracturing fluids often yields some residues (e.g., guar gum) in the formation that create difficulty in the post-fracturing process and results in formation damage. Single particle settling velocity, v_s in Newtonian fluids can be expressed as Eq. (10.3):

$$v_s = \frac{g\left(\rho_p - \rho_f\right)d_p^2}{18\mu_f},$$
(10.3)

where ρ_p is particle density, ρ_f is fluid density, d_p is proppant dimeter, μ_f is fluid viscosity, and g is acceleration of gravity. From Eq. (10.3) it is clear that the density difference between proppant and the fluid, the proppant diameter, and the fluid viscosity are important parameters. However, predicting proppant settling velocity based on Eq. (10.3) might be inadequate, because particle-particle and particle-wall interactions are not considered.

10.2.2 Proppant embedment

During the production life of a fractured well, the pressure inside the fractures decreases leading to an increase in the net effective stress as shown in Eq. (10.4).

$$\sigma_{eff} = \sigma_T - \alpha P_p, \tag{10.4}$$

where σ_{eff} is net effective stress, σ_T is overburden stress, α is effective stress coefficient which is also called Biot's poroelastic term, and P_p is pore pressure. An increase of effective stress results in mechanical interaction between proppants and fracture surface (Bennion, 2002; Reinicke et al., 2010). This stress-induced interaction between proppants and the fracture surface can lead to proppant embedment and crushing, breakdown, and consequent generation of fine particles.

Some of the factors affecting the level of embedment are proppant type, size, concentration, mineral composition, hardness of rock, reservoir temperature and pressure, type of fracturing fluid, and exposure time (Li et al., 2015; Guo and Liu, 2012). Palisch et al. (2010) expressed embedment depth, W_E in inches as a function of the median proppant dimeter, d_p, also in inches, Young's modulus of the formation, E, in MMpsi, and closure stress, σ_{eff} in psi as Eq. (10.5):

$$W_E = 2.5 d_p \left(1 + \frac{4000E}{\sigma_{eff}} \right)^{-1.5}, \tag{10.5}$$

From Eq. (10.5), it is clear that the larger the particle diameter, the more indentation would occur; moreover, through the role of the Young's modulus in Eq. (10.5), soft rocks lead to more indentation than hard rocks (Denney, 2012; Alramahi and Sundberg, 2012). However, in Eq. (10.5) impact of different proppant properties are not considered; whereas the degree of proppant embedment can be considerably different depending on the type of proppant. Ceramic proppant exhibits higher embedment than sand at the same confining stress; however, embedment depth of resin–coated sand is less than that of ceramic or sand proppants (Weaver et al., 2005). This observation can be explained because resin–coated sand proppants provide higher contact area which helps distribute stress load.

Ghosh et al. (2014) conducted experiments to analyze embedment depth for several proppant types. Based on their experimental results, the embedment depth of 20/40 mesh size sand was measured in the range of 30−80 μm at 4000 psi effective pressure for a Barnett shale sample. In the case of 30/50 mesh size resin-coated sand, indentation depth reduced to $20-60 \, \mu m$, whereas in the case of 40/70 mesh size ceramic proppant embedment depth increased to $50-80 \, \mu m$. Moreover, Zhang et al. (2015a) have shown that the severity of fracture conductivity loss due to proppant embedment can be different depending on the number of proppant layers. Given the same level of embedment, the fracture propped with multiple proppant layers experiences less conductivity reduction when compared with the fracture propped by a monolayer of proppants. For multiple layers of proppant, the fracture conductivity depends more on the internal path within proppants rather than flow areas adjacent to the fracture walls.

The skin factor resulting from proppant-embedment, s_p can be expressed as a function of damage depth, b_s, fracture half-length, x_f, initial permeability, K, and damaged permeability, K_s as shown in Eq. (10.6) (Jackson and Rai, 2012; Civan, 2014):

$$s_p = \frac{\pi b_s}{2x_f} \left(\frac{K}{K_s} - 1 \right), \tag{10.6}$$

10.2.3 Proppant crushing

Sustainability of fracture connectivity strongly depends on the strength and the stiffness of the proppant (Han and Wang, 2014). Besides the proppant embedment issue, further increase in effective stress leads to proppant crushing (depending on proppant type and hardness of the rock). As illustrated in Fig. 10.1, permeability reduction can be up to 70% at 6000 psi effective stress owing to proppant crushing. Sand and resin-coated sand exhibit relatively similar permeability loss (68% at 6000 psi), whereas ceramic proppant exhibits comparably less permeability reduction, that is, close to 55% at 6000 psi effective stress.

10.2.4 Fines migration and plugging

Crushing of proppants generates fines particles that can later migrate and block the flow channels, consequently resulting in conductivity loss (Gidley et al., 1995; Lv, 2010; Kang et al., 2013). The amount and characteristic (size distribution) of fine particles are strongly related to the type of proppant; for example, significant amounts of fine particles are generated associated with sand crushing. Although resin-coated sand will crush in a similar fashion to sand, since the resin coating encapsulates crushed particles, it helps prevent fines migration. On the other hand,

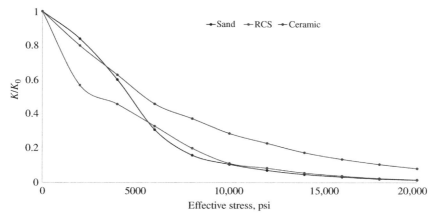

Figure 10.1 Permeability reduction for three proppant types (Sand, Resin-coated sand, and Ceramic) as a function of effective stress.

ceramic proppant crushing results in much larger size particles, which are more difficult to migrate and therefore are less likely to plug the fracture throats (Ghosh et al., 2014).

To illustrate fracture permeability reduction due to fines migration, experimental data for a Barnett shale sample is shown in Fig. 10.2 (Ghosh et al., 2014). As can be seen, in the case of a 40/70 mesh size sand, permeability reduction can be as much as 99%. On the other hand, ceramic proppant exhibits 70% drop after 10 days. In contrast, the, resin–coated sand shows no permeability reduction due to fines migration, since that type of proppant completely encapsulates the crushed particles.

Based on experimental work, Zhang et al. (2015a) suggested that reversal of flow direction can help to partially mitigate the negative effects of fines migration on permeability reduction. In their experiments, a fracture with initial conductivity of 87 mD-ft results in a reduction of conductivity to 4 mD-ft after a fracture-stimulation operation (including flow-back water). When the direction of the flow-back water is abruptly switched, conductivity is temporarily restored to 15 mD-ft. This is an indication of the capturing and release of fines due to flow reversal.

10.2.5 Gelling damage

To transport and properly distribute proppant, fracturing fluid with sufficient apparent viscosity is required, which can be achieved by adding cross-linked gel. After proppant placement is concluded, a breaker is injected to

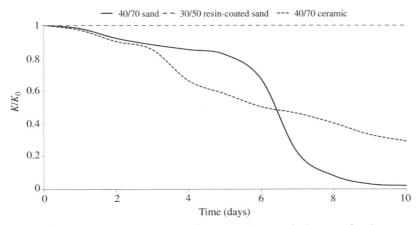

Figure 10.2 Permeability drop due to fines migration and plugging for three proppant types.

break–up the polymer networks and reduce fracturing fluid's apparent viscosity, which helps the flow-back efficiency. However, unbroken cross-linked gel molecules in fractures yield gel residue, which can significantly reduce fracture conductivity (Iqbal and Civan, 1993; Shaoul et al., 2011).

The skin factor associated with hydraulically created fractures, s_f, can be expressed as function of dimensionless fracture conductivity, C_f, fracture half-length, x_f, and well radius, r_w as shown in Eq. (10.7) (Valko et al., 1998):

$$s_f = \frac{1.65 - 0.32v + 0.116v^2}{1 + 0.18v + 0.064v^2 + 0.05v^3} - \ln\left(\frac{x_f}{r_w}\right), \qquad (10.7)$$

where $v = \ln(C_f)$.

Alternatively, Eq. (10.7) can be written if fracture connectivity is greater than 100, $C_f > 100$ as Eq. (10.8):

$$s_f \approx 0.7 - \ln(x_f/r_w) \qquad (10.8)$$

Eq. (10.8) indicates that the skin factor of fractures with high-conductivity only depends on fracture length, x_f, whereas the fracture permeability and fracture width have no effect.

On the other hand, for $C_f < 1$, Eq. (10.7) can be simplified to Eq. (10.9):

$$s_f \approx 1.52 + 2.31\log(r_w) - 1.55\log\left(\frac{K_f w}{K}\right) - 0.77\log(x_f) \qquad (10.9)$$

where K_f is fracture permeability, K is reservoir permeability, w is fracture width, x_f is fracture half-length, and r_w is well radius. Through comparison of the coefficients of the last two terms in Eq. (10.9), one can conclude that the skin factor of the fractured well with low-conductivity is more sensitive to the fracture permeability and the fracture width than the fracture length (Guo et al., 2007).

10.2.6 Multiphase and non-Darcy inertial flow

Large fluid velocities in fractures yield significant amounts of the energy loss, which leads to an additional pressure drop. Forchheimer's equation, which considers this extra inertial pressure loss can be expressed as Eq. (10.10) (Forchheimer, 1901; Teng and Zhao, 2000):

$$\frac{\Delta p}{x_f} = \frac{\mu_f}{K_f} \cdot u + \beta \rho_f u_f^2 \qquad (10.10)$$

where Δp is pressure gradient, x_f is fracture half-length, K_f is fracture permeability, u_f is fluid velocity, ρ_f is fluid density, μ_f is fluid viscosity, and β is inertial flow coefficient in $1/\text{ft}$.

Palisch et al. (2007) has shown that fracture conductivity loss owing to inertial flow can be in the range of 50%—85%. Moreover, experimental results demonstrate that having an additional phase in the fracture will also result in a significant amount of pressure drop. This is typically attributed to the inefficient flow regimes where gas and liquid molecules move through the proppant pack and associated momentum transfer between phases.

10.3 RESERVOIR FORMATION DAMAGE

Prediction of productivity and well deliverability requires a thorough understanding of how porosity and pore connectivity affect reservoir permeability. Although there are several models to predict hydraulic conductivity and intrinsic permeability (Pape et al., 2000; Civan, 2001), in most cases the interplay between pore compressibility and connectivity loss (coordination number reduction) due to effective stress and/or pore plugging has been oversimplified. Thus, it is crucial to have permeability model which accounts for the complex nature of the shale-pore connectivity (Civan, 2011).

10.3.1 Intrinsic permeability—kozeny-carmen equation

One of the fundamental permeability models is Kozeny-Carmen (KC) equation (Kozeny, 1927; Carman, 1937), which assumes that the pores are bundle of cylindrical tubes and relates permeability to porosity (ϕ), tortuosity (τ), hydraulic pore radius (r_h), and constant term (c):

$$K = \frac{r_h^2}{c} \frac{\phi}{\tau}, \tag{10.11}$$

Many variations of the KC model have been proposed in the literature. Based on the assumption that the electrical field lines and the fluid stream lines are identical, Paterson (1983), and Walsh and Brace (1984) proposed the equivalent channel model, which is expressed as Eq. 10.12:

$$K = \frac{r_h^2}{c} \frac{1}{F}, \tag{10.12}$$

where r_h is hydraulic radius, F is electrical formation factor, and c is constant term.

Later, Katz and Thompson (1986, 1987) applied critical-path analysis to determine permeability. They recognized that fluid flow and electrical conductance through porous media are percolation processes, and permeability can be related to electrical conductivity and critical pore diameter. The Katz and Thompson model is expressed by Eq. (10.13):

$$K = \frac{1}{c} \frac{\sigma_b}{\sigma_w} r_c^2 = \frac{r_c^2}{c} \frac{1}{F}, \tag{10.13}$$

where σ_b is bulk electrical conductivity, σ_w is saturating fluid electrical conductivity, r_c is critical pore radius, and c is a constant. Here r_c is the critical pore radius, defined as the largest value of r, for which an interconnected path may exist from one side of a system to the other. Katz and Thompson (1986) argued that the critical pore radius can be estimated from mercury-intrusion porosimetry data, and the inflection point on the mercury-intrusion curve corresponds to this critical pore radius.

If the effective hydraulic pore radius, r_h is substituted by the surface area per unit of grain volume, which it is related to grain radius, r_g and porosity, then Eq. (10.11) can be rewritten as Eq. (10.14):

$$K = \frac{r_g^2}{c} \left(\frac{\phi}{1-\phi}\right)^2 \frac{\phi}{\tau}, \qquad (10.14)$$

In Eq. (10.14), $\phi/(1-\phi)$ represents pore to grain volume and permeability can be estimated as a function of porosity and tortuosity. Although Eq. (10.14) is used extensively, it has many limitations, thus has been frequently modified for accurate estimation of permeability.

10.3.2 Power—law permeability equation

Civan (2001) suggested that the KC equation cannot properly address the gate /valve effect and predict permeability when pore throats are blocked. The blockage of pore throats creates isolated pores; therefore, the KC equation needs to be modified to include an interconnectivity parameter, Γ, as expressed by Eq. (10.15):

$$k = \Gamma\phi\left(\frac{\phi}{1-\phi}\right)^{2\eta} \qquad (10.15)$$

where η is the exponent and Γ is a measure of the pore space connectivity which represents the valve effect of the pore throats. The valve effect controls the pore connectivity to pore spaces in an interconnected network (Civan, 2011). The number of pore throats is generally denoted by z, called the coordination number. The values $z = 0$ and $z = 1$ represent isolated pores and dead—end pores, respectively, making $z = 2$ the lowest value of the coordination number associated with a conductive pore body is. Thus, the interconnectivity parameter is a strong function of average coordination number and it becomes zero when all existing pore throats are blocked due to mechanisms like fine migration, deposition of precipitates including gels, waxes, and asphaltenes, and/or collapse of pore throats under mechanical stresses. Consequently, Eq. (10.15) predicts that permeability can become zero even if the porosity is nonzero.

10.3.3 Permeability as function of coordination number

Alternatively, the Katz-Thompson model [Eq. (10.13)] can be modified based on fractal and percolation theories. Following Daigle (2016), the critical pore radius can be expressed in terms of percolation threshold in Eq. (10.16):

$$r_c = r_{\text{max}}(1-p_c)^{\frac{1}{3-D}}, \tag{10.16}$$

where r_{max} is maximum accessible pore radius, D is fractal dimension, and p_c is critical percolation threshold.

Additionally, the formation factor, F which is the ratio of electrical conductivity to the electrical conductivity of the fluid-saturated pore, also may be expressed in terms of total porosity, ϕ, percolation threshold, p_c, and exponent, m as Eq. (10.17) (Ghanbarian et al., 2014):

$$\frac{1}{F} = \left[\frac{\phi(1-p_c)}{1-\phi p_c}\right]^m, \tag{10.17}$$

Moreover, Revil (2002) has suggested that in clay-bearing rocks, $1 - \phi p_c \approx 1$, and based on that assumption, the formation factor can be expressed as Eq. (10.18):

$$\frac{1}{F} = [\phi(1-p_c)]^m, \tag{10.18}$$

It should be noted that Eq. (10.18) reduces to the empirical Archie model, $1/F = \phi^m$, when p_c approaches zero. On the basis of these arguments, if Eqs. (10.16) and (10.18) are substituted into Eq. (10.13), then permeability can be estimated by Eq. (10.19):

$$K = \frac{r_{\text{max}}^2}{c} \phi^m (1-p_c)^{\frac{2}{3-D}+m}, \tag{10.19}$$

Critical percolation threshold, p_c can be expressed in terms of the coordination number as $p_c = 1.5/z$, and if c is assumed to be equal to 8, then Eq. (10.19) can be rewritten as Eq. (10.20):

$$K = \frac{r_{\text{max}}^2}{8} \phi^m \left(\frac{z-1.5}{z}\right)^{\frac{2}{3-D}+m} \tag{10.20}$$

When Eq. (10.20) is compared with Civan's model [Eq. (10.15)], it is clear that $r_{max}^2 \phi^2/8$ is the maximum achievable permeability and $[(z-1.5)/z]^{\frac{2}{3-D}+m}$ represents the interconnectivity term (Γ). Also, Eq. (10.20) is essentially a modified version of KT model [Eq. (10.13)], which includes the interconnectivity term implicitly. However, by analyzing Eq. (10.20), the impact of pore volume reduction and connectivity

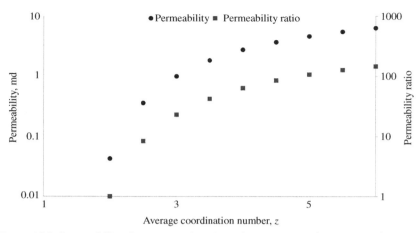

Figure 10.3 Permeability change as a function of average coordination number.

loss can be distinguished and evaluated separately. To illustrate the importance of connectivity, permeability reduction with decreasing coordination number based on the proposed model [Eq. (10.20)] is shown in Fig. 10.3. The results of that analysis indicate that the reduction in permeability can be as much as 150 times when the average pore coordination number is decreased from 6 to 2. This clearly indicates the importance of pore connectivity and how crucial it might be to lose critical bond connections. Note that the based on proposed model [Eq. (10.20)], permeability reduction is also related to fractal dimension, D, which is in range of $2 \leq D < 3$ for three-dimensional system and it is considered as 2.5 in these calculations (Krohn, 1988; Liu et al., 2015).

10.3.4 Intrinsic permeability reduction by effective stress

Various experimental studies have been performed to study permeability variation with respect to pressure for shale formations (Tinni et al., 2012; Metwally and Sondergeld, 2011; Dong et al., 2010). Experimental results from the samples have shown a nonlinear reduction in permeability with an increase in effective pressure, where this reduction might be as much as two orders of magnitude. This severe reduction in permeability cannot be explained solely by pore volume reduction caused by compressibility, rather it can be better explained by the combination of two main phenomena; pore shrinkage and connectivity loss due to bond breakage between interconnected pores (Lan et al., 2017; Davudov et al., 2016; Davudov and Moghanloo, 2016).

10.3.4.1 Empirical models

To describe the relationship between severe permeability reduction and slight porosity change under effective stress, researchers have suggested several empirical models that describe permeability change as function of effective stress. Shi and Wang (1986) suggested that the relationship between effective pressure and rock permeability should follow a power law, which can be expressed as Eq. (10.21):

$$\frac{K_e}{K_o} = \left(\frac{P_e}{P_o}\right)^{-p}, \tag{10.21}$$

where K_e denotes the permeability under the net effective stress P_e, K_o represents the reference permeability under atmospheric pressure P_o and p is a constant.

On the other hand, David et al. (1994) and Evans et al. (1997) suggested an exponential relationship to model permeability change as a function of effective stress expressed as Eq. (10.22):

$$\frac{K_e}{K_o} = \exp[-\omega(P_e - P_o)], \tag{10.22}$$

where ω is a constant value.

Kwon et al. (2001) found that the reduction in permeability with increasing effective stress, P_e can be described by cubic law function as expressed in Eq. (10.23):

$$\frac{K_e}{K_o} = \left[1 - \left(\frac{P_e}{P*}\right)^q\right]^3, \tag{10.23}$$

where K_e denotes the permeability under the net effective stress P_e, K_o represents the reference permeability at $P_e = 0$, and parameters q and $P*$ are constants associated with geometry and pore surface topography. Although, the "cubic law" model is one of the most used equation to explain permeability decrease caused by the effective stress, the flaw in this relationship is that it yields a negative K when P_e is larger than $P*$.

Since permeability is a function of porosity and pore structure, researchers have also studied the permeability—porosity relationship under effective stress. David et al. (1994) proposed a power law relationship to describe the permeability—porosity relationship induced by mechanical compaction as expressed by Eq. (10.24):

$$\frac{K_e}{K_o} = \left(\frac{\phi}{\phi_o}\right)^{\gamma}, \tag{10.24}$$

where γ is a constant called the porosity sensitivity exponent. Based on Dong et al. (2010) experimental results, porosity sensitivity exponents for sandstone, range from 3 to 6, where these values ranged from 25 to 55 for the tested silty-shale samples. The porosity sensitivity exponent for the silty-shale is considerably higher than that of the sandstone, which clearly indicates that lightly decreased porosity potentially causes dramatic decreases in permeability for shale formations. Kwon et al. (2004) suggested that such large and nonrecoverable decreases observed in permeability of shale samples are due to the closure of critical pore links in the network and permanent reductions in connected pore space, while the small recoverable changes in permeability represent the elastic response of the pore space.

10.3.4.2 Effect of pore compressibility and connectivity loss

Alternatively, based on the permeability model described in Eq. (10.19), the effects of both pore shrinkage and bond breakage on permeability reduction can be analyzed using Eq. (10.25):

$$\frac{K_e}{K_o} = \frac{r_{max}^2}{r_{max_o}^2} \left(\frac{\phi}{\phi_o}\right)^m \frac{\Gamma}{\Gamma_o}, \tag{10.25}$$

If it is assumed that $r_{max}^2 / r_{max_o}^2 = \frac{\phi}{\phi_o}$, Eq. (10.25) becomes simplified to Eq. (10.26):

$$\underbrace{\frac{K_e}{K_o}}_{\text{Permeability Reduction Rate}} = \underbrace{\left(\frac{\phi}{\phi_o}\right)^{m+1}}_{\text{Pore Volume Shrinkage}} * \underbrace{\frac{\Gamma}{\Gamma_o}}_{\text{Connectivity Loss}} \tag{10.26}$$

where first and second terms on the right side of Eq. (10.26) express permeability reduction due to pore volume shrinkage/pore compressibility and connectivity loss, respectively (Davudov and Moghanloo, 2018). Alternatively, an average coordination number as a function of effective stress can be estimated by combining Eqs. (10.20) and (10.26) in the form of Eq. (10.27):

$$\frac{\Gamma}{\Gamma_o} = \frac{\frac{K_e}{K_o}}{\left(\frac{\phi}{\phi_o}\right)^{m+1}} = \frac{\left(\frac{z-1.5}{z}\right)^{\frac{2}{3-D}+m}}{\left(\frac{z_o-1.5}{z_o}\right)^{\frac{2}{3-D}+m}} \tag{10.27}$$

where z_o is initial coordination number and it can be estimated as a function of initial porosity (Doyen, 1988; Bernabe et al., 2010) by Eq. (10.28):

$$z_o = A + B \log(\phi), \tag{10.28}$$

where both A and B are constants. Bernabe et al. (2010) suggested that for a two-dimensional system A is 10.4 and B is equal to 6.25. Average coordination number decrease might be minimal and therefore can be neglected in conventional formations, however, in formations like shale, intrinsic permeability reduction is mainly due to bond breakage and reduction of the connected pore volume network.

To evaluate the impact of both pore compressibility and bond breakage on permeability reduction, two sets of experimental data from sandstone and shale samples are selected from the literature (Dong et al., 2010). To measure porosity and intrinsic permeability, experiments were conducted that gradually increase the confining pressure from 435 to 725 psi, then to 1450, and finally (in 1450 psi increments) up to 10000 psi, while keeping pore pressure constant. Table 10.1 and Table 10.2 detail experimental data for sandstone and shale samples, respectively.

Fig. 10.4 illustrates porosity and porosity ratio, $\left(\phi/\phi_o\right)$ reduction under effective stress for both sandstone and shale samples in Tables 10.1 and 10.2. Results indicate that when effective stress reaches 8500 psi, porosity reduction in sandstone is close to 15%, whereas this value is around 10% for the shale sample. On the other hand, when permeability reduction rates are compared for the two samples it can be observed that,

Table 10.1 Porosity and permeability data under effective stress for a sandstone sample from Cholan formation in western Taiwan

Net stress, psi	Porosity, %	Net stress, psi	Permeability, md
0	18.5	0	76.0
659.7	18.1	580.3	73.0
1389.9	17.4	1358.4	66.4
2853.2	16.7	2791.3	60.9
4287.1	16.5	4329.3	57.8
5720.9	16.2	5718.2	56.5
7207.3	16.0	7143.5	55.2
8623.7	15.8	8641.0	54.4

Adopted from Dong et al., 2010.

Table 10.2 Porosity and permeability under effective stress data for shale sample from Chinshui shale formation in western Taiwan

Net stress, psi	Porosity, %	Net stress, psi	Permeability, md	Calculated Z
0	10.8	0	0.513	3.59
297.2	10.3	600.1	0.366	3.15
527.5	10.2	1335.2	0.168	2.80
1262.2	10.1	2032.5	0.072	2.56
2678.6	10.0	2809.1	0.044	2.36
4141.0	10.0	4208.0	0.022	2.08
5565.6	9.9	5646.8	0.015	1.87
7071.5	9.9	7125.0	0.010	1.71
8495.5	9.8	8642.1	0.008	1.57

Adopted from Dong et al., 2010.

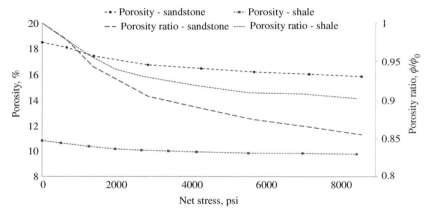

Figure 10.4 Porosity as a function of effective stress comparing sandstone and shale samples (Experimental data adopted from Dong et al., 2010).

permeability for shale sample is almost 98% less than its initial value when effective stress is increased to 8500 psi; whereas for sandstone sample, this reduction is less than 30% (Fig. 10.5). These two observations clearly indicate the importance of bond breakage and connectivity loss in shale formations.

In order to understand connectivity loss under effective stress, the interconnectivity ratio (Γ/Γ_o) is calculated based on Eq. (10.27). For the sandstone sample, results indicate that permeability reduction is only due to pore volume shrinkage where exponent m is around 1.2, and connectivity loss is negligible. On the other hand, for the shale sample, if initial coordination number is estimated based on Eq. (10.28) to be 4.3, fractal dimension, D is assumed to be equal to 2.5 and formation

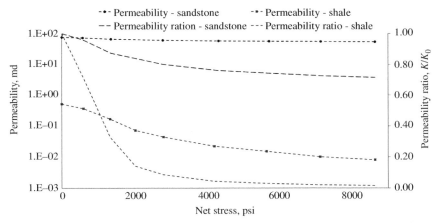

Figure 10.5 Permeability as a function of effective stress (A) Sandstone sample (B) Shale sample (Experimental data adopted from Dong et al., 2010).

factor exponent, m is considered as a universal exponent equal to 2 (Clerc et al., 2000; Stauffer and Aharony, 1992), then the interconnectivity parameter and coordination number drastically decrease. Fig. 10.6 and Table 10.3 summarize results of change in the interconnectivity parameter and reduction of the coordination number with respect to their initial values as a function of pressure. Results show that when effective pressure reaches 10,000 psi, the initial value of coordination number reduces by about 50% from 4.3 to 2.3. Thus, it can be concluded that for conventional formations the KC equation can be sufficient to model permeability, since permeability reduction can be solely due to pore volume shrinkage/pore compressibility. However, for tight formations with very small pore throats, one of the major reasons of permeability decrease can be attributed to connectivity/coordination number loss. Please note that in these calculations, the impact of micro-crack closure on permeability has not been considered, although as shown in Davudov and Moghanloo (2018) the closure effect can play a significant role at early stages.

Moreover, when pore diameters are in the order of micro–nano meters this will cause a violation of the Darcy's law to describe fluid flow in shale formations (Davudov et al., 2017b; Ghanbarnezhad Moghanloo and Javadpour, 2014). Depending on the pore size and gas properties, non–Darcy flow mechanisms such as slip-flow, molecular diffusion, and Knudsen diffusion can affect the matrix permeability. Thus, it is essential to distinguish and incorporate all important physical

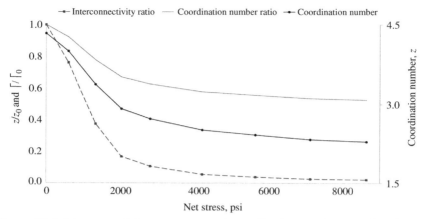

Figure 10.6 Interconnectivity parameter and coordination number as a function of effective stress for shale sample.

Table 10.3 Porosity and permeability data for shale sample under effective stress

Net stress, psi	Porosity ratio	Permeability ratio	Interconnectivity ratio	Coordination number ratio	Coordination number
0	1.00	1.00	1.00	1.00	4.3
600.1	0.98	0.713	0.763	0.92	4.0
1335.2	0.95	0.327	0.377	0.78	3.4
2032.5	0.94	0.140	0.168	0.67	2.9
2809.1	0.93	0.086	0.106	0.63	2.7
4208.0	0.92	0.043	0.055	0.58	2.5
5646.8	0.91	0.029	0.038	0.56	2.4
7125.0	0.91	0.019	0.026	0.54	2.3
8642.1	0.90	0.016	0.021	0.53	2.3

Adopted from Dong et al., 2010.

parameters accordingly to estimate permeability evolution under effective stress. During production, as pore pressure decreases, the effective stress will increase and that can potentially lead to intrinsic permeability reduction because of pore shrinkage and connectivity loss. At the same time, due to the effect of non–Darcy flow regimes, apparent permeability will increase (Fig. 10.7). Depending on the pore type, size, and topography as well as pore pressure, one of the above effects may dominate matrix permeability (Davudov et al., 2017b). However, non–Darcy flow regimes and their effect on permeability is beyond the scope of the present chapter and is not discussed further here.

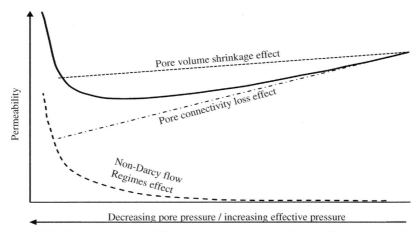

Figure 10.7 Shale gas permeability considering the combined effects of non-Darcy flow regimes, pore volume shrinkage and pore connectivity loss (solid line).

10.4 FLUID DAMAGE

During hydraulic fracturing, a significant amount of water is utilized, where only 25%—60% can be reproduced back to the surface. Jung et al. (2013) emphasized the sensitivity of shale formations to water, where the major concern is its interaction with water-sensitive clays (clay swelling) and its penetration into the formation because of high capillary pressure. Water imbibition in the shale formation can causes water blockage by increasing of the water saturation in the vicinity of fracture interface.

The water transport in shale formations can occur through a combination of various mechanisms, including convection, capillary suction/imbibition, and osmosis (Bader and Heister, 2006; Al-Bazali et al., 2009). Fakcharoenphol et al. (2013) have stated that imbibition and osmotic are the most dominant factors affecting the transfer of water from natural fractures into the shale matrix.

Holditch (1979) performed extensive studies to investigate formation-damage mechanisms in fractured wells; he concluded that capillary pressure and relative permeability are the most important factors controlling the success of well cleanup. Moreover, he suggested that water blockage effects can be negligible if the pressure drawdown is greater than the capillary pressure in the formation. Based on numerical simulations, Gdanski et al. (2005) confirmed the results reported by Holditch (1979). They have concluded that fracture-face matrix damage caused by clay swelling and water blockage may result in up to 90% loss of the original permeability which may have a significant impact on the production rate and the formation deliverability.

10.4.1 Clay swelling

Water imbibition from fracture network into matrix pore body will cause the embedded clay structure to swell. Clay swelling will result in significant reduction of fracture conductivity and reservoir permeability (Shaoul et al., 2011; Alkouh et al., 2014).

Based on results obtained from experiments with water-based fracturing fluids, Bazin et al. (2010) have shown that matrix permeability and formation deliverability can be reduced by up to 70% due to illite and montmorrilinite clays swelling and pore plugging because of kaolinite clay migration. Moreover, based on experimental studies, Zhang et al. (2015b) have observed that, for the Barnett shale samples, the initial fracture conductivity was reduced by 80% because of clay swelling, whereas for the Eagle Ford shale sample with 12% clay content, this value was around 20%.

The major factors affecting shale swelling are the initial water content W (%), the difference between salinity of the injected water and the bond water, clay content C (%), in-situ confined pressure P (kpa), and plasticity index (PI). Numerous models have been developed to predict shale swelling. Erol and Dhowian (1990) used regression analyses and proposed a model to predict swelling volume S (%) of shale as a function of initial water content, W, liquid limit, LL (%), plasticity index, PI, and clay percent, CL, as expressed in Eq. (10.29):

$$S = 0.925(0.43LL - W)^{0.51} + 1.19PI^{0.4} + 0.74CL^{0.25} - 4.14 \quad (10.29)$$

Based on experimental results from 30 shale samples Sabtan (2005) suggested a linear empirical equation as a function of the plasticity index, initial water content, and clay percent as expressed in Eq. (10.30):

$$S = 1.0 + 0.06(CL + PI - W), \quad (10.30)$$

Lyu et al. (2015) suggested that clay swelling is a strong function of in-situ pressure, P (kpa) and based on multiple linear regression of experimental data obtained from 45 samples, they proposed the empirical Eq. (10.31):

$$S = 30.02 - 0.27W + 0.046CL - 9.18 \log(P),   \tag{10.31}$$

Maghrabi et al. (2013) suggested that initial swelling occurs rapidly; however, it stops when the maximum swelling limit is attained. They predicted swelling volume as a function of time applying fitting constants A, B, and C in Eq. (10.32):

$$S(t) = A\left(1 - \frac{1}{\exp(Bt) + C\sqrt{t}}\right),   \tag{10.32}$$

where A represent saturation swelling volume, which will be reached after sufficient time.

In addition, shale swelling is also affected by the salinity of the water. Maghrabi et al. (2013) have shown that increasing salt concentration from zero to 10% may result in up to 60% less swelling by volume.

Although additives can be used to prevent clay swelling, the effectiveness of additives can diminish with time because of mixing/dilution with the resident water in the formation (Alkouh et al., 2014).

10.4.2 Water phase trapping and blockage

Another major reason for a long-lasting cleanup phase (initial period in which hydrocarbon production rate increases gradually over time, rather than starting with peak production rate) in shale reservoirs is due to water blockage in the fracture-face matrix. If capillary pressure exceeds drawdown pressure, water trapping will occur in water-wet pores, which will result in a dramatic reduction of the relative permeability of hydrocarbons (Holditch 1979; Alvarez et al., 2007; Zitha et al., 2013). Alternatively, invasion of oil-based fluids into zones of low oil saturation can result in oil phase trapping. Shaoul et al. (2011) has suggested that entrapped water can be produced back during the cleanup period only if drawdown pressure can overcome capillary pressure.

Iglauer et al. (2011) determined that the capillary-trapping capacity, C_{nw-max} (a product of residual saturation and porosity) can be correlated with the porosity. Civan (2014) proposed an empirical correlation to estimate maximum nonwetting phase capillary-trapping capacity as a function of porosity as expressed in Eq. (10.33):

$$C_{nw-\max} = \phi S_{nw-\max} = \phi \left[A\log\left(\frac{C}{\phi} - 1\right) + B \right], \tag{10.33}$$

where $A = 0.616$, $B = -0.182$, and $C = 1.7$.

10.5 SUMMARY

Production from shale resources is strongly affected and controlled by the reduction of fracture conductivity and reservoir matrix deliverability due to formation-damage mechanisms. In the context of shale resources, formation damage can be characterized as a combination of various phenomena that jeopardize flow deliverability of the stimulated reservoir volume (SRV). In general, it can be classified as fracture conductivity loss, reservoir matrix permeability loss, and fluid damage.

The conductivity of hydraulic fractures is often significantly reduced owing to formation damage induced by various factors, such as proppant transport and placement, proppant embedment and crushing, fine migration and plugging, gelling damage, and non–Darcy and multiphase flow effect. When multiple formation-damage parameters are considered together, the overall conductivity can diminish by 90% and thus should be considered as part of the fracture-stimulation design. The impact of damaging factors may also vary depending upon reservoir properties and heterogeneities, proppant type, or proppant concentration.

Reservoir matrix permeability in shale formations is strong function of effective stress; during production, as pore pressure decreases, the effective stress will increase, which potentially lead to up to two orders of intrinsic permeability reduction. This severe reduction can be explained by combined effect of pore shrinkage (pore compressibility) and connectivity loss (coordination number reduction).

In this chapter we propose intrinsic permeability model, which is function of pore radius, porosity, and coordination number, and based on proposed model the impact of pore volume reduction and connectivity loss are distinguished and evaluated separately. Two sets of experimental data one from each sandstone and shale samples are selected from the literature. Permeability reduction and impact of pore compressibility and connectivity loss are evaluated for both samples. Results have shown that, for studied sandstone sample connectivity loss is negligible and permeability reduction is only function of pore volume compressibility. On the other hand, for shale sample, connectivity loss is the dominant mechanism; when pressure reaches 8600 psi coordination number reduces by about 50% from 4.3 to 2.3. Although it has not been discussed in detail here, it should be noted that, decreasing pore pressure also have positive effect on reservoir permeability due to effect of non-Darcy flow regimes.

Other major formation damage can be caused by clay swelling and water phase blockage due to water imbibition from fracture to matrix pore system during hydraulic fracturing. Clay swelling can result in severe permeability reduction, but fortunately additives can be used to prevent its impact. On the other hand, if capillary pressure is high enough to exceed drawdown pressure, then water trapping will occur which will lead to long-lasting cleanup period.

In general, it can be summarized that, during drill-in and completion stage, fracturing fluid loss is the most significant cause of formation damage leading to clay swelling and water trapping and blockage. Moreover, well distributed and uniform proppant placement is essential for creating high conductive fractures. After production starts, increasing effective stress will cause to proppant embedment, crushing, and fine migration and plugging which will lead to significant fracture conductivity loss. At the same time, increasing effective stress will reduce reservoir matrix permeability because of pore volume shrinkage and connectivity loss (Fig. 10.8).

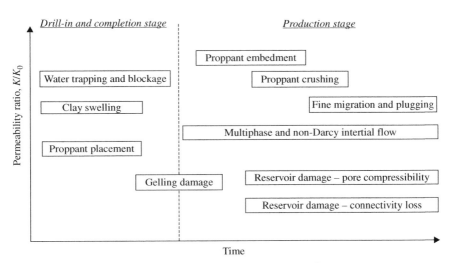

Figure 10.8 Influence diagram of all formation damage mechanism.

NOMENCLATURE

A	constant in Eqs. (10.28), (10.32), and (10.33)
B	constant in Eqs. (10.28), (10.32), and (10.33)
b_s	damage depth
C	constant in Eqs. (10.32) and (10.33)
c	constant term in Eq. (10.11)
C_f	dimensionless form fracture connectivity
CL	clay percent
C_{nw-max}	capillary-trapping capacity
D	fractal dimension
d_p	proppant dimeter
$\mathbf{d_{50}}$	the median particle size (mm)
$\mathbf{d_{90}\text{-}d_{50}}$	difference between 90th and 50th percentile particle size (mm)
E	Young's modulus of the formation (MMpsi)
F	electrical formation factor
K	reservoir permeability
K_e	permeability under the net effective stress
$\mathbf{K_f}$	fracture permeability
K_o	reference permeability under atmospheric pressure
K_s	damaged permeability
LL	liquid limit
p	exponent in Eq. (10.21)
P	in-situ pressure
$P*$	constant in Eq. (10.23)
PI	plasticity index
p_c	critical percolation threshold
P_e	net effective stress
P_o	atmospheric pressure
P_p	pore pressure
q	constant in Eq. (10.23)
r_c	critical pore radius
r_g	grain radius
r_h	hydraulic pore radius
r_{max}	maximum accessible pore radius
r_w	well radius
S	swelling volume in percent
S(t)	clay swelling as a function of time
S_{nw-max}	residual saturation
s_p	skin factor resulting from proppant-embedment
s_f	skin factor associated with hydraulically fractures
t	time
u_f	fluid velocity
W	initial water content
W_E	embedment depth (inches)
w	fracture width

$\mathbf{x}_f$	fracture half-length
$\mathbf{z}$	pore coordination number
α	Biot's poroelastic term
ϕ	porosity
ν_s	particle settling velocity
ρ_p	particle density
ρ_f	fluid density
μ_f	fluid viscosity
g	acceleration of gravity
σ_{eff}	net effective stress
σ_T	overburden stress $\nu = ln\left(C_f\right)$
Δp	pressure gradient
β	inertial flow coefficient (1/ft)
τ	tortuosity
σ_b	bulk electrical conductivity
σ_w	saturating fluid electrical conductivity
η	the exponent in Eq. (10.15)
Γ	interconnectivity term
ω	constant in Eq. (10.22)
γ	porosity sensitivity exponent in Eq. (10.24)

REFERENCES

Al-Bazali, T.M., Zhang, J., Chenevert, M.E., Sharma, M.M., 2009. An Experimental Investigation on the Impact of Capillary Pressure, Diffusion Osmosis, and Chemical Osmosis on the Stability and Reservoir Hydrocarbon Capacity of Shales, Paper SPE 121451-MS, Offshore Europe, 8-11 September, Aberdeen, UK.

Alkouh, A., Mcketta, S., Wattenbarger, R.A., 2014. Estimation of effective-fracture volume using water-flowback and production data for shale-gas wells. J. Can. Pet. Technol 53 (5), 290−303.

Alramahi, B., Sundberg, M., 2012. Proppant Embedment and Conductivity of Hydraulic Factures in Shales. Presented at the 46th US Rock Mechanics/Geomechanics Symposium, Chicago, Illinois, 24-27 June. ARMA-2012-291.

Alvarez, A., Hime, A., Bedrikovetsky, P., 2007. The inverse problem of determining the filtration function and permeability reduction in flow of water with particles in porous media. Transp. Porous Media 70 (1), 43−62.

Bader, S., Heister, K., 2006. The effect of membrane potential on the development of chemical osmotic pressure in compacted clay. J. Colloid Interface Sci. 297 (1), 329−340. May 1.

Bahrami, H., Rezaee, R., Clennell, B., 2012. Water blocking damage in hydraulically fractured tight-sand gas reservoirs: an example from Perth Basin, Western Australia. J. Pet. Sci. Eng. 88−89, 100−106.

Bazin, B., Bekri, S., Vizika, O., Herzhaft, B., Aubry, E., 2010. Fracturing in tight gas reservoirs: application of special-core-analysis methods to investigate formation damage mechanisms. SPE J 15 (4), 975−982.

Bennion, D.B., 2002. An overview of formation damage mechanisms causing a reduction in the productivity and injectivity of oil and gas producing formations. J. Can. Pet. Technol 41 (11), 29−36.

Bernabé, Y., Li, M., Maineult, A., 2010. Permeability and pore connectivity: a new model based on network simulations. J. Geophys. Res. Solid Earth 115 (B10).

Carman, P., 1937. Fluid flow through a granular bed. Trans. Inst. Chem. Eng 15, 150–167.

Civan, F., 2001. Scale effect on porosity and permeability: Kinetics, model, and correlation. AIChE J. 47 (2), 271–287.

Civan, F., 2010. Effective correlation of apparent gas permeability in low-permeability porous media. Transport in Porous Media 82 (2), 375–384.

Civan, F., 2011. Porous Media Transport Phenomena. John Wiley & Sons, Hoboken, NJ.

Civan, F., 2014. Analyses of processes, mechanisms, and preventive measures of shale-gas Gas reservoir fluid, completion, and formation damage. In: SPE 168164, Presented at the SPE International Symposium and Exhibition on Formation Damage Control Held in Lafayette, Louisiana, USA, 26–28 February.

Clerc, J.P., Podolskiy, V.A., Sarychev, A.K., 2000. Precise determination of the conductivity exponent of 3D percolation using exact numerical renormalization. Eur. Phys. J. B 15, 507–516.

Daigle, H., 2016. Application of critical path analysis for permeability prediction in natural porous media. Adv. in Water Resour. 96, 43–54.

David, C., Wong, T.F., Zhu, W., Zhang, J., 1994. Laboratory measurement of compaction-induced permeability change in porous rocks: implications for the generation and maintenance of pore pressure excess in the crust. Pure Appl. Geophys. 143 (1–3), 425–456.

Davudov, D., Moghanloo, R.G., 2016. Upscaling of Pore Connectivity Results from Lab-Scale to Well-Scale for Barnett and Haynesville Shale Plays. Society of Petroleum Engineers. Available from: https://doi.org/10.2118/181433-MS.

Davudov, D., Moghanloo, R.G., Yuan, B., 2016. Impact of Pore Connectivity and Topology on Gas Productivity in Barnett and Haynesville Shale Plays. Presented at the Unconventional Resources Technology Conference, San Antonio, Texas, 1–3 August. Available from: URTEC-2461331-MS. https://doi.org/10.15530/URTEC-2016-2461331.

Davudov, D., Moghanloo, R.G., Lan, Y., 2017a. Non-Darcy Flow Regimes Coupled with Pore Compaction in Shale Gas Formations. Unconventional Resources Technology Conference, Austin, Texas, 24–26 July 2017: pp. 2935–2954.

Davudov, D., Moghanloo, R.G., Lan, Y. et al., 2017b. Investigation of Shale Pore Compressibility Impact on Production with Reservoir Simulation. Presented at the SPE Unconventional Resources Conference, Calgary, 15–16 February. SPE-185059-MS. Available from: https://doi.org/10.2118/185059-MS.

Davudov, D., Moghanloo, R.G., 2018. Impact of pore compressibility and connectivity loss on shale permeability. Int. J. Coal Geol. 187, 98–113.

Denney, D., 2012. Fracturing-fluid effects on shale and proppant embedment. J. Pet. Technol. 64 (03), 59–61. Available from: https://doi.org/10.2118/0312-0059-JPT. SPE-0312-0059-JPT.

Dong, J.J., Hsu, J.Y., Wu, W.J., Shimamoto, T., Hung, J.H., Yeh, E.C., et al., 2010. Stress-dependence of the permeability and porosity of sandstone and shale from TCDP Hole-A. Int. J. Rock Mech. Min. Sci. 47 (7), 1141–1157.

Doyen, P.M., 1988. Permeability, conductivity, and pore geometry of sandstone. J. Geophys. Res. Solid Earth 93 (B7), 7729–7740.

Erol, A.O., Dhowian, A., 1990. Swell behavior of arid climate shales from Saudi Arabia. Q. J. Eng. Geol. Hydrogeol 23 (3), 243–254.

Evans, J.P., Forster, C.B., Goddard, J.V., 1997. Permeability of fault-related rocks, and implications for hydraulic structure of fault zones. J. Struct. Geol. 19 (11), 1393–1404.

Fakcharoenphol, P., Torcuk, M., Bertoncello, A., Kazemi, H., Wu, Y.-S., Wallace, J., et al., 2013. Managing Shut-in Time to Enhance Gas Flow Rate in Hydraulic Fractured Shale Reservoirs: A Simulation Study, Paper SPE 166098-MS, SPE Annual Technical Conference and Exhibition, 30 September–2 October, New Orleans, Louisiana, USA.

Forchheimer, P., 1901. Wasserbewegung durch Boden. Zeitschrift des Vereines Deutscher Ingenieuer, 45th edition.

Gdanski, R.D., Weaver, J.D., Slabaugh, B.F., Walters, H.G., Parker, M.A., 2005. Fracture Face Damage - It Matters. Society of Petroleum Engineers. Available from: http://dx.doi.org/10.2118/94649-MS.

Ghanbarian, B., Hunt, A.G., Ewing, R.P., Skinner, T.E., 2014. Universal scaling of the formation factor in porous media derived by combining percolation and effective medium theories. Geophys. Res. Lett. 41 (11), 3884−3890.

Ghanbarnezhad Moghanloo, R., Javadpour, F., 2014. Applying method of characteristics to determine pressure distribution in 1D shale-gas samples. Soc. Pet. Eng. 19 (3). Available from: https://doi.org/10.2118/168218-PA.

Ghanbarnezhad Moghanloo, R., Javadpour, F., Davudov, D., 2013. Contribution of methane molecular diffusion in kerogen to gas-in-place and production. Soc. Pet. Eng. Available from: https://doi.org/10.2118/165376-MS.

Ghosh, S., Rai, C.S., Sondergeld, C.H., Larese, R.E., 2014. Experimental Investigation of Proppant Diagenesis. Presented at the SPE/CSUR Unconventional Resources Conference−Canada, Calgary, Alberta, 30 September−2 October. SPE-171604-MS. Available from: https://doi.org/10.2118/171604-MS.

Gidley, J.L., Penny, G.S., McDaniel, R.R., 1995. Effect of proppant failure and fines migration on conductivity of propped fractures. SPE Prod. Facilities 10 (01), 20−25.

Guo, J.C., Liu, Y.X., 2012. Modelling of proppant embedment: elastic deformation and creep deformation. In: SPE 157449, Presented at the SPE International Production and Operations Conference and Exhibition Held in Doha Qatar, 14−16 May.

Guo, B., Lyons, W., Ghalambor, A., 2007. Petroleum Production Engineering, 7/89-90. Gulf Professional Publishing, Oxford.

Guo, B.Y., Gao, D.L., Wang, Q., 2011. The role of formation damage in hydraulic fracturing shale gas wells. In: SPE 148778, Presented at SPE Eastern Regional Meeting Held in Columbus, Ohio, USA, 17−19 August.

Han, J., Wang, J.Y., 2014. Fracture Conductivity Decrease Due to Proppant Deformation and Crushing, a Parametrical Study. Presented at the SPE Eastern Regional Meeting, Charleston, WV, 21-23 October. SPE-171019-MS.

Holditch, S.A., 1979. Factors affecting water blocking and gas flow from hydraulically fractured gas wells. J. Pet. Technol. 31 (12), 1515−1524.

Hu, Y., Devegowda, D., Striolo, A., Phan, Van, Van Anh, A., Ho, T.A., et al., 2015. Microscopic dynamics of water and hydrocarbon in shale-kerogen pores of potentially mixed wettability. SPE J. 20 (1), 112−124.

Iglauer, S., Wülling, W., Pentland, C.H., Al-Mansoori, S.K., Blunt, M.J., 2011. Capillary-trapping capacity of sandstones and sandpacks Paper SPE 120960-PA SPE J. 16 (4), 778−783.

Iqbal, M.M., Civan, F., 1993. Simulation of Skin Effects and Liquid Cleanup in Hydraulically Fractured Wells", Paper SPE 25482, Proceedings of the SPE Production Operations Symposium, Oklahoma City, OK, 03/21-23, pp. 665−677.

Jackson, G., Rai, R., 2012. The Impact of Completion Related Pressure Losses on Productivity in Shale Gas Wells, 162213-MS, Abu Dhabi International Petroleum Conference and Exhibition, 11−14 November, Abu Dhabi, UAE.

Javadpour, F., Fisher, D., Unsworth, M., 2007. Nanoscale gas flow in shale gas sediments. J. Can. Pet. Technol 46 (10), 55−61.

Javadpour, F., 2009. Nanopores and apparent permeability of gas flow in mudrocks (shales and siltstone). J. Can. Petrol. Technol 48 (08), 16−21.

Jung, C.M., Zhou, J., Chenevert, M.E., Sharma, M.M., 2013. The Impact of Shale Preservation on the Petrophysical Properties of Organic-Rich Shales, Paper SPE 166419-MS, SPE Annual Technical Conference and Exhibition, 30 September−2 October, New Orleans, Louisiana, USA.

Kang, Y.L., Yang, B., You, L.J., 2013. Damage evaluation of oil-based drill-in fluids to shale reservoirs. Nat. Gas. Ind. 33 (12), 99−104.

Katz, A.J., Thompson, A.H., 1986. Quantitative prediction of permeability in porous rock. Phys. Rev. B 34 (11), 8179−8181.

Katz, A.J., Thompson, A.H., 1987. Prediction of rock electrical conductivity from mercury injection measurements. J. Geophys. Res. 92 (B1), 599−607.

Kelkar, M., 2008. Natural Gas Production Engineering. PennWell Books, Tulsa, Oklahoma, p. 214.

Kozeny, J., 1927. Uber kapillare Leitung der Wasser in Boden. Sitzungsber. Akad. Wiss. Wien 136, 271−306.

Krohn, C.E., 1988. Fractal measurements of sandstones, shales, and carbonates. J. Geophys. Res.: Sol. Earth 93 (B4), 3297−3305.

Kwon, O., Kronenberg, A.K., Gangi, A.F., Johnson, B., 2001. Permeability of Wilcox shale and its effective pressure law. J. Geophys. Res.: Sol. Earth 106 (B9), 19339−19353.

Kwon, O., Herbert, B.E., Kronenberg, A.K., 2004. Permeability of illite-bearing shale: 2. Influence of fluid chemistry on flow and functionally connected pores. J. Geophys. Res.: Sol. Earth 109 (B10).

Lan, Y., Ghanbarnezhad Moghanloo, R., Davudov, D., 2017. Pore compressibility of shale formations. SPE J. 22 (6).

Li, J., Guo, B., Gao, D., Ai, C., 2012. The effect of fracture-face matrix damage on productivity of fractures with infinite and finite conductivities in shale-gas reservoirs, paper SPE 143304-PA. SPE Drill Completion 27 (3), 347−353.

Li, K.W., Gao, Y.P., Lyu, Y., Wang, M., 2015. New mathematical models for calculating proppant embedment and fracture conductivity. SPE J. 20 (3), 496−507.

Liu, X., Xiong, J., Liang, L., 2015. Investigation of pore structure and fractal characteristics of organic-rich Yanchang formation shale in central China by nitrogen adsorption/desorption analysis. J. Nat. Gas Sci. Eng. 22, 62−72.

Loucks, R.G., Reed, R.M., Ruppel, S.C., Hammes, U., 2012. Spectrum of pore types and networks in mudrocks and a descriptive classification for matrix-related mudrock pores. AAPG Bulletin 96 (6), 1071−1098.

Lv, K.H., 2010. Technology of Formation Damage Control in Oil and Gas Reservoir. China University of Petroleum Press, Dongying, China.

Lyu, Q., Ranjith, P.G., Long, X., Kang, Y., Huang, M., 2015. A review of shale swelling by water adsorption. J. Nat. Gas Sci. Eng. 27, 1421−1431.

Maghrabi, S., Kulkarni, D., Teke, K., Kulkarni, S.D., Jamison, D., 2013. Modeling of Shale-Swelling Behavior in Aqueous Drilling Fluids, Paper SPE 164253-MS, 18th Middle East Oil & Gas Show and Conference (MEOS), Mar 10−13, Bahrain International Exhibition Centre, Manama, Bahrain.

Metwally, Y.M., Sondergeld, C.H., 2011. Measuring low permeability of gas-sands and shales using a pressure transmission technique. Int. J. of Rock Mech. Min. Sci. 1135−1144.

Palisch, T.T., Duenckel, R.J., Bazan, L.W., Heidt, J.H., Turk, G.A., 2007. Determining realistic fracture conductivity and understanding its impact on well performance-theory and field examples. SPE Hydraulic Fracturing Technology Conference. Society of Petroleum Engineers.

Palisch, T.T., Vincent, M., Handren, P.J., 2010. Slickwater fracturing: food for thought. SPE Prod. Oper. 25 (03), 327−344. Available from: https://doi.org/10.2118/115766-PA. SPE-115766-PA.

Pape, H., Clauser, C., Iffland, J., 2000. Variation of permeability with porosity in sandstone diagenesis interpreted with a fractal pore space model. Pure Appl. Geophys 157, 603−619.

Passey, Q.R., Bohacs, K., Esch, W.L., et al., 2010. From oil-prone source rock to gas- producing shale reservoir e geologic and petrophysical characterization of unconventional shale gas reservoirs. In: SPE 131350, Presented at the International Oil and Gas Conference and Exhibition Held in Beijing, China, 8−10 June.

Paterson, M.S., 1983. The equivalent channel model for permeability and resistivity in fluid-saturated rocks − a reappraisal. Mech. Mater 2, 345−352.

Pattamasingh, P., 2016. Fracturing Optimization Based on Dynamic Conductivity. the University of OklahomaMSc. Thesis. Available from: https://shareok.org/handle/11244/34647.

Reinicke, A., Rybacki, E., Stanchits, S., Huenges, E., Dresen, G., 2010. Hydraulic fracturing stimulation techniques and formation damage mechanisms—implications from laboratory testing of tight sandstone−proppant systems. Chemie der Erde-Geochem. 70, 107−117.

Revil, A., 2002. The hydroelectric problem of porous rocks: thermodynamic approach and introduction of a percolation threshold. Geophys. J. Int 151, 944−949.

Romero, D.J., Valko, P.P., Economides, M.J., 2002. The Optimization Of The Productivity Index And The Fracture Geometry Of A Stimulated Well With Fracture Face And Choke Skins. Presented at the International Symposium and Exhibition on Formation Damage Control, Lafayette, Louisiana, 20-21 February. SPE-73758-MS. Available from: https://doi.org/10.2118/73758-MS.

Roy, S., Raju, R., Chuang, H.F., Cruden, B.A., Meyyappan, M., 2003. Modeling gas flow through microchannels and nanopores. J. Appl. Phy. 93 (8), 4870−4879.

Sabtan, A.A., 2005. Geotechnical properties of expansive clay shale in Tabuk, Saudi Arabia. J. Asian Earth Sci 25 (5), 747−757.

Saldungaray, P.M., Palisch, T.T. 2012. Hydraulic fracture optimization in unconventional reservoirs. Presented at the SPE Middle East Unconventional Gas Conference and Exhibition, Abu Dhabi, UAE, 23-25 January. SPE-151128-MS. Available from: https://doi.org/10.2118/151128-MS.

Schubarth, S., Milton-Tayler, D., 2004. Investigating how proppant packs change under stress. Presented at the SPE Annual Technical Conference and Exhibition, Houston, Texas, 26-29 September. SPE-90562-MS. Available from: https://doi.org/10.2118/90562-MS.

Shaoul, J., van Zelm, L., Pater, C.J., et al., 2011. Damage mechanisms in unconventional gas well stimulation-a new look at an old problem. SPE Prod. Oper 26 (4), 388−400.

Shi, T., Wang, C.Y., 1986. Pore pressure generation in sedimentary basins: overloading versus aquathermal. J. Geophys. Res. 91 (B2), 2153−2162.

Stauffer, D., Aharony, A., 1992. Introduction to Percolation Theory, second ed. Taylor and Francis, London.

Teng, H., Zhao, T.S., 2000. An extension of Darcy's law to non-stokes flow in porous media. Chem. Eng. Sci 55 (14), 2727−2735.

Tinni, A., Fathi, E., Agarwal, R., Sondergeld, C.H., Akkutlu, I.Y., Rai, C.S., 2012. Shale Permeability Measurements on Plugs and Crushed Samples. Society of Petroleum Engineers. Available from: http://dx.doi.org/10.2118/162235-MS.

Valko, P.P., Oligney, R.E., Economides, M.J., 1998. High permeability fracturing of gas wells. Pe. Eng. Int. 71 (1).

Vincent, M.C., 2009. Examining Our Assumptions--Have Oversimplifications Jeopardized Our Ability to Design Optimal Fracture Treatments? Presented at the SPE Hydraulic Fracturing Technology Conference, The Woodlands, Texas, 19-21 January. SPE-119143-MS.

Walsh, J.B., Brace, W.F., 1984. The effect of pressure on porosity and the transport properties of rock. J. Geophys. Res. Sol. Earth 89, 9425−9431.

Weaver, J.D., Nguyen, P.D., Parker, M.A., van Batenburg, D.W., 2005. Sustaining Fracture Conductivity. Presented at the SPE European Formation Damage

Conference, Sheveningen, The Netherlands, 25–27 May. SPE-94666-MS. Available from: https://doi.org/10.2118/94666-MS.

Xu, T.T., Xiong, Y.M., Kang, Y.L., 2010. Technology of Formation Damage Control in Oil and Gas Reservoir. Petroleum Industry Press, Peking, China.

You, Z., Bedrikovetsky, P., Badalyan, A., Hand, M., 2015. Particle mobilization in porous media: temperature effects on competing electrostatic and drag forces. Geophys. Res. Lett 42, 2852–2860.

Yuan, B., Moghanloo, R.G., 2016. Analytical evaluation of nanoparticles application to reduce fines migration in porous media. SPE J. 21 (06), 2317–2332.

Yuan, B., Su, Y., Moghanloo, R.G., 2015a. A new analytical multi-linear solution for gas flow toward fractured horizontal well with different fracture intensity. J. Nat. Gas Sci. Eng. 23, 227–238.

Yuan, B., Wood, D.A., Yu, W., 2015b. Stimulation and hydraulic fracturing technology in natural gas reservoirs: theory and case study (2012–2015). J. Nat. Gas Sci. Eng. 26, 1414–1421.

Yuan, B., Moghanloo, R.G., Shariff, E., et al., 2016. Integrated investigation of dynamic drainage volume (DDV) and inflow performance relationship (transient IPR) to optimize multi-stage fractured horizontal wells in shale oil. J. Ener. Res. Technol. 138 (5), 052901–052909.

Yuan, B., Moghanloo, G.R., Zheng, D., 2017. A novel integrated production analysis workflow for evaluation, optimization and predication in shale plays. Int. J. Coal Geol. 180, 18–28.

Zhang, J., Ouyang, L., Zhu, D., Hill, A.D., 2015a. Experimental and numerical studies of reduced fracture conductivity due to proppant embedment in the shale reservoirs. J. Pet. Sci. Eng. 130 (1), 37–45.

Zhang, J., Zhu, D., Hill, A.D., 2015b. Water-induced fracture conductivity damage in shale formations. In: SPE 173346, Presented at the SPE Hydraulic Fracturing Technology Conference Held in the Woodlands, Texas, USA, 3–5 February.

Zheng, D., Yuan, B., Moghanloo, R.G., et al., 2016. Analytical modeling dynamic drainage volume for transient flow towards multi-stage fractured wells in composite shale reservoirs. J. Pet. Sci. Eng. 149, 756–764.

Zitha, P., Frequin, D., Bedrikovetsky, P., 2013. CT scan study of the leak-off of oil- based drilling fluids into saturated media. In: SPE 165193 Presented at the SPE Formation Damage Conference and Exhibition Held in Njirwjik, Netherlands, 5–9 June.

FURTHER READING

Davudov, D., Ghanbarnezhad Moghanloo, R., Dadmohammadi, Y., Curtis, M., Javadpour, F., 2016b. Impact of Pore Topology on Gas Diffusion and Productivity in Barnett and Haynesville Shale Plays. ASME 2016 35th International Conference on Ocean. Offshore and Arctic Engineering.

Kang, Y.L., Xu, C.Y., You, L.J., Yu, H.F., Zhang, B.J., 2014. Comprehensive evaluation of formation damage induced by working fluid loss in fractured tight gas reservoir. J. Nat. Gas Sci. Eng 18 (1), 353–359.

Palisch, T.T., Duenckel, R.J., Chapman, M.A., Woolfolk, S., Vincent, M.C., 2009. How to Use and Misuse Proppant Crush Tests--Exposing the Top 10 Myths. Presented at the SPE Hydraulic Fracturing Technology Conference, The Woodlands, Texas, 19–21 January. SPE-119242-MS. Available from: https://doi.org/10.2118/119242-MS.

Weaver, J., Liang, F., Schultheiss, N., 2015. Assessment of fracturing-fluid cleanup by use of a rapid-gel-damage method. SPE Prod. Oper. 30 (01), 69–75. Available from: https://doi.org/10.2118/165086-PA. SPE-165086-PA.

CHAPTER ELEVEN

A Special Focus on Formation Damage in Offshore and Deepwater Reservoirs

Xingru Wu

The University of Oklahoma, Norman, OK, United States

Contents

11.1 INTRODUCTION

Deepwater is usually defined for an offshore region where the water of depth, from the sea floor (mud line) to the sea level, is between 1000 ft and 5000 ft, and regions with water depth above 5000 ft are usually referred to as ultra–deepwater. In this chapter, rather than distinguishing between deepwater and ultra-deep water, we specify an offshore field with a water depth greater than 1000 ft as a deepwater field. Fig. 11.1 shows the number of wells drilled in the Gulf of Mexico from the early 1950s to 2010. From the figure we can see that the drilling depth increased gradually by 2000 ft, and the number of wells referenced in the deepwater category have become much more numerous over time, which is attributed to the progression of the subsea and associated engineering technologies. In order to achieve this, managing a deepwater asset is not only the responsibility for the petroleum engineer, but also an integrated

Formation Damage during Improved Oil Recovery.
DOI: https://doi.org/10.1016/B978-0-12-813782-6.00011-7

© 2018 Elsevier Inc.
All rights reserved.

417

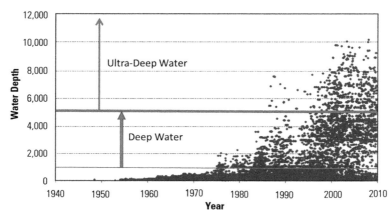

Figure 11.1 Number of wells in the Gulf of Mexico and classification of deepwater and ultra-deepwater. *Modified from the National Commission on the BP Deepwater Horizon Oil Spill (2011).*

team effort from multiple disciplines such as subsea engineering, process engineering, flow assurance engineering, and production chemistry engineering.

Developing deepwater hydrocarbon resources is one of the frontier activities in the petroleum industry, not only because it needs many advanced subsea and surveillance technologies, but also because it has a significant amount of subsurface uncertainties and operational risks that need to be fully addressed before or during production. Fig. 11.2 shows one of many offshore development concepts, in which well heads are placed on the sea floor (mud line) and produced and injected streams pass through a series of flow lines and manifolds. The platform can be fixed on the sea floor or it can be mobile/floating on the sea surface, depending on the water depth and pay load on the platforms. A simplified diagram is shown in Fig. 11.3 for illustrative purpose.

Deepwater wells are usually much more costly than their counterparts, onshore or shallow water wells, and this partially results in large well spacing in the typical development plan. A large well spacing usually indicates that there remains a high level of uncertainty in subsurface reservoir characterization that can lead to surprises for field operations and production teams. Typical uncertainties in reservoir characterization include reservoir compartmentalization, the presence of aquifer and reservoir energy, pore pressure variation, reservoir deformation, and others. Therefore, prevention and mitigation plans should be in place to fully address the potential

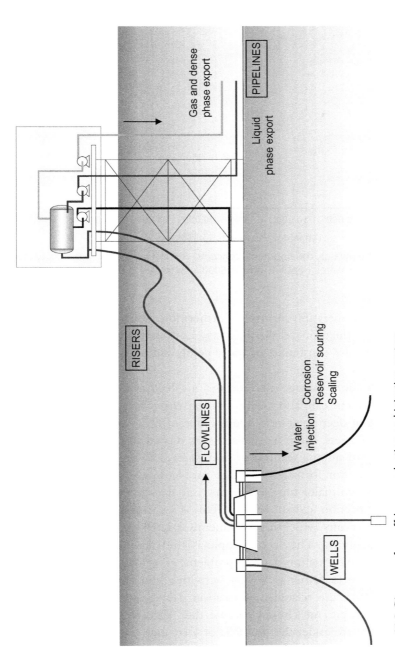

Figure 11.2 Diagram of an offshore production and injection system.

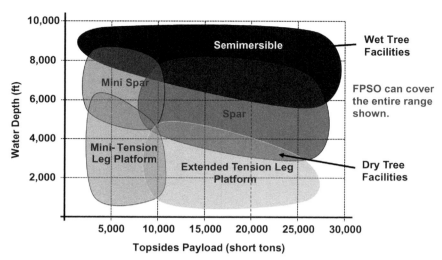

Figure 11.3 Deepwater development concepts for different payloads and water depths. When the water depth is above 5000 ft, wet tree production facilities are often used.

production and operation surprises. Furthermore, some mature technologies such as production well logging may not be applicable to deepwater wells owing to the constraints of water depth and operational difficulties/ costs.

Any formation damage that occurs in shallow water reservoirs or onshore reservoirs can also occur in deepwater formations. Since the development of deepwater resource is different from the development of other types of assets, some particular issues related to formation damage, remediation and unique features of deepwater development are discussed in this chapter. For formation damage problems, we are going to focus on discussing two major topics. The first one is formation damage due to reservoir compaction and subsidence. This is a common phenomenon in deepwater reservoirs, especially for unconsolidated sandstone formations or chalk formations. The second topic is the surveillance technologies used to diagnose formation damage problems as well as formation damage remediation efforts. Other issues and activities associated with deepwater development that are related to formation damaged are also briefly discussed. These topics are chosen because they feature deepwater development, and they are under the umbrella of formation damage.

Challenges of developing deepwater hydrocarbon resources come from a number of sources. In contrast to onshore reservoir development, the operator of a deepwater development needs to handle many offshore geohazards, such as shallow water flow from overpressurized zones and gas hydrates that pose tremendous challenges for the drilling and completion of deepwater wells. Additionally, metocean environment like a Tsunami and hurricane can directly impact the integrity of production facilities, well tie-backs, and drilling activities. These challenges from subsea and offshore environments put additional constraints on the development planning. For instance, due to safety and environmental concerns, many field development concepts feature subsea installations of well heads and fluid conditioning facilities on the mud line. These concepts typically create many operational challenges that limit access for well workover's and disable some applications of traditional surveillance tools. Wu et al. (2016) reviewed technologies and challenges of applying subsea fluid processing and conditioning facilities for deepwater operations. More importantly, the high development cost of a deepwater project is one of the key drivers for many development decisions. Because of the high cost of drilling and completing wells in the deepwater environment, the expected well productivity needs to be high. Therefore, any possible cause leading to formation damage or s increase in skin damage should be fully understood and minimized.

11.2 PRESSURE AND TEMPERATURE OF DEEPWATER RESERVOIRS

Distinct from onshore environments, many offshore operational problems are associated with the dramatic changes of temperature from the reservoir to the sea floor and to the surface. Husson et al. (2008) gave a depth dependent sea floor temperature relationship with 30°C sea level temperature expressed as Eq. (11.1):

$$T_{surf} = 4 + 26exp\left(-\frac{z_s}{300}\right) \tag{11.1}$$

where z_s = depth below the sea level inmeters.

Analyzing 8500 subsea temperature data points in six zones of the Gulf of Mexico, Husson et al. (2008) reported a 20−23°C/km geothermal gradient for shallow formations (<3000 m from sea level). Fig. 11.4 shows the temperature variation as a function of water depth and formation depth.

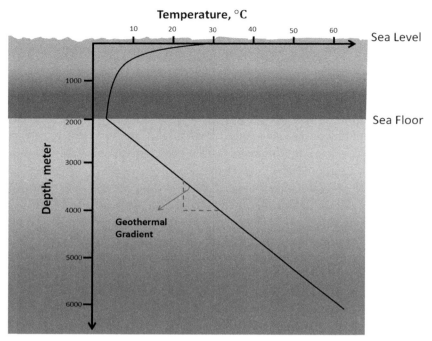

Figure 11.4 Temperature profile in ocean and subsurface with a constant geothermal gradient. Assuming water depth is 2000 m, and geothermal temperature gradient of 0.02°C/m.

Most of the deepwater reservoirs are overpressured, which leads to a significant challenge in drilling deepwater wells (Marland et al., 2007, Willson et al., 2003). Fig. 11.5 shows the illustrative relationship of different pressure variations with depth from the mud line, and we can see that because of the over pressure, the pressure gradient difference between the overburden and pore pressure is small. In an overpressured reservoir, in order to drill to the designed depth, a very precise control on the drilling pressure and drilling mud density are needed, given the small window between the pore pressure and the fracturing pressure. Furthermore, since deepwater reservoir formations are relatively young, reservoir consolidation can be poor, since the subsurface systems probably have not reached hydraulic equilibria. Therefore, it is possible to have dramatic pore pressure differences at the same vertical depth in nearby regions. It is often the case that the actual drilling duration is much longer than the planned drilling time because of the pore pressure uncertainty, particularly for drilling exploration or appraisal wells. Willson et al. (2003) reviewed and

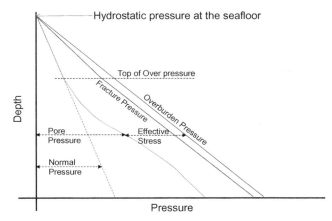

Figure 11.5 Pressures below the mud line in an offshore environment.

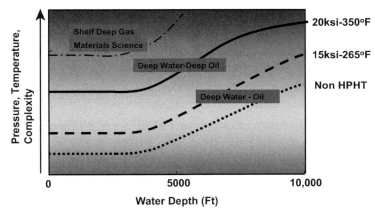

Figure 11.6 Illustrative diagram of reservoir pressure, temperature, and complexity in the deepwater environment.

summarized drilling uncertainties and challenges in drilling deepwater wells.

As the formation depth gets deeper and deeper, both the reservoir temperature and pressure increase toward current operational limits. To develop yet deeper formations, new technologies and significant progressions on facilities, equipment, new material, new development concepts and engineering processes are needed for the high pressure and high temperature (HP/HT) reservoirs. Fig. 11.6 shows an illustrative diagram of reservoir pressure and temperature, and boundaries of fluids. As an example, current subsea and offshore engineering capacities allow an operator to develop hydrocarbon resources from the Miocene formations in the Gulf of Mexico when the pore pressure is below 20,000 psi and the temperature is below 350°F, but developing deeper/older formations like the Wilcox pose tremendous challenges.

11.3 COMPLEX WELL STRUCTURE WITH INTELLIGENT WELL COMPLETION

Because of over pressure and the small drilling mud window between the fracturing pressure and pore pressure, the well bore structure in deepwater is usually very complex since multiple layers of casing have to be used in order to drill to the total depth of the design. Fig. 11.7 shows a diagram of casing configuration designed to provide an acceptable environment for the subsurface equipment in producing wells. For example, nine layers of casings and liners were used in the Macondo well in order to drill to the planned total depth, as shown in Fig. 11.8. The complicated wellbore structure creates challenges for pressure and temperature management in the annuli, especially for the annulus between the production casing and the production tubing, and the annulus between the production casing and immediate outer casing or liner. During production, the fluids in the confined space of the annuli expand because of the increasing temperature. If the inner pressure of the annulus

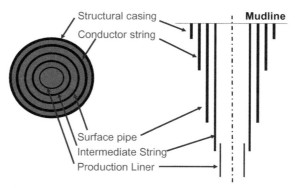

Figure 11.7 Multiple layers of casing are typically required from surface conductor to production casing or liner for deepwater well designs and completion. The specific number of casing layers required is determined by the contrast of pore pressure and fracturing pressure in the section to be drilled.

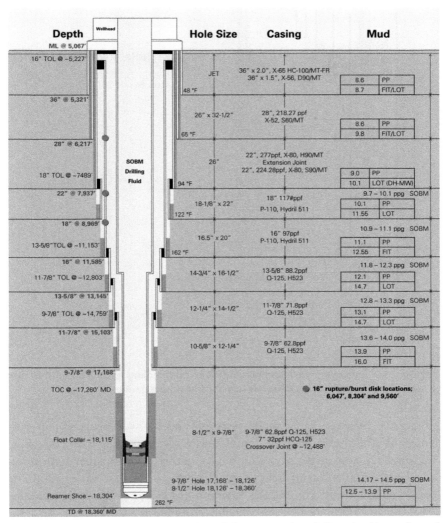

Figure 11.8 The casing structure and mud density used in drilling the Macondo well (BP, 2010).

is too high, the outer casing can burst or deform. Therefore, the annulus pressure buildup needs to be accurately managed, especially during transient production periods, such as production ramp up.

Given that more and more well heads are installed on the mud line due to safety and environmental concerns, operators prefer to equip these deepwater wells with more "intelligent" control devices and sensors to

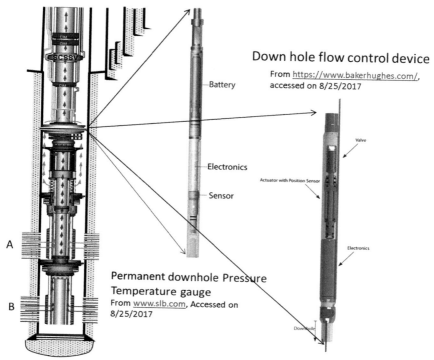

Figure 11.9 Intelligent well completions for commingling production from dual zones with downhole flow control device and downhole gauges.

obtain the required data for the reservoir and for well management purposes. Typically an intelligent completion is a system capable of collecting, transmitting, and analyzing wellbore production data, together with reservoir and completion-integrity data; thereby, enabling remote action to better monitor and control the reservoir, well and production processes (Robinson, 2003). Initially, intelligent completions were based on conventional wireline operated sleeve valves. Operators can increase hydrocarbon production by commingling production from multiple zones by using the inflow control devices such as the one shown in Fig. 11.9. Intelligent completions enable selective zonal control, thus increasing the efficiency of reservoir development.

The intelligent wells are also commonly equipped with sensors from which significant data can be collected, filtered, and analyzed. The collected data include pressure and temperature at the wellhead and downhole, acoustic signals for sand monitoring, and some time rate metering and flow logics. Fig. 11.9 shows an example of downhole permanent pressure and temperature gauges (PDG) that are used for deepwater wells. These gauges are installed inside the wellbore, usually placed close to the perforation. PDG has more and more applications significantly in popularity, partially due to the increasing survivability of PDGs because of optimized design in recent years. As of 2007, more than 10,000 PDGs were installed worldwide (Horne, 2007). Today, most offshore operators accept PDG as a part of regular wellbore design. PDGs provide a continuous data source a recording downhole real-time data every second and over periods of years. Such a frequency of measurement could generate 32 million data points per year (Horne, 2007).

The collected data from these sensors are transmitted through fiber optic lines connected to a communications network to data processing systems from where engineers can diagnose production problems and/or monitor production processes. Fig. 11.10 illustrates the major devices and

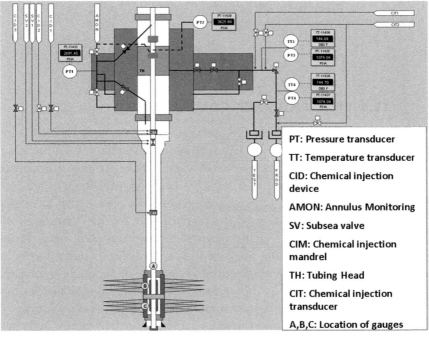

Figure 11.10 Major component of an intelligent well completion. Pressure and temperature transducers are installed to measure the real-time pressure and temperature at these locations; chemical mandrels and devices are used to inject chemicals for flow assurance; and subsea valves are mainly used for production control.

controls used in an intelligent well completion. The pressure data collected from transducers can be used for real-time surveillance (Wu et al., 2012, 2013a,b, 2014). Furthermore, the resolution and accuracy of the PDG enables users to record subtle changes and use them innovatively. For example, some quartz gauges have a pressure resolution of $\pm$ 0.005 psi at a 1 s sampling rate (Schlumberger, 2013).

In addition to PDGs, operators sometime install the Fiber Optical Distributed Temperature Sensor to provide temperature logs in real-time. The Distributed Temperature Sensor (DTS) system uses the light source from a laser to record temperature traces, deriving the temperature information distinguishing when and where different fluid phases enter the wellbore or when and where flow rates change (Hadley and Kimish, 2008). The DTS systems transmit continuous, precise, and accurate temperature measurements up to 15 km with 1.0 m measured interval and 0.1°C temperature resolution (Costello et al., 2012). When properly analyzed the data obtained in the wellbore temperature profiles from DTS system help to monitor wellbore conditions and identify inflow performance.

11.4 FRAC-AND-PACK COMPLETION

Frac-and-pack is often used as a well completion for deepwater wells, especially when the formation is unconsolidated and sand production would be expected without active sand control. It may also be used when the reservoir matrix has a high permeability, to sustain a high production rate. This form of well completion can reduce the formation damage and even stimulate production. Furthermore, since frac-and-pack provides short but wide fractures bypassing any damaged zones during drilling and completion, the frac-and-pack completion is often needed in reservoirs with the following features: (1) Poorly or weakly consolidated reservoir having sand migration problems. A high conductive fracture reduces drawdown pressure and near wellbore fluid flow velocity. (2) Laminated sand/shale sequences where vertical conductivity is low. Hydraulic fracturing eliminates vertical flow barriers in the near well bore region. In these reservoirs, sand migration is unavoidable, especially after water breakthrough and flowing the well at high rates or at high pressure drawdown.

When a well is in the pseudo-steady state production regime, the oil production rate can be calculated from Eq. (11.2):

$$q_o = \frac{4\pi k h k_{ro}}{\mu B_o \left[\ln\left(\frac{4A}{\gamma C_A r_w^2}\right) + 2S\right]} (\bar{p} - p_{wf}) \tag{11.2}$$

where k is the formation permeability, h is net pay thickness, k_{ro} is relative permeability of oil, μ is the oil viscosity, B_o is the oil volume factor, A is the drainage area, C_A is the dietz shape factor determined by well location and drainage shape, $\gamma = 1.78$ which is Euler's constant, r_w is the wellbore radius and S is the skin factor. $\bar{p}$ and p_{wf} are average reservoir pressure and flow bottom hole pressure, respectively. If the fractures bypass the damaged zone, all formation skin factor within the damaged zone can be removed. Additionally, if we assume the fracture half-length is x_f and the fracture conductivity is more than 10, the effective well bore radius is expressed by Eq. (11.3):

$$r_w' = \frac{x_f}{2} \tag{11.3}$$

The fracture skin is defined as Eq. (11.4)

$$s_f = \ln\left(\frac{r_w}{r_w'}\right) \tag{11.4}$$

Since x_f is usually much larger than the diameter of the wellbore, the fracture skin would be negative. For a negative fracture skin, at the same pressure drawdown, the actual production rate is higher than the unstimulated well. The magnitude of improvement can be quantified by the flow efficiency which is defined as Eq. (11.5):

$$FE = \frac{\ln\left(\frac{4A}{\gamma C_A r_w^2}\right)}{\ln\left(\frac{4A}{\gamma C_A r_w^2}\right) + 2S_f} \approx \frac{7.5}{7.5 + S_f} \tag{11.5}$$

When the skin is negative, the flow efficiency is more than 100%. Fig. 11.11 shows a comparison of frac-and-pack with other well completion schemes in the term of flow efficiency, and it clearly shows the superiority of this well completion method. Fig. 11.12 shows the life span expectancy of frac-and-pack for the same group of well completions.

Frac-and-pack is definitely not risk free, and it is the most costly and complex well completion scheme so far deployed in the deepwater

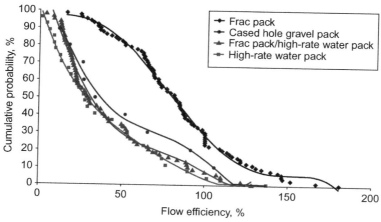

Figure 11.11 A comparison of well productivities with different well completion schemes. From Sanchez and Packing (2007).

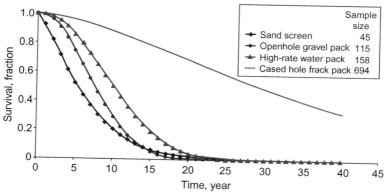

Figure 11.12 The comparison of completion life span for frac-and-pack design compared with other completion designs. From (Sanchez and Packing, 2007).

environment. Other than frac-and-pack completions, there are a few other alternative completions used for sand control in the deepwater environment. High Rate Water Pack (HRWP) is often used when the interval is less than 50 m thick and the formation transmissibility is less than 10 D-ft/cp. In HRWP, premature screen out is possible especially when the formation permeability is low, which results in the fractures terminating prematurely within the damaged zone.

There are several key advantages of frac-and-pack completion over alternative well completions. Firstly, it stimulates the well by fracturing the formation to bypass the damaged zone and lowers the near well pressure drop for the designed production rate. Lowering the pressure drop can directly mitigate other formation damage impacts, such as onset of asphaltene deposition. Secondly, the frac-and-pack completion can effectively control sand migration and production through gravel packing. Lastly, the life span of frac-and-pack well completion is longer than the alternative completion designs (Fig. 11.12), which reduces the frequency of well workover.

11.5 FORMATION DAMAGE DUE TO RESERVOIR COMPACTION

Reservoir compaction is one of the factors driving formation damage mechanisms, but care needs to be applied when assessing this in the deepwater environment. The overburden stress at depth H can be determined by Eq. (11.6):

$$\sigma_V = g \int_0^H \rho_r dh \tag{11.6}$$

where ρ_r is the density of the formation overlaying the target reservoir. Assuming an average formation density is used in lbm/ft³ and the depth is in ft, then the overburden stress can be derived by Eq. (11.7):

$$\sigma_V = \frac{\rho H}{144} \tag{11.7}$$

where σ_V is the vertical stress from the overburden in psi. In the case of a porous medium, the weight of overburden is supported by the rock skeleton and the fluid within the pore space, the effective stress, σ'_V is defined by Eq. (11.8):

$$\sigma'_V = \sigma_V - \alpha p \tag{11.8}$$

where α is Biot's poroelastic constant (Biot, 1941), and p is the pore pressure from fluid within the pore space. When the reservoir is being depleted, the pore pressure gradually declines, which lead to an increase of effective stress. When the effective stress is sufficiently high, the pore structure will ultimately collapse, which results in irreversible reductions

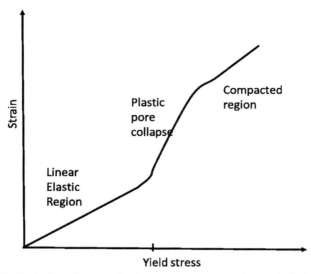

Figure 11.13 Typical rock stress/strain curve showing the rock behavior under stress. Elastic, plastic, and pore collapse, and compacted rock hardening. From Ruddy et al. (1989).

in permeability and porosity. If the reservoir pressure is too low, the reservoir formation may be compacted, resulting in casing deformation damage impacting the well completion, potentially leading to sand control problems, which would result in the loss of future reservoir drillability. Examples of reservoir compaction and subsidence are available in literature (Ruddy et al., 1989; Li et al., 2003).

Fig. 11.13 shows the type of rock behavior typically associated with an increase of net stress. In the early reservoir development, the reservoir pressure is high and net stress is low, the strain and stress may have a linear elastic behavior in which the porosity change is reversible. When the pore pressure is further reduced, the rock strain and stress is no longer a linear relationship, and the change on pore structure is irreversible. Since the pressure drop in the vicinity of the wellbore is much higher than the pressure drop farther away from the wellbore, the compaction or volumetric strain near the wellbore region could be significant. If the strains experienced by the well completion and sand control devices are different from the strain of the near-wellbore region, the well tubulars or sand control configurations can be destroyed.

The principle variables determining compaction in a reservoir are the initial porosity, formation thickness and pore pressure depletion.

Assuming the rock grains to be incompressible, the compaction strain can be expressed as Eq. (11.9):

$$\varepsilon = \phi_i - \phi \tag{11.9}$$

where ϕ_i is the initial porosity, and ϕ is current porosity at a specific pressure. In near wellbore region, the sandface pressure and flowing bottom hole pressure are much lower than the initial reservoir pressure when a high rate of production is being extracted; and a positive skin factor exacerbates the pressure drawdown. The consequence of the compact strain in this region can cause significant damage to the casing and sand control completion, such as frac-and-pack, because different Poisson ratios exist for casing, cement, gravel pack, and formation.

The pore volume change with reservoir depletion can be characterized by the pore volume compressibility which is defined by Eq. (11.10):

$$c_{pV} = \frac{1}{V_p}\frac{dV_p}{dp} \tag{11.10}$$

where V_p is the pore volume of reservoir, and p is pore pressure. In normally pressured reservoirs, the pore volume change with pressure can be assumed to be minimal, because the overburden stress is mainly supported by the skeleton of rock. However, in overpressured reservoirs, the pore volume change with pressure can become significant as the reservoir is progressively depleted. Fig. 11.14 shows the pore volume compressibility determined from uniaxial strain compaction based on the inferred reservoir pressure of a deepwater reservoir. In this measurement, the bulk compressibility was determined first from the derivative of the strain-stress slope, and then the pore volume compressibility was calculated from the bulk compressibility, over discrete segments, as a function of the effective mean stress by assuming a unit Biot's coefficient. At the initial reservoir pressure of 9160 psi, all samples have about the same initial porosity of 31%. When the reservoir pressure varies between 6000 and 9000 psi, the pore volume compressibility is about the same at a value of $5.5 \times 10^{-6} psi^{-1}$. However, when the reservoir pressure drops below 6000 psi, the pore volume compressibility increases in an accelerated fashion to the value of $50 \times 10^{-6} psi^{-1}$.

Yilmaz et al. (1994) presented a correlation to correlate the permeability change with reservoir pressure, and the correlation has a form identical to compressibility, expressed as Eq. (11.11).

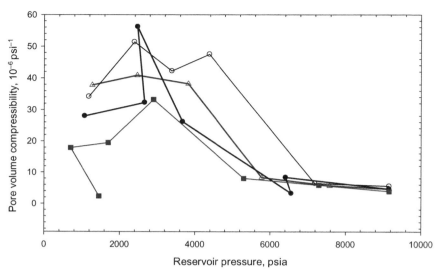

Figure 11.14 Pore volume compressibility (each line represents a sample, and the points are for data measured in laboratory) determined from the change in effective mean stress for an unconsolidated deepwater formation.

$$\gamma_k = \frac{1}{k}\frac{dk}{dp} \qquad (11.11)$$

where γ_k is referred to as the permeability modulus. If the initial pressure is p_i, which corresponds to the initial permeability of k_i, and these values are used as reference conditions, then the permeability at an arbitrary condition is given by Eq. (11.12):

$$k = k_i \exp\left[-\gamma(p_i - p)\right] \qquad (11.12)$$

From Eq. (11.12), we can see that the permeability reduction for the formation can be significant when the permeability modulus is large.

From above analysis, a deepwater field under development with limited aquifer support, experiences rapid reservoir pressure decline as production progresses. If the reservoir pressure depletion is not arrested, some catastrophic consequence can occur. (1) Reservoir compaction would lead to a reduction in permeability such as to reduce the productivity of production wells. Reservoir compaction provides additional natural energy to production by reducing the porosity. When reservoir depletion pressure exceeds a threshold, laboratory experiments show that the compressibility of pore volume would increase dramatically. (2) For existing wells, if the reservoir compaction is greater than the tolerable strain for well completion and sand control scheme, the well can fail prematurely. (3) Reservoir depletion will affect the drillability of the field for future wells. During the drilling process, the drilling fluid pressure should be within the window of formation fracture pressure and formation pore pressure. As reservoir pressure decreases, so will the formation fracturing gradient. The drilling window will get smaller and smaller, which results in more significant potential drilling problems for new wells drilled. Shutting in the surrounding producers can boost the drillability of a new well, but this is usually a costly option. (4) For near wellbore formation damage, when the bottom hole pressure of a well is below the fluid asphaltene onset pressure, asphaltenes would dropout and potentially permanently damage the formation, as well as the wellbore. Once the asphaltene precipitates in the formation, there is no proven and economic remediation method to remove the precipitants.

11.6 DEEPWATER SURVEILLANCE

Surveillance technologies are critical in deepwater development and production as they provide hard data for decision making on reservoir and well management. Capitalizing on fiber optic and sensor technologies of the intelligent completion systems, the engineer uses real-time surveillance technologies to quickly identify formation damage symptoms and make prompt decision to remedy the problems. So far many mature technologies, such as the alarm and event system, classical pressure transient analysis, and the "virtual flow meter" use real-time pressure and temperature. Readers can refer to external sources for these applications, and this section is mainly dedicated to some recent advances in reservoir and well surveillance.

Since temperature is not a direct driving force for fluid flow, there are many uncertainties and challenges in quantitative usage of measured temperature without complex modeling (Wu et al., 2012) or using and interpreting data from distributed temperature sensors (Li and Zhu, 2010; Hoang et al., 2012; Wang et al., 2008; Yoshioka et al., 2007) and production/injection well logging (Witterholt and Tixier, 1972; Steffensen and Smith, 1973). Temperature variation in the wellbore during flow periods is usually caused by subtle energy effects such as Joule–Thomson expansion, viscous dissipative heating, and thermal transfer. Currently temperature utilization for flow rate and phase estimation mainly use empirical or semi-analytical models. Wu and his coworker utilized temperature data as an indicator of formation damage. Given that temperature is one of the most important factors influencing fluid mobility, it can be utilized together with pressure data in modeling flow direction.

When the distance between the perforations and the downhole temperature gauge location is relatively short, the heat loss from the wellbore to the surrounding formation can be neglected, which satisfies the Joule–Thomson adiabatic assumption. The Joule–Thomson effect describes the temperature variation when fluid travels through a restrained device such as a well perforation and near wellbore porous media. Depending on the pressure magnitude and flow fluid properties, the resulting temperature from the Joule–Thomson effect can be subjected to either a heating or a cooling effect. Mathematically the Joule–Thomson coefficient is defined as Eq. (11.13):

$$\mu_{JT} = \left(\frac{\partial T}{\partial p}\right)_H \tag{11.13}$$

where T is temperature and p is pressure. Eq. (11.14) expresses the above equation in a finite difference form:

$$\mu_{JT} = \frac{\Delta T}{\Delta p} = \frac{T_r - T_g}{p_r - p_{wf}} \tag{11.14}$$

where T_r is the temperature of fluid entering the perforation interval, p_r is the average reservoir pressure, T_g is the temperature at the downhole gauge, and p_{wf} is the pressure measured by the downhole gauge. A more general discussion about the derivation can be found in the literature (Abdel Rasoul et al., 2011). Assuming Darcy flow in an isothermal reservoir in pseudo steady-state regime, the well production rate can be calculated. Wu et al. (2014) derived the quantitative relationship between the skin factor from formation damage and temperature difference expressed as Eq. (11.15):

$$T_g(t) = T_r - a(b + s_T)q(t) \tag{11.15}$$

where s_T denotes the skin factor determined from the temperature data. The relationship between temperature change and skin factor is given by Eq. (11.16),

$$f(s) = a(b + s) \qquad (11.16)$$

In both Eqs. (11.15) and (11.16), the constants a and b are in SI units system and are derived as follows.

$$a = \mu_{JT} \frac{\mu}{2\pi kh}$$

$$b = \frac{1}{2} \ln \frac{4A}{C_A \gamma r_w^2}$$

where μ is fluid viscosity, k is formation permeability, h is the net pay thickness, C_A is shape factor, r_w is wellbore radius, and A is the well drainage area. In a constant skin factor period, one can select two different rates and their corresponding gauge temperatures, and reconfigure Eq. (11.16) to be expressed as Eq. (11.17):

$$f(s) = a(b + s) = \frac{T_{g,2} - T_{g,1}}{q_1 - q_2} \qquad (11.17)$$

Using Eq. (11.17), multiple surveillance objectives can be achieved by integrating the temperature data with pressure transient analysis, petrophysical study, and production data for different reservoir and well conditions. For a well with a constant skin factor, f(s) derived by Eq. (11.17) will also be a constant. For a reservoir with a constant temperature, Tr, temperature data can be used to determine the change in skin factor.

Fig. 11.15 shows a deepwater well production rate (curve) and recorded downhole temperature (dots). The well had an asphaltene precipitation issue in the near wellbore region because the sandface flowing pressure is less than the onset pressure for asphaltene precipitation. In the production history, several xylene treatments were conducted to dissolve the precipitates. Monitoring the skin change over the treatment was necessary to evaluate the effectiveness of the treatments. The curve in Fig. 11.16 shows the skin factor estimated using pressure transient analysis. The variations of skin factor in the first 900 days of production, especially for the skin within the ellipse, were mainly caused by well cleanup operations, rate dependent skin changes, and possibly inaccurate rate allocations to this well. After 900 days of production, a significant event led the well to be fully opened and to blow down the reservoir, as the oil rate went up from 3200 STB/D to nearly 7000 STB/D. The dramatic skin change during the blowdown was because of asphaltene deposition and xylene treatments. Two independent methods were used to analyze

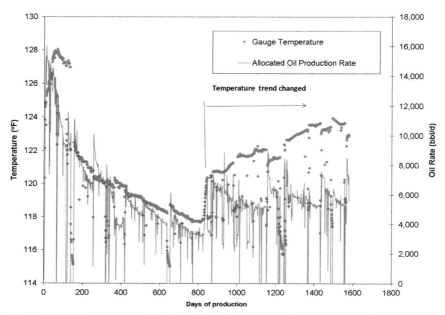

Figure 11.15 Temperature (dots) and production rate history (curve) for producing well B in the field.

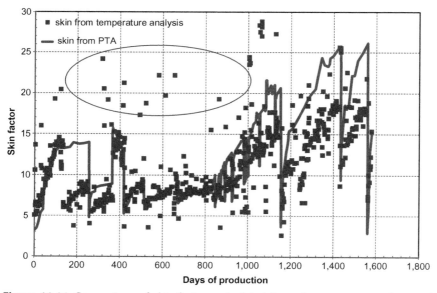

Figure 11.16 Comparison of skin factors calculated from the temperature data and the pressure transient analysis applying Eq. (11.17).

the skin factor change. One method used the downhole pressure data, and the other used the downhole temperature data. The pressure method here was not only using pressure build-up data, but also matching the downhole pressure gauge data during production using time superposition applying Eq. (11.18) in field units.

$$p_{wf}(t) = P_i - \frac{141.2B\mu}{kh} \left[\sum_{j=1}^{N} \left(q_j - q_{j-1} \right) p_D \left(t_D - t_{jD} \right) + q_N s_p \right] \quad (11.18)$$

where p_i is the initial reservoir pressure, psi, B is oil volume factor, μ is oil viscosity, cp; k is formation permeability, md, h is the formation net pay thickness, ft. q_j is oil production rate at the moment of j, and p_D is pressure response function in dimensionless form. Matching p_{wf} with the gauge data by changing the skin factor will yield a skin history (denoted by s_p in the equation). A detailed discussion concerning Eq. (11.18) is provided in the literature (Horne, 2000). The above example presents methods to determine the skin factors from both pressure and temperature during the production stage, which has a clear advantage over classical pressure build up (PBU) technique historically used to determine the skin factor. The real-time data of pressure and temperature enable the determination of the skin history from which formation damage can be identified. Additionally, since the temperature approach is based on a completely different mechanism to the PBU approach, confirmation, and cross checking using both methods can increase the confidence about skin factor changes and their causes. For the above temperature model, with little additional effort, it can be extended to monitor the productivity changes caused by the permeability changes, reservoir compaction, and fluid composition changes.

The temperature model is based upon assumptions of single phase flow and no heat loss to the formation. The second assumption indicates that the production rate has to be high and the distance between the production interval to the downhole gauge location has to be small. Violating these two fundamental assumptions may yield inappropriate values for the input variables for the temperature model.

When the flow rate is low or the distance between the perforation interval to the downhole temperature gauge is large, we can't neglect the heat losses from the wellbore to the surrounding media. In this case, a tangible temperature difference can be obtained from the recorded temperature data from which the fluid production rate can be calculated.

The temperature at the gauge has a nonlinear relationship with fluid flow rate in the wellbore as determined by Eq. (11.19) (Prats, 1986):

$$T(D, t) = T_{Geo}(D) + G_g \rho_f q_f C_p R_h$$
$$+ \left[T_p - T_{perf} - G_g \rho_f q_f C_p R_h \right] \exp \left(\frac{T_{Geo}(D) - T_{perf}}{G_g \rho_f q_f C_p R_h} \right) \quad (11.19)$$

where T is the measured temperature at the gauge, and T_{Geo} is the geothermal temperature at the gauge location, G_g is the geothermal gradient, ρ_f is fluid density, C_p is fluid heat capacity, R_h is the heat resistance of formation, and T_{perf} is the fluid temperature flowing out of the perforation. In this model, the volumetric flow rate, q_f, can be estimated implicitly using numerical iterations. For practical purposes, the combined entity $(G_g \rho_f C_p R_h)$ can be determined as one variable from well rate test result. This method is applicable when the production rate is sufficiently high and stable for providing quantitative analysis; otherwise the method should be used in the qualitative sense as the temperature change is small and variation of the rate corresponding to the small temperature change can be large.

Pressure data can also be used to monitor formation compressibility change if a high resolution and accurate downhole pressure gauge is used. Wu et al. (2013b) presented the procedure to estimate the total compressibility change based on the tidal signal collected during a long term well shut-in. The procedure and result are unique in the terms of scale of investigation for the parameters. Tide height at sea level generates a periodic pressure signal on the sea floor, and the pressure variation on the sea floor will be transmitted to the reservoir formation as the overburden pressure changes, which in turn results in pore pressure changes. Assuming the tidal loading is uniaxial in the vertical direction only and the grain compressibility is negligible compared with pore volume compressibility, the effective stress caused by the tidal pressure in the reservoir formation is expressed as Eq. (11.20):

$$\Delta \sigma_{eff} = \Delta \sigma_v - \Delta P_p \quad (11.20)$$

where $\Delta \sigma_v$ is the change of overburden stress and ΔP_p is the change of pore pressure. Reconfiguring Eq. (11.20) yields Eq. (11.21):

$$\frac{\Delta \sigma_{eff}}{\Delta P_p} = \frac{\Delta \sigma_v}{\Delta P_p} - 1 \quad (11.21)$$

Defining the uniaxial pore volume compressibility as Eq. (11.22):

$$c_{PV} = \frac{1}{\varnothing} \frac{\partial \varnothing}{\partial \sigma_{eff}} \quad (11.22)$$

The fluid compressibility, c_f, in the formation can be expressed as Eq. (11.23):

$$c_f = \frac{1}{\varnothing} \frac{\partial \varnothing}{\partial P_p} \quad (11.23)$$

The fluid compressibility can be determined for a given fluid and conditions using Equations of State or correlations. The ratio of the fluid compressibility to the pore volume compressibility is derived using Eq. (11.24):

$$\frac{c_f}{c_{PV}} = \frac{\Delta \sigma_v}{\Delta P_p} - 1 \quad (11.24)$$

Defining the ratio of the pore pressure change to the vertical stress change as the tidal transmission efficiency, T can be calculated using Eq. (11.25):

$$T = \frac{\Delta P_p}{\Delta \sigma_v} = \frac{c_f}{c_f + c_{PV}} \quad (11.25)$$

Applying the Fast Fourier Transform (FFT) to the pressure data collected from the downhole gauge during a well shut-in period, we can extract the tidal signal as shown in Fig. 11.17, which is based on pressure data of about 3500 h. Manipulating the transmission coefficient (T) to match the real tidal data with the pressure extract data, one can determine the change of fluid compressibility or the pore volume compressibility.

11.7 SUMMARY OF DEEPWATER DEVELOPMENT

Developing hydrocarbon resources from deepwater environments is an expensive and challenging activity. The challenges come from metocean conditions, geohazards, subsea engineering, and subsurface surprises. The cost of developing deepwater fields calls for high production rates of oil or gas to make an economically viable project. Therefore, any

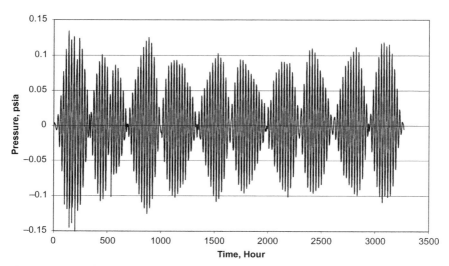

Figure 11.17 Tidal signal filtered from the downhole pressure gauge applying a Fast Fourier Transform (FFT) to the pressure data recorded for the well.

potential cause of formation damage should be fully understood and minimized, and advanced well completions such as frac-and-pack are used to further enhance the well productivity index.

Overpressure is a double-edged sword in offshore development. First, it provides high pore pressure and energy for initial well production. At the same time, the pore pressure causes challenges in drilling and completing wells as the pore pressure varies significantly at the same depth. Furthermore, with high depletion, the fracture pressure gradient is lowered and becomes close to the pore pressure gradient, which causes drillability problems for in-fill drilling. When the overpressured reservoir is being depleted, the reservoir compaction may result in irreversible permeability and porosity reduction.

Deepwater wells feature intelligent well completions with sensors and inflow control devices. The intelligent completion enables real-time surveillance for reservoir and production management. Pressure and temperature data are most often used for flow rate allocation, formation testing, and well performance monitoring. In addition to classical rate and phase determination and event and alarm systems, some innovative applications such as skin factor monitoring using downhole pressure and temperature well data, and pore volume compressibility enable formation damage to be better determined and quantified.

In recent years, we have made significant progresses on deepwater technologies and have been pushing the frontier boundaries of deepwater development, however, our current technologies can still only be deployed in regions of water depth less than 10,000 ft, reservoir pressure less than 20,000 psia, and reservoir temperature is less than 300°F. To develop deeper formations, new development focused on materials sciences, innovative tools, and engineered processes are needed.

NOMENCLATURES

A	drainage area, L^2
a	a mathematical parameter, tT/L^3
B	oil formation volume factor, *dimensionless*
B_o	oil formation volume factor, *dimensionless*
b	a mathematical parameter, *dimensionless*
C_A	dietz shape factor, *dimensionless*
C_{pV}	pore volume compressibility, $M^{-1}Lt^2$
C_p	fluid heat capacity, $M^2Lt^{-2}T$
C_{pv}	uniaxial pore volume compressibility, $M^{-1}Lt^2$
C_f	fluid compressibility, $M^{-1}Lt^2$
FE	flow efficiency, *dimensionless*
G_g	geothermal gradient, T/L
H	depth, L
h	net pay thickness, L
k	formation permeability, L^2
k_i	initial permeability, L^2
k_{ro}	relative permeability of oil, *dimensionless*
p	pore pressure, $ML^{-1}t^{-2}$
P_p	pore pressure, $ML^{-1}t^{-2}$
p_D	dimensionless pressure, *dimensionless*
p_i	initial pressure, $ML^{-1}t^{-2}$
$\bar{p}$	average reservoir pressure, $ML^{-1}t^{-2}$
p_r	average reservoir pressure, $ML^{-1}t^{-2}$
p_{wf}	flowing bottom hole pressure, $ML^{-1}t^{-2}$
q	rate, L^3/t
q_f	volumetric rate, L^3/t
q_o	oil production rate, L^3/t
q_j	production rate at moment j, L^3/t
R_h	heat resistivity radius, $\tau M^{-1}L^{-2}t^3$
r_w	wellbore radius, L
r_w'	effective wellbore radius, L

S skin factor, *dimensionless*

s_f fracture skin, *dimensionless*

s_p matching skin factor, *dimensionless*

s_T skin factor, *dimensionless*

T tidal transmission efficiency, *dimensionless*

T_g fluid temperature at the downhole gauge, T

T_{Geo} geothermal temperature at the gauge location, T

T_{perf} fluid temperature flowing out of the perforation, T

T_r fluid temperature entering the perforation interval, T

T_{surf} surface temperature, T

t time, t

t_D dimensionless time, *dimensionless*

Vp pore volume, L^3

x_f fracture half-length, L

z_s depth below the sea level, L

α Biot's poroelastic constant, *dimensionless*

ε strain, *dimensionless*

μ oil viscosity, $ML^{-1}t^{-1}$

μ_{JT} Joule–Thomson coefficient, $TM^{-1}Lt^2$

γ Euler's constant, 1.78, *dimensionless*

γ_k permeability modulus, $M^{-1}Lt$

ρ_r density of the formation overlaying the target reservoir, M/L^3

ρ_f fluid density, M/L^3

σ_{eff} effective stress, $ML^{-1}t^{-2}$

σ_V vertical overburden stress, $ML^{-1}t^{-2}$

σ'_V effective vertical stress, $ML^{-1}t^{-2}$

ϕ current porosity at a specific pressure, *dimensionless*

ϕ_i initial porosity, *dimensionless*

REFERENCES

Abdel Rasoul, R., Salah, A., Daoud, A., 2011. Production allocation in multi-layers gas producing wells using temperature measurements (By Genetic Algorithm). SPE Middle East Oil and Gas Show and Conference, 25-28 September 2011 Manama, Bahrain.

BP, 2010. Deepwater Horizon Accident Investigation Report. BP.

Biot, M.A., 1941. General theory of three-dimensional consolidation. J. Appl. Phys. 12 (2), 155–164.

Costello, C., Sordyl, P., Hughes, C.T., Figueroa, M.R., Balster, E.P., Brown, G., 2012. Permanent distributed temperature sensing (DTS) technology applied in mature fields-a forties field case study. SPE Intelligent Energy International. Society of Petroleum Engineers.

Hadley, M., Kimish, R., 2008. Distributed temperature sensor measures temperature resolution in real time. SPE Annual Technical Conference and Exhibition. Society of Petroleum Engineers, January.

Hoang, H., Mahadevan, J., Lopez, H., 2012. Injection profiling during limited-entry fracturing using distributed-temperature-sensor data. Soc. Pet. Eng. J. 17 (3), 752–767.

Horne, R.N., 2000. Modern Well Test Analysis: A Computer-Aided Approach. Petroway, Inc., CA.

Horne, R.N., 2007. Listening to the reservoir—interpreting data from permanent down-hole gauges. J. Pet. Technol. 59 (12), 78—86.

Husson, L., Henry, P., Le Pichon, X., 2008. Thermal regime of the NW shelf of the Gulf of Mexico. Part A: Thermal and pressure fields. Bull. Soc. Geol. Fr. 179, 129—137.

Li, X., Mitchum, F., Bruno, M., Pattillo, P., Willson, S., 2003. Compaction, subsidence, and associated casing damage and well failure assessment for the Gulf of Mexico shelf Matagorda Island 623 field. SPE Annual Technical Conference and Exhibition, 5-8 October 2003 Denver, Colorado. Society of Petroleum Engineers.

Li, Z., Zhu, D., 2010. Predicting flow profile of horizontal well by downhole pressure and distributed-temperature data for waterdrive reservoir. Soc. Pet. Eng. Prod. Oper. 25 (3), 296—304.

Marland, C.N., Nicholas, S.M., Cox, W., Flannery, C., Thistle, B., 2007. Pressure prediction and drilling challenges in a deepwater subsalt well from Offshore Nova Scotia, Canada. Soc. Pet. Eng. Drill. Completion 22 (3), 227—236.

Prats, M., 1986. Thermal Recovery. Society of Petroleum Engineers, New York.

Robinson, M., 2003. Intelligent well completions. J. Pet. Technol 55 (08), 57—59.

Ruddy, I., Andersen, M.A., Pattillo, P., Bishlawi, M., Foged, N., 1989. Rock compressibility, compaction, and subsidence in a high-porosity chalk reservoir: a case study of Valhall field. J. Pet. Technol. 41 (7), 741—746.

Sanchez, M., Packing, F., 2007. Fracturing for Sand Control. Middle East and Asia Reservoir Review 37—49.

Schlumberger., 2013. Signature Quartz Gauges [Online]. Available from: http://www.slb.com/~/media/Files/testing/product_sheets/pressure/signature_quartz_gauge_ps.pdf (accessed 16.04.2013).

Spill, N. C. O. T. B. D. H. O, 2011. Deep Water: The Gulf Oil Disaster and the Future of Offshore Drilling. Perseus Distribution Digital.

Steffensen, R., Smith, R., 1973. The importance of Joule-Thomson heating (or Cooling) in temperature log interpretation. Fall Meeting of the Society of Petroleum Engineers of AIME, 30 September-3 October 1973 Las Vegas, Nevada.

Wang, X., Lee, J., Vachon, G., 2008 Distributed temperature sensor (DTS) system modeling and application. SPE Saudi Arabia Section Technical Symposium, 10-12 May 2008 Al-Khobar, Saudi Arabia.

Willson, S., Edwards, S., Heppard, P., Li, X., Coltrin, G., Chester, D., et al., 2003. Wellbore stability challenges in the deep water, Gulf of Mexico: case history examples from the pompano field. SPE Annual Technical Conference and Exhibition, 2003. Society of Petroleum Engineers.

Witterholt, E.J., Tixier, M.R. Temperature logging in injection wells. Fall Meeting of the Society of Petroleum Engineers of AIME, 8—11 October 1972 San Antonio, Texas.

Wu, X., Humphrey, K. & Liao, T. Enhancing production allocation in intelligent wells via application of models and real-time surveillance data. SPE International Production and Operations Conference & Exhibition, 14-16 May 2012 Doha, Qatar.

Wu, X., Jiang, Y., Sui, W., 2013a. Innovative applications of downhole temperature data. SPE Middle East Intelligent Energy Conference and Exhibition, 2013a. Society of Petroleum Engineers.

Wu, X., Ling, K., Liu, D., 2013b. Deepwater-reservoir characterization by use of tidal signal extracted from permanent downhole pressure gauge. Soc. Pet. Eng. Reserv. Eval Eng. 16, 390—400.

Wu, X., Sui, W., Jiang, Y., 2014. Semiquantitative applications of downhole-temperature data in subsurface surveillance. Soc. Pet. Eng. Prod Oper. 29, 323—328.

Wu, X., Babatola, F., Jiang, L., Tolbert, B.T., Liu, J., 2016. Applying subsea fluid-processing technologies for deepwater operations. Oil and Gas Facilities 5.

Yilmaz, Ö., Nolen-Hoeksema, R.C., Nur, A., 1994. Pore pressure profiles in fractured and compliant rocks. Geophys. Prospect. 42, 693–714.

Yoshioka, K., Zhu, D., Hill, A., Dawkrajai, P., Lake, L., 2007. Prediction of temperature changes caused by water or gas entry into a horizontal well. Soc. Pet. Eng. Prod. Oper. 22, 425–433.

CHAPTER TWELVE

Formation Damage Challenges in Geothermal Reservoirs

Zhenjiang You, Alexander Badalyan, Yulong Yang and Pavel Bedrikovetsky

The University of Adelaide, Adelaide, SA, Australia

Contents

12.1 INTRODUCTION

The primary condition for successful geothermal field exploitation is high rock permeability that provides sufficient well productivity and economic efficiency. Therefore, the permeability decline during inflow

Formation Damage during Improved Oil Recovery.
DOI: https://doi.org/10.1016/B978-0-12-813782-6.00012-9

© 2018 Elsevier Inc.
All rights reserved.

447

performance is a significant negative factor in geothermal energy development (Ghassemi and Zhou, 2011; Aragón-Aguilar et al., 2013). One of the main reasons for permeability reduction during geothermal exploitation is fines migration and straining (Baudracco, 1990; Rosenbrand et al., 2012, 2013, 2014, 2015). Therefore, the assessment of the formation damage caused by mobilization and migration of fines is a vital part of the feasibility study for geothermal energy projects.

The temperature dependencies of rock properties significantly affect the energy recovery from geothermal reservoirs. Seismic properties of rocks along with their electrical conductivity and porosity are highly temperature sensitive (Jaya et al., 2010; Kristinsdottir et al., 2010; Milsch et al., 2010). The temperature sensitivity of rock permeability is explained by induced fines migration (Rosenbrand et al., 2015).

Similar processes of fines migration and permeability damage occur in the so-called enhanced geothermal projects, where hot water is produced, and the reservoir pressure is maintained by the cold-water injection and during the seasonal storage of hot water in geothermal aquifers.

Fines-migration-induced permeability reduction in rocks is a well-known phenomenon occurring in both geothermal reservoirs (Priisholm et al., 1987; Baudracco, 1990; Baudracco and Aoubouazza, 1995; Rosenbrand et al., 2012, 2013, 2014, 2015) and conventional oil and gas fields (Khilar and Fogler, 1998; Civan, 2007; Tiab and Donaldson, 2012; Marquez et al., 2014). This permeability damage is caused by mobilization of fine particles after disturbance to the balance between moments of attaching electrostatic force and detaching drag and lifting forces occurring at reduced ionic strengths and elevated velocities of geothermal fluids (Bradford et al., 2013). The detached particles migrate in a porous rock and block narrow pore throats leading to permeability decline (Muecke, 1979; Lever and Dawe, 1984; Sarkar and Sharma, 1990) (Figs. 12.1 and 12.2). The detachment of particles from grain surfaces yields some insignificant increase of permeability while straining and blocking of pore throats result in significant permeability decrease. The permeability decline caused by mobilization, movement, and straining of fines occurs since the attached particles are mobilized and become strained (Muecke, 1979; Khilar and Fogler, 1998; Tiab and Donaldson, 2012). The effects of salinity, velocity, and pH on fines migration and permeability impairment have been investigated by laboratory experiments and mathematical modeling (Lever and Dawe, 1984; Kia et al., 1987a, b; Sharma and Yortsos, 1987b; Sarkar and Sharma, 1990; Miranda and Underdown, 1993; Khilar and Fogler, 1998; Ochi and Vernoux, 1998;

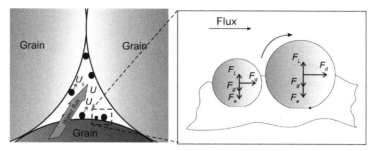

Figure 12.1 Cross section of a pore throat and forces acting on particles attached at a grain surface.

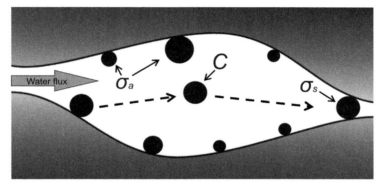

Figure 12.2 Mobilization, movement, and straining of fines in porous media.

Rosenbrand et al., 2015). Yuan et al. (2015, 2016, 2017) and Yuan and Moghanloo (2017) provide a series of mathematical works to confirm the effectiveness of nanofluids to control problems of fines migration and mitigate formation damage. Qualitative effects of temperature on fines migration have been investigated by Schembre and Kovscek (2005), Rodríguez and Araujo (2006), and Rosenbrand et al. (2012, 2013, 2014, 2015). To the best of our knowledge, the quantitative temperature effects necessary for reservoir modeling of the geothermal energy development have not been reported in the literature. In the present work, we predict mobilization of fine particles at high geothermal temperatures and the consequent formation damage using mathematical modeling adjusted to experimental coreflood data.

The well productivity index (the ratio of flow rate to the pressure drawdown) is used to measure the ability of a reservoir to deliver fluids to the wellbore. This index for wells located in geothermal sandstone reservoirs is significantly affected by the mobilization of clay particles due to high fluid velocities and/or low ionic strength of reservoir water. On the one side, dielectric permittivity of solution decreases with the increase of temperature, thus reducing the electrical double layer (EDL) thickness (Rosenbrand et al., 2014) and weakening the repulsion between clay particles and sand surface. On the other side, magnitudes of negative surface charges of clay particles and sand increase with temperature (Ramachandran and Somasundaran, 1986; Rosenbrand et al., 2014), which enhances particle–grain repulsion. The dominance of the latter effect over the former one indicates that temperature increase results in stronger electrostatic repulsion and fines mobilization. A particle can be mobilized by a flowing fluid, if the hydrodynamic force exceeds the resulting Derjaguin–Landau–Verwey–Overbeek (DLVO) attractive electrostatic force exerted on the particle attached to the surface of a rock (Jiao and Sharma, 1994). Rock permeability reduction due to particle capture is caused by retention of the mobilized particles by straining (Khilar and Fogler, 1998; Bedrikovetsky, 2008; Yuan et al., 2013), bridging at the pore entrances (Muecke, 1979; Rousseau et al., 2008) and internal cake formation (Nabzar and Chauveteau, 1997; Rousseau et al., 2008).

Reservoir simulation can determine well productivity and technological efficiency of enhanced geothermal projects when fine particles migrate in geothermal reservoirs. Several laboratory studies evaluate how temperature, salinity, and velocity affect the geothermal reservoir permeability under fines migration (Baudracco, 1990; Baudracco and Aoubouazza, 1995; Schembre and Kovscek, 2005; Rosenbrand et al., 2012, 2013, 2014, 2015). However, experimental investigations of fines migration in geothermal fields with the goal to provide the necessary data for reservoir simulation are not available in the literature.

The present laboratory study investigates the effect of variations of the geothermal fluid's ionic strength and velocity on migration of fine particles and consecutive permeability damage. The investigated core is analogous to the rock of the Salamander geothermal field (Otway Basin, Australia). The mobilized fines of the core contain predominantly kaolinite and illite/chlorite minerals, causing the formation damage. We show that low ionic strength or high velocity of the reservoir water is responsible for significant fines mobilization and permeability impairment of the core. The coreflood tests show that periods of stabilization for rock permeability are significantly greater than one pore volume (one PV) injected, whereas if one assumes that velocities of a fine particle and water are equal, stabilization of core permeability is achieved after injection of one PV. The delayed permeability stabilization observed in coreflood tests is explained by a slow transport of fine particles near pore walls.

For the first time, the laboratory translation procedure between temperature and velocity dependencies of fines mobilization is established. The new methodology is developed based on the mechanical equilibrium of particles on rock surfaces. It allows substituting multiple cumbersome laboratory experiments with high-velocity variation by a single feasible one with ionic strength variation. The derived formula for the maximum retention function (MRF) has been calculated accounting for information available in the literature on temperature variation of Hamaker constant, zeta potential, water viscosity, etc. It is found that the MRF declines with the increase of temperature for the parameter ranges investigated and the type of core used. Consequently, geothermal reservoirs are more susceptible to formation damage due to fines migration than the conventional aquifers and oil reservoirs.

Design and planning of geothermal processes accompanied by suspension transport significantly depend on the results of mathematical modeling of deep bed filtration caused by mobilization of particles, their straining in narrow pore throats, and consequent blocking of the rock porous network. Straining of mobilized particles is considered as the major formation damage physical mechanism during suspension transport in porous media. The jamming ratio, defined as the ratio between mean particle size and mean pore throat size, determines size exclusion particle capture at conditions of mutual particle–grain repulsion. Therefore, microscale models considering size distributions of particles and pores are sufficient for prediction of permeability damage during migration of fines (Fig. 12.1 shows the processes of particle mobilization, migration, and capture by straining). The microscale random walks models (Cortis et al., 2006; Shapiro, 2007; Yuan and Shapiro, 2011), population balance models (Sharma and Yortsos, 1987a, b; You et al., 2013), and Boltzmann's physical kinetics equation (Shapiro and Wesselingh, 2008) thoroughly describe the process of fines migration considering size distributions of particles and pores. However, no experimental data on size distribution of particles and pores during suspension transport in porous media are available in literature. For this reason, to predict and assess the consequences of fines migration, we used the averaged equations which operate with concentrations of attached, suspended, and retained particles.

The mass balance equation for solute transport with the sink term accounting for the retention of fine particles and the source term accounting for the displacement of a particle from the rock surface is the most widely applied mathematical model for particle detachment, migration, and capture (Schijven and Hassanizadeh, 2000; Logan, 2001; Cortis et al., 2006; Tufenkji, 2007; Shapiro and Yuan, 2013):

$$\frac{\partial}{\partial t}(\phi c + \sigma) + U\frac{\partial c}{\partial x} = D\frac{\partial^2 c}{\partial x^2} \tag{12.1}$$

$$\frac{\partial \sigma}{\partial t} = \lambda(\sigma)cU - k_{det}\sigma \tag{12.2}$$

where c and σ are volumetric concentrations of suspended and strained particles, respectively; U is the flow velocity and D is the diffusion coefficient.

The first term in the right-hand side of Eq. (12.2) is called the capture term. The proportionality of this term to the advective particle flux is through the filtration coefficient, λ. The detachment rate coefficient, k_{det}, establishes the proportionality between the detachment term (the second term in the right-hand side of Eq. 12.2) and the concentration of particles retained in a porous medium. The system of equations (12.1, 12.2) and microscale modeling-based formula for coefficient λ are referred to as the classical filtration theory. Advanced theories determine the relationship between the filtration coefficient and particle−particle/particle−grain interactions, suspension velocity, sedimentation under gravity, and Brownian diffusion (Nabzar and Chauveteau, 1997; Chauveteau et al., 1998; Tufenkji and Elimelech, 2004; Rousseau et al., 2008; Yuan and Shapiro, 2012), while k_{det} is usually calculated by tuning guided by experimental data. This is a limitation of the advective−diffusive attachment−detachment model with kinetics of the particle detachment based on Eqs. (12.1), (12.2).

The instant release of fines due to sudden increase of the pressure gradient or decrease of fluid ionic strength (Miranda and Underdown, 1993; Khilar and Fogler, 1998) is in contradiction with the asymptotic stabilization of the concentration of retained particles and rock permeability when time tends to infinity; this is another limitation of the above-mentioned advective−diffusive attachment−detachment model. Core permeability immediately responds to the instant increase of the fluid velocity during coreflood tests (Ochi and Vernoux, 1998; Bedrikovetsky et al., 2012; Oliveira et al., 2014).

Disturbance of the mechanical equilibrium of a particle located on the internal filter cake leads to detachment of this particle (Schechter, 1992; Rahman et al., 1994; Civan, 2007). Electrostatic, drag, lifting, and gravitational forces act on a particle located in the internal filter cake. Usually, one can neglect lifting and gravitational forces due to their very small values compared with electrostatic and drag forces, as was shown for the case of ambient and geothermal reservoir conditions (You et al., 2015). For this reason, Fig. 12.1 shows only drag and electrostatic forces. Some studies examine the balance between the drag force acting on the particle from the bypassing fluid and the friction force with an empirical Coulomb coefficient (Civan, 2007). Other authors consider the momentum balance of forces (Jiao and Sharma, 1994; Freitas and Sharma, 2001):

$$F_d(U, T, r_s)l(r_s) = F_e(\gamma, T, r_s), l = l_d/l_e \tag{12.3}$$

where F_d and F_e are drag and electrostatic forces, respectively, l_d and l_e are corresponding lever arms, l is the lever arm ratio, U is flow velocity, γ is the ionic strength of the reservoir brine, and r_s is the particle radius.

A revised Stokes' formula is derived for a spherical particle retained on the wall of a pore, which allows the calculation of the drag force via the radius of a particle and its velocity (Jiao and Sharma, 1994; Ochi and Vernoux, 1998; Bradford et al., 2013). A formula for the drag force includes the shape factor accounting for deformation of a particle on the rock surface by electrostatic attracting forces, the roughness of the rock surface, and the form of a particle.

Differentiation of the total interaction potential energy by the distance between a particle and wall results in the total electrostatic force. The total interaction potential energy is the sum of London—van der Waals (LW), EDL, and Born interaction potential energies. The DLVO theory gives explicit equations for these three interaction potential energies (Derjaguin and Landau, 1941; Verwey and Overbeek, 1948; Israelachvili, 2011). Expressions for dependencies of these three respective forces on fluid velocity and ionic strength, particle, and pore radii (Khilar and Fogler, 1998) are applied to migration of fines in porous media in the present chapter.

The two approaches explained here are mathematically equivalent. The mechanical equilibrium of a single particle does not affect the kinetic detachment term in Eq. (12.2), and the advective—diffusion equation with kinetic detachment term does not reflect the mechanical equilibrium of a particle.

The electrostatic force in Eq. (12.3) varies with fluid ionic strength, radius of a particle, and fluid/rock temperature, whereas particle radius and fluid velocity affect the drag force. Therefore, it is possible to calculate particle radius r_s from transcendental equation (12.3): $r_{scr} = r_s(U, \gamma, T)$. The critical radius of the retained particles, r_{scr}, is a monotonic function which decreases with U and T, and increases with γ for Eq. (12.3). Therefore, we can identify the critical radius of the retained particles as the minimum size of particles mobilized by increased fluid velocity and temperature, and by decreased ionic strength of the fluid. It means that, initially larger particles are mobilized followed by smaller ones. As a result, the total concentration of particles remaining on the rock surface is equal to the sum of concentrations for all particles smaller than r_{scr}:

$$\sigma_a = \sigma_{cr}(U, \gamma, T),$$

$$\sigma_{cr}(U, \gamma, T) = \int_0^{r_{scr}(U, \gamma, T)} \Sigma_a(r_s) dr_s \tag{12.4}$$

Here $\Sigma_a(r_s)$ is the size distribution of particles attached to the rock surface. Eq. (12.4) is called the MRF. This empirical function depends on the properties of porous media and the flowing-through particles.

The MRF as an instant function of fluid ionic strength, velocity, and temperature is used in the improved mathematical model describing the mobilization of particles (Bedrikovetsky et al., 2011, 2012). The MRF either decreases or increases if any of the above three parameters change at point (x, t), resulting in either instant fines detachment or timely attachment of new fines to the rock surface, respectively. If fluid ionic strength monotonically decreases and fluid velocity and temperature monotonically increase, then the MRF declines, and the particles attached to the rock surface are mobilized.

Therefore, the MRF is calculated via the torque balance of the particle located on the surface of a pore of the internal cake, Eq. (12.3). The kinetics of particle release in the classical attachment—detachment model, Eq. (12.2), is replaced by the phenomenological function, Eq. (12.4). Therefore, the modified model, Eqs. (12.3), (12.4), is free of the above-mentioned shortcomings.

The MRF is in analogy to an adsorption isotherm. The difference between them is that the MRF is a function of fluid velocity, which is not a thermodynamic parameter (Bedrikovetsky, 1993). Eq. (12.4) for the dependency of the MRF on fluid ionic strength, velocity, and temperature does not correspond to an energy minimum, since the drag force in Eq. (12.3) does not have potential. Generalization of Eq. (12.4) can be done in a way analogous to nonequilibrium sorption.

Electrostatic attraction deforms both the matrix surface and the particle attached to it. High salinity fluid forms a very thin film between them. Injection of low-ionic-strength fluid results in diffusion of salt ions from high-concentration fluid film to low-concentration bypassing fluid, leading to particle detachment. The kinetics of this process are modeled by Bradford et al. (2012, 2013) by the introduction of time delay into the MRF equation (12.4).

According to the assumption of the conventional model, Eqs. (12.1), (12.2), velocities of the mobilized particles and the carrier fluid are equal. Considering this assumption and instant mobilization of fine particles results in zero concentration of particles at the core outlet after injection of one PV. This is because the particle mobilized at the core inlet reaches the core outlet after dimensionless time equal to one PV. Accordingly, stabilization of the pressure drop across the core length occurs also after the injection of one PV. But, stabilization of pressure drop/permeability with considerable time delay was observed in numerous laboratory studies (Lever and Dawe, 1984; Ochi and Vernoux, 1998; Oliveira et al., 2014). Yang et al. (2016) explain such particle behavior by its slow motion near the surface of the rock grains. Several experimental studies observed slow drift of mobilized particles. The microscale modeling using the Navier—Stokes equation was used to study slow particle motion in the vicinity of the asperities of the pore walls (Sefrioui et al., 2013). A delayed arrival of particles was observed from the breakthrough concentration curves (Yuan and Shapiro, 2011; Bradford et al., 2013). The breakthrough concentrations were successfully matched using the proposed two-speed model. However, it is not possible to determine six empirical modeling kinetic coefficients of mass transfer between particles attached to the wall, slowly moving near the wall, and fast moving in the bulk fluid only from the breakthrough concentration history. The complete characterization of the above model can be achieved using experimental data from significantly more sophisticated laboratory tests. Also, due to the existence of mass exchange between fluxes of fast and slowly moving particles, the core scale concentrations of particles may be equal causing propagation of the overall particle ensemble with a low average speed. Therefore, the single-velocity model accounting for particle speed, which is lower than the velocity of the fluid, is used in the present work.

In the current study, the observed extended periods for stabilization of core permeability are explained by slow movement of mobilized fine particles along the rock surface. Therefore, water flow velocity U in basic equations (12.1), (12.2) is substituted by the particle velocity $U_s < U$. Further improvement of the system of equations (12.1), (12.2) is the introduction of the MRF for a monolayer of size-distributed fines. This explains the nonconvex form of the MRF. The exact solution to the improved system of basic equations (12.1), (12.2) for one-dimensional (1D) flow with piecewise constant velocity increase is derived. The analytical model matches with the experimental data on pressure drop along the core length during coreflood. It is shown that the velocity of mobilized fines is significantly lower than the velocity of the carrier water, i.e., $U_s \ll U$. The proposed model accounting for a slow motion of mobilized fine particles along the rock surface is validated by a good agreement between the laboratory data and modeling results.

The historical production data for the Salamander geothermal well (Otway Basin, Australia) were used to test laboratory-based analytical models accounting for the MRF dependency on temperature. Well impedance growth, calculated according to the model, closely agrees with that calculated from field historical data with the coefficient of determination $R^2 = 0.99$. According to the model, geothermal wells are more susceptible to fines migration than conventional oil and artesian wells.

The structure of this chapter is as follows. Section 12.2 presents the physics of particle mobilization and straining in porous media. Section 12.3 reports laboratory investigations of fines migration in geothermal fields, including materials, experimental setup, procedures, and electrostatic particle—rock interaction analysis. Section 12.4 presents the system of basic equations for 1D suspension—colloidal transport accompanied by particle detachment, migration, and capture together with analytical model for injection of fluid with piecewise constantly decreasing ionic strength. The coefficients of this model are tuned using the laboratory coreflood data as shown in Section 12.5. Section 12.6 provides the governing equations and analytical model for axisymmetric flow toward a wellbore, accounting for fines detachment, migration, capture in rocks, and resultant permeability damage. It is followed by the application of the analytical model to the treatment of Salamander geothermal field data in Section 12.7. The main conclusions drawn in Section 12.8 finalize the chapter.

12.2 PHYSICS OF PARTICLE MOBILIZATION AND STRAINING IN POROUS MEDIA

The mobilization of fine particles is determined by the acting forces, which are presented in Fig. 12.1. The primary detaching forces are the drag force, F_d, and the lifting force, F_L. Formulae for these forces can be found in the works by Altmann and Ripperger (1997) and Bradford et al. (2013). From these expressions, it follows that the primary dependencies of these two forces are on the velocity, U, the particle radius, r_s, the ionic strength, γ, and the temperature, T. The ranges of these variables relevant to this study are fluid velocities between 10^{-5} and 10^{-3} m/s, particle radii between 1 and 5 μm, ionic strength values between 0.025 and 0.20 M for NaCl solutions, and temperatures between 25°C and 130°C. For these conditions, the ratio F_L/F_d does not exceed 0.0004. As such, the lifting force, F_L, is negligible and can be removed from further calculations. The gravitational force, which also acts on attached particles, can similarly be discarded as the ratio F_g/F_d lies below 0.02. Not shown in Fig. 12.1 is the Brownian diffusive force which arises from collisions between particles and water molecules. This force is modeled as Gaussian noise (Kim and Zydney, 2004). For the conditions described earlier, the ratio between the Brownian force and the drag force is less than 0.05, which means that the Brownian force can be neglected. The primary attaching force is the electrostatic force, $F_{\bar{e}}$. Equations for this force are given by the DLVO theory, expressions of which can be found in the work by Yang et al. (2016).

The torque balance conditions for particle mobilization thus consists of only two forces, the attaching electrostatic force, F_e, and the detaching drag force, F_d. The final expression is given in Eq. (12.3), where l is the lever arm ratio. This ratio is determined by the geometry of the attached particle on the rock surface. The parameter l is calculated either from an asperity on the rock surface or from the extent of particle deformation of the rock surface (Freitas and Sharma, 2001; Bradford et al., 2013). Calculations using the latter apply Hertz contact theory to incorporate both particle and rock deformation resulting from the attaching forces (Bradford et al., 2013; Kalantariasl and Bedrikovetsky, 2014). Typical calculations for illite/chlorite particles in sandstone rocks give $l = 0.0021$.

Eq. (12.3) is a transcendental equation with respect to the particle radius, r_s. At any value of the flow velocity, ionic strength, and temperature, this equation enables the calculation of the critical particle size, r_{scr}, which satisfies the condition of torque equilibrium. All particles larger than this size have detaching torques exceeding the attaching torques, and thus detach from the rock surface. The critical particle radius can be expressed as the function $r_{scr} = r_s(U, \gamma, T)$, which monotonically decreases with U and monotonically increases with γ. It follows that during particle detachment induced by increasing the velocity, or by decreasing the salinity, the attached particles detach in the order of decreasing size. Variation of the attached concentration with alteration to the injection conditions is expressed by the MRF function, σ_{cr}, given in Eq. (12.4). Taking the derivative of this expression with respect to the particle size yields the size distribution of the attached particles, $\Sigma_a(r_s)$. This facilitates a laboratory method for determining the size distribution of attached fines on rock surface.

Fig. 12.3 presents typical forms of the MRF as functions of the salinity and velocity, as calculated from Eq. (12.4). The initial attached concentration σ_{ai} is shown by the point I. Particle detachment is induced when the attached concentration lies above the maximum retention concentration at given conditions. At 25°C, an increase in velocity from zero to U_B (or equivalently, a reduction of ionic strength from γ_1 to γ_2) will result in no particle detachment, as the value of the MRF at point B is exactly the initial attached concentration. A further increase in velocity to U_C (or decrease in ionic strength to γ_3) will result in a decrease in the attached concentration by $\Delta\sigma$.

The MRF function also varies considerably with changes in temperature. This dependency is determined by the dependencies of the water viscosity and the electrostatic constants on the temperature. These parameters influence the drag and electrostatic forces. Fig. 12.3 demonstrates the resulting effect on the MRF. The figure shows that higher temperatures correspond to a lower value of the MRF, and hence a greater number of detached fines, which leads to a severer permeability decline.

12.3 LABORATORY STUDY ON FINES MIGRATION IN GEOTHERMAL FIELDS

This section outlines the preparation of a core for experiments (Section 12.3.1), describes the laboratory setup (Section 12.3.2), and experimental procedure (Section 12.3.3) for a laboratory investigation of fines migration in geothermal fields.

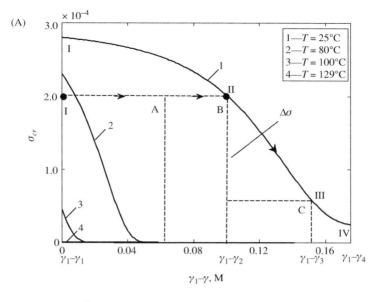

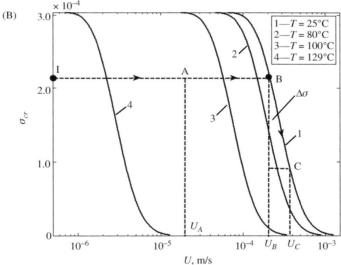

Figure 12.3 Dependency of the maximum concentration of attached particles σ_{cr} on (A) fluid ionic strength (γ) ($\gamma_1 = 0.2$ M NaCl) at various temperature values T; (B) flow velocity (U) at various temperature values T.

12.3.1 Materials

The goal of the laboratory study is the prediction of permeability damage due to fines migration in the geothermal well Salamander-1 (Otway Basin, South Australia, Australia). The absence of cores inhibits the analysis of the Salamander-1 well performance. However, cores are available from the nearby Ladbroke Grove-1 well from the Pretty Hill Sandstone Formation (Otway Basin, South Australia, Australia), with identified analogous mineral and rock characteristics.

Thin section, SEM-EDX (scanning electron microscope coupled with the thin film energy dispersive X-ray analysis) and XRD (X-ray powder diffraction) analyses provide the following mineral components for the Salamander-1 rock sample, which consist of quartz, albite, phengite, kaolinite, calcite, clinochlore, microcline, biotite, illite/muscovite, muscovite, and siderite; and for the Ladbroke Grove-1, which consist of quartz, albite, kaolinite, biotite, siderite, k–feldspar, illite/chlorite, illite/muscovite, calcite, dolomite, and muscovite.

Permeabilities and porosities for Salamander-1 and Ladbroke Grove-1 rock samples are as follows: 4.90 mD and 10.6%, and 5.46 mD and 17.2%, respectively. Pore size distribution has not been determined for Salamander-1 and Ladbroke Grove-1 rock samples. Instead, we evaluated mean pore throat sizes for these rocks. Using the formula $r_p = 0.5\sqrt{k/(4.48\phi^2)}$ proposed by Katz and Thompson (1986), we obtained the following values of mean pore throat size: 3.21 and 4.93 μm for the Salamander-1 and Ladbroke Grove-1 samples, respectively. Densities of the Salamander-1 and Ladbroke Grove-1 rock samples are 2144 and 2163 kg/m^3, respectively.

Similarities in mineral composition and mean pore throat size for these two rock samples allow us to use the rock from Ladbroke Grove-1 well for evaluation of Salamander-1 well behavior. A core plug was drilled in a horizontal direction of a Ladbroke Grove-1 core slab sample coming from the depth of 2557.12 m. The core plug diameter and length are as follows: 3.92 and 6.33 cm, respectively. This core plug was evacuated at 1.5 Pa for 24 hours and saturated with 0.6 M NaCl solution prepared from the degassed ultrapure deionized water (MilliQ water, Merck Millipore, United States).

12.3.2 Experimental setup

The schematic and photograph of the laboratory setup are shown in Fig. 12.4A and B, respectively. A core 1 is placed inside a Viton sleeve 2

(A)

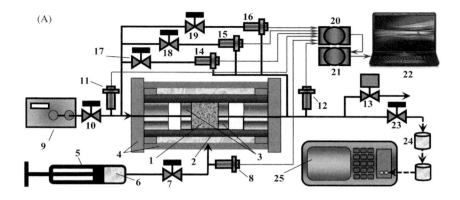

(B)

Figure 12.4 Experimental setup for fines migration studies: (A) schematic and (B) photo. 1—core; 2—Viton sleeve; 3—stainless steel stoppers; 4—high-pressure core holder; 5—manual pressure generator; 6—distilled water; 7, 10, 17, 18, 19, 23—two-way manual valves; 8—11, 12—PA-33X PTs; 9—HPLC pump; 13—back pressure regulator; 14—16—Validyne DPTs; 20—ADAM-4019+ data acquisition module; 21—ADAM-5060 signal converter; 22—personal computer; 24—beakers for effluents; 25—PAMAS particle counter/sizer.

which is fixed by inlet and outlet fluid distributors 3. The core arrangements are placed inside the standard Hassler-type core holder 4 (model TEMCO RCHR-1-1.5″-series, CoreLab, United States). Radial overburden pressures of up to 1000 psi are applied to cores and controlled within ±0.2% of the reading by a manual pressure generator 5 (model 87-6-5, High Pressure Equipment Company, United States). Distilled water 6 is fed into an overburden space via a manual valve 7. A pressure transmitter (PT) 8 (model PA-33X, KELLER AG fur

Druckmesstechnik, Switzerland) measures overburden pressure. The HPLC pump 9 (model Prep-36, Scientific Systems, Inc., Lab Alliance, United States) delivers NaCl solutions through cores and fragments by a manual valve 10. The inlet and outlet pressures are measured by PA-33X PTs 11 and 12. Located at the outlet of the core holder the backpressure regulator 13 is used to control backup pressure and maintain stable fluid flow rates through the core. Pressure difference between inlet and outlet of the core is measured by differential pressure transmitters (DPTs) *14−16* (model Validyne Engineering, United States) having three measuring ranges: $0−1.25$, $0−12.5$, and $0−125$ psi. Manual valves $17−19$ connect these DPTs to a flow-through system. Electrical signals from PTs and DPTs are transmitted to a real-time data acquisition system incorporating the inlet data acquisition module 20 (model ADAM-4019 + , ADVANTECH, Taiwan) and the RS-232/RS-485/RS-422 signal conditioner 21 (model ADAM-5060, ADVANTECH, Taiwan). Custom-built data acquisition software (ADAMView Ver. 4.25 application builder, ADVANTECH, Taiwan) records all experimental parameters (pressure, differential pressure, and time) in real time using a stand-alone computer 22. A manual valve 23 is used for fine backup pressure control. Effluent suspensions are collected in plastic beakers 24. The suspended particle concentrations and particle size distribution are measured by a portable particle counter 25 (model PAMAS S4031 GO, PAMAS GmbH, Germany).

12.3.3 Experimental procedures

The laboratory procedure includes the following steps: (1) injection of water with piecewise constant decreasing ionic strength with measurements of the differential pressure across the core; (2) measurements of particle concentrations in the effluent fluid; (3) collection of the breakthrough particles and their SEM-EDX analyses for identification of clay minerals; and (4) particle zeta potential measurements for various fluid ionic strengths. The procedure determines the conditions favorable for particle mobilization, fines removal capacity, and their effects on formation damage.

Fines mobilization and permeability decline for the core as a function of fluid ionic strength varying from 0.6 to 1.28×10^{-4} M NaCl is studied at Darcy velocity of 1.38×10^{-4} m/s. The ionic strength I is calculated as $I = 0.5 \sum c_i z_i^2$, where c_i is molar concentration of "i"-th dissolved ion

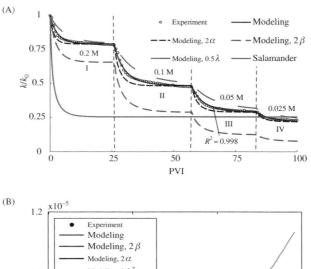

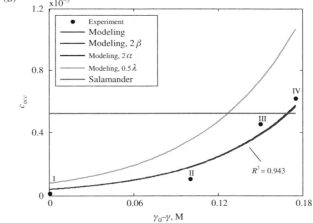

Figure 12.5 Experimental data from coreflood tests with varying fluid ionic strength tuned by the analytical model and well behavior prediction for the Salamander geothermal well: (A) decrease of the core permeability during the test with piecewise decreasing fluid ionic strength and (B) cumulative breakthrough concentration (c_{acc}) at various fluid ionic strengths.

expressed in mol/L (M), and z_i is valence of "i"-th dissolved ion. For monovalent salts such as NaCl, the concentration of the salt in solution is equal to its ionic strength according to the above formula. Rock permeability values are measured (black circles in Fig. 12.5A) and effluent samples are collected (black points in Fig. 12.5B) until stabilization of core permeability is achieved. Matching the above experimental data using the mathematical model for fines migration with particle straining is presented by You et al. (2016).

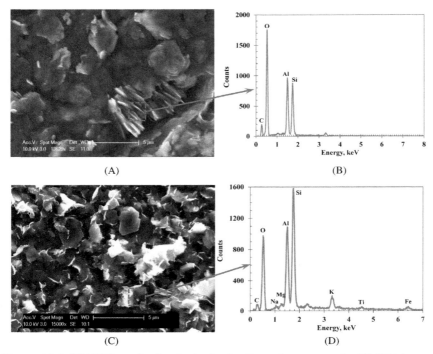

Figure 12.6 SEM-EDX results for the Ladbroke Grove-1 core sample: (A) SEM image for kaolinite; (B) EDX spectra for kaolinite; (C) SEM image for illite/chlorite; and (D) EDX spectra for illite/chlorite.

Concentration and size distribution of the effluent fines are measured by the particle counter PAMAS. Volumes of all the effluent samples are measured and used to calculate the total volume of collected fines as $V_{tot}^{fines} = 1.08 \times 10^{-9}$ m^3. The effluent fluid is filtered through a 0.45-μm nylon filter and dried in the atmospheric oven at 60°C for 12 hours. A Philips XL30 Scanning Electron Microscopes coupled with the thin film EDX is used for imaging of dried fines and X-ray analyses for the identification of minerals (Fig. 12.6).

Rock samples are saturated in vacuum with 0.6 M NaCl solution (pH value of 7.2) for porosity measurements, before the flow-through experiments. After the tests, zeta potentials for collected fines are measured by a Zetasizer Nano Z (model ZEN3600, Malvern Instruments Ltd, United Kingdom) at various ionic strengths of suspensions and pH ≈ 7.2 (Fig. 12.7A). Verification of the Zetasizer was carried out by measuring electrophoretic mobility of zeta potential transfer standard with

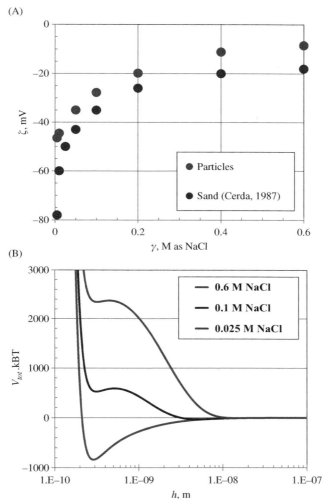

Figure 12.7 Zeta potentials (A) and the DLVO total interaction potential energy (B) between the fine particle and sand grain at 25°C.

-42.0 mV $\pm$ 4.2 mV (Malvern Instruments Ltd, United Kingdom). The measured zeta potential value of -38.9 mV agrees with the standard within the reported uncertainty. The sizes of particles collected from the effluent streams vary between 1.23 and 1.63 μm. Particle concentrations vary in the range $0.8-1.7$ ppm (volumetric concentrations). Conversion of electrophoretic mobilities into zeta potentials was carried out using the Smoluchowski model (Hunter, 1981). Zeta potentials for sand are adopted from Cerda (1987) with $pH \approx 7.0$. The values of zeta potentials are then used for the calculation of electrostatic forces between fines and rock.

12.3.4 Electrostatic particle—rock interaction analysis

This section presents the results of the experimental study described earlier and the calculations of particle—rock interactions. It includes the effects of fluid ionic strength on the fines release capacity and rock permeability (Section 12.3.4.1), and SEM-EDX analyses for the identification of clay minerals in the collected breakthrough fines (Section 12.3.4.2). Total interaction potential energy between fines and sand matrix is calculated in Section 12.3.4.3 by DLVO theory.

12.3.4.1 Effect of fluid ionic strength on rock permeability

Decrease of the core permeability is observed at the interval of fluid ionic strengths from 0.2 to 0.025 M NaCl as shown in Fig. 12.5A (black circles). Sequential corefloods are performed for four ionic strengths of NaCl solutions in the decreasing order of their ionic strengths. The largest permeability reduction from about 3.18 down to 0.93 mD is observed in the range of ionic strength from 0.2 to 0.05 M NaCl (stages I—III). Further reduction of ionic strength (interval IV) has a lower effect on core permeability.

12.3.4.2 SEM-EDX analyses for released fines

The results of SEM-EDX analyses for the released fines are shown in Fig. 12.6. Kaolinite particles show characteristic plate-like pseudo-hexagonal geometry as shown in the SEM image (Fig. 12.6A). Relative molar proportions of elements were determined by the so-called peak height ratio from EDX spectrum data (see Fig. 12.6B) and used for the identification of kaolinite, $Al_2(Si_2O_5)(OH)_4$. Flakes of clay minerals are visible in the SEM image as shown in Fig. 12.6C. Although their morphology indicates the presence of chlorite $(Mg,Al,Fe)_{12}[(Si,Al)_8O_{20}](OH)_{16}$, their EDX spectra are more indicative of illite $K_{1-1.5}Al_4[Si_{7-6.5}Al_{1-1.5}O_{20}](OH)_4$ (see Fig. 12.6D). However, the moderate iron content is an indication of the formation of chlorite (Vortisch et al., 2003). Therefore, the observed clay particles can be identified as the mixed-layer illite/chlorite mineral. A double-stick electrically conductive carbon tape produces carbon peaks in both EDX spectra in Fig. 12.6B and D.

The SEM analysis alone might not be definitive enough for precise determination of fines mineralogy, and XRD analysis would be required. However, low quantities of collected fines did not allow for XRD analyses in this study.

12.3.4.3 DLVO interaction between the particles and pore matrix

Interaction between particles and porous matrix (sand) is quantitatively described by the DLVO theory (Derjaguin and Landau, 1941; Verwey and Overbeek, 1948). The total DLVO interaction energy potential is the sum of the interaction potential energies arising from the attractive long-range ($10 \, nm < h < 100 \, nm$, see Israelachvili (2011)) LW forces, the short-range ($0.2 \, nm < h < 10 \, nm$, see Capco and Chen (2014)) attractive/repulsive EDL and Born repulsion forces. The positive sign of the DLVO total interaction potential energy is an indication of the domination of the repulsive EDL and Born forces over the attractive LW forces resulting in particle—sand repulsion and mobilization of fines (Sen and Khilar, 2006). Attraction of particles to sand occurs if the attractive LW forces dominate over the EDL and Born forces, and the sign for the DLVO total interaction potential energy changes to negative. The mobilization of fines starts below the so-called critical salt concentration which corresponds to the case where both the total interaction potential energy and the resultant force acting on particles are equal to zero (Khilar and Fogler, 1998). For particle—sand attraction, particles with sufficient energy can overcome the interaction potential energy barrier, approach the surface of a porous medium at the separation distance of a few nanometers, and be irreversibly attached to the surface in the primary energy minimum. The presence of a secondary energy minimum at a larger separation distance can lead to reversible particle capture provided the particles have insufficient energy to escape (Kuznar and Elimelech, 2007).

The measured values for zeta potentials of collected fines are shown in Fig. 12.7A; they increase with fluid ionic strength. The values of zeta potentials for collected fines and sand are used for the calculation of the DLVO total interaction potential energy between fines and sand (Elimelech et al., 1998). Variation of the DLVO total interaction potential energy with separation distance between fines and sand at various ionic strengths is shown in Fig. 12.7B. The presence of a negative primary energy minimum of $-838 \, k_B T$ at separation distance $h = 0.29 \, nm$, the absence of potential barriers, and a secondary energy minimum imply the attraction between the clay particles and sand at $0.6 \, M$ NaCl ionic

Table 12.1 Temperature effects on the parameters in the DLVO interaction energy model

Parameter	Temperature effect	References
λ	N/A	Gregory (1981)
ε_1	Table 12.2	Leluk et al. (2010)
ε_2	Negligible if $T < 170°C$ (Fig. 1 in reference)	Stuart (1955)
ε_3	Table 12.2	Marshall (2008)
n_1	N/A	Egan and Hilgeman (1979)
n_2	Interpolation from Fig. 1 in reference	Leviton and Frey (2006)
n_3	Eq. (8) in reference	Aly and Esmail (1993)
ζ_s	Eq. (9) in reference	Schembre and Kovscek (2005)
ζ_{pm}	Eq. (9) in reference	Schembre and Kovscek (2005)
σ_c	N/A	Elimelech et al. (1998)

strength. Reduction of fluid ionic strength from 0.1 M NaCl down to 0.025 M significantly changes the particle—sand interaction: it is not only that the sign of primary energy minima changes from negative to positive (indicating particle—sand repulsion), but also the values of the primary minima increase from 520 to 2343 k_BT at about $h = 0.31$ nm separation distance, and the energy barriers rise from 585 to 2365 k_BT at a separation distance of $h = 0.51$ nm. Overall, the reduction of fluid ionic strength has a significant effect on particle—sand interaction, and, consequently on fines mobilization and formation damage.

The temperature effects on all the parameters in DLVO interaction potential energy expressions are summarized in Table 12.1. The data on temperature dependencies of the characteristic wavelength of interaction λ, refractive index of clay n_1, and collision diameter σ_c are unavailable in the literature. In the present work, sensitivity analysis of these parameters on the particle—rock interaction energy is performed (Fig. 12.8). The results show that increase of λ (Fig. 12.8A) and n_1 (Fig. 12.8B) enhances the particle—rock attraction, while larger σ_c weakens the attractive potential (Fig. 12.8C). Following Schembre and Kovscek (2005), Schembre et al. (2006a, b), and Lagasca and Kovscek (2014), we assume the above-mentioned parameters to be constant. The values of these constants are taken from works by Gregory (1981), Egan and Hilgeman (1979), and Elimelech et al. (1998).

Attractive LW potential and repulsive Born potential are functions of Hamaker constant A_{132}, which depends on the dielectric constants of clay ε_1, quartz ε_2, and brine ε_3 (Israelachvili, 2011). These dielectric

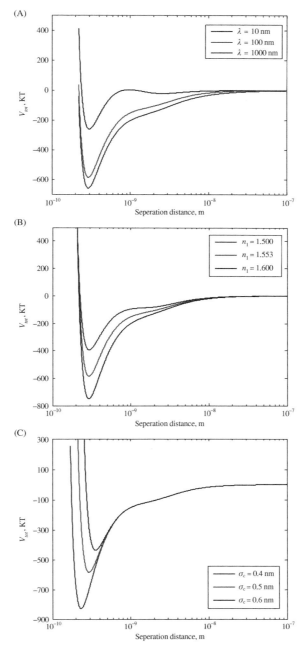

Figure 12.8 Sensitivity analysis of the DLVO parameters on the particle—rock interaction energy for $r_s = 1\ \mu m$, $\gamma = 0.2\ M$: (A) characteristic wavelength of interaction λ; (B) refractive index of clay n_1; and (C) collision diameter σ_c.

Table 12.2 Dielectric constants at different temperatures

T, °C	ε_1	ε_2	ε_3
25	6.66	3.80	72.75
80	6.26	3.80	55.74
100	6.12	3.80	50.64
129	5.90	3.80	44.11

parameters are temperature dependent. Gradients of ε_1, ε_2, and ε_3 with respect to temperature are $-0.007°C$ (Leluk et al., 2010), $0.00°C$ (Stuart, 1955), and $-0.32°C$ (for 0.6 M NaCl solution) (Buchner et al., 1999; Marshall, 2008), respectively. Values of dielectric constants at different temperatures are presented in Table 12.2. The calculated Hamaker constant shows that temperature rise from 25°C to 129°C increases A_{132} by 23% for different ionic strengths (Table 12.3).

Electrostatic double layer potential is a function of the following parameters: Debye–Hückel parameter κ, and zeta potentials ζ_s and ζ_{pm}, all of which are temperature dependent. Inverse Debye length κ^{-1} characterizes the thickness of the double layer (Ohshima, 2006). Increase of fluid temperature from 25°C to 129°C decreases κ^{-1} by 10.5% in 0.6 M NaCl solution, from 0.38 to 0.34 nm (Table 12.3). The reduced double layer thickness leads to weakened EDL repulsion between particles and porous matrix. Zeta potentials for kaolinite and quartz decrease with temperature (Ramachandran and Somasundaran, 1986; Schembre and Kovscek, 2005; Rodríguez and Araujo, 2006). Applying measured zeta potentials at ambient conditions (Cerda, 1987) to the formula by Schembre and Kovscek (2005) provides the zeta potentials at different temperatures (Table 12.3).

Although the particle size is present in DLVO potentials, thermal expansion of kaolinite particles and quartz can be neglected, due to negligible increase of their linear dimensions ($\sim 0.06\%$) with a temperature rise from 25°C to 130°C (McKinstry, 1965).

12.4 ANALYTICAL MODELING OF FINES MIGRATION IN LABORATORY COREFLOODING

In the current section, first we provide the assumptions used for deriving the analytical model (Section 12.4.1). Then the governing equations for suspension flow with slow drift of detached particles and

Table 12.3 Variation of the DLVO parameters with temperature at different ionic strengths

Ionic strength, M NaCl	$t, °C$	25	80	100	129
0.6	κ^{-1}, nm	0.38	0.36	0.35	0.34
	A_{132}, J	1.100E − 20	1.214E − 20	1.266E − 20	1.356E − 20
	ζ_s, mV	− 10.30	− 20.1	− 23.6	− 28.7
	ζ_{pm}, mV	− 18.0	− 35.0	− 41.1	− 50.1
0.2	κ^{-1}, nm	0.67	0.64	0.63	0.61
	A_{132}, J	1.104E − 20	1.219E − 20	1.272E − 20	1.363E − 20
	ζ_s, mV	− 14.5	− 28.1	− 33.1	− 40.3
	ζ_{pm}, mV	− 26.2	− 50.8	− 59.7	− 72.7
0.07	κ^{-1}, nm	1.15	1.09	1.07	1.04
	A_{132}, J	1.105E − 20	1.221E − 20	1.274E − 20	1.365E − 20
	ζ_s, mV	− 28.50	− 55.40	− 65.10	− 79.30
	ζ_{pm}, mV	− 39.10	− 75.90	− 89.20	− 108.6
0.025	κ^{-1}, nm	1.93	1.84	1.80	1.75
	A_{132}, J	1.105E − 20	1.222E − 20	1.275E − 20	1.365E − 20
	ζ_s, mV	− 41.90	− 81.30	− 95.70	− 116.5
	ζ_{pm}, mV	− 50.20	− 97.50	− 114.7	− 139.6

consequent straining downstream are presented in Section 12.4.2. The 1D analytical solution of the governing system is derived in Section 12.4.3.

12.4.1 Model assumptions

The following list provides major assumptions applied in deriving the theoretical model for suspension flow with slow drift of fines and their straining:

1. Fluid is incompressible, and particles are nondeformable.
2. Rock porosity variation is negligible due to low concentrations of attached and strained fines.
3. Particle drift velocity differs from that of the carrier fluid.
4. Particle dispersion is negligible.
5. Particle detachment is instant according to the maximum retention condition, Eqs. (12.3), (12.4).
6. Particle straining obeys linear kinetics, i.e., filtration and formation damage coefficients are constant.
7. There are no strained particles initially.
8. Once strained, particles cannot be detached.
9. Increase of permeability due to fines detachment is negligible.

Applying these assumptions, we establish the system of governing equations for fines migration in the next section.

12.4.2 System of governing equations

The equation of population balance is valid for suspended, strained, and attached particles:

$$\frac{\partial(\phi c + \sigma_s + \sigma_a)}{\partial t} + \alpha U \frac{\partial c}{\partial x} = 0 \qquad (12.5)$$

where σ_a stands for attached particle concentration and σ_s refers to strained particle. The drift delay factor α indicates the slow motion of migrated particles.

Following Herzig et al. (1970), the rate of particle capture is proportional to the flux of suspended particles carried by water:

$$\frac{\partial \sigma_s}{\partial t} = \lambda \alpha U c \qquad (12.6)$$

We define this proportionality coefficient λ as the filtration coefficient. It indicates the probability of particle retention per unit length.

Permeability reduction following particle straining is evaluated from the Darcy equation as:

$$U = -\frac{k(\sigma_s)}{\mu}\frac{\partial p}{\partial x}, \quad k(\sigma_s) = \frac{k_0}{1 + \beta\sigma_s} \tag{12.7}$$

where the permeability k is dependent on the strained concentration of particles, k_0 is the initial permeability, β is the formation damage coefficient, μ is the viscosity of suspension, and p is the pressure.

The initial and boundary conditions posed for the scenario of water injection into a clean core are:

$$c(x,0) = 0, \quad c(0,t) = 0,$$
$$\sigma_a(x,0) = \sigma_{a0}, \quad \sigma_s(x,0) = 0 \tag{12.8}$$

In Fig. 12.9, the route $I \rightarrow A \rightarrow B \rightarrow C$ shown by black arrows indicates the transition which happens when water with ionic strength γ_3 is injected into the core with water of γ_1. Fig. 12.9 clearly shows the dependence of the attached concentration on ionic strength. The dot I $(0, \sigma_{a0})$ indicates the initial concentration of attached fines. The section $I \rightarrow A \rightarrow B$ illustrates the water injection process with less ionic strength but without fines detachment. The dot B corresponds to the start point of fines release, where the ionic strength reaches γ_2. Further reduction of the ionic strength from γ_2 to γ_3 results in the total number of $\Delta\sigma$ of fines detachment. The released fines then immediately become suspended in the carrying water with a concentration of $\Delta\sigma/\phi$.

Consider the process of water injection with piecewise constant ionic strength values $\gamma_i (i = 1, 2, \ldots)$. Suspended concentration resulting from fines detachment with ionic strength alteration from γ_{i-1} (stage $i - 1$) to γ_i (stage i) is evaluated from the reduction of the attached concentration:

$$\Delta c(x, t_i) = \phi^{-1}\{\sigma_{cr}[\gamma(x, t_{i-1})] - \sigma_{cr}[\gamma(x, t_i)]\} = \phi^{-1}\Delta\sigma_a(\gamma_{i-1}, \gamma_i) \tag{12.9}$$

where t_i in Eq. (12.9) corresponds to the moment at which ionic strength changes from γ_{i-1} to γ_i.

In Fig. 12.9A, the number of detached fines $\Delta\sigma$ represents the difference between the values of the MRF corresponding to ionic strengths from γ_{i-1} to γ_i.

Fines with distributed sizes are detached in different stages, according to the variation of water ionic strength. Consequently, the drift delay factor $\alpha_i (i = 1, 2, \ldots)$ varies from stage to stage. The same rule applies to the filtration and formation damage coefficients.

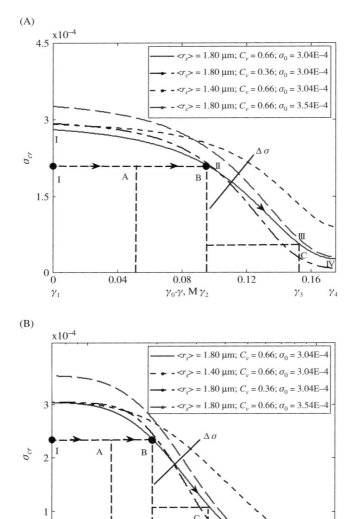

Figure 12.9 Sensitivity analysis of mean particle radius, variance coefficient, and maximum concentration value on the critical retention concentration, as a function of (A) fluid ionic strength and (B) flow velocity.

The dimensionless variables are defined and applied to the governing equations (12.3), (12.5)−(12.8):

$$S_a = \frac{\sigma_a}{\phi}, S_s = \frac{\sigma_s}{\phi}, \lambda_D = \lambda L,$$

$$t_D = \frac{1}{\phi L}\int_0^t U(\gamma)d\gamma, x_D = \frac{x}{L}, \alpha = \frac{U_s}{U} \quad (12.10)$$

where L is the core length and t_D is the injected water volume, noted as PVI.

The dimensionless system of governing equations for the suspended, strained, and attached particle concentrations c_i, S_{si}, and S_{ai} during stage i are derived by substituting the expressions (12.10) into Eqs. (12.3), (12.5)−(12.8):

$$\frac{\partial(c_i + S_{si})}{\partial t_D} + \alpha_i \frac{\partial c_i}{\partial x_D} = 0 \quad (12.11)$$

$$\frac{\partial S_{si}}{\partial t_D} = \alpha_i \lambda_{Di} c_i \quad (12.12)$$

$$S_{ai}(x_D, t_D) = S_{cr}(\gamma_i) \quad (12.13)$$

$$c_i(x_D, t_{Di}) = c_{i-1}(x_D, t_{Di}) + \Delta S_a(\gamma_{i-1}, \gamma_i),$$
$$c_i(0, t_D) = 0, S_{si}(x_D, t_{Di}) = S_{s,i-1}(x_D, t_{Di}) \quad (12.14)$$

The initial condition for c_i (the first equation in Eq. 12.14) implies that the number of particles, detached at t_{Di} when water salinity changes from γ_{i-1} to γ_i, adds to the suspended concentration of fines inherited from the injection of water with salinity γ_{i-1}.

The system of four governing equations (12.7), (12.11)−(12.13) together with the initial and boundary conditions, Eq. (12.14), determines unknowns c_i, S_{si}, and S_{ai}, as well as pressure during stage i. The equations for three concentrations (12.11)−(12.13) decouple from Eq. (12.7) for pressure. Therefore, the solution of pressure can be obtained from Eq. (12.7) separately, after solving the system of equations (12.11)−(12.13).

12.4.3 Analytical solution

Instant particle detachment with the consequent mobilization and straining is a linear hyperbolic problem, which allows for an exact solution. For the case $\alpha = 1$, Bedrikovetsky et al. (2011, 2012) reported explicit formulae for suspended and strained particle concentrations as well as pressure drop along the core. For $\alpha < 1$, the analytical solution is briefly presented here.

The suspended particle concentration can be derived from Eqs. (12.11), (12.12) by applying the method of characteristics (Courant and Hilbert, 1962; Polyanin et al., 2002):

$$c_i(x_D, t_D) = \begin{cases} 0, & x_D \leq \alpha_i(t_D - t_{Di}) \\ \left[c_{i-1}(x_D, t_{Di}) + \Delta S_a(\gamma_{i-1}, \gamma_i) \right] e^{-\alpha_i \lambda_{Di}(t_D - t_{Di})}, & x_D > \alpha_i(t_D - t_{Di}) \end{cases}$$

$$(12.15)$$

where $\alpha_i = dx_D/dt_D$ is the speed of front propagation (Fig. 12.10A). Eq. (12.15) is substituted into Eq. (12.12), which is then integrated with respect to time and results in the solution of strained fines concentration (the second line in Table 12.4 provides the strained fines concentration at the first stage S_{s1}). The suspended fines concentration is zero behind the front. The moment $1/\alpha_1$ corresponds to the breakthrough of the "last" detached particle. Fig. 12.10B demonstrates the suspended concentration profiles at three moments: $t_D = 0$, t_{Da} (before the concentration front breakthrough), and t_{Db} (after the concentration front breakthrough). The initial suspended concentration is ΔS_{a1}. Before the front breakthrough at the outlet, the suspended concentration is zero behind the front and constant ahead of the front. The constant decreases with time (Eq. 12.15). The suspended concentration drops to zero after the concentration front breakthrough, because all the suspended particles are either retained in the core or carried through the outlet by water. The strained particle concentration profile at three moments 0, t_{Da}, and t_{Db} are shown in Fig. 12.10C. The strained particle concentration increases with time until the front arrival and is invariant subsequently. The larger is the distance x_D, the longer is the stabilization time and the greater is the maximum value of the accumulated strained concentration; therefore, the strained particle concentration rises with x_D. This profile is constant ahead of the front, because the advective flux of suspension and particle retention probability are constant.

The suspended particle concentration during stage i is the total concentration of all particles inherited from the previous stage. Applying Eq. (12.15) to each stage and taking the sum yields:

$$c_{i,j}(x_D, t_D) = \begin{cases} 0, & j = 1 \\ \displaystyle\sum_{j=2}^{i+1} \left(\Delta S_a(\gamma_{i-j+1}, \gamma_{i-j+2}) \prod_{n=i-j+2}^{i} Q_n \right), & j > 1 \end{cases} \quad (12.16)$$

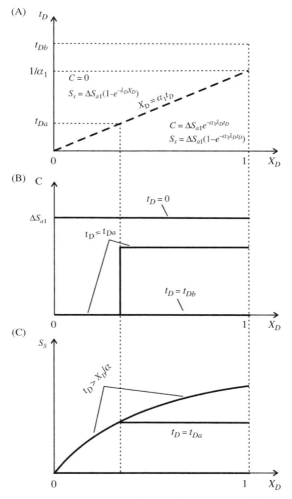

Figure 12.10 Solution of the flow problem with migration of fine particles at elevated velocity: (A) trajectory of concentration front in (x_D, t_D) plane; (B) profiles of concentrations of suspended particles at three moments; and (C) profiles of concentrations of retained particles at three moments.

where the function Q_n is defined as:

$$Q_n(t_D) = e^{-\alpha_n \lambda_{Dn}(t_D - t_{Dn})} \qquad (12.17)$$

According to the population balance equation (12.11), the total concentration of strained particles equals the initial suspended and strained

Table 12.4 Analytical solutions for fines concentrations and permeability evolution

Line	Term	Notation	Expression
1	Suspended fines concentration during stage 1	$c_1(x_D, t_D)$	$\begin{cases} 0, & x_D \le \alpha_1(t_D - t_{D1}) \\ \Delta S_a(\gamma_0, \gamma_1)\, e^{-\alpha_1 \lambda_{D1}(t_D - t_{D1})}, & x_D > \alpha_1(t_D - t_{D1}) \end{cases}$
2	Strained fines concentration during stage 1	$S_{s1}(x_D, t_D)$	$\begin{cases} \Delta S_a(\gamma_0, \gamma_1)\left(1 - e^{-\lambda_{D1} x_D}\right), & x_D \le \alpha_1(t_D - t_{D1}) \\ \Delta S_a(\gamma_0, \gamma_1)\left(1 - e^{-\alpha_1 \lambda_{D1}(t_D - t_{D1})}\right), & x_D > \alpha_1(t_D - t_{D1}) \end{cases}$
3	Permeability during stage 1	$k_1(t_D)$	$\dfrac{k_0}{1 + \beta_1 \phi \int_0^1 S_{s1}(x_D, t_D)\, dx_D}$
4	Total suspended fines concentration during stage $i-1$	$\int_0^1 c_{i-1}(x_D, t_{Di})\, dx_D$	$\displaystyle\sum_{j=2}^{i} \left[(x_{D(i-1)j} - x_{D(i-1)j-1}) \Delta S_a(\gamma_{i-j}, \gamma_{i-j+1}) \prod_{n=i-j+1}^{i-1} Q_n \right]$
5	Cumulative effluent fines concentration during stage i	$\int_{t_{Di}}^{t_D} \alpha_i c_i(1, t_D)\, dt_D$	$\dfrac{1}{\lambda_{Di}} \left\{ 1 - \exp\left[-\alpha_i \lambda_{Di}(t_D - t_{Di}) \right] \right\} \times$ $\left\{ \Delta S_a(\gamma_{i-1}, \gamma_i) + \displaystyle\sum_{j=3}^{i+1} \left[\Delta S_a(\gamma_{i-j+1}, \gamma_{i-j+2}) \prod_{n=i-j+2}^{i-1} Q_n \right] \right\}$
6	Total suspended fines concentration during stage i	$\int_0^1 c_i(x_D, t_D)\, dx_D$	$\displaystyle\sum_{j=2}^{i+1} \left[(x_{Dij} - x_{Dij-1}) \Delta S_a(\gamma_{i-j+1}, \gamma_{i-j+2}) \prod_{n=i-j+2}^{i} Q_n \right]$
7	Total strained fines concentration during stage i	$\int_0^1 \left[S_{si}(x_D, t_D) - S_{s,i-1}(x_D, t_{Di}) \right] dx_D$	$\Delta S_a(\gamma_{i-1}, \gamma_i) + \int_0^1 c_{i-1}(x_D, t_{Di})\, dx_D - \left[\int_{t_{Di}}^{t_D} \alpha_i c_i(1, t_D)\, dt_D + \int_0^1 c_i(x_D, t_D)\, dx_D \right]$
8	Permeability during stage i	$k_i(t_D)$	$k_0 \left\{ \displaystyle\prod_{n=1}^{i} \left[1 + \beta_n \phi \int_0^1 \left(S_{sn}(x_D, t_D) - S_{s,n-1}(x_D, t_{Dn}) \right) dx_D \right] \right\}^{-1}$

concentrations at the beginning of stage i minus the concentration of breakthrough particles and suspended particles. Thus, the total concentration of strained particles can be expressed as:

$$\int_0^1 S_{si}(x_D,t_D)dx_D = \underbrace{\Delta S_a\left(\gamma_{i-1},\gamma_i\right) + \int_0^1 c_{i-1}(x_D,t_{Di})dx_D + \int_0^1 S_{s,i-1}(x_D,t_{Di})dx_D}_{\text{initial particles}}$$

$$- \left[\underbrace{\int_{t_{Di}}^{t_D} \alpha_i c_i(1,t_D)dt_D}_{\text{breakthrough particles}} + \underbrace{\int_0^1 c_i(x_D,t_D)dx_D}_{\text{suspended particles}}\right]$$

(12.18)

where

$$\int_0^1 c_{i-1}(x_D,t_{Di})dx_D = \sum_{j=2}^{i}\left[\left(x_{D(i-1),j} - x_{D(i-1),j-1}\right)\Delta S_a(\gamma_{i-j},\gamma_{i-j+1}) \prod_{n=i-j+1}^{i-1} Q_n\right]$$

(12.19)

stands for the suspended particle concentration along the core at the end of stage $i-1$. This term combined with the mobilized particle concentration at the end of stage $i-1$ determines the initial suspended particle concentration for stage i. The cumulative particle breakthrough concentration during stage i and the total suspended particle concentration along the core during stage i are obtained, if Eq. (12.16) is substituted into the last two terms of Eq. (12.18):

$$\int_{t_{Di}}^{t_D} \alpha_i c_i(1,t_D)dt_D = \frac{1}{\lambda_{Di}}\left\{1 - \exp\left[-\alpha_i\lambda_{D_i}(t_D - t_{Di})\right]\right\}$$

$$\left\{\Delta S_a(\gamma_{i-1},\gamma_i) + \sum_{j=3}^{i+1}\left[\Delta S_a(\gamma_{i-j+1},\gamma_{i-j+2}) \prod_{n=i-j+2}^{i-1} Q_n\right]\right\}$$

(12.20)

$$\int_0^1 c_i(x_D,t_D)dx_D = \sum_{j=2}^{i+1}\left[\left(x_{Di,j} - x_{Di,j-1}\right)\Delta S_a(\gamma_{i-j+1},\gamma_{i-j+2}) \prod_{n=i-j+2}^{i} Q_n\right]$$

(12.21)

The permeability evolution during stage i is obtained from the second equation of Eq. (12.7), considering permeability decline history during each stage:

$$k_i(t_D) = k_{i-1}\left\{1+\beta_i\phi\int_0^1\left[S_{si}(x_D,t_D)-S_{s,i-1}(x_D,t_{Di})\right]dx_D\right\}^{-1}$$
$$= k_0\left\{\prod_{n=1}^i\left[1+\beta_n\phi\int_0^1\left(S_{sn}(x_D,t_D)-S_{s,n-1}(x_D,t_{Dn})\right)dx_D\right]\right\}^{-1}$$

(12.22)

where $\int_0^1\left[S_{si}(x_D,t_D)-S_{s,i-1}(x_D,t_{Di})\right]dx_D$ corresponds to the total strained particle concentration along the core during stage i, which can be obtained from Eq. (12.18).

Table 12.4 provides the list of analytical solutions for fines concentrations and permeability evolution. The suspended particle concentration during stage 1 is c_1, which can be obtained from Eq. (12.15) by letting $i=1$ (first line, Table 12.4). Substituting c_1 into Eq. (12.12) and integrating with respect to t_D lead to strained particle concentration S_{s1} during stage 1 (second line, Table 12.4). Permeability during stage 1, k_1, can be obtained from Eq. (12.22) (third line, Table 12.4). Integration of the suspended particle concentration over the core length $\int_0^1 c_{i-1}(x_D,t_{Di})dx_D$ at the end of stage $i-1$ yields the total inherited suspended concentration from stage $i-1$ (fourth line, Table 12.4). Accumulated breakthrough particle concentration during stage i is calculated from Eq. (12.20) as $\int_{t_{Di}}^{t_D}\alpha_ic_i(1,t_D)dt_D$ (fifth line, Table 12.4). The total suspended particle concentration during stage i is obtained from Eq. (12.21) as $\int_0^1 c_i(x_D,t_D)dx_D$ (sixth line, Table 12.4). The seventh line in Table 12.4 presents the total strained concentration during stage i, $\int_0^1\left[S_{si}(x_D,t_D)-S_{s,i-1}(x_D,t_{Di})\right]dx_D$, resulting from Eq. (12.18). Permeability k_i from Eq. (12.22) during stage i is provided in eighth line of Table 12.4.

12.5 HISTORY MATCHING OF THE LABORATORY COREFLOOD TEST RESULTS

The analytical model presented in Section 12.4.3 is used for history matching of experimental data from the laboratory coreflood test with piecewise decreasing ionic strength. Availability of only sandstone frag-ments from the Salamander-1 geothermal well (Australia) inhibits the study of formation damage in this well due to fines migration. Therefore, a sandstone core from Ladbroke Grove-1 well (Australia) located in the same geological formation where rocks have mineral composition similar to those from Salamander well (Badalyan et al., 2013, 2014) is chosen for experimental study. The main parameters of the core sample are as fol-lows: porosity 17.2%, permeability 5.46 mD, and length 6.33 cm. The experimental procedure is as follows:

- Initially, the core is placed in the desiccator, which is evacuated.
- After the residual pressure 1.5 Pa is reached, 0.6 M NaCl solution is introduced into the desiccator, and the core is saturated.
- The saturated core is transferred to a Viton sleeve, and this arrange-ment is placed inside the core holder operating at pressures of up to 5000 psi.
- A manual pressure generator developed overburden pressure 1000 psi by compressing distilled water in the volume surrounding the Viton sleeve.
- Coreflood tests start from pumping 0.6 M NaCl solution through the core with velocity of 1.4×10^{-4} m/s.
- Effluents were collected in sampling tubes with predetermined time intervals until rock permeability is stabilized.
- The particle counter PAMAS measures concentrations and size distri-butions of particles in the collected effluent samples.
- After rock permeability stabilization, another solution with lower ionic strength 0.4 M NaCl is injected into the core, and the above procedure is repeated.
- The following fluid ionic strengths were used in the present study: 0.2, 0.1, 0.05, and 0.025 M NaCl.

Fig. 12.5 shows the obtained experimental coreflood data. Permeability data obtained during corefloods with 0.6 and 0.4 M NaCl solutions are not shown in Fig. 12.5A, since no change in core perme-ability was observed at these fluid ionic strengths. This core permeability was adopted as initial/undamaged permeability k_0. The model was tuned using experimental data of core permeability (black circles in Fig. 12.5A) and effluent fines concentrations (black points in Fig. 12.5B). During model tuning, we assumed that the particle size distribution has a lognor-mal form. The MRF for a monolayer of size-distributed particles is calcu-lated according to the method developed by Yang et al. (2016). Using a torque balance equation the minimum radius of a particle, r_{scr} mobilized by the flowing fluid with ionic strength γ, is calculated. Therefore, particles which remain immobilized on the surface of a rock should have radii smaller than r_{scr}. One arrives at the conclusion that the critical retention concentration is the total concentration of fines with radii smaller than $r_{scr}(\gamma)$ originally attached to the rock surface.

There are five model parameters to be tuned during data treatment: the average particle radius $\langle r_s \rangle$, the variance coefficient of size distribution of particles C_v, the particle drift delay factor α, the filtration coefficient λ, and the formation damage coefficient β. Smaller injected ionic strength leads to finer particles mobilized. During each stage, the size of detached particles is assumed to be constant. Hence, the parameters α, λ, and β are all constant during each stage. The Levenberg–Marquardt minimization scheme has been applied to optimizing the model parameters (Marquardt, 1963; Chalk et al., 2012; You et al., 2013). Table 12.5 presents the results of tuning parameters. The dark gray curves (only Modeling) in Fig. 12.5A and B show the permeability and effluent concentration curves from model optimization, respectively. It is observed that modeling results match well with experimental data. The coefficient of determination is $R^2 = 0.998$ for permeability and $R^2 = 0.943$ for particle effluent concentration. The good agreement validates the analytical model developed.

Table 12.5 Model parameters tuned from the coreflood test data

Parameter	Value
r_s, µm	1.80
C_v	0.66
σ_0	3.04E − 4
α_1	4.10E − 3
α_2	2.96E − 3
α_3	2.81E − 3
α_4	2.74E − 3
β_1	9793
β_2	7631
β_3	7391
β_4	7158
λ_{D1}	67.14
λ_{D2}	53.79
λ_{D3}	51.11
λ_{D4}	50.13

A parameter sensitivity study is applied to the three main parameters in the model: the drift delay factor α, filtration coefficient λ, and formation damage coefficient β. The larger the value of α, the shorter the stabilization time, because of the higher velocity of particles (dashed curve (Modelling, 2α) in Fig. 12.5A). Decrease of λ reduces the particle capture probability, which leads to the extension of stabilization time (dashed curve (Modelling, 0.5λ) in Fig. 12.5A). The larger value of β, the more permeability reduction with time (dashed curve (Modeling, 2β) in Fig. 12.5A). Among the three model parameters, the effluent particle concentration is sensitive to the filtration coefficient λ only, since it determines the capture probability of detached particles (Fig. 12.5B). The lower the value of λ, the greater the effluent particle concentration (curve (Modelling, 0.5λ) in Fig. 12.5B). The effect of α on effluent particle concentration is insignificant, because the concentration of breakthrough fines is much smaller than the detached particle concentration $\Delta\sigma_i$ in this test (curve (Modelling, 2α) in Fig. 12.5B). The effluent particle concentration is not affected by β (curve (Modeling, 2β) in Fig. 12.5B).

The curve (Salamander) in Fig. 12.5A and B is generated based on the temperature of the Salamander geothermal field, while all the other curves are obtained based on the laboratory data.

Table 12.5 provides that the obtained values of λ and β fall within the common ranges of these parameters reported in the literature (Nabzar and Chauveteau, 1997; Pang and Sharma, 1997; Sharma et al., 2000; Civan, 2007).

The dependence of ionic strength on the critical retention concentration σ_{cr} is shown in Fig. 12.9A, obtained based on the model of monolayer size-distributed particles. The solid gray curve represents the optimized values of the average radius $<r_s>$ and the variance coefficient C_v of the size distribution of the attached fines (see Table 12.5). It is observed that the critical retention concentration reduces with the ionic strength.

According to its definition, the lower the value of C_v indicates the narrower size distribution of attached fines, thus, the smaller fraction of large particles in the overall particle size distribution. As a result, the decrease of C_v causes the increase of the critical retention concentration at large ionic strength (black short and long dashed curve in Fig. 12.9A). The critical retention concentration of fines reduces with C_v at small ionic strength.

Reduction in $<r_s>$ leads to the increase in the fraction of small particles in the overall size distribution. It yields a greater critical retention concentration (short dashed curve in Fig. 12.9A). Increase in σ_0 results in a greater critical retention concentration at the given ionic strength (long dashed curve in Fig. 12.9A).

Assuming fine particles with varied sizes attach on the rock surface in a single layer, the size distribution of the attached fines can be determined from the MRF function. Differentiating Eq. (12.4) yields the distribution function of the attached fines concentration in terms of MRF. Fig. 12.11 presents the resulting size distribution function versus particle size from the proposed model. The obtained parameters of size distribution function are $<r_s> = 1.80 \, \mu m$ and $C_v = 0.66$ (Table 12.5). The four blue (gray) points in Fig. 12.11 indicate the critical sizes of mobilized particles in the laboratory coreflood where four ionic strength values are applied. It is shown that the decrease of ionic strength from stage 1 to stage 4 reduces the critical sizes of detached fines. Consequently, all the three parameters α, λ, and β decrease from stage 1 to stage 4 (Table 12.5).

Based on Eq. (12.7), the permeability reduction resulting from fines migration is calculated using the MRF function. The results of dimensionless permeability as functions of flow velocity and ionic strength at different temperature values are presented in Fig. 12.12A and B, respectively. The significant effect of temperature rise on the diminution of the electrostatic attaching force yields substantial reduction of the critical retention concentration of fines. As a result, the lower permeability

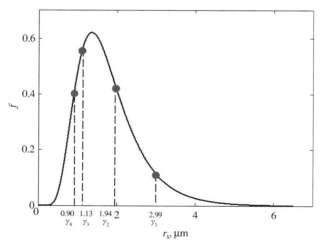

Figure 12.11 Size distribution of detachable fines calculated from the experimental data tuning using the analytical model.

corresponds to higher temperature, as shown in Fig. 12.12A and B. Particularly, at the temperature $T = 129°C$ and the ionic strength γ below 0.2 M NaCl in the Salamander geothermal reservoir, the attaching force exerting on particles is so weak that nearly no fines can stay attached on the rock surface. The critical retention concentration in this case tends to zero, resulting in the permeability decline to a constant level in the entire range of $0 \leq \gamma \leq 0.2$ M NaCl (black dashed curve in Fig. 12.12B).

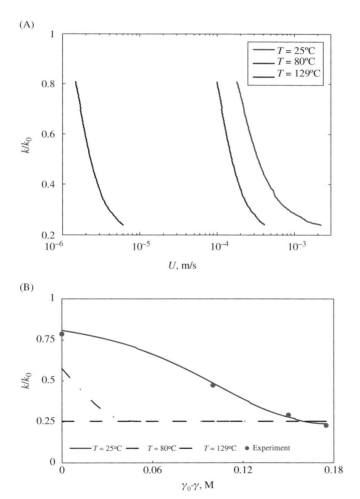

Figure 12.12 Effect of temperature on rock permeability due to migration of fine particles at various: (A) flow velocities and (B) fluid ionic strengths.

12.6 MATHEMATICAL MODEL FOR FINES MIGRATION DURING FLOW TOWARD WELL

12.6.1 System of equations

The following list provides the main assumptions applied in the proposed model for fines detachment, mobilization, and migration during flow toward a well:

1. Particle drift speed is significantly lower than the carrier fluid velocity.
2. Reservoir is homogeneous.
3. Water compressibility is small.
4. Salt and particle diffusions are negligible.
5. Particle detachment is instant (without delay).
6. Particle straining obeys linear kinetics, i.e., filtration and formation damage coefficients are constant.
7. Increase of permeability resulting from fines detachment is negligible.

Applying these assumptions, we establish the governing system of equations for fluid flow accompanied by fines migration in rocks. The system includes mass balance equations for water, ions, and particles, captures rate equation for particle retention, equation of state for water, and Darcy equation considering permeability impairment due to size exclusion:

$$\frac{\partial}{\partial t}\left[(\phi - \sigma_a - \sigma_s)(1-c)\rho_w\right] + \nabla \cdot \left[(1-c)\rho_w \boldsymbol{U}\right] = 0 \tag{12.23}$$

$$\frac{\partial}{\partial t}\left[(\phi - \sigma_a - \sigma_s)(1-c)\gamma\rho_w\right] + \nabla \cdot \left[(1-c)\gamma\rho_w \boldsymbol{U}\right] = 0 \tag{12.24}$$

$$\frac{\partial}{\partial t}\left[(\phi - \sigma_a - \sigma_s)c + \sigma_a + \sigma_s\right] + \nabla \cdot (c\alpha\boldsymbol{U}) = 0 \tag{12.25}$$

$$\frac{\partial \sigma_s}{\partial t} = \lambda_s c\alpha|\boldsymbol{U}|, \rho_w = \rho_{w0}e^{c_w(p-p_0)}, \boldsymbol{U} = -\frac{k_0}{\mu(1+\beta_s\sigma_s)}\nabla p \tag{12.26}$$

where ϕ is porosity; c, σ_a, and σ_s are suspended, attached, and strained concentrations of fines, respectively; ρ_w is water density; $\boldsymbol{U}$ is flow velocity; p is pressure; α is particle drift delay factor; λ_s is filtration coefficient; β_s is formation damage coefficient; c_w is water compressibility; μ is water viscosity; and k_0 is the initial permeability. The governing system of equations (12.23)−(12.26) is solved in Section 12.6.2, and the solution is applied to the radial flow toward a wellbore in a field application in Section 12.7.

12.6.2 Analytical model

We consider the problem of radial flow toward a wellbore. In this case, the initial conditions are as follows:

$$t = 0 : \sigma_a = \sigma_{ai}, \sigma_s = 0,$$

$$c = c_i(r) = c_0 + \frac{1}{\phi} \left[\sigma_{ai} - \sigma_{cr} \left(\frac{q}{2\pi r} \right) \right] \tag{12.27}$$

The initial suspended particle concentration is the sum of the initial suspended concentration in porous rock c_0 and the concentration of fines detached from the grain surface, under the flow velocity $q/2\pi r$. The initial strained particle concentration is zero, since particle straining is accounted for in the initial permeability k_0. The initial concentration of attached fines is given as σ_{ai}, which can be evaluated from core samples.

The boundary condition of zero rate is defined at the reservoir boundary $r = r_e$. Either the pressure p_w or the rate q at the wellbore $r = r_w$ is provided as an inner boundary condition. If the ionic strength is constant, Eq. (12.24) is trivial and can be removed.

Fines are detached at large velocities $U > U_i$ (point B in Fig. 12.3B) within the zone $r < r_i = q/2\pi U_i$, $\sigma_{ai} = \sigma_{cr}(U_i, \gamma)$, where U_i is the maximum flow velocity below which fines remain attached on the rock surface. This domain is also called the damaged zone, where incompressible water flow is assumed. The domain $r_i < r < r_e$ is referred to as the undamaged zone, since no fines migration occurs. The flow of compressible water is governed by the pressure diffusivity equation, which is a linear parabolic equation.

The flow problem in the damaged zone, Eqs. (12.4)−(12.25), allows for exact solution by using the method of characteristics:

$$c(r, t) = c_i(r) e^{\lambda_s \left(r - \sqrt{r^2 + 2Bt} \right)}, B = \frac{q}{2\pi\alpha\phi} \tag{12.28}$$

$$\sigma_s(r, t) = \frac{\phi}{r} c_i(r) \left[r + \lambda_s^{-1} - e^{\lambda_s \left(r - \sqrt{r^2 + 2Bt} \right)} \left(\sqrt{r^2 + 2Bt} + \lambda_s^{-1} \right) \right] \tag{12.29}$$

Subsequently, the pressure profile in the damaged zone $r_w < r < r_i$ can be obtained by integrating Eq. (12.26) with respect to r. The pressure profile in the undamaged zone $r_i < r < r_e$ can be derived from the Fourier series (Polyanin, 2007). The analytical solution can be applied to the calculation of the well impedance J, which is defined as the pressure drop from the reservoir boundary to the wellbore (Δp) per unit of rate (q):

$$J(t) = \Delta p(t)q(t = 0)/\Delta p(t = 0)q(t) \quad (12.30)$$

The finite difference method is applied to the system of governing equations (12.23)−(12.26). First-order backward difference scheme is applied to time derivatives, and second-order central difference scheme is applied to spatial derivatives (Iserles, 2009). The implicit approach is used for solving the suspended concentration c and the flow velocity U; the explicit approach is used for other unknowns p, ρ_w, σ_s, and σ_a. In the case study of the Salamander geothermal reservoir, the deviation between analytical and numerical results in wellbore pressure and well impedance is 2.5%.

12.7 FIELD APPLICATION

The mathematical model developed in Section 12.6 is applied to the Salamander geothermal field in this section. The discharging time of the well was $t_0 = 5$ hours. The constant volumetric flow rate $q = 15.5$ L/s was maintained during the production time. Pressure drawdown grew from 20 bars at the beginning to 55 bars at the end of discharge. The well data of impedance history are shown as black points in Fig. 12.5.

The DLVO parameters for electrostatic particle−rock interaction have been determined in the experimental study in Section 12.3: the Salamander reservoir temperature $T = 129°C$, permeability $k_0 = 6.9$ mD, and porosity $\phi = 0.1$. The dynamic viscosity of the brine was taken from Al-Shemmeri (2012). Three model parameters for optimization were fines drift delay factor α, filtration and formation damage coefficients λ_s and β_s. The treatment of field data using the analytical model (12.28)−(12.30) was performed by the method of Leverberg−Marqardt optimization to minimize the difference between the field data and modeled results. The optimized values of three parameters included: $\alpha = 0.1$, $\lambda_s = 10$ m^{-1}, and $\beta_s = 9900$. A good match is observed between the field

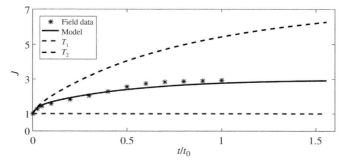

Figure 12.13 Field data on well impedance history, modeling results at reservoir temperature 129°C, and two other temperatures for sensitivity study ($T_1 = 25$°C, $T_2 = 180$°C).

impedance data (black points in Fig. 12.13) and modeled results (solid curve), with a coefficient of determination $R^2 = 0.99$. The optimal values of parameters lie within the common ranges encountered in field applications (Khilar and Fogler, 1998).

Fig. 12.13 also presents the result of a sensitivity study with respect to the temperature. The analytical model equations (12.28)−(12.30) with the parameter values optimized from the treatment of the Salamander field data was applied to the well impedance prediction. The well impedance values versus time from the model prediction at 25°C and 180°C are shown as gray and black dashed curves, respectively. Temperature rise yields reduction in the MRF and increase in detached fines concentration during flow toward a well. However, it also leads to reduction in water viscosity and decrease in detaching drag force. The overall effect shows that higher temperature results in more significant formation damage, i.e., the effect of temperature on electrostatic force is dominant if compared with that of water viscosity. Consequently, geothermal fields are more susceptible to productivity reduction and formation damage induced by fines migration, than conventional oilfields and aquifers.

In some other field cases with high permeability and small fines, particle straining is rare, therefore fines migration may not induce productivity decline. For example, the Celsius geothermal reservoir (Cooper Basin, Australia) was investigated. Laboratory coreflood tests resulted in permeability improvement during flow with fines migration, which yields enhancement of well productivity during exploitation.

12.8 CONCLUSIONS

The laboratory study and mathematical modeling of formation damage induced by mobilized fines in geothermal reservoirs lead to the following main conclusions.

Under geothermal field conditions, the drag and electrostatic forces acting on fine particles are usually two to four orders of magnitude larger than the gravitational and lifting forces. Torque balance of particles attached to the rock surface is mainly determined by drag and electrostatic forces. Two tuning parameters can be eliminated by neglecting the gravitational and lifting forces.

Laboratory-based model predictions under high-temperature geothermal field conditions show that the electrostatic attaching force decreases with temperature increase, and the detaching drag force decreases with water viscosity decrease. The former effect is much more significant, which leads to the reduction of the maximum retention concentration with temperature increase. Consequently, geothermal fields are more susceptible to fine particle mobilization than conventional oilfields and aquifers.

The laboratory-based "velocity—ionic strength" translation methodology together with the particle torque balance model lead to a new method for determining the velocity dependency of the MRF from the coreflood test with changing ionic strength.

Temperature dependency of the MRF is obtained for specific conditions of geothermal reservoirs.

Kaolinite and illite/chlorite are the main clay minerals present in the released fines from coreflood experiments in the present study and are responsible for the formation damage observed.

The corefloods performed exhibit an instant permeability response to abrupt salinity alteration, indicating that a substantial fraction of fines attached on the rock surface is detached without delay (almost instantaneously).

The analytical model for fine particle mobilization, migration, and size exclusion in 1D flow results in explicit formulae for suspended and strained particle concentrations, and the permeability profile.

Long permeability stabilization periods during flow with fines migration are interpreted by slow drift of detached rolling or sliding fines. History matching of permeability is achieved by introducing the drift delay factor for fines.

Comparison between the measured data and model prediction shows a close match to the permeability history and qualitative agreement with the effluent particle concentration data.

The MRF for a monolayer of multisized particles is expressed by an explicit formula. This formula can be applied by determining the size distribution of attached fines on the rock surface from coreflood data.

The analytical model for axisymmetric flow toward a wellbore with fines mobilization, migration, and straining derives explicit formulae for suspended and strained fines concentrations and for well impedance.

The laboratory-based mathematical model facilitates the prediction of the well impedance dynamics.

NOMENCLATURE

A_{132} Hamaker constant, $M L^2 T^{-2}$

c volumetric concentration of suspended particles, L^{-3}; molar concentration of ions, $N L^{-3}$

c_w water compressibility, $M^{-1} T^2 L$

D diffusion coefficient, $L^2 T^{-1}$

e elementary electric charge, $I T$

F force, $M L T^{-2}$

h separation distance, L

h_c thickness of internal cake, L

I ionic strength, $N L^{-3}$

k permeability, L^2

k_B Boltzmann constant, $M L^2 \Theta^{-1} T^{-2}$

k_{det} coefficient of detachment rate, T^{-1}

l lever, L

n concentration of ions, L^{-3}; refractive index (for clay, quartz, and NaCl solution)

p pressure, $M T^{-2} L^{-1}$

Q intermediate function

q well rate per unit of production layer thickness, $L^2 T^{-1}$

r radius, L

S concentration of retained particles, dimensionless

T temperature, Θ

t time, T

U flow velocity, $L T^{-1}$

U_t Fluid velocity at the center of particle, $L T^{-1}$

V potential energy, $M L^2 T^{-2}$

x distance, L

z valence of electrolyte solution

Greek letters

α particle drift delay factor
β formation damage coefficient
γ ionic strength
ε erosion ratio (ratio between the detaching and attaching torques); static dielectric constant (for clay, quartz, and NaCl solution)
ε_0 dielectric permittivity of vacuum, $I^2\,T^4\,M^{-1}\,L^{-3}$
ζ zeta potential, $M\,L^2\,I^{-1}\,T^{-3}$
κ Debye–Hückel parameter, L^{-1}
λ filtration coefficient, L^{-1}; characteristic wavelength of interaction, L
μ dynamic viscosity, $M\,L^{-1}\,T^{-1}$
ν_e absorption frequency, T^{-1}
ρ_w water density, $M\,L^{-3}$
Σ concentration distribution of captured particles, L^{-4}
σ total volumetric concentration of captured particles, L^{-3}
σ_c collision diameter, L
ϕ porosity
ω drag force coefficient

Super/Subscripts

a attachment
B Born
cr critical
D dimensionless
d drag
EDL electrical double layer
e drainage (for reservoir radius), electrostatic (for force)
g gravitational
i initial
LW London–van der Waals
L lifting
m maximum value (for velocity)
p pore
pm porous matrix
s particle
scr critical radius (for captured particles)
0   initial value
1   clay
2   quartz
3   electrolyte solution

ACKNOWLEDGMENTS

Drs. Ulrike Schacht, Themis Carageorgos, Alex Musson and Chris Matthews (University of Adelaide) are gratefully acknowledged for cooperation and support. Special thanks are extended to David Wood (DWA Energy Ltd) for his assistance in the proofreading.

REFERENCES

Al-Shemmeri, T., 2012. Engineering Fluid Mechanics. Ventus Publishing ApS.

Altmann, J., Ripperger, S., 1997. Particle deposition and layer formation at the crossflow microfiltration. J. Memb. Sci. 124, 119—128.

Aly, K.M., Esmail, E., 1993. Refractive index of salt water: effect of temperature. Optic. Mater 2, 195—199.

Aragón-Aguilar, A., Barragán-Reyes, R.M., Arellano-Gómez, V.M., 2013. Methodologies for analysis of productivity decline: a review and application. Geothermics 48, 69—79.

Badalyan, A., Carageorgos, T., You, Z., Schacht, U., Bedrikovetsky, P., Matthews, C., et al., 2013. Analysis of field case in Salamander-1 geothermal well. In: Proceedings of the Sixth Annual Australian Geothermal Energy Conference, Brisbane, Queensland, Australia.

Badalyan, A., Carageorgos, T., You, Z., Schacht, U., Bedrikovetsky, P., Matthews, C., et al., 2014. A new experimental procedure for formation damage assessment in geothermal wells. Proceedings, Thirty-Ninth Workshop on Geothermal Reservoir Engineering. Stanford University, Standford, CA.

Baudracco, J., 1990. Variations in permeability and fine particle migrations in unconsolidated sandstones submitted to saline circulations. Geothermics 19, 213—221.

Baudracco, J., Aoubouazza, M., 1995. Permeability variations in Berea and Vosges sandstone submitted to cyclic temperature percolation of saline fluids. Geothermics 24, 661—677.

Bedrikovetsky, P.G., 1993. Mathematical Theory of Oil & Gas Recovery. Kluwer Academic Publishers, London—Boston—Dordrecht, p. 600.

Bedrikovetsky, P., 2008. Upscaling of stochastic micro model for suspension transport in porous media. Transp. Porous Media 75, 335—369.

Bedrikovetsky, P., Siqueira, F.D., Furtado, C.A., Souza, A.L.S., 2011. Modified particle detachment model for colloidal transport in porous media. Transp. Porous Media 86, 353—383.

Bedrikovetsky, P., Zeinijahromi, A., Siqueira, F.D., Furtado, C.A., de Souza, A.L.S., 2012. Particle detachment under velocity alternation during suspension transport in porous media. Transp. Porous Media 91, 173—197.

Bradford, S.A., Torkzaban, S., Kim, H., Simunek, J., 2012. Modeling colloid and microorganism transport and release with transients in solution ionic strength. Water Resou. Res. 48, W09509.

Bradford, S.A., Torkzaban, S., Shapiro, A.A., 2013. A theoretical analysis of colloid attachment and straining in chemically heterogeneous porous media. Langmuir 29, 6944—6952.

Buchner, R., Hefter, G.T., May, P.M., 1999. Dielectric relaxation of aqueous NaCl solutions. J. Phys. Chem. A 103, 8—9.

Capco, D.G., Chen, Y. (Eds.), 2014. Nanomaterial: Impacts on Cell Biology and Medicine. Springer, Dordrecht, The Netherlands.

Cerda, C.M., 1987. Mobilization of kaolinite fines in porous media. Colloids Surf. 27, 219—241.

Chalk, P., Gooding, N., Hutten, S., You, Z., Bedrikovetsky, P., 2012. Pore size distribution from challenge coreflood testing by colloidal flow. Chem. Eng. Res. Design 90, 63—77.

Chauveteau, G., Nabzar, L., Coste, J.P., 1998. Physics and modeling of permeability damage induced by particle deposition. In: Proceedings of the SPE International Symposium on Formation Damage Control, pp. 409—419.

Civan, F., 2007. Reservoir formation damage: fundamentals, modeling, assessment, and mitigation, second ed Gulf Publishing Company, Houston, TX.

Cortis, A., Harter, T., Hou, L., Atwill, E.R., Packman, A.I., Green, P.G., 2006. Transport of *Cryptosporidium parvum* in porous media: Long-term elution experiments and continuous time random walk filtration modeling. Water Resour. Res. 42, W12S13.

Courant, R., Hilbert, D., 1962. Methods of Mathematical Physics, Volume II. Wiley-Interscience, New York, London.

Derjaguin, B.V., Landau, L., 1941. Theory of the stability of strongly charged lyophobic sols and of the adhesion of strongly charged particles in solutions of electrolytes. Acta Phys. Chem. (URSS) 14, 633–662.

Egan, W.G., Hilgeman, T.W., 1979. Optical Properties of Inhomogeneous Materials: Applications to Geology, Astronomy, Chemistry, and Engineering. Academic Press, New York.

Elimelech, M., Jia, X., Gregory, J., Williams, R., 1998. Particle deposition and aggregation: measurement, modelling and simulation. Butterworth-Heinemann, Oxford.

Freitas, A.M., Sharma, M.M., 2001. Detachment of particles from surfaces: an AFM study. J. Colloid Interf. Sci. 233, 73–82.

Ghassemi, A., Zhou, X., 2011. A three-dimensional thermo-poroelastic model for fracture response to injection/extraction in enhanced geothermal systems. Geothermics 40, 39–49.

Gregory, J., 1981. Approximate expressions for retarded van der Waals interaction. J. Colloid Interf. Sci. 83, 138–145.

Herzig, J.P., Leclerc, D.M., Legoff, P., 1970. Flow of suspensions through porous media—application to deep filtration. Indus. Eng. Chem. 62, 8–35.

Hunter, R.J., 1981. Zeta Potential in Colloid Science: Principles and Applications. Academic Press, London.

Iserles, A., 2009. A First Course in the Numerical Analysis of Differential Equations. Cambridge University Press, Cambridge, UK.

Israelachvili, J.N., 2011. Intermolecular and Surface Forces, third ed Elsevier, Amsterdam, The Netherlands.

Jaya, M.S., Shapiro, S.A., Kristinsdottir, L.H., Bruhn, D., Milsch, H., Spangenberg, E., 2010. Temperature dependence of seismic properties in geothermal rocks at reservoir conditions. Geothermics 39, 115–123.

Jiao, D., Sharma, M.M., 1994. Mechanism of cake buildup in crossflow filtration of colloidal suspensions. J. Colloid Interf. Sci. 162, 454–462.

Kalantariasl, A., Bedrikovetsky, P., 2014. Stabilization of external filter cake by colloidal forces in "well-reservoir" system. Indus. Eng. Chem. Res. 53, 930–944.

Katz, A.J., Thompson, A.H., 1986. Quantitative prediction of permeability in porous rock. Phys. Rev. B 34, 8179–8181.

Khilar, K.C., Fogler, H.S., 1998. Migrations of Fines in Porous Media. Kluwer Academic Publishers, Dordrecht, The Netherlands.

Kia, S.F., Fogler, H.S., Reed, M.G., 1987a. Effect of pH on colloidally induced fines migration. J. Colloid Interf. Sci. 118, 158–168.

Kia, S.F., Fogler, H.S., Reed, M.G., Vaidya, R.N., 1987b. Effect of salt composition on clay release in Berea sandstones. SPE Prod. Eng. 2, 277–283.

Kim, M., Zydney, A.L., 2004. Effect of electrostatic, hydrodynamic, and Brownian forces on particle trajectories and sieving in normal flow filtration. J. Colloid Interf. Sci. 269, 425–431.

Kristinsdottir, L.H., Flovenz, O.G., Arnason, K., Bruhn, D., Milsch, H., Spangenberg, E., et al., 2010. Electrical conductivity and P-wave velocity in rock samples from high-temperature Icelandic geothermal fields. Geothermics 39, 94–105.

Kuznar, Z.A., Elimelech, M., 2007. Direct microscopic observation of particle deposition in porous media: role of the secondary energy minimum. Colloids Surf. A: Physicochem. Eng. Aspects 294, 156–162.

Lagasca, J.R.P., Kovscek, A.R., 2014. Fines migration and compaction in diatomaceous rocks. J. Petrol. Sci. Eng. 122, 108—118.

Leluk, K., Orzechowski, K., Jerie, K., Baranowski, A., Słonka, T., Głowiński, J., 2010. Dielectric permittivity of kaolinite heated to high temperatures. J. Phys. Chem. Solids 71, 827—831.

Lever, A., Dawe, R.A., 1984. Water-sensitivity and migration of fines in the Hopeman sandstone (Scotland). J. Petrol. Geol. 7, 97—107.

Leviton, D.B., Frey, B.J., 2006. Temperature-dependent absolute refractive index measurements of synthetic fused silica. In: Astronomical Telescopes and Instrumentation, International Society for Optics and Photonics, pp. 62732K-62711.

Logan, J.D., 2001. Transport Modeling in Hydrogeochemical Systems. Springer, New York.

Marquardt, D., 1963. An algorithm for least-squares estimation of nonlinear parameters. J. Soc. Indus. Appl. Math. 11, 431—441.

Marquez, M., Williams, W., Knobles, M., Bedrikovetsky, P., You, Z., 2014. Fines migration in fractured wells: integrating modeling with field and laboratory data. SPE Prod. Oper. 29, 309—322.

Marshall, W.L., 2008. Dielectric constant of water discovered to be simple function of density over extreme ranges from −35 to +600°C and to 1200 MPa (12000 Atm.), believed universal. Nature Precedings. Available from: http://dx.doi.org/10.1038/npre.2008.2472.1.

McKinstry, H.A., 1965. Thermal expansion of clay minerals. Am. Miner. 50, 212—222.

Milsch, H., Kristinsdottir, L.H., Spangenberg, E., Bruhn, D., Flovenz, O.G., 2010. Effect of the water-steam phase transition on the electrical conductivity of porous rocks. Geothermics 39, 106—114.

Miranda, R.M., Underdown, D.R., 1993. Laboratory measurement of critical rate: A novel approach for quantifying fines migration problems. SPE Production Operations Symposium, 21-23 March. Society of Petroleum Engineers, Oklahoma City, Oklahoma.

Muecke, T.W., 1979. Formation fines and factors controlling their movement in porous media. J. Petrol. Technol. 31, 144—150.

Nabzar, L., Chauveteau, G., 1997. Permeability damage by deposition of colloidal particles. SPE European Formation Damage Conference. The Hague, The Netherlands.

Ochi, J., Vernoux, J.F., 1998. Permeability decrease in sandstone reservoirs by fluid injection: hydrodynamic and chemical effects. J. Hydrol. 208, 237—248.

Ohshima, H., 2006. Theory of Colloid and Interfacial Electric Phenomena. Academic Press, Amsterdam, The Netherlands.

Oliveira, M., Vaz, A., Siqueira, F., Yang, Y., You, Z., Bedrikovetsky, P., 2014. Slow migration of mobilised fines during flow in reservoir rocks: laboratory study. J. Petrol. Sci. Eng. 122, 534—541.

Pang, S., Sharma, M.M., 1997. A model for predicting injectivity decline in water-injection wells. SPE Formation Eval. 12, 194—201.

Polyanin, A.D., 2007. Handbook of Mathematics for Engineers and Scientists. Chapman & Hall/CRC, Boca Raton, FL.

Polyanin, A.D., Zaitsev, V.F., Moussiaux, A., 2002. Handbook of First Order Partial Differential Equations. Taylor & Francis, London.

Priisholm, S., Nielsen, B.L., Haslund, O., 1987. Fines migration, blocking, and clay swelling of potential geothermal sandstone reservoirs, Denmark. SPE Formation Evaluation 2, 168—178.

Rahman, S.S., Arshad, A., Chen, H., 1994. Prediction of critical condition for fines migration in petroleum reservoirs. SPE Asia Pacific Oil and Gas Conference. Society of Petroleum Engineers, Melbourne, Australia.

Ramachandran, R., Somasundaran, P., 1986. Effect of temperature on the interfacial properties of silicates. Colloids Surf. 21, 355—369.

Rodríguez, K., Araujo, M., 2006. Temperature and pressure effects on zeta potential values of reservoir minerals. J. Colloid Interf. Sci. 300, 788—794.

Rosenbrand, E., Fabricius, I.L., Yuan, H., 2012. Thermally induced permeability reduction due to particle migration in sandstones: the effect of temperature on kaolinite mobilisation and aggregation. Proceedings of the Thirty-Seventh Workshop on Geothermal Reservoir Engineering. Stanford University, Stanford, CA.

Rosenbrand, E., Fabricius, I.L., Kets, F., 2013. Kaolinite mobilisation in sandstone: pore plugging vs suspended particles. Proceedings of the Thirty-Eighth Workshop on Geothermal Reservoir Engineering. Stanford University, Stanford, CA.

Rosenbrand, E., Haugwitz, C., Jacobsen, P.S.M., Kjøller, C., Fabricius, I.L., 2014. The effect of hot water injection on sandstone permeability. Geothermics 50, 155—166.

Rosenbrand, E., Kjøller, C., Riis, J.F., Kets, F., Fabricius, I.L., 2015. Different effects of temperature and salinity on permeability reduction by fines migration in Berea sandstone. Geothermics 53, 225—235.

Rousseau, D., Hadi, L., Nabzar, L., 2008. Injectivity decline from produced-water reinjection: New insights on in-depth particle—deposition mechanisms. SPE Prod. Oper. 23, 525—531.

Sarkar, A.K., Sharma, M.M., 1990. Fines migration in two-phase flow. J. Petrol. Technol. 42, 646—652.

Schechter, R.S., 1992. Oil Well Stimulation. Prentice Hall, Englewood Cliffs, NJ.

Schembre, J.M., Kovscek, A.R., 2005. Mechanism of formation damage at elevated temperature. J. Energ. Res. Technol. Transact. ASME 127, 171—180.

Schembre, J.M., Tang, G., Kovscek, A.R., 2006a. Wettability alteration and oil recovery by water imbibition at elevated temperatures. J. Petrol. Sci. Eng. 52, 131—148.

Schembre, J.M., Tang, G., Kovscek, A.R., 2006b. Interrelationship of temperature and wettability on the relative permeability of heavy-oil in diatomaceous rocks. SPE Reserv. Eval. Eng. 9, 239—250.

Schijven, J.F., Hassanizadeh, S.M., 2000. Removal of viruses by soil passage: overview of modeling, processes, and parameters. Crit. Rev. Environ. Sci. Technol. 30, 49—127.

Sefrioui, N., Ahmadi, A., Omari, A., Bertin, H., 2013. Numerical simulation of retention and release of colloids in porous media at the pore scale. Colloids Surf. A Physicochem. Eng. Aspects 427, 33—40.

Sen, T.K., Khilar, K.C., 2006. Review on subsurface colloids and colloid-associated contaminant transport in saturated porous media. Adv. Colloid Interf. Sci. 119, 71—96.

Shapiro, A.A., 2007. Elliptic equation for random walks. Application to transport in microporous media. Phys. A Stat. Mech. Appl. 375, 81—96.

Shapiro, A.A., Wesselingh, J.A., 2008. Gas transport in tight porous media. Gas kinetic approach. Chem. Eng. J. 142, 14—22.

Shapiro, A.A., Yuan, H., 2013. Application of stochastic approaches to modeling suspension flow in porous media. In: Skogseid, A., Fasano, V. (Eds.), Statistical Mechanics and Random Walks: Principles, Processes and Applications. Nova Science Publishers, Inc, New York, pp. 1—38.

Sharma, M.M., Yortsos, Y.C., 1987a. Transport of particulate suspensions in porous media: model formulation. AIChE J. 33, 1636—1643.

Sharma, M.M., Yortsos, Y.C., 1987b. Fines migration in porous media. AIChE J. 33, 1654—1662.

Sharma, M.M., Pang, S., Wennberg, K.E., Morgenthaler, L.N., 2000. Injectivity decline in water-injection wells: an offshore gulf of Mexico case study. SPE Prod. Facil. 15, 6—13.

Stuart, M.R., 1955. Dielectric constant of quartz as a function of frequency and temperature. J. Appl. Phys. 26, 1399−1404.

Tiab, D., Donaldson, E.C., 2012. Petrophysics: Theory and Practice of Measuring Reservoir Rock and Fluid Transport Properties, third ed Elsevier/Gulf Professional Pub., Amsterdam, The Netherlands.

Tufenkji, N., 2007. Colloid and microbe migration in granular environments: a discussion of modelling methods. In: Frimmel, F.H., von der Kammer, F., Flemming, H.-C. (Eds.), Colloidal Transport in Porous Media. Springer, Berlin, Germany, pp. 119−142.

Tufenkji, N., Elimelech, M., 2004. Correlation equation for predicting single-collector efficiency in physicochemical filtration in saturated porous media. Environ. Sci. Technol. 38, 529−536.

Verwey, E.J.W., Overbeek, J.Th.G., 1948. Theory of the Stability of Lyophobic Colloids. Elsevier, Amsterdam, The Netherlands.

Vortisch, W., Harding, D., Morgan, J., 2003. Petrographic analysis using cathodoluminescence microscopy with simultaneous energy-dispersive X-ray spectroscopy. Mineral. Petrol 79, 193−202.

Yang, Y., Siqueira, F.D., Vaz, A., You, Z., Bedrikovetsky, P., 2016. Slow migration of detached fine particles over rock surface in porous media. J. Nat. Gas Sci. Eng. 34, 1159−1173.

You, Z., Badalyan, A., Bedrikovetsky, P., 2013. Size-exclusion colloidal transport in porous media—stochastic modeling and experimental study. SPE J. 18, 620−633.

You, Z., Bedrikovetsky, P., Badalyan, A., Hand, M., 2015. Particle mobilization in porous media: temperature effects on competing electrostatic and drag forces. Geophys. Res. Lett. 42, 2852−2860.

You, Z., Yang, Y., Badalyan, A., Bedrikovetsky, P., Hand, M., 2016. Mathematical modelling of fines migration in geothermal reservoirs. Geothermics 59, 123−133.

Yuan, B., Moghanloo, Rouzbeh G., 2017. Analytical modeling improved well performance by nanofluid pre-flush. Fuel 202, 380−394.

Yuan, B., Su, Y., Moghanloo, R.G., 2015. A new analytical multi-linear solution for gas flow toward fractured horizontal well with different fracture intensity. J. Nat. Gas Sci. Eng. 23, 227−238.

Yuan, B., Moghanloo, R.G., Zheng, D., 2016. Analytical evaluation of nanoparticles application to reduce fines migration in porous media. SPE J. 21 (06), 2317−2332.

Yuan, B., Moghanloo, G.R., Zheng, D., 2017. A novel integrated production analysis workflow for evaluation, optimization and predication in shale plays. Int. J. Coal Geol. 180, 18−28.

Yuan, H., Shapiro, A.A., 2011. A mathematical model for non-monotonic deposition profiles in deep bed filtration systems. Chem. Eng. J. 166, 105−115.

Yuan, H., Shapiro, A.A., 2012. Colloid transport and retention: recent advances in colloids filtration theory. In: Ray, P.C. (Ed.), Colloids: Classification, Properties and Applications. Nova Science Publishers, Inc, New York, pp. 201−242.

Yuan, H., You, Z., Shapiro, A., Bedrikovetsky, P., 2013. Improved population balance model for straining-dominant deep bed filtration using network calculations. Chem. Eng. J. 226, 227−237.

CHAPTER THIRTEEN

Formation Damage in Coalbed Methane Recovery

Jack C. Pashin[1], Sarada P. Pradhan[2] and Vikram Vishal[3]

[1]Oklahoma State University, Stillwater, OK, United States
[2]Indian Institute of Technology Roorkee, Roorkee, Uttarakhand, India
[3]Indian Institute of Technology Bombay, Mumbai, India

Contents

13.1 INTRODUCTION

Coal is a rock composed of $>50\%$ organic matter and thus has many unique properties as a natural gas reservoir because of its low density, adsorptivity, natural fracture networks, and compressibility. These properties facilitate production of large volumes of methane and associated gases at shallow depth and also require care to minimize formation damage when drilling, completing, and producing wells. This paper provides a summary of common types of formation damage in coalbed methane (CBM) wells, as well as some insight on how damage can be minimized during development. Coal is classified as a continuous–type unconventional reservoir that acts as both source rock and sink (reservoir) for natural gas (e.g., Kim, 1977; Scott, 2002; Pashin, 2008; Seidle, 2011; Moore, 2012). Adsorption is the dominant mechanism of hydrocarbon

Formation Damage during Improved Oil Recovery.
DOI: https://doi.org/10.1016/B978-0-12-813782-6.00013-0

© 2018 Elsevier Inc.
All rights reserved.

storage in coal, and diffusion is the dominant flow mechanism within the organic rock matrix. Lowering reservoir pressure by dewatering is the principal mechanism by which adsorbed natural gas is liberated from the coal matrix. Reservoir coal seams contain closely spaced natural fractures called cleats, and these fractures facilitate commercial flow rates and the transport of water and natural gas to the wellbore (Fig. 13.1). Coal is an exceptionally stress–sensitive rock type because of the high organic content, and permeability generally decreases exponentially with reservoir depth (McKee et al., 1988; Vishal et al., 2013a,b). Near-surface coal can have high permeability, measured on the scale of Darcies, whereas permeability can be lower than 1 mD at depths greater than about 700 m (Fig. 13.2).

Although highly permeable coal can be produced without stimulation, most coal seams are hydraulically fractured to achieve commercial flow rates. Numerous hydraulic fracturing techniques are employed using a variety of proppants, completion fluids, and treatment pressures, and these are reviewed in detail by Lambert et al. (1980) and Seidle (2011). Horizontal and multilateral directional drilling is increasingly common in CBM reservoirs and has been used successfully to achieve high gas

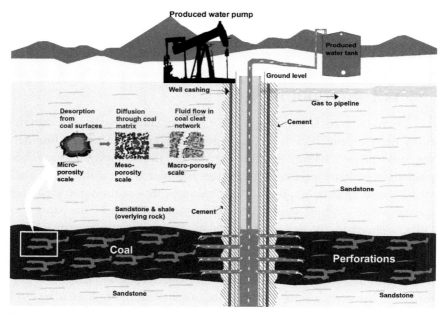

Figure 13.1 Schematic of CBM production site.

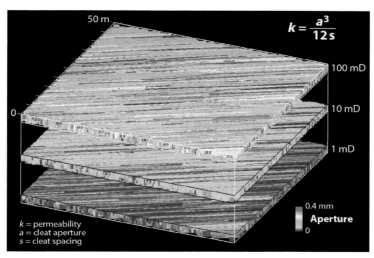

Figure 13.2 Discrete fracture network model of coal showing decreasing permeability and cleat aperture with depth.

recovery in coal seams as thin as 1 m without stimulation, and these wells have opened up major reserves in coal with permeability lower than 1 mD (e.g., von Schoenfeldt et al., 2004; Al-Jubori et al., 2009).

Coal formation damage constitutes impairment of reservoir permeability and reduction of reservoir quality due to CBM well development and recovery. Formation damage may be caused by many factors, including physical, chemical, chemical, biological, hydrodynamic, and thermal interactions of the formation, particles, and fluids, and the mechanical deformation of a coal formation under stress and fluid shear. These processes are triggered during drilling, hydraulic fracturing, production, and workover operations. Coal beds are heterogeneous and weak; therefore many broken fragments and fine particles may be produced during drilling and completion. These fragments do not allow the water and gas to flow toward the well and reduce the permeability, which can significantly hamper the production rate if suitable precautions are not taken. Improper dewatering at the initial stages of production can cause damage and changes in coal seam properties. Improper dewatering refers to the suboptimal water production rate and drainage time. Liu et al. (2013) found that the drainage system has substantial influence on productivity of CBM. Incorrect drainage may cause significant changes in pressure and drainage velocity throughout the CBM reservoir.

The following is a list of various damage mechanisms (after Bishop, 1997), and more detailed insights are given in subsequent section:

1. Fines migration, such as the internal movement of fine particles within a rock's pore structure, resulting in bridging and plugging of the fractures and pore throats.
2. The incompatibilities between fluids; such as physico-chemical inter-action between drilling mud and formation water.
3. Interaction of rock with fluid; for example, the interaction of smectite or kaolinite clay with water-based mud, which may increase the skin and significantly reduce permeability.
4. Fines migration, such as the internal movement of fine particles within a rock's pore structure resulting in bridging and plugging of the fractures and pore throats.
5. Phase trapping/blocking; for example, the invasion and entrapment of water-based fluids adjacent to the wellbore of a gas well.
6. Alteration of chemical adsorption/wettability.
7. Biological activity, such as the introduction of microbial agents into the formation during drilling and the subsequent generation of bio-films that reduce permeability.

13.2 FORMATION DAMAGE IN COALBED METHANE RESERVOIRS

This section describes and discusses the major types of formation damage that commonly occur when developing CBM reservoirs. Four basic types of damage are covered. First is formation damage associated with drilling, and second is formation damage associated with comple-tion, and specifically hydraulic fracturing. The discussion continues with a summary of formation damage due to shrinkage and swelling of the coal matrix, and concludes with a summary of production-related reser-voir failure.

13.2.1 Drilling-related formation damage

Coal is a weak rock type, and due to its fragile nature is vulnerable to damage during drilling, completion, and gas production. Drilling wells can result in formation damage related to caving and cavitation of coal seams, drilling-induced fractures and associated borehole breakouts, coal fines, and infiltration of particulate matter into the cleat networks. Closely spaced cleats and shear fractures can make coal highly friable and prone to caving and cavitation (Fig. 13.3). The degree of caving and cavitation is readily apparent in caliper logs, which provide a direct measure of borehole expansion in caved zones. Strong, blocky coal seams may show little or no evidence of caving, whereas caving of highly friable coal seams may cause borehole expansion that is beyond the reach of borehole caliper tools. In extreme cases, cavitation of the coal may cause wellbore instability through collapse or heaving of the bounding roof and floor strata, and cavitation may affect the effectiveness of hydraulic fracture treatments. In some cases, however, cavitation is not only desirable, but is used as a completion technique, for example, in the main production fairway of the San Juan Basin of Colorado and New Mexico (e.g. Saulsberry et al., 1996; Jahediesfanjani and Civan, 2005). Where these completions are effective, hydraulically induced shear failure of the near-wellbore coal mass is used to make a cavity that helps the wellbore access a large reservoir volume (Palmer et al., 2005).

When drilling horizontal wells, the integrity and strength of coal is an important consideration. Borehole collapse is a significant risk in coal that is weak or friable, and Palmer et al. (2005) have discussed ways in which coal strength is proportional to coal rank, correlating specifically with the Hardgrove grindability index (HGI). Thus, coal of medium volatile bituminous to low volatile bituminous rank has low HGI and is this best suited for drilling long-length horizontal well segments without casing or

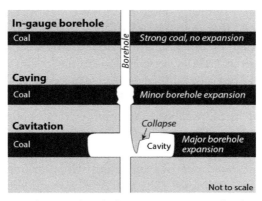

Figure 13.3 Diagram showing borehole expansion in coalbed methane reservoirs related to caving and cavitation.

sleeving. Collapse of a segment of a directional well equates to an effective loss of borehole length, and so care must be taken to design wells with coal strength and heterogeneity as key parameters. Where coal is weak and prone to collapse, parts or all of the horizontal legs may need to be sleeved to maintain effective borehole length.

Some mechanical damage can also take place due to contact of the drill bit with the wellbore wall. Drilling wells commonly induces fractures and borehole breakouts, which can introduce irregularities in borehole geometry. In extreme cases during bridging of the borehole, it may pose difficulty for reentry, completion, wellbore maintenance, and production operations. Most commonly, drilling-induced fractures and borehole breakouts are of limited consequence for development. Instead, these features provide critical information on tectonic and hoop stresses affecting the wellbore (e.g., Plumb and Hickman, 1985; Zoback et al., 2003). In the Black Warrior Basin of Alabama, Groshong et al. (2003) found that wellbore anomalies associated with drilling-induced fractures and borehole breakouts can be identified in caliper logs and are most common in the shale and sandstone layers between the coal seams (Figs. 13.2 and 13.3). Many of the borehole breakouts analyzed by Groshong et al. (2003) occur in fault zones.

Other types of drilling-related formation damage are caused by particulate matter derived from the formation and/or from drilling fluids (e.g., Park et al., 2014). For example, intraformational coal fines, sand, and clay pulverized by the bit may plug cleats when fluid flows from the wellbore into the reservoir coal seams, and chemically charged fines may make this type of formation damage difficult to address (Baltoiu et al., 2008). Drilling mud may similarly plug cleats near the wellbore. Park et al. (2014) determined that the presence of solid particles in the drilling fluid results in blocking of pores and fractures of the coal seam leading to potential damage in CBM reservoirs. To minimize this type of damage, CBM wells are typically drilled underbalanced (e.g., Johnson, 1995; Seidle, 2011); that is, the wells are drilled so that formation pressure is higher than that in the wellbore. Accordingly, flow potential is from the formation into the wellbore, which minimizes the risk of plugging the cleats with fines. Where possible, CBM wells tend to be drilled with air rigs that facilitate rapid rate of penetration while drilling underbalanced. But there are some additional risks associated with underbalanced drilling, such as pressure kicks and increased chances of blow out. Another possibility for reducing fluid-related damage during drilling is by introducing silica nanoparticles to foamed drilling fluid (Cai et al., 2016). However, during casing and cementing, flow pathways may be blocked by cement, mud solids pushed ahead of the cement, or adverse interaction between chemicals pumped ahead of cement and reservoir particles and fluids. Invasion of cement can result in clay slaking, fines migration, and silica dissolution.

13.2.2 Formation damage associated with hydraulic fracturing

Hydraulic fracturing is required to achieve economic production rates in any vertical CBM wells, and a broad range of treatment fluids and proppants have been used (Lambert et al., 1980; Puri et al., 1991; Chen et al., 2006; Seidle, 2011). CBM wells are generally fractured using viscous fluids and proppants. Fluid volume, proppant volume and type, and treatment pressure can vary greatly depending on the coal and subsurface stress conditions. Common treatment fluids deployed in CBM wells include water, slickwater, cross-linked gel, and nitrogen foam, and each fluid offers a distinct set of pros and cons (Holditch et al., 1989). Water, for example, is inexpensive but is limited in its ability to transport proppant deep into the formation. Slickwater, nitrogen foam, and cross-linked gel are more expensive but have superior proppant transport capability. Fracturing fluids containing anionic surfactants can minimize formation damage because they do not interfere with desorption of gas and facilitate reduced fluid contact angles (You et al., 2015).

Fracturing fluids cause formation damage by water-block, solids invasion associated with fluid leakoff, and clay hydration in the near-fracture formation. Therefore, it is important to use compatible fluids and fluid-loss additives (Keelan and Koepf, 1977). Filtrate invasion with gelled fluids during fracturing operations has the potential to cause formation damage. Damage to coal permeability caused by gelled fracture fluids, friction reducing agents, and other chemicals used during a hydraulic fracture treatment was studied by Puri et al. (1991), who attempted to determine the magnitude of change in coal permeability by contact with fracturing fluids. Gelled fluids can block cleats, which are the principal source of permeability in coal. Cross-linked gel requires a breaker to reduce viscosity and facilitate flowback after treatment. Breaking is most effective in deep, warm reservoirs. In reservoirs where the breaker does not fully activate, cleat surfaces may remain coated with gel, thus constituting a major source of formation damage (Holditch et al., 1989). In some reservoirs, restimulation with nitrogen foam has been used as a remedial measure to improve production of wells with significant gel damage (Saulsberry, 2007). Palmer et al. (1991) suggested that leakoff of treatment fluid from hydraulic fractures into the formation can constitute between 15% and 30% of the total treatment fluid volume. Accordingly, fluid-related formation damage could be exacerbated in situations where a large proportion of the treatment fluid penetrates the reservoir and does not flowback to the well following stimulation. Accumulation of coal fines and rock chips can further contribute to formation damage during hydraulic fracture treatments (Jones et al., 1987). Water blocking and reactivity of stimulation fluid with the target coal seams also are common sources of formation damage (Huang et al., 2015).

Screenout of proppant at the entry of hydraulic fractures has been a significant problem since the early days of hydraulic fracturing in coal (Lambert et al., 1980). One way to mitigate screenout is to select treatment fluids with superior proppant transport capabilities. Another method is using smaller proppant sizes, and there has been a general progression in the United States from 40 to 80 mesh and even finer grained proppant. The aforementioned exponential decrease in the permeability of coal with depth limits proppant transport into deep, tight coal seams. In the Black Warrior Basin, coal seams are typically stimulated through perforations, but the deepest commercial coal seam is typically completed in open hole to maximize the surface area exposed to treatment fluids and reduce the potential for screenout (Pashin et al., 2015). Further, high in situ stresses and complex stress fields in deep coal reservoirs suppress the development of hydraulic fractures, leading to reduced permeability (Lu et al., 2015).

13.2.3 Changes in coal matrix

One of the more remarkable properties of coal is that the organic matrix shrinks during desorption and swells during adsorption, thereby changing cleat aperture (Harpalani and Chen, 1995; Palmer and Mansoori, 1998; Pekot and Reeves, 2003; Vishal and Singh, 2015; Vishal et al., 2013b). The volumetric strain associated with this is proportional to the amount of gas sorbed and correlates well with the Langmuir isotherm. Sorption capacity and strain are least with small molecules like hydrogen, which form a thin Langmuir monolayer, and increase proportionally with Langmuir monolayer thickness (Chikatamarla et al., 2009) (Fig. 13.4). Of the three major gases that occur naturally in coal, nitrogen forms the

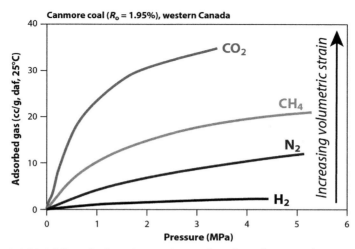

Figure 13.4 Variability of adsorption isotherms for selected gases in bituminous coal. Coal volume changes as gas is adsorbed or desorbed, and volumetric strain is directly proportional to sorbed gas volume. *After Chikatamarla et al. (2009).*

thinnest monolayer, methane forms an intermediate monolayer, and carbon dioxide forms a thick monolayer.

Sorption-induced swelling of the coal matrix can also occur by interaction with the filtered fracturing fluids and can cause formation damage. Efforts were made to restore the coal permeability by prolonged water and acid flushing but were not very successful (Puri et al., 1991). Even low concentrations of friction reducing compounds in water could reduce the permeability of coal samples. Reucroft and Patel (1986) hypothesized that the maximum swelling occurs when the solubility parameter of the molecular probe (fluid phase) is equal to the solubility parameter of the macromolecular network. This could help explain the higher swelling of coal due to adsorption of carbon dioxide over methane. Shrinkage of coal matrix can result in an increase in permeability greater than an order of magnitude during the productive life of a well (Harpalani and Chen, 1995; Palmer and Mansoori, 1998). The magnitude of this effect across a broad range of reservoir conditions is highly variable and depends on gas content, reservoir stresses, and the mechanical properties of coal (Palmer and Mansoori, 1998; Pekot and Reeves, 2003). The shrinkage effect caused by production (i.e., reduction of sorbed gas content), is generally thought to be advantageous, and the most tangible downside may be minor formation damage related to remobilization of proppant and fines as the matrix shrinks and the fracture aperture increases.

Injection of carbon dioxide can enhance the recovery of CBM significantly (Gentzis, 2000; Vishal et al., 2013c, 2015b). Carbon dioxide and nitrogen have been used for enhanced CBM recovery commercially in the San Juan basin and experimentally in many basins (e.g., Reeves, 2001; Pashin, 2016). Matrix swelling is considered a significant issue in all CO_2-enhanced CBM recovery experiments, although it seems that it can be managed operationally by limiting injection rate (Reeves, 2001). Flooding coal with nitrogen, by contrast, will tend to shrink the coal matrix, and investigators have explored the utility of CO_2-N_2 mixes to optimize recovery and limit swelling-related formation damage (Lin and Kovoscek, 2014). However, Vishal (2017a,b), using laboratory experiments, showed that significant matrix swelling occurs during liquid/supercritical CO_2 injection and only a small part swelling is reversed by N_2 injection.

Swelling clays are abundant in shale reservoirs but are much less common in CBM reservoirs. Thus, damage associated with swelling clay is of less concern than fine particle migration. Swelling of the coal matrix due to water adsorption can significantly reduce permeability. In a study conducted by Zhang et al. (2016) on the effect of swelling in two types of identified cleats (i.e., cleats in the coal and those in the mineral partings), approximately 80% of the coal cleats closed upon water adsorption, while the cleats in the mineral partings were not affected. This cleat closure by water adsorption dramatically reduced porosity and permeability. When a fresh-water filtrate invades the reservoir rock, it can cause the clay to swell and thus reduces or totally blocks the pore throats. Filtering liquids invade into the coal through parallel particles, bedding, joints, and cleats of the coal, causing damage if the liquids are incompatible with the coal. Clay minerals in micropores will hydrate once in contact with water, leading to flow path and particulate output reduction (Huang et al., 2015). To minimize such damage, clay stabilizers or surfactants and salts like KCl are added to the fracture fluid.

13.2.4 Production-related reservoir failure

Permeability and porosity are important parameters which affect the production of gas from deep coal beds. During production, it is important to make sure that these parameters are not diminished. Due to the changes occurring during production, decreased productivity may make it uneconomical to extract the gases. Variable water production rates and frequently changing of wellhead pressure and the production system often leads to pressure decline as the flow paths in the coal seam get blocked by sand. This may cause permanent damage to permeability of coal bed methane reservoir and productivity as well (Spencer, 1989). Initiation of fines movement during initial drill stem tests by using excessive draw-down pressure or inorganic/organic scaling through an abrupt shift in thermodynamic conditions also leads to formation damage.

As mentioned in Section 13.2.3, a significant change that commonly takes place is improvement of permeability as a response to matrix shrinkage and an associated expansion of cleat aperture. However, Harpalani et al. (2015) noted examples of reservoir failure during the late stages of CBM production in the northern San Juan Basin. During the life of the wells that underwent failure, production and permeability enhancement related to matrix shrinkage appears to have occurred normally as pressure declined from about 10 to 4 MPa. At pressures below 4 MPa, pressure dropped rapidly to about 0.4 MPa, and permeability increased by a factor of about 400. This phenomenon was observed only in wells deeper than 900 m, and it is thought that coal entered the failure envelope at a pressure of about 0.4 MPa.

Harpalani et al. (2015) interpreted this phenomenon as a response to a major decrease of horizontal stress as the coal relaxed during desorption. This apparently resulted in a substantial differential between vertical and horizontal stresses, which in turn resulted in rapidly increasing permeability, perhaps resulting in the generation of new fractures. They further noted that coal failure typically results in the generation of coal fines and rubble, which can damage the formation, thereby temporarily reducing permeability and requiring workover operations to restore the well to a producing condition. It is not clear how similar types of reservoir failure may occur in other coal basins, although the potential for either major permeability enhancement or major formation damage necessitates care as pressure is depleted in deep CBM wells and overburden pressure approaches the failure envelope.

13.2.5 Diagnosis of formation damage and evaluation of damage potential

Discovering the reasons for formation damage is important for formulation of a mitigation strategy to improve well performance. This requires a systematic approach to research, planning, and evaluatingi all available information. The measurement and diagnosis of formation damage can be assessed by well testing, logging, evaluating production performance analysis, etc. A few approaches are listed below:

1. Quantification of the degree of existing damage.
2. Performing specific laboratory studies.
3. Performing downhole diagnostic tests that can be conducted in the field, which include well-test analysis, downhole imaging, and physical sampling of the produced fluids and solids in the production zone. Physical sampling can be done by packer testing, sidewall coring, and conducting special core analysis. After finding the cause for the formation damage, it is essential to quantify its extent.

The measurement of formation damage is quantified in terms of

1. Skin factor—the reduced in permeability of the reservoir due to damage of the formation around the wellbore.
2. Change in permeability.
3. Change in viscosity, effective fluid mobility, and flow rate and flow efficiency.

13.3 SUMMARY AND CONCLUSIONS

CBM reservoirs are characterized by low mechanical strength, presence of developed cleats, heterogeneity and anisotropy, and a larger surface area as compared to the conventional gas reservoirs. Therefore, potential damage in coal may occur during drilling, completion, and production. Additional damage occurs when there is interaction of various fluids (Vishal et al, 2015a). Such damage includes blocking of pores and fractures by solid particles in the drilling fluids containing produced coal cuttings, precipitation during certain reactions, due to stress sensitivity, hydration swelling, and dispersion of nonorganic materials like clays mingling in the coal beds.

We have focused on four major types of coal formation damage that include drilling-induced damage, hydraulic fracturing-induced damage, coal matrix deformation, and production-related reservoir failure. An advance understanding of the possible formation damage in coal can help minimize the chances of damage, facilitating smooth operations. Various methods to overcome such damage need to be studied further. Water blocking can be prevented by adding agents that adjust the coal–water wettability. Adding inhibitors, such as inorganic salts, reduces the working fluid inhibition. Surfactants are used in a water-based, working fluid that can reduce the capillary forces to adjust the coal-water wettability. The differential pressures and rate of discharge may be controlled to limit the stress sensitivity of the coal. Huang et al. (2015) showed that $CaCl_2$ could reduce water sensitivity and other viscosifiers, friction reducing agents, and fluid-loss additive operating liquids could minimize the damage to the permeability of CBM reservoirs. With the inclusion of CO_2-enhanced CBM recovery, new techniques and technologies will need to be developed to minimize formation damage in CBM reservoirs.

REFERENCES

Al-Jubori, A., Johnston, S., Boyer, C., Lambert, S.W., Bustos, O.A., Pashin, J.C., et al., 2009. Coalbed methane: clean energy for the world. Oilfield Rev 21 (2), 4−13.

Baltoiu, L.V., Warren, B.K., Natras, T.A., 2008. State-of-the-art in coalbed methane drilling fluids. SPE Drill. Complet. 23 (3), 250−257.

Bishop, S.R., 1997. The experimental investigation of formation damage due to the induced flocculation of clays within a sandstone pore structure by a high salinity brine. Proceedings of SPE European Formation Damage Conference, June 1997. Society of Petroleum Engineers. Available from: https://doi.org/10.2118/38156-MS.

Cai, J., Gu, S., Wang, F., Yang, X., Yue, Y., Xiaoming, Wu, et al., 2016. Decreasing coalbed methane formation damage using microfoamed drilling fluid stabilized by silica nanoparticles. J. Nanotechnol. 11. article 9037532.

Chen, Z., Khaja, N., Valencia, K.L., Rahman, S.S., 2006. Formation damage induced by fracture fluids in coalbed methane reservoirs. Society of Petroleum Engineers, p. 4, Paper SPE 101127.

Chikatamarla,, L.M., Bustin, R.M., Cui, X., 2009. CO_2 sequestration into coalbeds: insights from laboratory experiments and numerical modeling. AAPG Stud. Geol. 59, 457−474.

Gentzis, T., 2000. Subsurface sequestration of carbon dioxide—an overview from an Alberta (Canada) perspective. Int. J. Coal Geol. 43, 287−305.

Groshong, R.H., Jr., Cox, M.H., Pashin, J.C., McIntyre, M.R., 2003. Relationship between gas and water production and structure in southeastern Deerlick Creek coalbed methane field, Black Warrior basin, Alabama, Tuscaloosa, Alabama, University of Alabama College of Continuing Studies, 2003 International Coalbed Methane Symposium Proceedings, Paper 0306, p. 12.

Harpalani, S., Chen, G., 1995. Influence of gas production induced volumetric strain on permeability of coal. Geotec. Geol. Eng. 15 (4), 303−325.

Harpalani, S., Singh, V., Soni, A., 2015. Coal failure with continued depletion of coalbed methane reservoirs and resulting jump in permeability. Society of Exploration Geophysicists International Conference and Exhibition, Melbourne, Australia, pp. 209−209.

Holditch, S.A., Ely, J.W., Carter, R.H., 1989. Development of a coal seam fracture design manual. In: Proceedings of the 1989 Coalbed Methane Symposium, University of Alabama, Tuscaloosa, Alabama, pp. 299−320.

Huang, W., Lei, M., Qui, Z., Leong, Y.−K., Zhong, H., Zhang, S., 2015. Damage mechanism and protection measures of a coalbed methane reservoir in the Zhenghuang block. J. Nat. Gas. Sci. Eng. 26, 683−694.

Jahediesfanjani, H., Civan, F., 2005. Damage tolerance of well-completion and stimulation techniques in coalbed methane reservoirs. Am. Soc. Min. Eng. Trans. 127, 248−256.

Johnson, P.W., 1995. Design techniques in air and gas drilling: cleaning criteria and minimum flowing pressure gradients. J. Canadian Petrol. Technol. 34, 18−26.

Jones, A.H., Bell, G.J., Morales, R.H., 1987. The influence of coal fines/chips on the behavior of hydraulic fracture simulation treatments. In: Proceedings of the 1987 Coalbed Methane Symposium, University of Alabama, Tuscaloosa, Alabama, pp. 93−102.

Keelan, D.K., Koepf, E.H., 1977. The role of cores and core analysis in evaluation of formation damage. SPEJ. Available from: https://doi.org/10.2118/5696-PA.

Kim, A.G., 1977. Estimating methane content of bituminous coals from adsorption data. U.S. Bureau of Mines Report of Investigations 2845, p. 22.

Lambert, S.W., Trevits, M.A., Steidl, P.F., 1980. Vertical borehole design and completion practices to remove methane gas from mineable coal beds. U.S. Department of Energy Report DOE/CMTC/TR-80/2, p. 163.

Lin, W., Kovoscek, A.R., 2014. Gas sorption and the consequent volumetric and permeability change of coal I: experimental. Transport Porous Med 105, 371−389.

Liu, H., Sang, S., Formolo, M., Li, M., Liu, S., Xu, H., et al., 2013. Production characteristics and drainage optimization of coalbed methane wells: a case study from low-permeability anthracite hosted reservoirs in southern Qinshui Basin, China. Energy. Sust. Dev. 17 (5), 412−423.

Lu, Y., Yang, Z., Li, X., Han, J., Guofa, J., 2015. Problems and methods for optimization of hydraulic fracturing of deep coal beds in China. Chem. Technol. Fuels Oils 51, 41−48.

McKee, C.R., Bumb, A.C., and Koenig, R.A., Stress-dependent permeability and porosity of coal and other geologic formations: Society of Petroleum Engineers Formation Evaluation, 1988, p. 81−91.

Moore, T.A., 2012. Coalbed methane: a review. Int. J. Coal. Geol. 101, 36−81.

Palmer, I., Mansoori, J., 1998. How permeability depends on stress and pore pressure in coalbeds: a new model. Society of Petroleum Engineers Reservoir Evaluation and Engineering, paper SPE 52607, pp. 539−544.

Palmer, I.D., Fryar, R.T., Tumino, K.A., Puri, R., 1991. Comparison between gel-fracture and water-fracture stimulations in the Black Warrior basin. Coalbed Methane Symposium Proceedings. University of Alabama, Tuscaloosa, Alabama, pp. 233−242.

Palmer, I.D., Moshovidis, Z., Cameron, J., 2005. Coal failure and consequences for coalbed methane wells. Society of Petroleum Engineers, paper SPE 9682.

Park,, S., Song,, H., Park,, J., 2014. Selection of suitable aqueous potassium amino acid salts: CH_4 recovery in coal bed methane via CO_2 removal. Fuel Process. Technol. 120, 48−53.

Pashin, J.C., 2008. Coal as a petroleum source rock and reservoir rock. In: Ruiz, I.S., Crelling, J.C. (Eds.), Applied Coal Petrology—The Role of Petrology in Coal Utilization. Elsevier, Amsterdam, pp. 227−262.

Pashin, J.C., 2016. Geologic considerations for CO_2 storage in coal. In: Singh, T.N. (Ed.), Geologic Carbon Sequestration: Understanding Reservoir Behavior. Springer, Berlin, pp. 137−159.

Pashin, J.C., Clark, P.E., McIntyre-Redden, M.R., Carroll, R.E., Esposito, R.A., Oudinot, A.Y., et al., 2015. SECARB CO_2 injection test in mature coalbed methane reservoirs of the Black Warrior Basin, Blue Creek Field, Alabama. Int. J. Coal Geol. 144, 71−87.

Pekot, L.J., Reeves, S.R. 2003. Modeling the effects of matrix shrinkage and differential swelling on coalbed methane recovery and carbon sequestration. Tuscaloosa, 2003 Int. Coalbed Methane Symposium Proc., paper per 0328, p. 16.

Plumb, R.A., Hickman, S.H., 1985. Stress-induced borehole elongation: a comparison between the four-arm dipmeter and the borehole televiewer in the Auburn Geothermal Well. J. Geophys. Res. Solid Earth 90 (B7), 5513—5521.

Puri, R., King, G.E., Palmer, I.D., 1991. Damage to coal permeability during hydraulic fracturing. Society of Petroleum Engineers Paper SPE 21813.

Reeves, S.R., 2001. Geological sequestration of CO_2 in deep, unmineable coalbeds: an integrated research and commercial-scale field demonstration test. Society of Petroleum Engineers, paper 71749, p. 11.

Reucroft, P.J., Patel, H., 1986. Gas-induced swelling in coal. Fuel 65, 816—820.

Saulsberry, J.L., 2007. A case history: Oak Grove Field, Black Warrior Basin, USA; CBM well stimulations, in Coalbed methane. Durango, Society of Petroleum Engineers Regional Technology Workshop Proceedings, p. 32.

Saulsberry, J.L., Schafer, P.S. Schraufnagel, R.A., 1996. A Guide to Coalbed Methane Reservoir Engineering. Gas Research Institute Report GRI-94/0397, Chicago.

Scott, A.R., 2002. Hydrogeological factors affecting gas content distribution in coal beds. Int. J. Coal Geol. 50, 363—387.

Seidle, J., 2011. Fundamentals of Coalbed Methane Reservoir Engineering. PenWell Corporation, Tulsa, p. 401.

Spencer, C.W., 1989. Review of characteristics of low-permeability gas reservoirs in West United State. AAPG Bull. 73 (5), 613—619.

Vishal, V., 2017a. In-situ disposal of CO_2: liquid and supercritical CO_2 permeability in coal at multiple down-hole stress conditions. J. CO_2 Util. 17, 235—242.

Vishal, V., 2017b. Saturation time dependency of liquid and supercritical CO_2 permeability of bituminous coals: implications for carbon storage. Fuel 192, 201—207.

Vishal,, V., Singh,, T.N., 2015. A laboratory investigation of permeability of coal to supercritical CO_2. Geotech. Geol. Eng. 33 (4), 1009—1016.

Vishal,, V., Ranjith,, P.G., Pradhan,, S.P., Singh,, T.N., 2013a. Permeability of sub-critical carbon dioxide in naturally fractured Indian bituminous coal at a range of down-hole stress conditions. Eng. Geol. 167, 148—156.

Vishal,, V., Ranjith,, P.G., Singh,, T.N., 2013b. CO_2 permeability of Indian bituminous coals: implications for carbon sequestration. Int. J. Coal Geol. 105, 36—47.

Vishal,, V., Singh,, L., Pradhan,, S.P., Singh, T.N., Ranjith, P.G., 2013c. Numerical modeling of Gondwana coal seams in India as coalbed methane reservoirs substituted for carbon dioxide sequestration. Energy 49, 384—394.

Vishal,, V., Ranjith,, P.G., Singh,, T.N., 2015a. An experimental investigation on behaviour of coal under fluid saturation, using acoustic emission. J. Nat. Gas Sci. Eng. 22, 428—436.

Vishal,, V., Singh,, T.N., Ranjith,, P.G., 2015b. Influence of sorption time in CO_2-ECBM process in Indian coals using coupled numerical simulation. Fuel 139, 51—58.

von Schoenfeldt, H., Zupanic, J., Wight, D.R., 2004. Unconventional drilling methods for unconventional reservoirs in the US and overseas. Tuscaloosa, Alabama, University of Alabama College of Continuing Studies, 2004 International Coalbed Methane Symposium Proceedings, Paper 0441, p. 10.

You, Q., Wang, C., Ding, Q., Zhao, G., Fang, J., Liu, Y., et al., 2015. Impact of surfactant in fracturing fluid on the adsorption-desorption processes of coalbed methane. J. Nat. Gas Sci. Eng. 26, 35—41.

Zhang, Y., Lebedev, M., Sarmadivaleh, M., Barifcani, A., Rahman, T.,, Iglauer, S., 2016. Swelling effect on coal micro structure and associated permeability reduction. Fuel 182, 568−576.

Zoback, M., Barton, C., Brudy, M., Castillo, D., Finkbeiner, T., Grollimund, B., et al., 2003. Determination of stress orientation and magnitude in deep wells. Int. J. Rock Mec. Min Sci. 40, 1049−1076.

FURTHER READING

Civan, F., 2000. Reservoir formation damage: fundamentals, modeling, assessment and mitigation, second ed. Gulf Publishing Company, Houston, Texas.

CHAPTER FOURTEEN

Special Focus on Produced Water in Oil and Gas Fields: Origin, Management, and Reinjection Practice

Yu Liang[1], Yang Ning[2], Lulu Liao[3] and Bin Yuan[4]
[1]The University of Texas at Austin, Austin, TX, United States
[2]University of Houston, Houston, TX, United States
[3]Sinopec Research Institute of Petroleum Engineering, Beijing, China
[4]The University of Oklahoma, Norman, OK, United States

Contents

14.1 ORIGIN, CHARACTERISTIC, AND PRODUCTION OF PRODUCED WATER

14.1.1 Origin of produced water

With global energy demand continually growing, oil and gas plays an increasingly important role in supporting the development of society.

Formation Damage during Improved Oil Recovery.
DOI: https://doi.org/10.1016/B978-0-12-813782-6.00014-2

© 2018 Elsevier Inc.
All rights reserved.
515

The global petroleum daily consumption has increased from 80 million barrels in 2000 to 98 million barrels in 2017, indicating that every day a large amount of oil and gas is produced from conventional and unconventional fields. However, the largest produced stream is not oil and gas; the average global water and oil production ratio is 3:1, indicating that every day approximately 300 million barrels of water is also brought up to the surface together with oil and gas. Produced water usually contains toxic pollutants, posing a great threat to environment and increasing field cost. In addition, both production and composition of produced water vary significantly from field to field, determined by many complicated factors. Produced water can potentially pose many serious issues to oil and gas exploration and production activities, while the industry has very limited knowledge about it.

In hydrocarbon-bearing reservoirs, saline water usually exists and resides below oil and gas zones due to its higher density compared to those hydrocarbons, as conceptually depicted in Fig. 14.1. Generally, there are two primary sources of saline water that are naturally permeated with hydrocarbons into a formation, including flow from the same hydrocarbon zone due to hydrocarbon production and flow from other hydrocarbon zones due to hydrocarbon migration (Veil et al., 2004). The saline water in these two categories is defined as "formation water," and it becomes produced water when it is brought up to the surface together with oil and gas as a mixture.

During hydrocarbon production activities, in some cases, additional water may be injected into the formation in order to maintain the reservoir pressure and/or to facilitate various enhanced oil recovery (EOR) processes (Lake, 1989; Sheng, 2013; Yuan et al., 2016b; Zhao et al., 2011). Usually the water is injected from wells (called "injectors") into the target formation, pushing oil in the formation flowing toward other well (called "producers"), as shown in Fig. 14.2. At the breakthrough time, when either the injected water or the formation water reaches the producer, the producer starts to produce hydrocarbons as well as water, and this lifted water is defined as "produced water." At the surface, most of hydrocarbons are separated from produced water and sent to storage units for onward sales and shipment (Ekins et al., 2007); however, there is still a small amount of dissolved oil and dispersed oil remaining in the produced water after the separation process. Therefore, produced water is

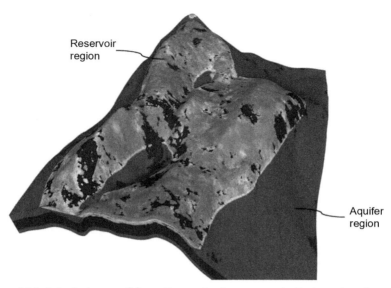

Figure 14.1 A typical case of formation water in a reservoir. Hydrocarbon (reservoir region) resides above formation water (aquifer region). *Reproduced with permission from Du, S., Fung, L.S., Dogru, A.H., 2017. Dual grid method for aquifer acceleration in parallel implicit field-scale reservoir simulation. SPE Reservoir Simulation Conference. Society of Petroleum Engineers, Tulsa, OK.*

a mixture of formation water, injected water, liquid and gaseous hydrocarbons, dissolved or suspended solid, produced sand and chemical additives, generated during drilling, completion, production, and separation processes in oil and gas industry (Fakhru'l-Razi et al., 2009; Igunnu and Chen, 2012; Reed and Johnsen, 2012). Depending upon its location, produced water can also be classified as: (1) oil field produced water, (2) gas field produced water, (3) coalbed methane produced water, and (4) unconventional field produced water.

14.1.2 Characteristics of produced water

Understanding the characteristics of produced water is very important to guide us to take the appropriate measures to handle it. On one hand, knowledge of its physical and chemical properties can significantly facilitate petroleum exploration and production activities. For example, by knowing resistivity of produced water in a pay zone, based on Archie's law reservoir engineers can determine hydrocarbon saturation via well

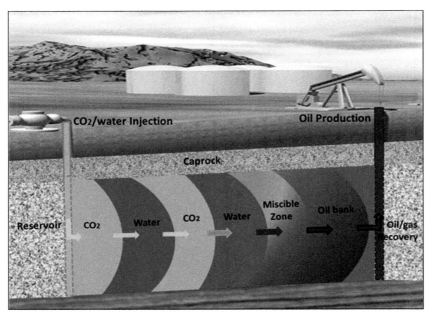

Figure 14.2 Water/CO_2 flooding process: water and CO_2 is injected into a formation to enhance oil recovery, pushing oil from injectors to producers. *Reproduced with permission from Dai, Z., Viswanathan, H., Middleton, R., Pan, F., Ampomah, W., Yang, C., et al., 2016. CO_2 accounting and risk analysis for CO_2 sequestration at enhanced oil recovery sites. Environ. Sci. Technol. 50, 7546–7554.*

logging methods (Katz and Thompson, 1985). On the other hand, flow of produced water usually contains toxic pollutants, posing a great potential threat to the environment and hindering production activities. It is essential to take actions to treat or reuse the produce water for environmental protection purposes. However, it should be noted that the characteristics of produced water change significantly depending on many factors, including mixture components, geological formation, geological location, lifetime of a reservoir, types of the production process, and other factors (Henderson et al., 1999). The great variances in chemical components of produced water are qualitatively observed in Tables 14.1–14.3. Therefore, produced water treatment has to be adjusted from site to site according to its properties in order to achieve the best removal results.

Table 14.1 Treatment chemicals in produced water (Hansen and Davies, 1994; Igunnu and Chen, 2012)

Chemical concentration	Oil field	Gas field
	Range (mg/L)	Range (mg/L)
Corrosion inhibitor	2—10	2—10
Demulsifier	1—2	—
Glycol	—	500—2000
Methanol	—	1000—15000
Polyelectrolyte	0—10	—
Scale inhibitor	4—30	—

Table 14.2 Constituents of produced water from oil fields (Fakhru'l-Razi et al., 2009; Igunnu and Chen, 2012; Tibbetts et al., 1992)

Parameter	Range	Parameter	Range
Aluminum	310—410	Manganese	0.004—175
Ammoniacal nitrogen	10—300	Mercury	0.001—0.002
Arsenic	0.005—0.3	pH (—)	4.3—10
Barium	1.3—650	Phenols	0.009—23
Beryllium	0.001—0.004	Potassium	24—4300
Bicarbonate	77—3990	Silver	0.001—0.15
Boron	5—95	Sodium	132—97000
Cadmium	0.005—0.2	Strontium	0.02—1000
Calcium	13—25800	Sulfate	2—1650
Chemical oxygen demand	1220	Sulfite	10
Chloride	80—200000	Surface Tension (dynes/cm)	43—78
Chromium	0.02—1.1	Titanium	0.01—0.7
Copper	0.002—1.5	Total organic carbon (TOC)	0—1500
Density (kg/m^3)	1014—1140	Total oil	2—565
Higher acids	1—63	Total polar	9.7—600
Iron	0.1—100	Total suspended solids (TSS)	1.2—1000
Lead	0.002—8.8	Volatile	0.39—35
Lithium	3—50	Volatile fatty acids	2—4900
Magnesium	8—6000	Zinc	0.01—35

The unit is in mg/L otherwise stated specifically.

14.1.2.1 Major compounds of produced water

Produced water consists of organic materials and inorganic materials. In general, major compositions of produced water include: (1) dissolved and dispersed oil, (2) dissolved minerals, (3) treatment chemicals, (4) produced solids, and (5) dissolved gases (Hansen and Davies, 1994). Produced water may also contain some minor compositions. Bacteria and scales are examples of those minor compositions, and they can cause severe pipeline blockage issues and add a significant economic burden to field operations (Neff, 2002). The large variations in composition of produced water also result in a great uncertainty regarding environmental protection requirements and hydrocarbon production optimization. The distribution of major compounds in produced water can be accurately analyzed by thermodynamic models and algorithms at different testing conditions, e.g., isobaric—isothermal and isobaric—isenthalpic (Zhu and Okuno, 2015a, b, 2016). During the online production life of a well, the compounds of produced water may change over time if water from other sources enters the production zones, such as via waterflooding or produced water reinjection (PWRI) processes (Sheng, 2013).

- *Dissolved and dispersed oil*: The compounds of dissolved and dispersed oil in produced water include PAHs (polyaromatic hydrocarbons), BTEX (benzene, toluene, ethylbenzene, xylenes), and phenols. Dissolved oil is defined as water-soluble organics in produced water, including BTEX, phenols, aliphatic hydrocarbons, carboxylic acid, and low molecular weight aromatic compounds. Dispersed oil typically occurs as small droplets of organics suspends in produced water, including PAHs and alkyl phenols. The dissolubility of oil mainly depends on temperature, oil composition, pH, total dissolved solids, salinity, oil/water ratio, oil field chemicals, and stability compounds (Veil et al., 2004). In produced water, most of hydrocarbons are in dispersed oil form, while dissolved oil still contains a high concentration of toxic components (Shannon et al., 2008). Generally, nonpolar organics in produced water are consistently toxic; aromatics and phenols in the dissolved hydrocarbons are the main contributor to acute toxicity (Fakhru'l-Razi et al., 2009; Gunatilake and Bandara, 2017).
- *Minerals*: The compounds of dissolved minerals include anions and cations, heavy metals, and radioactive materials (Arthur et al., 2005). The majority of chemical/physical properties of produced water, such as buffering capacity, salinity, and scale potential, are highly related to the concentration of those anions and cations, including Na^+, Ca^{2+},

Fe^{2+}, Cl^-, and so on. Especially, salinity of a formation is determined by Na^+ and Cl^- while conductivity is determined by all anions and cations. In addition, lead, nickel, zinc, cadmium, and copper commonly exist as heavy metal in oil and gas field produced water, and their concentration is proportional to the age of the formation (Roach et al., 1993). ^{226}Ra and ^{228}Ra are the most abundant radioactive materials while Barium sulfate is the most common scale coprecipitated (Fard et al., 2017; Pichtel, 2016; Vegueria et al., 2002).

• *Treatment chemicals*: Treatment chemicals are additives to facilitate oil and gas production, such as corrosion inhibitor and scale inhibitor, potentially posing the greatest concerns for toxicity of produced water (Henze et al., 2001). According to field operations, concentrations of treatment chemicals vary from field to field, and sometimes it can reach very high concentrations and cause disastrous results. Therefore, steps to decrease toxicity are necessary before any disposal or injection activity can be conducted. The common compounds of treatment chemicals detected in produced water from oil and gas fields are summarized in Table 14.1.

• *Produced solids*: These are suspended solids from wellbore and producing formations, including carbonates, clays, corrosion products, proppants, and sands. The amounts of those solids in produced water vary significantly from site to site, depending on the condition of the wellbore and formation. Produced solid is an important factor in determining the properties of produced water. For example, the concentration of fine grains can decrease efficiency of a water/oil separator, resulting in a higher oil and grease limits in the produced water. In addition, when produced water is reinjected (PWRI) into a reservoir, formation damage is mainly determined by the retention of the suspended particles (Barkman and Davidson, 1972). Most PWRI formation damage mathematical models focus on the transport of suspended solids in porous media, which will be discussed in detail in this chapter.

• *Dissolved gas*: The most common dissolved gases in produced water are oxygen, hydrogen sulfide, and carbon dioxide, affecting the physical properties of produced water such as density and viscosity (Osborn et al., 2011).

14.1.2.2 Compounds contained in produced water from conventional oil fields

Table 14.2 demonstrates the major constituents and corresponding characteristics of produced water from conventional oil fields. The organics and inorganics in the produced water exist in a variety of physical forms including emulsion, solution, suspension, particulates, and adsorbed particles. It is commonly observed that in addition to natural components such as formation water and produced solids, produced water can also contain treatment chemicals, injected water, and bacterial, which can greatly affect its physical and chemical properties. The commonly used chemicals in oil field produced water include coagulants, corrosion inhibitors, scale inhibitors, and emulsion breakers (McCormack et al., 2001).

14.1.2.3 Compounds contained in produced water from conventional gas fields

In gas fields, the major components of produced water are formation water and condensed water since water is usually not injected back into gas-bearing formations, in contrast to oil fields. Table 14.3 summarizes the constituents and corresponding characteristics of produced water from gas fields. Since gas contains more volatile aromatic hydrocarbon, produced water from gas fields, which contain a large volume of BTEX, is

Table 14.3 Constituents of produced water from gas fields (Fakhru'l-Razi et al., 2009; Igunnu and Chen, 2012; Shepherd et al., 1992; Silva et al., 2017)

Parameter	Range	Parameter	Range
Alkalinity	0−285	Lithium	18.6−235
Aluminum	n.d.−83	Magnesium	0.045−4300
Arsenic	0.004−151	Nickel	n.d.−9.2
Barium	n.d.−1740	Oil/grease	2.3−60
Benzene	0.01−10.3	pH (−)	3.1−7.0
Biochemical oxygen demand	75−2870	Potassium	149−3870
Boron	n.d.−56	Silver	0.047−7
Bromide	150−1149	Sodium	520−120000
Cadmium	n.d.−1.21	Strontium	6200
Calcium	n.d.−25000	Sulfate	n.d.−47
Chloride	1400−190000	Surfactants	0.08−1200
Chromium	n.d.−0.03	Tin	n.d.−1.1
Chemical oxygen demand	2600−120000	Total organic carbon (TOC)	67−38000
Conductivity (umhos/cm)	4200−586000	Toluene	0.01−18
Copper	n.d.−5	Total dissolved solid	2600−360000
Iron	n.d.−1100	TSS	8−5484
Lead	0.2−10.2	Zinc	n.d.−5

The unit is in mg/L otherwise stated specifically.

considered more toxic than produced water from oil fields, although its volume is lower than oil field produced water, resulting in a lower total impact on the environment. The commonly used chemicals in gas field produced water include dehydration chemicals, hydrogen sulfite remover, hydrate inhibitors, and mineral acids (Jacobs et al., 1992).

14.1.2.4 Compounds contained in produced water from coalbed methane fields

In order to produce gas from coalbeds, a "dewatering" process is commonly applied. The first step is to lift a large amount of formation water from an underground coal seam to the surface, and lifted formation water is regarded as produced water. As more and more formation water is produced from a coal seam, the methane production increases due to the desorption effect caused by the depletion of reservoir pressure. The produced water at the surface is either reinjected into the formation or discharged to the surface after proved treatments (Nghiem et al., 2011). In conventional oil and gas reservoirs, formation water has been in contact with oil and gas for centuries and reached a chemical equilibrium for component exchanges. However, in coalbed methane fields, the formation water contact is mainly in contact with the coal seams (solid phase) in fractures, and gas adsorption and desorption processes in its dual-porosity system significantly affect the chemical equilibrium (He et al., 2016; Liang et al., 2017; Ning et al., 2014, 2015a; Zhou et al., 2018). Therefore, the compounds of produced water from coalbed methane fields differ significantly from produced water from conventional oil and gas fields, as given in Table 14.4.

Table 14.4 Constituents of produced water from coalbed methane fields (Ayers Jr., 2002; Veil et al., 2004)

Parameter	Range (mg/L)	Mean (mg/L)
Barium	0.1−8	0.6
Calcium	5.9−200	36
Chloride	3−119	13
Iron	0.02−15.4	0.8
Magnesium	1.6−46	16
Sodium	110−800	305
Sodium adsorption ratio	5.7−32	11.7
Sulfate	0.01−17	2.4
Total dissolved solid	270−2010	862

14.1.2.5 Compounds contained in produced water from shale/tight gas fields

Since 2009, exploration and production from unconventional gas reservoirs, including tight gas and shale gas reservoirs, has become an increasingly important part of global hydrocarbon production (Yuan, Zheng, et al., 2017). In terms of chemical compounds, there is no significant difference between unconventional gas and conventional gas; the possible difference lies in the formation rocks that the hydrocarbons remain in contact with for centuries (Law and Curtis, 2002; Ning et al., 2015b). In addition, since the permeability of shale or tight gas reservoirs is usually much smaller than conventional reservoirs, additional technology and chemical additives are applied during hydrocarbon exploration and production, resulting in a great variance in the chemical compounds contained in the produced water from shale/tight gas fields. For example, Rowan et al. (2011) reported that produced water from the Marcellus shale contains more radium and a lower radium isotopic ratio (0.3) than other reservoirs. Consequently, the radium isotopic ratio can be used as an indicator of radium sources in an unknown region.

Fig. 14.3 shows a standard procedure to produce gas from shale/tight reservoirs. First, a well (either horizontal or vertical) is drilled with some injection of a mixture of drilling fluids, solids, and chemicals into the formations. The purpose of the injection is to keep the bottom pressure balanced in case of hydrocarbon kicks, as shown in Step 1. Then, by injecting a large amount of hydraulic fracturing fluids, which is a mixture of water, proppant, acids, surfactants, and other chemical additives at high pressure, hydraulic fracturing is achieved to create well-connected narrow paths to increase permeability in the target formation, as shown in Step 2. According to the geological and operational conditions, substances added to hydraulic facture fluids vary from site to site (Wu and Sharma, 2017). Ferrer and Thurman (2015) discussed the chemical components of hydraulic fracturing fluids in detail, and Table 14.5 illustrates some commonly applied chemicals, additives, related chemical compounds, and corresponding functions. Pump trucks inject the hydraulic fracturing fluids downhole at high pressure to break the subsurface rocks and establish networks of pathways through which gas is transported toward the wellbore. Usually a well is fractured in multiple stages in order to obtain optimal results (Yuan et al., 2015a). It is typical to require larger amounts of fracturing fluids and sands for large-scale, multistage fracturing in

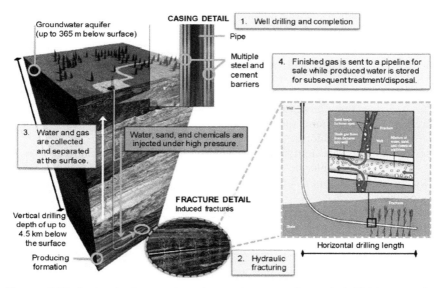

Figure 14.3 A standard procedure to produce gas from shale/tight reservoirs. *Reproduced with permission from Silva, T.L., Morales-Torres, S., Castro-Silva, S., Figueiredo, J.L., Silva, A.M., 2017. An overview on exploration and environmental impact of unconventional gas sources and treatment options for produced water. J. Environ. Manag. 200, 511–529.*

unconventional reservoirs. Hence, the recycling of such large amounts of drilling and/or hydraulic fracturing fluids becomes a huge challenge.

Hydraulic fracturing fluids, with a small amount of formation water, drilling, and fracturing chemical additives and hydrocarbons, usually flow back to the surface following the termination of the hydraulic fracturing operation, as shown in Steps 1 and 2. This period is commonly defined as the "flowback" period by most of operators, taking up to about 1–2 months. During this period, the characteristics of the flowback fluids are closer to the hydraulic fracturing fluids and the generated water is defined as "flowback water." Steps 3 and 4 demonstrate how the underground gas is brought up to the surface, separated from the produced water, and delivered to a final sale point after the flowback period. Water produced from a well after the flowback period is commonly defined as "produced water." As more and more hydrocarbons and water are produced from a well (usually after the first month of production), the chemical composition of produced water becomes closer to the formation water, indicating that less hydraulic fracturing fluids remain in the formation. Flowback

Table 14.5 Common chemical additives in hydraulic fracturing fluids (Ferrer and Thurman, 2015; Silva et al., 2017)

Additives	Compounds	Functions
Acids	Acetic acid, hydrochloride acid	Dissolving solids during injection
Biocides	Glutaraldehyde, quaternary ammonium compounds	Prevent grow of bacteria
Breakers	Sodium chloride, potassium chloride	Reduce viscosity of fluid
Clay stabilizers	Potassium chloride, sodium chloride	Prevent swelling of clays
Corrosion inhibitor	Amines, amides, amino amides	Prevent rust in pipelines
Cross-linkers	Borate salts, potassium hydroxide, ethylene glycol	Main viscosity of fluid
Friction reducers	Polyacrylamide, acrylamide, methanol	Minimize friction in pipeline
Gellants	Guar gum, propylene glycol, ethylene glycol	Increase fluid viscosity
pH controller	Sodium carbonate, potassium carbonate, acetic acid	Adjust pH of fluid
Proppants	Sand, ceramic beads, zirconium oxide, graphite	Keep fractures open
Scale inhibitors	Citric acid, sodium polycarboxylate	Prevent buildup of mineral scale
Surfactants	Ethylene glycol, isopropanol, phosphate esters, alcohol polyethoxylates	Increase viscosity of injected fluid, reduce surface tension

water can be regarded as a special type of produced water. In many studies, flowback water production and produced water production are combined together without a distinction, even though flowback water and produced water have different chemical and physical properties (Jackson et al., 2013; Ning et al., 2016a; Yuan et al., 2015b; Yuan and Wood, 2015). Table 14.6 provides the chemical composition of produced water from shale gas and tight gas fields.

14.1.3 Produced water volume and affecting factors

During oil and gas exploration and production activities, produced water is the largest volume of the waste stream. In 2014, globally about 220 million barrels of water was produced from onshore fields every day,

Table 14.6 Constituents of produced water from unconventional gas fields (Abousnina et al., 2015; Cluff et al., 2014; Hayes et al., 2012; Silva et al., 2017)

Parameter	Produced water constituents range	
	Shale gas fields	Tight gas fields
Alkalinity	169—188	1424
Aluminum	n.d.—5290	—
Ammonium	—	2.74
Arsenic	—	0.17
Barium	n.d.—4370	—
Boron	0.12—24	—
Bromine	n.d.—10600	—
Cadmium	—	0.37
Calcium	0.65—83950	3—74185
Chlorine	48.9—212700	52—216000
Chromium	—	0.265
Conductivity (umhos/cm)	—	24400
Copper	n.d.—15	0.539
Fluorine	n.d.—33	—
Iron	n.d.—2838	0.015
Lithium	n.d.—611	—
Magnesium	1.08—25340	2—8750
Manganese	n.d.—96.5	0.525
Nickel	—	0.123
Nitrate	n.d.—2670	—
Oil and grease	—	42
pH (—)	1.21—8.86	5—8.6
Phosphate	n.d.—5.3	—
Potassium	0.21—5490	5—2500
Sodium	10.04—204302	648—80000
Strontium	0.03—1310	—
Sulfate	n.d.—3663	12—48
Zinc	n.d.—20	0.076

The unit is in mg/L otherwise stated specifically.

compared to about 90 million barrels per day for offshore produced water production. In contrast, the global oil production was about 78 million barrels per day, yielding an average water—oil ratio (WOR) of 3.0. For every one barrel of oil produced, there are three barrels of produced water brought up to surface simultaneously (Fig. 14.4), requiring a large portion of space at the surface dedicated to produced water management (Clark and Veil, 2009). Consequently, produced water management accounts for a high percentage of the total operation costs. Fig. 14.4

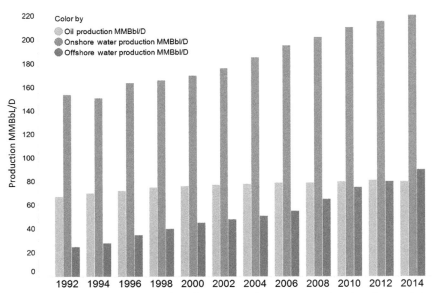

Figure 14.4 Global oil and produced water production. *Reproduced with permission from Fakhru'l-Razi, A., Pendashteh, A., Abdullah, L.C., Biak, D.R.A., Madaeni, S.S., Abidin, Z.Z., 2009. Review of technologies for oil and gas produced water treatment. J. Hazard. Mater. 170, 530-551; Khatib, Z., Verbeek, P., 2003. Water to value-produced water management for sustainable field development of mature and green fields. J. Pet. Technol. 55, 26-28; Zheng, J., Chen, B., Thanyamanta, W., Hawboldt, K., Zhang, B., Liu, B., 2016. Offshore produced water management: a review of current practice and challenges in harsh/Arctic environments. Mar. Pollut. Bull. 104, 7−19.*

shows global oil, offshore produced water, and onshore produced water production from 1992 to 2014, indicating a progressive increase in the cost for produced water management each year. It is notable that offshore produced water volume has increased at a significant rate over the past decades. Moreover, the average WOR of an onshore well in the United States is close to 7.5, which is higher than the global average WOR (Clark and Veil, 2009).

14.1.3.1 Typical features of the produced water volume for different types of reservoirs

The produced water volume and corresponding WOR change over time in producing fields. For conventional oil and gas wells, the volume of produced water increases through their production lives, leading to a lower hydrocarbon recovery and higher waste management cost each

Figure 14.5 Typical oil and water production for an oil well in West Louisiana.

year. Fig. 14.5 shows a typical conventional oil well in west Louisiana: at the beginning of production, produced water only accounts for a small percentage of the fluids brought up to the surface. As oil production decreases over time, the produced water production tends to increase, resulting in a higher WOR and lower profit margin. For some oil wells, at the end of their production lives, the water cut can be as high as 98% (Khatib and Verbeek, 2003). Onshore wells usually produce more produced water than offshore wells, while conventional oil wells typically produce higher volumes of produced water than conventional gas wells. On the other hand, unconventional oil and gas resources, because of the large amount of hydraulic fluids injected into the formation and brought up to the surface later, produce more produced water (including flowback water) than conventional resources.

Unconventional reservoirs usually include coalbed methane, shale oil/ gas, and tight oil/gas reservoirs. Produced water production in coalbed methane fields is highly sensitive to formation permeability, which is an inverse bell-shaped curve (Fig. 14.6) when plotted as a function of pore pressure (Liang et al., 2017). Therefore, coalbed methane fields usually generate a large amount of produced water from their coal seams at the beginning, which decreases dramatically as production advances. Shale gas/oil reservoirs and tight gas/oil also produce a large amount of

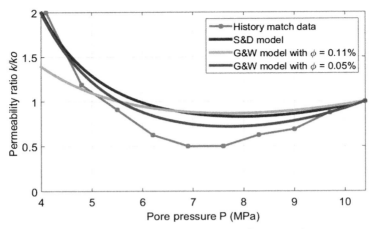

Figure 14.6 Permeability in coal seam changes as a function of pore pressure. The inverse bell curve is predicted by models. *Reproduced with permission from Liang, Y., 2017. Scaling of Solutal Convection in Porous Media. The University of Texas at Austin, Austin, TX.*

produced water (including flowback fluids) initially, the composition of which is close to the injected hydraulic fracturing fluids, but the volume of produced water drops significantly after the flowback period (Liang and Yuan, 2017a; Yuan et al., 2016a). If waterflooding is applied to heavy or light oil reservoirs, the volume of produced water increases dramatically after the breakthrough time.

14.1.3.2 Factors affecting produced water volume

The volume of produced water, similar to its chemical compositions, also differs significantly depending on many factors. For a conventional oil and gas well, generally an increasing produced water production through its production life is inevitable, resulting in a lower profit margin as the field ages because of higher waste management cost and lower hydrocarbon recovery. Therefore, understanding factors that influence produced water volume is essential to extend a field's economic life. Reynolds and Kiker (2003) summarized major factors that can affect the volume of produced water, including location of a well, type of well, type of completion, type of separation facilities, EOR, well integrity, new perforations, and subsurface communication.

- *Location of a well*: A well can produce more produced water, or start to produce water earlier than expected, if the well is improperly located within the reservoir structure/formation.
- *Type of well*: For a water-drive reservoir, a horizontal well can produce at higher rates or at lower pressure drawdown, delaying the break-through of produced water as a result, under the assumption that the impact of flowback water is not taken into consideration.
- *Type of completion*: In a pay zone, a perforated completion can offer better control of a well than an open-hole completion. If produced water volume at a specific interval is too high, significantly impedes oil production, and negatively affects the profit margin, it can be plugged and/or isolated in order to elongate the well's economic life.
- *Type of separation facilities*: Depending on different technologies, pro-duced water can be separated at either the surface or downhole. If downhole water/oil separation equipment or downhole water/gas sep-aration equipment is used, then the produced water can be directly disposed of into nonproducing formations below or above the produc-tion zone. A typical downhole separation assembly includes a separa-tion tool to separate water from hydrocarbon, a pump to pressurize water, a heavy-duty motor to operate the separation tool, and miscel-laneous pipe and valve equipment.
- *EOR method*: If waterflooding is applied to a reservoir, water is injected from an injector well, and pushes/sweeps remaining oil toward a producer well. Initially more oil can be produced; as the water reaches the producer well (the breakthrough point), the volume of the produced water increases significantly in the producer well, resulting in lower hydrocarbon recovery.
- *Well integrity*: Loss of well integrity can result in flow of fluids, includ-ing formation water, from other zones or aquifers into wellbore. Many factors can cause well integrity issue, including excessive pressure, cor-rosion, poorly cemented casing and formation deformation (Feng and Gray, 2016a, b).
- *Re-perforations*: When the initial completion zones of a well are depleted, new pay zones at other depths are sometimes available to be perforated in order to generate more hydrocarbon recovery and cash flow for profit maximization. Opening a new zone may result in a higher produced water volume at the surface if the initial water satura-tion in the new zone is high.

- *Subsurface communication*: Subsurface communication problems, including near-wellbore communication problems (completion near water, barrier breakdowns, and channels behind pipe) and reservoir communication problem (coning, cresting, channeling through more permeable zones), can result in higher produced water volumes in specific wells.

14.1.3.3 Produced water volume from onshore, offshore, and unconventional oil fields

Due to the difficulties in data collection, the latest reliable data is published by Argonne National Laboratory. In the report, Clark and Veil (2009) summarized, by state, the US onshore and offshore oil, gas, and produced water production for 2007, as given in Table 14.7. According to the report, the total volume of produced water generated in the United States in 2007 was 21 billion barrels with an average production of 57.4 million barrels per day; Texas made up the largest percentage of total produced water in the United States, accounting for 35% and followed by California (12%), Wyoming (11%), Oklahoma (11%), and Kansas (6%). These states represented approximately 75% of the total US produced water production. Based on the production–weighted averaged method, the national average offshore WOR is 5.3 and the national average onshore WOR is 7.6. Fig. 14.7 shows a treemap chart of the US produced water volume data. The size of the blocks represents the annual crude oil production and the color intensity indicates corresponding WORs. The treemap highlights that while both Texas and Alaska are two major producing states, Alaska has significantly lower WOR than other major oil producing states.

There are large amounts of offshore oil and gas fields in the Arctic region where produced water can result in more severe environmental issues due to its complexity. Table 14.8 provides production data from fields in the Arctic region. The produced water production, gas production, and oil production vary from site to site, while the WOR of those fields are approximately 3.0, which is consistent with the global average WOR. More attention needs to be paid in those regions because the extra-cold environment could exacerbate the potential environmental issues associated with produced water handling (Noble et al., 2013).

The implementation of multistage fractures or fracture clusters in horizontal wells has enabled the commercial production from such naturally

Table 14.7 2007 US oil, gas, and produced water volume

State	Annual oil production (MMbbl)	Annual gas production (Bcf)	Annual produced water production (MMbbl)
Alabama	5.0	285.0	119.0
Alaska	263.6	3498.0	801.3
Arizona	0.0	1.0	0.1
Arkansas	6.1	272.0	166.0
California	244.0	312.0	2552.2
Colorado	2.4	1288.0	383.8
Florida	2.1	2.0	50.3
Illinois	3.2	0.0	136.9
Indiana	1.7	4.0	40.2
Kansas	36.6	371.0	1244.3
Kentucky	3.6	95.0	24.6
Louisiana	52.5	1382.0	1149.6
Michigan	5.2	168.0	114.6
Mississippi	20.0	97.0	330.7
Missouri	0.1	0.0	1.6
Montana	34.7	95.0	182.3
Nebraska	2.3	1.0	49.3
Nevada	0.4	0.0	6.8
New Mexico	59.1	1526.0	665.7
New York	0.4	55.0	0.6
North Dakota	44.5	71.0	135.0
Ohio	5.4	86.0	6.9
Oklahoma	60.8	1643.0	2195.2
Pennsylvania	1.5	172.0	3.9
South Dakota	1.7	12.0	4.2
Tennessee	0.4	1.0	2.3
Texas	342.1	6878.0	7376.9
Utah	19.5	385.0	148.6
Virginia	0.0	112.0	1.6
West Virginia	0.7	225.0	8.3
Wyoming	54.1	2253.0	2355.7
Federal Offshore	467.2	2787.0	587.4
Tribal Lands	9.5	297.0	149.3
Total	1750.5	24374.0	20995.2

Source: Based on the data from Clark, C.E., Veil, J.A., 2009. Produced Water Volumes and Management Practices in the United States. Argonne National Laboratory (ANL), Washington, DC.

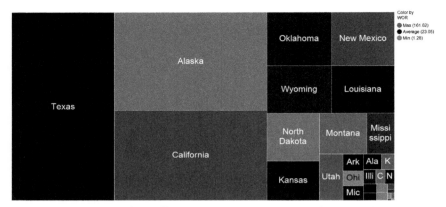

Figure 14.7 Treemap of US offshore oil production (size) and WOR (color intensity). *Reproduced with permission from Clark, C.E., Veil, J.A., 2009. Produced Water Volumes and Management Practices in the United States. Argonne National Laboratory (ANL), Washington, DC.*

Table 14.8 Volume of produced water from oil fields in the Arctic region (Clark and Veil, 2009; Harsem et al., 2011; Zheng et al., 2016)

Region	Field/ country	Annual oil production (MMbbl)	Annual gas production (Bcf)	Annual water production (MMbbl)	WOR
Alaska	Cook Inlet	96.9	137.3	368.0	3.8
North Atlantic	Hibernia	262.9	166.0	761.1	2.9
North Sea	Denmark	71.1	302.2	215.1	3.0
North Sea	Norway	528.3	2189.5	1258.0	2.4

"impermeable" zones, while produced water production from fracturing (flowback water) and production (produced water) stages has created some new issues associated with waste treatment and water handling for operators (Moghaddam et al., 2012; Ning et al., 2016b; Yuan et al., 2016c). Kondash et al. (2017) studied the quantity of flowback and produced water from unconventional resources. Flowback and produced water volume ranges from 0.5 to 3.8 million gallons/well over the first 5—10 years of their production lives, varying from field to field as is shown in Fig. 14.8. For a typical well in unconventional fields, 20%—50% of the total flowback and produced water is produced within the first 6 months.

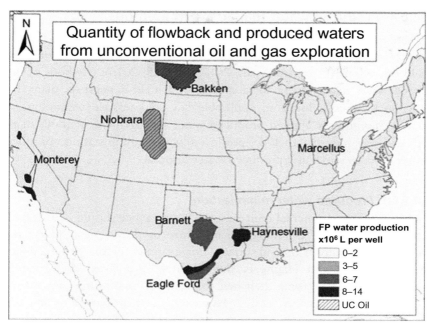

Figure 14.8 Geomap of flowback and produced water volume from unconventional fields. *Kondash, A.J., Albright, E., Vengosh, A., 2017. Quantity of flowback and produced waters from unconventional oil and gas exploration. Sci. Total Environ. 574, 314–321.*

14.2 PRODUCED WATER MANAGEMENT METHODS AND TECHNIQUES

In this section, major produced water management methods, including produced water minimization, produced water reuse/recycle, and produced water disposal, are discussed in detail. No matter which option is selected to manage produced water, a series of national and local guidelines, permits, regulations, and laws need to be adhered to by operators. Therefore, legal frameworks, policies, and regulations related to produced water from onshore and offshore oil and gas fields are considered in detail. In order to meet those requirements, produced water needs to be treated before any management operation. Consequently, this section also introduces commonly applied produced water treatment technologies applied in oil and gas industry.

14.2.1 Produced water management methods

Produced water is regarded as an oil field waste because of its negative impacts on the environment, generating a high economic burden on oil field operations (Liang and Yuan, 2017b; Pichtel, 2016). Methods of produced water management for a specific well/field depend on many factors, such as cost, location, applicable law, and feasible technology. In general, commonly applied produced water management methods in the oil and gas industry include produced water minimization, produced water recycle/reuse, and produced water disposal.

14.2.1.1 Produced water minimization

In terms of environmental concerns, the best practice should be the application of advanced technology to minimize produced water production being extracted from the subsurface. For example, before drilling a new well, geologists and engineers should carefully analyze all formations along the wellbore, trying their best to avoid perforating into water zones or high-water saturation zones. However, as hydrocarbon is depleted from a reservoir, a rising oil—water contact is typically inevitable, resulting in a dramatic increase in produced water volume, as shown in Fig. 14.5. Table 14.9 summarizes some common reservoir and production technologies that can restrict produced water from entering a production wellbore.

14.2.1.2 Produced water recycling and reuse

Produced water can be recycled and reused if its quality and composition meet national and local regulations and laws, and then its categorization is switched from "waste" to "resource." In most cases, produced water from oil and gas fields cannot meet those requirements; therefore, produced water treatments are mandatory before any recycling activities can commence. Coalbed methane is an exception; in some coalbed methane fields, the produced water naturally meets local requirements and no further activities are needed. Common produced water recycling and reuse options include PWRI, consumption by farm animals, irrigation, and industrial use.

Produced water can be reinjected into virgin or depleted formations, and this process is defined as PWRI. The most important objectives of PWRI are improved/enhanced oil recovery (IOR/EOR) and produced water disposal. These two objectives do not counteract each other. In

Table 14.9 Commonly used produced water minimization technologies in the oil and gas industry

Options	Technology facts
Mechanical blocking devices	Including packers, bridge plugs, sand plugs, cements, and other devices. This method is effective in blocking casing leaks and flow behind pipe, while it is ineffective in other produced water issues.
Water shutoff chemicals	Injection of polymer gels into formations effectively blocks the path of water to the wellbore while it has less impact on the path of hydrocarbons. Seright et al. (2001) reported that improved profits are obtained after application of water shutoff chemicals.
Dual-completion wells	Swisher (2000) reported a dual-completion well in Louisiana produced 55 bpd of oil while a nearby single-zone completion wells produced only 16 bpd, indicating a good produced water minimization result. However, the capital investment of dual-completion costs twice as much.
Downhole oil/water separators (DOWS)	DOWS separates water from oil within the wellbore and directly inject the produced water into the formation (Veil et al., 1999).
Downhole gas/water separators (DGWS)	DGWS separates water from gas within the wellbore. Veil et al. (1999) showed that the gas production rate increased 57% with some application of DGWS. However, 43% of tests failed due to poor injectivity issues.
Subsea separator	Subsea separation separates water from oil at the sea floor. The capacity of subsea separation is much higher than DOWS but is more expensive. Von Flatern (2003) reported that this technology enabled the Troll platform to produce an additional 2.5 million barrels of oil during the first year of its trail.

some cases, produced water is just injected into permeable formations only for disposal purpose. In most cases, produced water is injected into a target formation from an injector as a resource for IOR waterflooding and EOR steam flooding, sweeping more oil to producer wells. Once hydrocarbons are depleted in a production zone, the produced water can then be reinjected into that zone gradually filling the space originally occupied by the produced hydrocarbons. After injection of produced water, in some cases, there is the potential for supercritical CO_2 to also be injected into that produced water saturated formation for the purpose of CO_2 sequestration and/or EOR (Liang, 2014; Liang et al., 2015; Yuan, Zing, et al., 2017). It should be noted that treatments are also required for reinjected water at the surface to prevent wellbore/reservoir blockages and system failures. For onshore fields, PWRI is a common practice while it only makes up a small percentage in offshore fields due to higher costs and technical difficulties. Clark and Veil (2009) report that 98% of onshore produced water was reinjected for EOR and disposal purpose while only 9% of offshore produced water was injected in the US 2007. There are also very promising examples in unconventional reservoirs; 87% of produced water is reused in Marcellus shale (Zheng et al., 2016). However, the deposition of suspended particles during rejection may block pores and pore throats, resulting in reduced permeability (increased impedance, i.e., the inverse of injectivity) and thus the formation damage near the wellbore, as conceptually depicted in Fig. 14.9.

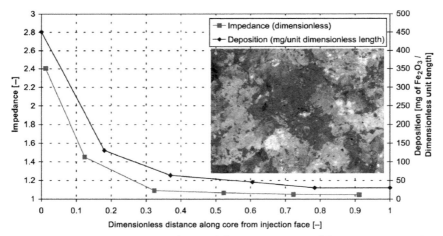

Figure 14.9 Microscope image of cores following injection of fluids with suspended particles (gray color). As water with solids is injected into the core, deposition of particles occupy pore spaces and block transport of fluid, contributing to a reduction in permeability (increasing impedance, defined as the inverse of injectivity) and formation damage near the wellbore. *Reproduced with permission from Al-Abduwani, F.A., Shirzadi, A., van den Brock, W., Currie, P.K., 2005. Formation damage vs. solid particles deposition profile during laboratory simulated PWRI. SPE J 10,138—151.*

With appropriate treatments, produced water can be used as drinking water for farm animals or for crop irrigation. According to a study by Veil et al. (2004), animals can drink water with total dissolved solids ranging from 1000 to 7000 ppm. The study also reports successful examples in Wyoming of animals being watered by produced water from oil fields nearby. Produced water also has a potential to be used as irrigation sources if its salinity, sodicity, and toxicity meet relevant requirements, but this process involves significant treatment.

As long as produced water satisfies industrial requirements, it can also be used as industrial water sources. The successful applications include oil field dust suppression, vehicle and equipment washing and power generation. It is reported that with certain treatments 360,000 bpd of produced water from a field in California is used as boiler feedwater in a cogeneration plant (Veil et al., 2004).

14.2.1.3 Produced water disposal

From the perspective of cost effectiveness, produced water disposal is the best method. Once separated from oil, the produced water is either discharged into surface watercourses or injected into the subsurface via wells. Due to the strict prohibition by laws, most US offshore wells reinject their produced water into the reservoir formation for either disposal or hydrocarbon enhancement; meanwhile most US onshore wells directly dispose their produced water into watercourses and rivers. Produced water can also be discharged into evaporation pits for natural evaporation, which is the case in some coalbed methane operations, but this can be costly if subsequent treatment/restoration is required.

According to the study from Argonne National Laboratory, 40% of offshore produced water was reinjected into formations for disposal purpose and 91% of onshore produced water was discharged to the ocean in the US 2007, as shown in Fig. 14.10. Inserts are corresponding percentage of each management practice. The data in the figure is based on the report by Clark and Veil (2009).

Directly disposing of produced water can present great threats to human, animals, the environment, and the whole ecosystem. Frost et al. (1998) state that many factors can determine the level of impacts, including the physical and chemical properties of constituents, the content of dissolved organic materials, the concentration of key elements, pH, chemical environment, temperature, metabolism, presence of other organic contaminants, and other factors.

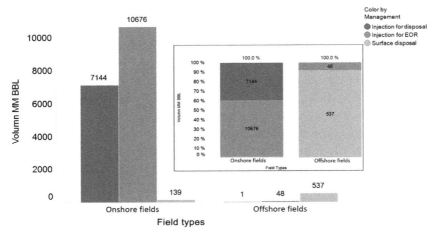

Figure 14.10 2007 US produced water volume by management practice. Insert graph shows corresponding percentages. *Reproduced with permission from Clark, C.E., Veil, J.A., 2009. Produced Water Volumes and Management Practices in the United States. Argonne National Laboratory (ANL), Washington, DC.*

In particular, the concentrations and corresponding characteristics of constituents are the most important factors in determining the impacts on the environment of offshore produced water disposal. A higher concentration of a more toxic compound can contribute to larger negative impacts on the environment than for compounds with lower concentrations and/or less toxicity. Georgie et al. (2001) summarized factors that affect the concentrations of constituents in seawater after disposal, including instantaneous and long-term precipitation, volatilization of low-weight hydrocarbon, physical/chemical reaction potential, adsorption, biodegradation, and dilution.

14.2.2 Legal framework, policy, and regulations

Once brought to the surface, produced water is separated from oil by gravity separators. However, the produced water may still contain additional emulsions and dissolved oil that are harmful to the receiving environments. Therefore, different guidelines, permits, regulations, and laws are cautiously developed and strictly enforced by local legislations and administrations, in order to minimize the potential impacts of produced water disposal and reinjection.

In the United States, produced water management is generally categorized into disposal and injection operations. Most of the onshore produced water is injected while most of the offshore produced water is discharged, regulated by different federal and state laws. At federal level, the Clean Water Act (CWA), Underground Injection Control (UIC), Safe Drinking Water Act (SDWA), and Effluent Limit Guidelines (ELGs) are highly relevant to oil and gas produced water management (Clark and Veil, 2009; Konschnik and Boling, 2014; Percival et al., 2015; Rebello et al., 2016). SDWA and UIC are used to regulate activities related to reinjection of produced water into formations. CWA regulates the disposal of produced water to the surface environment, stating that all discharges of pollutants at the surface must by authorized by the National Pollutant Discharge Elimination System, in order to prevent high-concentration toxic substances flowing into surface water bodies. ELGs are national technology-based minimum discharge requirements set by the Code of Federal Regulations (CFR). There are also different regional limits with respect to discharge rate, toxicity testing, and monitoring for certain harmful constituents, varying from state to state. In general, state programs are more stringent than federal requirements. The highlights of those guidelines, permits, regulations, and laws in the United States include:

- Onshore oil and gas operations may not discharge produced water into navigable waters with two specified exceptions according to Subpart C of 40 CRF Part 435: (1) facilities located west of the 98th meridian and (2) facilities that produced 10 barrels per day or less of crude oil.
- Oil and gas operations in coastal waters may not discharge produced water to the marine environment. Cook Inlet in Alaska is the only coastal location in the United States that allows disposal of produced water with the oil and grease limits of 42 mg/L daily maximum and 29 mg/L monthly average.
- Offshore oil and gas operations are allowed to discharge produced waters into sea, under the ELGs that set the limits for oil and grease at 42 mg/L daily maximum and 29 mg/L monthly average.
- Underground injection is grouped into five classes. Class I wells are used for the emplacement of hazardous, industrial, and municipal waters, and these wells are most strictly regulated by SDWA and Resource Conservation and Recovery Act. Class II includes wells with injection of brines and fluids associated with oil and gas. Class III includes wells with injection of fluids associated with minerals. Class IV includes wells with injection of hazardous or radioactive materials

into or above sources of drinking water, and this class is highly restricted. Class V includes the rest of injection wells. In general, the wells used to reinject produced water belong to Class II, with Class II−R for oil enhancement and Class II−D for disposal purposes.

• According to 40 CRF, all Class II−D wells are required to apply for a permit for the first year while Class II−R wells require authorization for the life of a well. Every new Class II−D and Class II−R well must be authorized before construction and injection can commence. Operators must demonstrate the internal and external integrity of the proposed injection wells. Operators must plug and abandon injection wells that have not been involved in any activity for 2 years. During any operation, the injection pressure cannot exceed the maximum injection pressure that may initiate new fractures. The owner must monitor and report conditions of injecting wells following the requirements by the guidance.

• The Bureau of Land Management in the US Department of the Interior has jurisdictions over onshore leasing, exploration, and production over federal lands while the Bureau of Safety and Environmental Enforcement (the BSEE) manages the operations on the Outer Continental Shelf.

Varying from country to country, there are different national limits with respect to discharge rate, toxicity testing, and monitoring for certain harmful constituents. Table 14.10 provides oil and grease limits for produced water from offshore fields worldwide.

14.2.3 Produced water treatment technologies

In order to meet guidelines, permits, regulations, and laws set by federal and local administrations, produced water needs to be treated before it can be reinjected or discharged. The general objectives of produced treatment include dispersed oil and greases removal, soluble organics removal, suspended sand removal, dissolved gas removal, radioactive materials removal, desalination, and disinfection. Several studies review technologies that are commonly applied in the produced water treatments (Fakhru'l-Razi et al., 2009; Igunnu and Chen, 2012; Silva et al., 2017; Zheng et al., 2016). In some cases, the identified technologies can be utilized separately. In other cases, a combination of different technologies can achieve better results. For instance, Barrufet et al. (2005) proposed a system to purify produced water into irrigation water, and the system

Table 14.10 Oil and grease limits for offshore produced water worldwide (Clark and Veil, 2009; Liang and Yuan, 2017b; Mahmoudi, 1997; Pavasovic, 1996; Tromp and Wieriks, 1994; Zheng et al., 2016)

Region/ country	Legislation	Oil and grease (mg/L) Daily maximum	Oil and grease (mg/L) Average
Alaska	40 CFR 435	42	29
Australia	—	50	30
Baltic Sea	HELCOM Convention	15	—
Brazil	—	20	—
Canada	Act RSC 1987	60	30
China	GB 4914-85	70	30−50
Indonesia	MD KEP 3/91; 42/97	100	75
Mediterranean	Barcelona Convention	100	40
North Sea	OSPAR Convention	—	30
Red Sea	Kuwait Convention	100	40
Thailand	NEQA 1992: Gov. Reg. 20/90	100	40
Vietnam	Decision no. 333/QB 1990	—	40

consists of sorption pellets, microfiltration (MF) membranes, and reverse osmosis (RO) membranes. This integrated system is more efficient than purification systems equipped solely with membranes.

14.2.3.1 Adsorption technology

Based on the feature that organic compounds and some heavy metals can adhere to carbon surfaces, adsorption technology is utilized to remove manganese, iron, total organic carbon (TOC), BTEX, and heavy metal from produced water while it can achieve almost 100% water recovery. The adsorption process is usually combined with other technologies such as membranes and media filtration. The commonly used adsorbents include activated carbon, activated alumina, and activated zeolites. The performance of adsorbents is affected by pH, temperature, suspended oil amounts, metal concentration, salinity, and dissolved contaminants. Janks and Cadena (1992) reported a successful case of BTEX removal in produced water by zeolites with an efficiency of 70%−85%. Spellman (2013) stated that over 80% of heavy metals could be removed by the adsorption technology. Chen et al. (2017) used Prussian blue and anion exchange to remove cesium. Both simulated rapid sand filter tests and field column tests demonstrate the effectiveness of cesium removal using the proposed

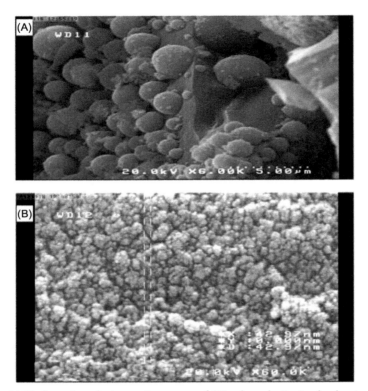

Figure 14.11 Scanning electron microscope of (A) ash and nFe-A adsorbents, which are used to remove manganese, iron, TOC, BTEX, and heavy metal from produced water. *Reproduced with permission from Ghasemi, M., Naushad, M., Ghasemi, N., Khosravi-Fard, Y., 2014. Adsorption of Pb (II) from aqueous solution using new adsorbents prepared from agricultural waste: adsorption isotherm and kinetic studies. J. Ind. Eng. Chem. 20, 2193–2199.*

adsorbents. Fig. 14.11 shows an example of scanning electron microscope images of surface morphology of ash and nFe-A adsorbents, which are used to remove manganese, iron, TOC, BTEX, and heavy metal from produced water. The adsorption process tends to be inefficient when the feed concentration is high, and it may generate additional waste.

14.2.3.2 Chemical oxidation technology
Based on oxidation and reduction reactions, chemical oxidation technology is used to remove color, odor, organic compounds, and inorganic compounds from produced water by applying a series of oxidants and

catalysts. The commonly used oxidants include chlorine, ozone, oxygen, peroxide, permanganate, and others. Even though oxidation processes involve many different reacting systems, they initiate similar chemical reactions, which is the production of OH radicals. Andreozzi et al. (1999) discussed the most popular oxidation processes in detail, including the Fenton processes (H_2O_2 and Fe^{2+}/Fe^{3+} as catalysts), photo-Fenton processes, photocatalysis, and ozone processes, concluding that oxidation technology is not more effective than other technologies. Moraes et al. (2004) reported a successful field application of the photo-Fenton process powered by solar energy in treating produced water, using ferrous ions (Fe^{2+}), hydrogen peroxide (H_2O_2), and ultraviolet (UV) irradiation as a source of hydroxyl radicals. The TOC removal efficiency reached 65%–80% in 4.5 hours. Aziz and Daud (2012) reported an application of the Penton process in oil fields effectively reduced 70% of the TOC and 98% of chemical oxygen demand (COD) from the produced water. Posada et al. (2014) reported catalytic wet air oxidation with a Cu/CeMnO catalyst at 160°C and 10 bar applied to remove organic pollutants from produced water, resulting in a TOC reduction of 80% (original TOC from 1249 to 1442 mg/L) in 40 minutess. The concept that dissolved oil can be removed by ozone was proposed by Morrow et al. (1999). Klasson (2002) conducted a series of experiments, showing that almost all extractable organic material could be destroyed by ozone following 3 days of exposure. McGuire (2016) proposed a pressurized system to reduce calcium carbonate in the produced water by ozone and electrochemicals.

14.2.3.3 Chemical precipitation technology

Coagulation and flocculation can be applied to de-emulsify, coalesce, aggregate, and remove suspended particles and large oil droplets from produced water. The commonly applied methods in the industry include lime softening, inorganic mixed metal FMA (Fe, Mg, and Al) polynuclear polymer, and other combinations of chemicals. Houcine (2002) reported successful applications of calcite and lime for heavy metals removal with efficiency of greater than 95%. Tansel and Pascual (2011) reported that applications of coagulant in the dissolved air flotation process can achieve an additional 5%–15% emulsified fuel oil removal, and its mechanism is conceptually depicted in Fig. 14.12. Chemical precipitation technology may generate a large volume of sludge (oil field waste) and its operational cost is high compared to other technologies.

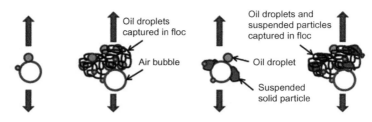

(A) Without coagulant   (B) With coagulant   (C) Without coagulant   (D) With coagulant
Brackish water with no suspended solids   Pond water with suspended solids

Figure 14.12 Removal of oil and suspended particles by air flotation with and without coagulation. *Reproduced with permission from Tansel, B., Pascual, B., 2011. Removal of emulsified fuel oils from brackish and pondwater by dissolved air flotation with and without polyelectrolyte use: pilot-scale investigation for estuarine and near shore applications. Chemosphere 85, 1182−1186.*

14.2.3.4 Electrodialysis and electrodialysis reversal technology

For the purpose of desalination, electrodialysis (ED) and electrodialysis reversal (EDR) involves a separation of dissolved ions from produced water via a series of ion exchange membranes. Only anions can pass through the membrane when it is positively charged, and vice versa. EDR includes the process of periodic reversal of polarity. Currently most ED and EDR studies are carried out on a laboratory scale, and its biggest challenge is high equipment and treatment costs. In addition, ED processes can generate HCl and NaOH as chemicals for desalination treatments.

Reig et al. (2016) proposed a combination of nanofiltration (NF) and electrodialysis with bipolar membranes for desalinating brines producing acids and bases under recirculation configuration, yielding a conversion percentage higher than 70% with energy consumption of 2.6 kWh/kg NaOH and current efficiency of 77%. Fig. 14.13 shows the proposed valorization process of seawater desalination brines.

Lopez et al. (2017) investigated ED and electrodeionization (EDI) to remove ions from brackish water. Results show that EDI is capable of removing contaminant ions at energy consumptions of $0.9-1.5$ kWh/m^3 water recovered with high-water productivity at $40-90$ L/ m^2. These results suggest that ED and EDI are feasible processes for brackish water desalination. Fig. 14.14 shows a schematic of the ED and EDI processes.

Hao et al. (2015) reported a system of combined coagulation with ED to treat produced water in order to meet local requirements. The integrated system has advantages of higher water recovery (85%) and longer

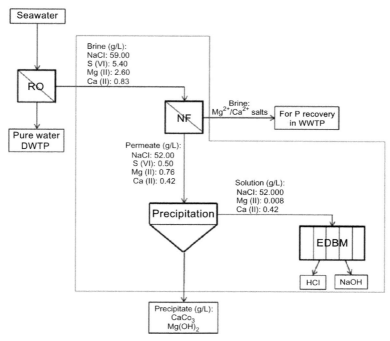

Figure 14.13 Valorization process of seawater desalination brines. *Reproduced with permission from Reig, M., Casas, S., Gibert, O., Valderrama, C., Cortina, J.L., 2016. Integration of nanofiltration and bipolar electrodialysis for valorization of seawater desalination brines: production of drinking and waste water treatment chemicals. Desalination 382,13–20.*

operation capacity. The results show that the coagulation process can effectively reduce 93% of COD, 94% of biochemical oxygen demand, and 99% of turbidity while the ED process can effectively reduce 91% of ion in produced water.

14.2.3.5 Evaporation technology
In order to prevent subsurface infiltration and downward migration, evaporation technology, including vertical tube, falling film, and vapor compression evaporation methods, can be applied to distill the produced water in an evaporation pit. For the purpose of desalination of high-salinity produced water, Onishi et al. (2017) proposed an optimization model for the single- or multiple-effect evaporation (SEE/MEE) systems with horizontal filling films, integrating mechanical vapor recompression (MVR) and heat recovery. In addition, a multiple-effect superstructure

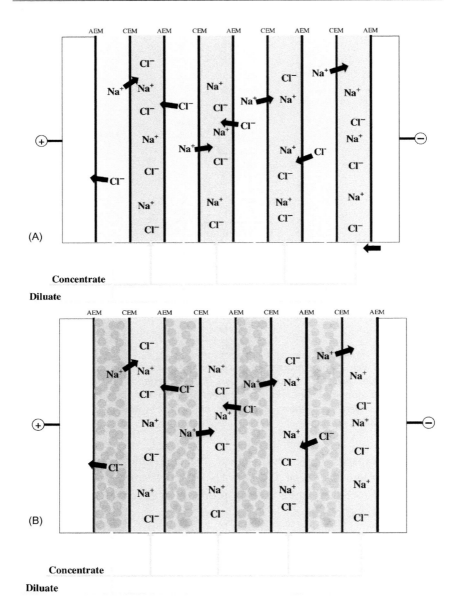

Figure 14.14 Schematic of (A) ED and (B) EDI. AEM is the anion exchange membrane and CEM is the cation exchange membrane. *Reproduced with permission from Lopez, A.M., Williams, M., Paiva, M., Demydov, D., Do, T.D., Fairey, J.L., et al., 2017. Potential of electrodialytic techniques in brackish desalination and recovery of industrial process water for reuse. Desalination 409, 108–114.*

with horizontal falling films, intermediate flashing tanks, and a distillate preheater are included in order to enhance the energy efficiency and reduce brine discharge. By comparison with the SEE—MVR system applied to a range of scenarios to desalinate produced water with different salinity, the horizontal falling film MEE—MVR system (Fig. 14.15) can save about 35% of the process cost and is therefore preferred; while the zero-liquid discharge condition is achieved in all cases, yielding a high freshwater recovery ratio of 0.77.

14.2.3.6 Gas flotation technology

While gas is injected into produced water, gas bubbles tend to migrate upward due to lower density. Suspended particles and oil droplets can adhere to the bubbles and rise to the top as a result. Based on the physics, flotation technology, including the dissolved gas flotation type and induced gas flotation type, is widely used to separate suspended particles, grease and oil, natural organic matter, and volatile organics, which are difficult for other technologies to cope with. Suspended particles as small as 25 μm can be removed by gas flotation. If pretreatment is properly applied, even contaminants up to 3 μm can be separated from produced water. Gas flotation can easily remove lighter and smaller particles while removal of dissolved oil requires chemical pretreatments. The gas flotation systems are usually compact and space efficient.

Saththasivam et al. (2016) reported an overview of gas flotation systems that are commonly used in the industry, as summarized in Table 14.11. Fig. 14.16 demonstrates a schematic of a conventional jet flotation system and a modified jet flotation system (Santander et al., 2011). Performance of gas flotation systems can be improved by pretreatment of coagulants, which destabilizes oil—water emulsion to promote droplet coalescence (Tansel and Pascual, 2011). The most commonly used coagulants include alum, aluminum sulfate, ferric sulfate, and ferric chloride, and studies show removal efficiency of 79%, 93%—99%, 72%—99%, and 73%—95%, respectively (Al-Shamrani et al., 2002; El-Gohary et al., 2010; Hoseini et al., 2015; Zouboulis and Avranas, 2000).

da Silva et al. (2015) studied the conjugation of flotation and the photo–Fenton processes to remove oil from produced water at the laboratory scale. A series of experiments were conducted in a column flotation and annular lamp reactor with a nonionic surfactant as a flotation agent. The results show a high oil removal efficiency of 99%, involving 10 minutes of the flotation process and 45 minutes of the photo–Fenton process.

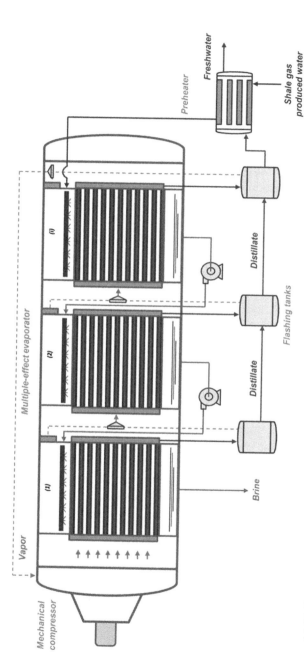

Figure 14.15 Superstructure of MEE—MVR system for desalination of high-salinity produced water. *Reproduced with permission from Onishi, V.C., Carrero-Parreno, A., Reyes-Labarta, J.A., Fraga, E.S., Caballero, J.A., 2017. Desalination of shale gas produced water: a rigorous design approach for zero-liquid discharge evaporation systems. J. Clean. Prod. 140, 1399–1414.*

Table 14.11 Summary of gas flotation technologies in produced water treatment (Saththasivam et al., 2016)

Model	System	Capacity (m^3/day)	Effluent oil concentration (mg/L)
AutoFlot—Mechanical	Horizontal mechanical induced air flotation	1248—27264	Not available
Epcon DualCFU	Centrifugal force and flotation	72—24000	<10
Hydrocell Hydraulic IAF	Horizontal hydraulic induced air flotation	556—15900	Not available
Quadricell	Mechanical induced gas flotation	<5000	<5
Revolift VS Flotation	Induced gas flotation systems	795—4372	<20
TST CFU	Vertical multistage separation with dissolved or induced gas flotation	120—16800	<10
Unicel	Vertical induced gas flotation	160—25400	5—10

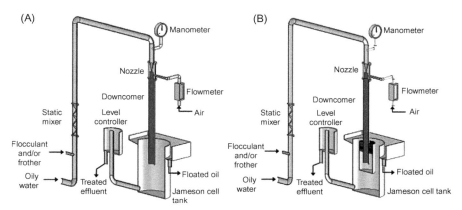

Figure 14.16 Schematic of (A) a conventional jet flotation system and (B) a modified jet flotation system. *Reproduced with permission from Santander, M., Rodrigues, R.T., Rubio, J., 2011. Modified jet flotation in oil (petroleum) emulsion/water separations. Colloids Surf. Physicochem. Eng. Asp. 375, 237—244.*

However, scaling this up to apply at an industrial scale requires further investigation. Jiménez et al. (2017) reported that a novel dissolved air flotation system, with glass microspheres and a coagulant of $FeCl_3$, can eliminate 90% of turbidity and oil and grease in produced water.

14.2.3.7 Hydrocyclone technology

Based on centrifugal and centripetal forces driven by density difference, hydrocyclones can separate solids from liquids in produced water with the removal capacity of $5-15$ μm particles and average effluent concentration of $20-30$ mg L (Lee and Frankiewicz, 2004). Generally, hydrocyclones have a long life span and no chemical additives are required for feedwater, but they generated a large amount of concentrated solids, which are also regarded as an oil field waste. Hydrocyclone systems are compact and capable of treating high concentrations of influent, but they are vulnerable to clogging and fouling issues. Fig. 14.17 shows a schematic of static hydrocyclone flow profile.

Young et al. (1994) carried out a series of experiments to study the impacts of hydrocyclone dimensions on oil and water separation, concluding that:

- The separation performance decreases as cylindrical length increases.
- The best separation performance is obtained with a cone angle of approximately 6 degrees.
- Underflow lengths up to 18 improve separation performance.
- Underflow size of 0.33 gives good separation performance over a range of flow rates.
- Feed size of 0.25 obtains the best separation performance.

14.2.3.8 Ion exchange technology

Ion exchange technology is used to remove monovalent and divalent ions and metals from produced water by ionizing the relevant minerals. The energy requirements are minimal while it is subjected to fouling and high chemical cost issues. Usually ion exchange technology is applied combined with other technologies such as adsorption to treat produced water. Pre- and posttreatments are necessary for the ion exchange process. Nadav (1999) suggested that barium could be removed by combination of ion exchange and RO. Fan et al. (2016) conducted a series of experiments involving natural zeolite and ion exchange resin, showing that clinoptilolite is a good option to treat produced water and its capacity for removing ^{226}Ra and Ba can be as high as 0.1 Gbg/ton.

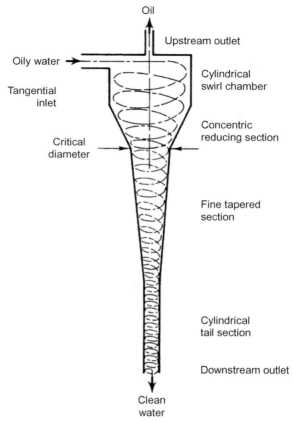

Figure 14.17 Schematic of a static hydrocyclone flow profile. *Reproduced with permission from Choi, M.S., 1990. Hydrocyclone produced water treatment for offshore developments. SPE Annual Technical Conference and Exhibition. Society of Petroleum Engineers, Tulsa, OK.*

14.2.3.9 Media filtration technology

Media filtration is extensively applied to separate dispersed oil and grease from produced water, and Drewes et al. (2009) reported that its efficiency could be as high as 90%. The common types of filtration media include anthracite, sand, gravel, and walnut shell. In particular, black walnut shell can easily filter oil to low concentrations due to its high oleophobicity (Srinivasan and Viraraghavan, 2008). The media filtration technology produces high-quality effluent and its efficiency can be further enhanced if coagulants are added to the feedwater, while it cannot remove soluble oil and may generate solid wastes after filtration (Rawlins, 2009). Yang et al. (2002) conducted produced water filtration experiments by applying hydrophilic fiber medium and walnut medium under the same experimental conditions. According to the experimental results, hydrophilic fiber is preferred because it can effectively filter and control oil concentration, suspended solid, and the d_{50} (the diameter of particle size of 50% by volume) in produced water to meet PWRI requirements for both low and high flow rate scenarios. Nair et al. (2016) reported that hydrocarbon removal efficiency of 96% was obtained by dual-media filtration, and this efficiency could be further improved by changing the bed depth and increasing the frequency of backwashing.

14.2.3.10 Membrane filtration technologies

Membranes are thin permeable films that selectively separate components from produced water driven by pressure difference during diffusive or convective mass transport (Padaki et al., 2015). They are usually compact units with a small footprint area. Depending on their pore sizes, in general membrane separation processes can be categorized as MF, ultrafiltration (UF), NF and RO filtration. Permeability of membranes decreases in the following order: MF > UF > NF > RO. MF membrane ($0.1-3$ µm pore size) is used to separate suspended particles and reduce turbidity, UF membrane ($0.01-0.1$ µm pore size) is used to separate macromolecules, NF membrane is used to filter multivalent ions, and RO membrane is used to separate dissolved and ionic components. Coday et al. (2014) stated that UF is one of the most effective methods to remove oil from produced water during oil exploration compared to conventional separation methods. Both NF and RO are pressure-driven processes; components are separated from produced water by membranes under a certain pressure gradient. They are very effective for produced water desalination, but they are significantly affected by inorganic scaling and particulate and biological organic fouling issues (Fig. 14.18), therefore feedwater pretreatments are necessary to achieve better filtration results and save long-term cost (Sutzkover-Gutman and Hasson, 2010). Materials of some polymeric membranes are subjected to degradation when temperature is higher than 50°C degrees Celsius. Dickhout et al. (2017) studied the relationship between membranes and emulsions in produced water from a colloidal perspective to better understand the membrane-fouling issue, as shown in Fig. 14.19.

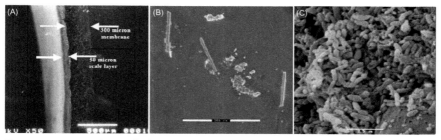

Figure 14.18 Membrane fouling. (A) Scaling on RO membrane. (B) Particulate fouling on RO membrane. (C) Biofouling on NF membrane. *Reproduced with permission from de Roever, E.W., Huisman, I.H., 2007. Microscopy as a tool for analysis of membrane failure and fouling. Desalination 207, 35—44; Ivnitsky, H., Katz, I., Minz, D., Shimoni, E., Chen, Y., Tarchitzky, J., et al., 2005. Characterization of membrane biofouling in nanofiltration processes of wastewater treatment. Desalination 185, 255—268; Semiat, R., Sutzkover, I., Hasson, D., 2003. Scaling of RO membranes from silica supersaturated solutions. Desalination 157, 169—191; Sutzkover-Gutman, I., Hasson, D., 2010. Feed water pretreatment for desalination plants. Desalination 264, 289—296.*

Direct-contact membrane distillation is one method used to dewater oil-in-water emulsions. Chew et al. (2017) studied the relationship between surfactant-stabilized oil-in-water emulsions and a polyvinylidene fluoride membrane surface, concluding that the membrane-fouling and wetting behaviors are significantly affected by the surfactant concentration and hydrophobicity. Especially, membrane hydrophobicity and fouling issues can be optimized by using surfactants with a lower hydrophilic lipophilic balance value, and this finding can be applied to make direct-contact membrane distillation an efficient and effective method to treat produced water.

Coday et al. (2014) reported that forward osmosis (FO) is an efficient and effective process to treat special types of produced water, which are difficult to filter at all scales. FO can be applied as a stand-alone desalination process, or it can be combined with other technologies to purify produced water. During the FO process, diffusion of water through a synthetic polymeric membrane is driven by osmotic pressure while almost all dissolved and suspended constituents are rejected. The commercial systems coupled with FO show net cost advantages within a range from 45% to 60%, compared to traditional methods. Some commercial FO systems need appropriate pretreatments to protect the membrane while others can operate without any pretreatments. Fig. 14.20 shows one commercial system coupled with FO and a thermolytic reconcentration system, which

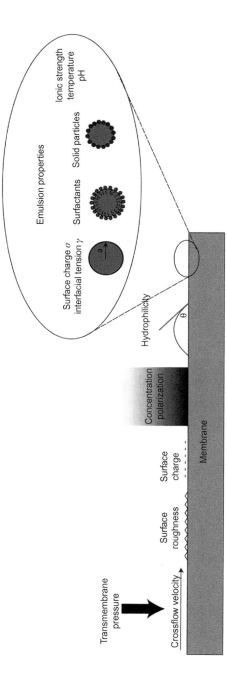

Figure 14.19 Interactions between membranes and emulsions in produced water. *Reproduced with permission from Dickhout, J.M., Moreno, J., Biesheuvel, P.M., Boels, L., Lammertink, R.G.H., de Vos, W.M., 2017. Produced water treatment by membranes: a review from a colloidal perspective. J. Colloid Interf. Sci. 487, 523–534.*

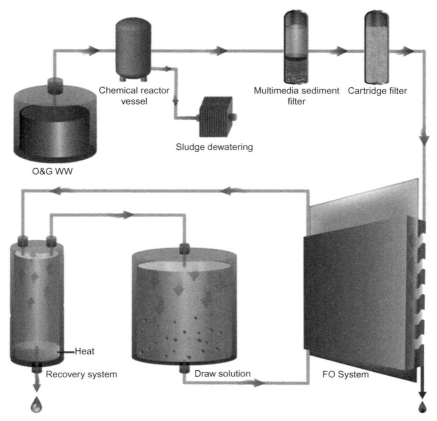

Figure 14.20 A commercial FO system with a thermolytic reconcentration system. Pretreatments technologies are used to reduce oil emulsions in produced water. *Reproduced with permission from Coday, B.D., Xu, P., Beaudry, E.G., Herron, J., Lampi, K., Hancock, N.T., et al., 2014. The sweet spot of forward osmosis: treatment of produced water, drilling wastewater, and other complex and difficult liquid streams. Desalination 333, 23—35.*

operates at a constant influent-draw solution concentration. For this system, pretreatments are required to filter oil emulsions in the produced water prior to entering the FO membrane.

14.2.3.11 Thermal technologies

Thermal separation treatments are applied for produced water desalination. During the process, a feed stream (produced water) is heated to form vapor, which can condense to high-quality water in later procedures. The major thermal desalination methods include multistage flash distillation (MSF), vapor compression distillation, and multieffect distillation (MED). Produced water recovery is about 20% for MSF and 45% for MED (Igunnu and Chen, 2012). The efficiency of thermal technology can be calculated by thermodynamic models and algorithms (Zhu et al., 2017; Zhu and Okuno, 2014a, b). The thermal technology is less effective for volatile organics removal since such compounds evaporate and condense together with produced water. However, due to the relatively large energy demands and competition by membrane technology, the market share of thermal separation technology has significantly decreased. Fig. 14.21 shows a typical process flow chart for an MSF plant.

14.2.3.12 Produced water treatments cost

When treating produced water, technological feasibility discussed earlier is one important factor; relevant cost and corresponding economic burden is another essential fact that operators must take into account. Based on published data, operating and capital cost for different produced water treatments is summarized in Table 14.12, providing indicative cost comparisons of technologies used for produced water treatments in the oil and gas industry.

14.3 FORMATION DAMAGE BY PWRI

PWRI is an important method for extending a field's economic life, complying with national and local regulations, and minimizing negative environment impacts. The most significant objectives of PWRI include oil recovery enhancement and produced water disposal (Aitkulov, 2014; Aitkulov et al., 2016, 2017; Aitkulov and Mohanty, 2016; Xiao et al., 2017). In addition, supercritical CO_2 can be injected into the formation previously injected by produced water in order to store captured CO_2 and achieve further EOR (Liang, 2014; Wu et al., 2017a, b, 2015; Wu and Bryant, 2014). In the United States, most of the offshore produced water is reinjected into the reservoir formations. However, the implementation of PWRI usually faces challenges with respect to safety and injectivity. During injection of produced water, the formation can be damaged by different mechanisms. Fig. 14.22 shows microscopic images of the deposition of hematite particles, as suspended particles colored in

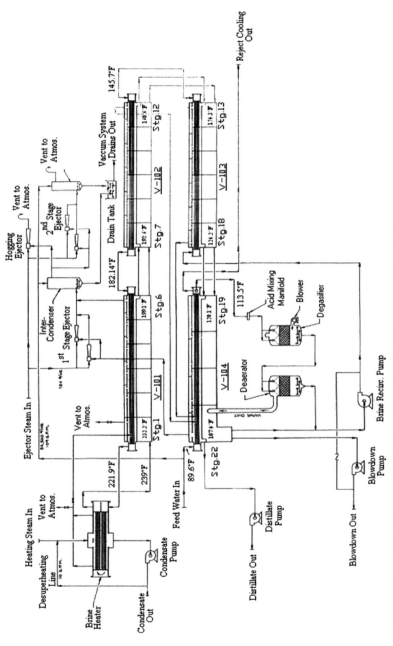

Figure 14.21 Process flow diagram of MSF plant. *Reproduced with permission from Khoshrou, I., Nasr, M.J., Bakhtari, K., 2017. New opportunities in mass and energy consumption of the multi-stage flash distillation type of brackish water desalination process. Sol. Energ. 153, 115−125.*

Table 14.12 Produced water treatments and cost (Al-Muntasheri et al., 2010; Fakhru'l-Razi et al., 2009; Igunnu and Chen, 2012; Jackson and Myers, 2003; Jing et al., 2015; Zheng et al., 2016)

Technologies	OPEX	CAPEX	Comments
Anoxic/aerobic granular activated carbon	$0.083 (bbl)	—	OPEX in 2003
Chemical oxidation	$0.026 (m³)	$2350	OPEX in 2013 CAPEX in 2013
Commercial water hauling	$0.01−5.50 (bbl)	—	OPEX in 2003
Constructed wetland	$0.001−2 (bbl)	—	OPEX in 2003
disposal wells	$0.05−2.65 (bbl)	—	OPEX in 2003
DOWS	$1603 (kg dissolved oil) $77 (kg dispersed oil)		OPEX in 2013 CAPEX in 2013
ED	$0.02−0.64 (bbl)	—	OPEX in 2003
Evaporation and flow lines	$1−1.75 (bbl)	—	OPEX in 2003
Evaporation pit	$0.01−0.8 (bbl)	—	OPEX in 2003
Freeze thaw evaporation	$2.65−5.00 (bbl)	—	OPEX in 2003
Hydrocyclone	$24−42 (kg dispersed oil)	$5000	OPEX in 2013 CAPEX in 2013
Induced air floatation	$0.05 (bbl)	$1568	OPEX in 2003 CAPEX in 2013
Membranes	Depends on membrane	$167000	OPEX in 2013 CAPEX in 2013
Secondary recovery	$0.05−1.25 (bbl)	—	OPEX in 2003
Shallow reinjection	$0.1−1.33 (bbl)	—	OPEX in 2003
Surface discharge	$0.01−0.8 (bbl)	—	OPEX in 2003
Water shutoff chemicals	$87−3705 (kg dissolved oil) $5.2−5557 (kg dispersed oil)	$942−3262	OPEX in 2013 CAPEX in 2013

dark gray, in a porous medium when water with suspended solids is injected into a core. The suspended particles in the fluid can block pores and pore throats of the porous medium, resulting in a decrease in the formation's permeability and thus reducing injectivity. This section discusses the formation damage associated with PWRI and its physics, and describes mathematical models that can be applied to explain and predict its effects.

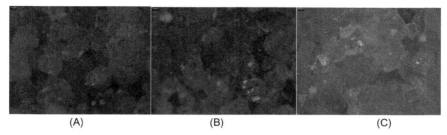

(A)	(B)	(C)

Figure 14.22 Microscopic images of deposition of hematite particles (suspended particles), shown as the dark gray color, in a porous medium at different distance from the injection face: (A) 0.74 inches, (B) 1.25 inches, and (C) 1.96 inches. *Reproduced with permission from Yerramilli, R.C., Zitha, P.L.J., Yerramilli, S.S., Bedrikovetsky, P., 2013. A novel water injectivity model and experimental validation using CT scanned core-floods. SPE European Formation Damage Conference & Exhibition. Society of Petroleum Engineers, Tulsa, OK.*

14.3.1 Formation damage mechanisms

Formation damage is commonly defined as the impairment of the permeability in formations. Barkman and Davidson (1972) state that there are four mechanisms contributing to formation damage, including invasion, perforation plugging, wellbore narrowing, and wellbore fill-up. When produced water is reinjected into a formation from an injector well, the suspended particles are deposited into the near-wellbore formations during the invasion process (internal filtration), and an external filter cake is formed on the surface of the well (external cake filtration), resulting in reductions in injectivity of the injector. This process is conceptually demonstrated in Fig. 14.23. The extent of formation damage is determined by the properties of rocks, such as pores and pore throat sizes, pore distribution and connectivity, as well as characteristics of injected produced water such as injection rate, particle size, particle distribution, surface charge, etc. (Yuan et al., 2016a).

In general, external filter cake inside the wellbore starts to build up (external cake filtration) after the near-wellbore formation is blocked by deposited particles from produced water (internal filtration). Therefore, at an early stage, the formation is mainly damaged by internal filtration, while external cake filtration plays an increasingly important role as it accumulates. Therefore, most studies divide the entire filtration process into two stages, the internal filtration and the external cake filtration, and these two stages are separated by a transition time, t^* (Pang and Sharma, 1995; Shutong and Sharma, 1997).

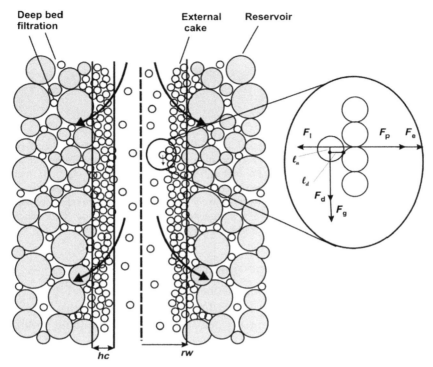

Figure 14.23 A formation is damaged by internal filtration and external cake. See explanations for the terms used in Eqs. (14.1)–(14.6) for the abbreviations used in this diagram. *Reproduced with permission from You, Z., Kalantariasl, A., Schulze, K., Storz, J., Burmester, C., Künckeler, S., et al., 2016. Injectivity impairment during produced water disposal into low-permeability Völkersen aquifer (compressibility and reservoir boundary effects). SPE International Conference and Exhibition on Formation Damage Control. Society of Petroleum Engineers, Tulsa, OK.*

14.3.2 Mathematical model for quantification of formation damage

14.3.2.1 Injectivity ratio and flow resistance in different damage zones

Some important concepts relevant to formation damage mathematical models are discussed here. The injectivity, I, describes how easy fluid can be injected into a well. When this value is high, it implies that under a certain pressure gradient, higher injection rate is achieved, which is preferable in terms of formation damage caused by PWRI. Injectivity can be expressed as:

$$I = \frac{q}{\Delta P} \tag{14.1}$$

Here q is the flow rate and ΔP is the pressure difference over a certain length. For one-dimensional (1D) flow, before particle deposition, the formation is homogeneous with an initial undamaged permeability, k_o. After invasion of particles, conceptually the formation can be divided into three parts: external cake filtration zone with an external cake damaged permeability, k_c; near-wellbore internal filtration zone with an internal filtration damaged permeability, k_f, and undamaged zone with the initial permeability, k_o, which is discussed in Section 14.3.1. Each damage zone has its own flow resistance (R), expressed as:

$$R_o = \frac{L}{k_o A} \tag{14.2}$$

$$R_c = \frac{h_c}{k_c A} \tag{14.3}$$

$$R_f = \frac{x_f}{\bar{k} A} \tag{14.4}$$

$$R_R = \frac{L - x_f}{k_o A} \tag{14.5}$$

where L is the total length of the medium, A is the cross-sectional area of the formation, h_c is the external cake thickness, x_f is the depth of the internal filtration, and R is the flow resistance. $\bar{k}$ is the average permeability. The subscription c denotes the external cake infiltration zone, f denotes the near-wellbore internal filtration zone, and the difference between internal and external filtrations is defined as the undamaged zone, as conceptually demonstrated in Figs. 14.23 and 14.24.

The injectivity ratio, α, is defined as the ratio of the instantaneous to the initial injectivitiy indices, while the reciprocal of the injectivity index, J, is referred to as the impedance index. A higher injectivity ratio indicates that the formation has higher permeability thus is less damaged after PWRI, which is preferable during hydrocarbon production activities. Substitutions of Eqs. (14.10) and (14.2)−(14.5) into Eq. (14.1) yields the expression of injectivity ratio as:

$$\alpha = \frac{I}{I_o} = \frac{q}{\Delta P} \bigg/ \left(\frac{q}{\Delta P}\right)_o = \frac{R_o}{R_c + R_f + R_R} \tag{14.6}$$

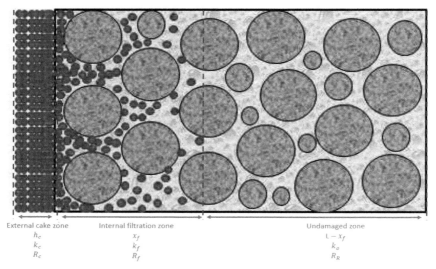

External cake zone   Internal filtration zone   Undamaged zone
h_c x_f $L - x_f$
k_c k_f k_o
R_c R_f R_R

Figure 14.24 Different formation damage zones during the PWRI process. Source: *Modified from the image by Yerramilli, R.C., Zitha, P.L.J., Yerramilli, S.S., Bedrikovetsky, P., 2013. A novel water injectivity model and experimental validation using CT scanned core-floods. In: SPE European Formation Damage Conference & Exhibition. Society of Petroleum Engineers, Tulsa, OK; Yerramilli, R.C., Zitha, P.L., Yerramilli, S.S., Bedrikovetsky, P., 2014. A novel water-injectivity model and experimental validation with CT-scanned corefloods. SPE J. 20, 1200—1211. See explanations for the terms used in Eqs. (14.1)— (14.6) for the abbreviations used in this diagram.*

14.3.2.2 Internal filtration model

When fluid with suspend particles is transported through a porous media, its pore and pore throat systems can filter the bypassing fluids: the granular rocks may block some suspended particles while the void spaces can allow transport of the fluid, similar to the case of suspended fluid through a membrane. If the porous media is assumed infinitely large and time is assumed infinitely long, then all particles will be separated from the fluid, either entrapped in the interstices along the fluid streamline or filtered at the surface. The nature of the internal filtration is actually the retention of the particles inside the pores.

The retention sites commonly include surface sites, crevice sites, construction sites, and cavern sites (Farajzadeh, 2004). Separation of suspended particles from fluids can change the properties of both the fluid and the porous medium, resulting in a permeability reduction in the formation.

The internal filtration process is commonly modeled by the deposition of particles in a 1D single-phase flow (Barkman and Davidson, 1972;

Bedrikovetsky et al., 2001; Civan, 2015; Herzig et al., 1970; Pang and Sharma, 1995; Sharma and Yortsos, 1987; Shutong and Sharma, 1997; Wennberg and Sharma, 1997). The assumptions constraining the internal filtration model include:

- The fluids and the solids are assumed to be incompressible.
- The density of the particles is assumed to be constant during the deposition process.
- The fluid viscosity is assumed to be constant.
- Effects of dispersion and diffusion in the porous media are neglected.

With these assumptions, the volumetric balance equation of the suspended and deposited particles for 1D linear flow is expressed as:

$$\frac{\partial(\phi c)}{\partial t} + u\frac{\partial c}{\partial x} + \frac{\partial \sigma}{\partial t} = 0 \tag{14.7}$$

where ϕ is the porosity, c is the volumetric fraction of suspended particles to unit liquid volume, σ is the volumetric concentration of deposited particles to unit filter volumes, and u is the volumetric flux. The particle deposition rate is described by Iwasaki et al. (1937) as:

$$\frac{d\sigma}{dt} = \lambda u c \tag{14.8}$$

and

$$\lambda = \lambda_o(1 + b\sigma) \tag{14.9}$$

where λ is defined as the filtration coefficient, b is an empirical constant, and λ_o is the filtration coefficient without particle deposition. In some models, λ is assumed constant while several studies consider it a function of fluid velocity, injected particle size, formation grain size, porosity, and ion concentration.

The 1D flow is described by Darcy's law as:

$$u = -\frac{k_o k(\sigma)}{\mu}\frac{\partial p}{\partial x} \tag{14.10}$$

where u is the volumetric flux, p is the pressure, μ is the fluid viscosity, and k_o is the undamaged formation permeability. $k(\sigma)$ is the formation damage function, and it is assumed to be a decreasing function of deposition concentration in this context. Physically it makes sense, as more particles deposit into pores, the pore throats have a higher chance of being blocked, resulting in a decrease in the formation's permeability. Similarly, the porosity decreases as the concentration of the particle deposits increases, and it is described as:

$$\phi = \phi_o - \sigma \qquad (14.11)$$

This model is constrained by the initial and boundary conditions as:

$$c = c_o(x), \sigma = \sigma_o(x), x > 0, \ t = 0 \qquad (14.12)$$

$$c = c_f(x), x = 0, \ t > 0 \qquad (14.13)$$

where subscript o denotes the initial state when $t = 0$ and subscript f denotes the state when $t > 0$.

The deposition of the particles in a 1D linear flow can be modeled by applying Eqs. (14.7)–(14.13) and many studies have presented analytical solutions for this model, as well as other models such as deposition in a radial flow. The most commonly used analytical solution was obtained under the following assumptions:

• Compared to initial porosity, the concentration of deposited particles, σ, is relatively minor and therefore this term can be neglected.
• The deposition coefficient, λ, is assumed to be a constant, $b = 0$ in Eq. (14.9).
• The fluid is injected at a constant rate, and the volumetric flux, u, is assumed to be a constant.

Based on these assumptions, the concentration of the suspended particles before the front is:

$$c(x, t) = c_o \exp\left(-\frac{\lambda u t}{\phi}\right) H\left(x - \frac{ut}{\phi}\right), \quad x > \frac{ut}{\phi} \qquad (14.14)$$

The concentration of the suspended particles after the front is:

$$c(x, t) = c_f \exp(-\lambda x) H\left(t - \frac{\phi x}{u}\right), \quad x < \frac{ut}{\phi} \qquad (14.15)$$

where H denotes a Heaviside unit step function.

For core flooding at laboratory scale, c_f is a constant, $c_o = 0$ and $\sigma_o = 0$, thus the solutions before and after the front can be simplified to:

$$c(x, t) = 0, \quad x > \frac{ut}{\phi} \qquad (14.16)$$

$$c(x, t) = c_f \exp(-\lambda x), \quad x < \frac{ut}{\phi} \qquad (14.17)$$

In particular, c is independent on time here since the transit period, when the first pore volume of injected fluids sweeps the system, is neglected. Substituting Eqs. (14.16) and (14.17) into Eq. (14.8) and integrating the equation derived, the volumetric concentration of deposited particles in a 1D linear flow is obtained as:

$$\sigma(x, t) = 0, \ x > \frac{ut}{\phi} \tag{14.18}$$

$$\sigma(x, t) = c_f \lambda ut \, \exp(-\lambda x), \ x < \frac{ut}{\phi} \tag{14.19}$$

Based on the assumption that permeability reduction results from pore throat clogging, the harmonic average permeability of the near-wellbore filtration damaged zone, $\bar{k}(t)$, is calculated by:

$$\frac{k_o}{\bar{k}(t)} = 1 + \frac{\beta c_f u}{x_f} \left(1 - e^{-\lambda x_f}\right) t + \frac{\beta c_f \phi}{\lambda x_f} \left(1 - e^{-\lambda x_f}\right) + \beta c_f \phi e^{-\lambda x_f} \tag{14.20}$$

where β is an empirical damage constant. At the laboratory scale, external cake filtration is not taken into consideration. Substitutions of Eqs. (14.2)−(14.6) into Eq. (14.19) yields the injectivity ratio, $\alpha(t)$, as:

$$\frac{1}{\alpha(t)} = 1 + \frac{\beta c_f u}{L} t + \frac{\beta c_f \phi}{\lambda L} \left(e^{-\lambda ut/\phi} - 1\right), \frac{ut}{\phi} < L \tag{14.21}$$

$$\frac{1}{\alpha(t)} = 1 + \frac{\beta c_f u}{L} \left(1 - e^{-\lambda L}\right) t + \frac{\beta c_f \phi}{\lambda L} \left(e^{-\lambda L} - 1\right) + \beta c_f \phi e^{-\lambda L}, \frac{ut}{\phi} > L \tag{14.22}$$

In some studies (Pang and Sharma, 1995; Wennberg and Sharma, 1997), the last two terms in Eq. (14.22) are neglected, yielding a 1D linear function of injectivity ratio with respect to time as:

$$\frac{1}{\alpha(t)} \cong 1 + \frac{\beta c_f u}{L} \left(1 - e^{-\lambda L}\right) t \tag{14.23}$$

According to Eqs. (14.17), (14.20), and Eq. (14.23), as more fluid is injected into the formation, the concentration front moves forward as a function of time, resulting into reductions in both the average permeability of the near-wellbore filtration damaged zone and the well injectivity ratio, and this result matches field experience of PWRI. It should be noted that in some studies, the inverse of injectivity is defined as impedance, commonly denoted by J, and relevant results are presented in terms of impedance instead of the inverse of injectivity (e.g., Fig. 14.29).

14.3.2.3 Transition time model

Internal filtration and external cake filtration are commonly assumed to be separated by a transit time t^* (Pang and Sharma, 1995). It is commonly assumed that at t^* the formation porosity reaches a critical value, ϕ^*, below which the suspended particles are blocked and accumulated at the wall of the wellbore (Pang and Sharma, 1995). Eq. (14.8) describes the deposition rate while the filtration coefficient is calculated by Eq. (14.9). A substitution of Eq. (14.8) into Eq. (14.9) yields:

$$\frac{d\sigma}{dt} = \lambda_o(1 + b\sigma)uc \tag{14.24}$$

With the initial condition

$$\sigma = 0, \; t = 0 \tag{14.25}$$

Integrating Eq. (14.24) with initial condition yields the following expressions:

$$\sigma = \lambda_o uct, \; b = 0 \tag{14.26}$$

$$\sigma = \left(\frac{1}{b}\right)(e^{\lambda_o buct} - 1), \; b \neq 0 \tag{14.27}$$

Within the transition time, the maximum volumetric concentration of the deposited particles is calculated as:

$$\sigma^* = \phi_o - \phi^* \tag{14.28}$$

Invoking Eq. (14.28) into Eqs. (14.26) and (14.27), the expressions to determine the transit time, t^*, are obtained as:

$$t^* = \frac{\phi_o - \phi^*}{\lambda_o uc}, \; b = 0 \tag{14.29}$$

$$t^* = \frac{\ln[b(\phi_o - \phi^*) + 1]}{\lambda_o ucb} b \neq 0 \tag{14.30}$$

14.3.2.4 External cake filtration model

The nature of the external cake filtrations is actually the separation of the particles at the wall of the wellbore. For the model discussed here, it is assumed the external cake starts to build up after the transit time, t^*. The external filtration cake thickness, h_c, is defined as the distance between the wall of wellbore to the surface of the external cake (see Fig. 14.24).

Therefore, before the transit time t^*, the external filtration cake thickness, h_c, is commonly assumed to be 0.

By volumetric balance of the suspended particles in the external cake, Sharma et al. (2000) proposed the cake thickness after transit time for a 1D linear flow as:

$$h_c = \frac{\int_{t^*}^{t} q(t)c_f(t)dt}{(1-\phi_c)A} \tag{14.31}$$

With the assumptions of constant rate injection and constant concentration of suspended particles at the surface, Eq. (14.31) simplifies as:

$$h_c = \frac{c_f q t}{(1-\phi_c)A} \tag{14.32}$$

The permeability of external cake damaged zone is a function of time and location, and the average permeability of external cake damaged zone is given by:

$$\frac{h_c}{\overline{k_c}(t)} = \int_0^{h_c} \frac{dx}{k_c(x,t)} \tag{14.33}$$

Substitution of Eqs. (14.33), (14.2), and (14.3) into Eq. (14.6) yields the injectivity ratio as:

$$\alpha(t) = \frac{L/k_o}{h_c/\overline{k_c}(t) + L/k_o} \tag{14.34}$$

With the assumption that external cake thickness is much smaller than its total length, the following expression is obtained indicating that the inverse of the injectivity ratio of external cake is a linear function of time (Sharma et al., 2000).

$$\frac{1}{\alpha(t)} \cong 1 + \frac{uk_o}{L\overline{k_c}}\left(\frac{c_f}{1-\phi_c}\right)t \tag{14.35}$$

According to Eqs. (14.32), (14.33), and (14.34), after the transit time t^*, external cake starts to build up on the wall of the wellbore; the external cake thickness increases as a function of time, resulting in reductions in both the average permeability of the external cake damaged zone and the well injectivity ratio. This model also matches field experience of PWRI.

Especially, in the models discussed earlier, the filtration coefficient, λ (Eq. 14.8), and the damage coefficient, β (Eq. 14.20), are assumed constant while Iwasaki et al. (1937) in the original work stated that the filtration coefficient and damage coefficient could be affected by previously deposited particles, concentration of suspension, velocity, particles size, gran size and many other factors. Many experiments and simulation works have been conducted to measure those two parameters to find corresponding dependence.

14.3.2.5 Other PWRI mathematical models

The classic models discussed through Sections 14.3.2.2—14.3.2.4 are for deposition of suspended particles in 1D linear singe-phase flow under certain assumptions. With the purpose of obtaining higher accuracy and broader applications, many studies were carried out to establish models with fewer restrictions. Liu and Civan (1996) developed a model applicable for a two-phase system, and that model couples the external cake and the internal filtration effects. Civan and Rasmussen (2005) reported the analytical solutions of internal filtration for radial flow with constant injection rates, as well as for radial flow with variable injection rates. In their studies, they also derived the analytical solutions of external cake for radial flow with constant injection rates. Yerramilli et al. (2013, 2014) developed a model that considers a nonlinear retention function, hydrodynamic dispersion, and suspension viscosity. Sacramento et al. (2015) proposed a mathematical model for cake formation with internal filtration for injection with two-size particle suspension, and the model was verified by laboratory core flooding results. Obe et al. (2017) developed a model incorporating the effects of geochemical rationing, adsorption kinetics, and the hydrodynamics of molecular transport. In particular, mechanical dispersion in porous media at high Peclet number, which was previously neglected, may play an important role in affecting transport and deposition of particles during the PWRI process, as shown in Fig. 14.25 (Liang, 2017; Liang et al., 2013, 2015). The effects of the mechanical dispersion have been studied and verified by numerical simulations (Wen, 2015; Wen et al., 2013, 2015; Wen and Chini, 2015).

14.3.3 Type curves for water injectivity test

Another application of the 1D linear models described in Section 14.3.2.2—14.3.2.4 is to interpret water injectivity tests. Based on

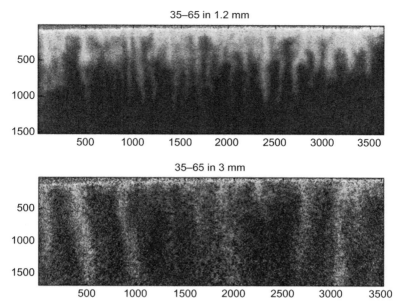

Figure 14.25 Two solute experiments with different Peclet numbers. The upper image is the experiment with a low Peclet number and the lower image is with a high Peclet number. Mechanical dispersion can greatly affect transport and deposition of particles during the PWRI process. *Reproduced with permission from Liang, Y., 2017. Scaling of Solutal Convection in Porous Media. The University of Texas at Austin, Austin, TX; Liang, Y., DiCarlo, D.A., Hesse, M.A., 2013. Experimental study of convective dissolution of carbon dioxide in heterogeneous media. In: AGU Fall Meeting Abstracts; Liang, Y., DiCarlo, D.A., Hesse, M.A., 2015. Experiment and simulation study of hydrodynamic dispersion and finger dynamics for convective dissolution of carbon dioxide. In: AGU Fall Meeting Abstracts.*

the experimental core flow data, Shutong and Sharma (1997) summarized four distinct type curves, as shown in Fig. 14.26.

- The shape of type curve 1 is a straight line, resulting from incompressible external cakes or internal filtration cake that is very close to the wellbore.
- The shape of type curve 2 is a curve with an increasing slope, resulting from compressible external filter cakes or internal filtration cakes. In the invasion zone, the retention of the particles blocks the pore throats, resulting in a reduction in permeability. However, the permeability changes more rapidly than the porosity. Therefore, type curve 2 shows an increasing slope while more fluids are injected into the formations.

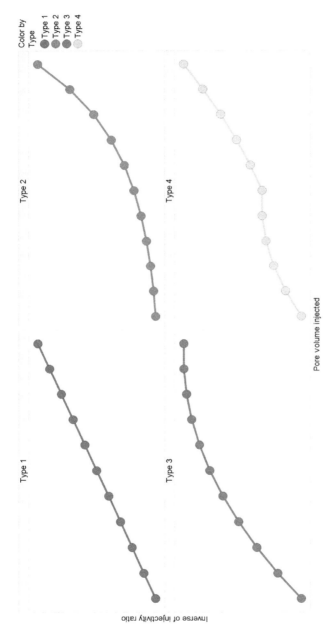

Figure 14.26 Type curves for water injectivity test. *Reproduced with permission from Pang, S., Sharma, M.M., 1995. Evaluating the performance of open-hole, perforated and fractured water injection wells. SPE European Formation Damage Conference. Society of Petroleum Engineers, Tulsa, OK.*

- The shape of type curve 3 is a curve with a decreasing slope, resulting from a deeper internal filtration invasion without external cakes. Pore filling is the main mechanism in this case since a smaller pore space can increase the interstitial velocity of injected fluids, contributing to less deposition of suspended particles. Therefore, the change of permeability slows down as a function of time.
- The shape of type curve 4 is an S-shaped curve, resulting from a combination of different filtration mechanisms simultaneously during the filtration process.

14.3.4 Case studies

The 1D linear model presented in Sections 14.3.2.2—14.3.2.4 can match experimental results and field data very well. In this section, two cases based on that 1D linear model are discussed in order to demonstrate how the mathematical models can help academia and industry understand and improve PWRI processes from core to field scales.

Based on that 1D linear model, Sharma et al. (2000) developed a simulator to predict injectivity decline and permeability changes near the wellbore formation. In addition, they evaluated the results from five offshore wells in Gulf of Mexico to test the simulator. The results show that the simulator works well when the water quality meets certain requirements. Fig. 14.27 shows one example from their cases study. They concluded that the decline rate of the injectivity ratio (the slope of the curve in Fig. 14.27) could be minimized with a large amount of treatments adding to the injected water. In addition, they also observed that fractured wells have a smaller decline rate of injectivity ratio than unfractured wells, implying that fractured wells have a better tolerance to injected water.

Another case discussed here is focused at the core scale. Kalantariasl et al. (2015) conducted a core flooding experiment using low permeability sandstone core sample ($k = 0.437$ md, $\phi = 0.123$). Water with suspended latex microsphere with radii of 0.505 μm was injected into the core sample. Fig. 14.28 shows the external cake, which has a higher permeability than the core sample, and was built up on the injection face. Fig. 14.29 shows the experimental results and the model predictions. As observed, the initially rapid increase of impedance before 40 pore volume injection (PVI) is due to the internal filtration; then the increasing slope of impedance becomes less steep, resulting from the buildup of external cake. The prediction by the 1D linear model, presented in Sections

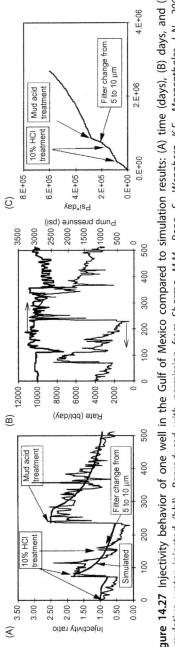

Figure 14.27 Injectivity behavior of one well in the Gulf of Mexico compared to simulation results: (A) time (days), (B) days, and (C) cumulative water injected (bbl). *Reproduced with permission from Sharma, M.M., Pang, S., Wennberg, K.E., Morgenthaler, L.N., 2000. Injectivity decline in water-injection wells: an offshore Gulf of Mexico case study. SPE Prod. Facil. 15, 6—13.*

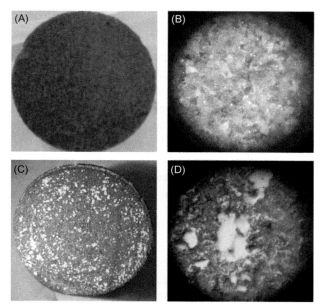

Figure 14.28 Injection face of the core sample: (A) before injection normal, (B) before injection magnification, (C) after injection normal, and (D) after injection magnification. *Reproduced with permission from Kalantariasl, A., Schulze, K., Storz, J., Burmester, C., Kuenckeler, S., You, Z., et al., 2015. Injectivity during PWRI and disposal in thick low permeable formations (laboratory and mathematical modelling, field case). SPE European Formation Damage Conference and Exhibition. Society of Petroleum Engineers, Tulsa, OK.*

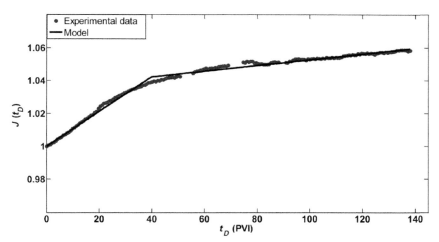

Figure 14.29 Experimental data and predicted model for impedance growth, where impedance is the inverse of injectivity. *Reproduced with permission from Kalantariasl, A., Schulze, K., Storz, J., Burmester, C., Kuenckeler, S., You, Z., et al., 2015. Injectivity during PWRI and disposal in thick low permeable formations (laboratory and mathematical modelling, field case). SPE European Formation Damage Conference and Exhibition. Society of Petroleum Engineers, Tulsa, OK.*

14.3.2.2–14.3.2.4, matches the experimental results very well, and this impedance curve matches the type curve 3 in Fig. 14.26. They concluded that the ratio between formation and external permeability could determine the degree of convexity of the impedance curve; the external cake plays a less important role if its permeability is higher than the formation. The turning (inflection) point on the impedance curve, 40 PVI for this example, confirms the existence of the transit time. They proposed that in a thick low permeable formation, water injection is still feasible since large reservoir thickness can offset the negative impacts of low permeability.

14.4 SUMMARY AND CONCLUSION

In this chapter, all facts about produced water in oil and gas fields are discussed in detail. The origin, the characteristic, and the production of produced water are introduced to facilitate the understanding of its physics, chemistry, affecting factors, and corresponding impacts. It is concluded that produced water in oil and gas fields plays an increasingly important role in affecting profit margins for oil and gas exploration and production activities, thus operators need to consider and examine it carefully to elongate well and field life spans and thus enhance profits.

The produced water can be treated by different methods to meet regulations and enhance profits. The major produced water management methods, including produced water minimization, produced water reuse/recycling and produced water disposal, as well as legal frameworks, policies, and regulations related to produced water from onshore and offshore oil and gas fields are described. The commonly applied produced water treatment technologies in the oil and gas industry, corresponding technical details, advantages and disadvantages are evaluated. According to the analysis, comparisons, and evaluations, produced water issues can be effectively mitigated by various produced water treatment technologies under certain legal frameworks, policies, and regulations.

PWRI is an important method for handling and utilizing produced water, but the reservoir formations can be severely damaged during the reinjection process if the process is not fully understood and adequately managed. Formation damage caused by PWRI, its physics, mathematical models, and case studies are described in detail in this chapter. Based on the data and fitting results at all scales, reductions in formation permeability and well injectivity during PWRI process can be accurately predicted and simulated by established mathematical models. A higher PWRI efficiency can be potentially achieved with appropriate produced water treatments before the reinjection process.

REFERENCES

Abousnina, R.M., Nghiem, L.D., Bundschuh, J., 2015. Comparison between oily and coal seam gas produced water with respect to quantity, characteristics and treatment technologies: a review. Desal. Water Treat 54, 1793—1808.

Aitkulov, A., 2014. Two-Dimensional ASP Flood for a Viscous Oil. The University of Texas at Austin, Austin, TX.

Aitkulov, A., Mohanty, K.K., 2016. Timing of ASP injection for viscous oil recovery. SPE Improved Oil Recovery Conference. Society of Petroleum Engineers, Tulsa, OK.

Aitkulov, A., Lu, J., Pope, G., Mohanty, K.K., 2016. Optimum time of ACP injection for heavy oil recovery. SPE Canada Heavy Oil Technical Conference. Society of Petroleum Engineers, Tulsa, OK.

Aitkulov, A., Luo, H., Lu, J., Mohanty, K.K., 2017. Alkali—cosolvent—polymer flooding for viscous oil recovery: 2D evaluation. Energy Fuels 31, 7015—7025.

Al-Abduwani, F.A., Shirzadi, A., van den Brock, W., Currie, P.K., 2005. Formation damage vs. solid particles deposition profile during laboratory simulated PWRI. SPE J. 10, 138—151.

Al-Muntasheri, G.A., Sierra, L., Garzon, F.O., Lynn, J.D., Izquierdo, G.A., 2010. Water shut-off with polymer gels in a high temperature horizontal gas well: a success story. SPE Improved Oil Recovery Symposium. Society of Petroleum Engineers, Tulsa, OK.

Al-Shamrani, A.A., James, A., Xiao, H., 2002. Destabilisation of oil—water emulsions and separation by dissolved air flotation. Water Res. 36, 1503—1512.

Andreozzi, R., Caprio, V., Insola, A., Marotta, R., 1999. Advanced oxidation processes (AOP) for water purification and recovery. Catal. Today 53, 51—59.

Arthur, J.D., Langhus, B.G., Patel, C., 2005. Technical Summary of Oil & Gas Produced Water Treatment Technologies. Consult. LLC, Tulsa OK.

Ayers Jr, W.B., 2002. Coalbed gas systems, resources, and production and a review of contrasting cases from the San Juan and Powder River basins. AAPG Bull 86, 1853—1890.

Aziz, A.A., Daud, W.M.A.W., 2012. Oxidative mineralisation of petroleum refinery effluent using Fenton-like process. Chem. Eng. Res. Des 90, 298—307.

Barkman, J.H., Davidson, D.H., 1972. Measuring water quality and predicting well impairment. J. Pet. Technol. 24, 865—873.

Barrufet, M.A., Burnett, D.B., Mareth, B., 2005. Modeling and operation of oil removal and desalting oilfield brines with modular units. SPE Annual Technical Conference and Exhibition. Society of Petroleum Engineers, Tulsa, OK.

Bedrikovetsky, P., Marchesin, D., Shecaira, F., Souza, A.L., Milanez, P.V., Rezende, E., 2001. Characterisation of deep bed filtration system from laboratory pressure drop measurements. J. Pet. Sci. Eng. 32, 167—177.

Chen, G.-R., Chang, Y.-R., Liu, X., Kawamoto, T., Tanaka, H., Parajuli, D., et al., 2017. Cesium removal from drinking water using Prussian blue adsorption followed by anion exchange process. Sep. Purif. Technol. 172, 147–151.

Chew, N.G.P., Zhao, S., Loh, C.H., Permogorov, N., Wang, R., 2017. Surfactant effects on water recovery from produced water via direct-contact membrane distillation. J. Membr. Sci. 528, 126–134.

Choi, M.S., 1990. Hydrocyclone produced water treatment for offshore developments. SPE Annual Technical Conference and Exhibition. Society of Petroleum Engineers, Tulsa, OK.

Civan, F., 2015. Reservoir Formation Damage. Gulf Professional Publishing, Oxford.

Civan, F., Rasmussen, M.L., 2005. Analytical models for porous media impairment by particles in rectilinear and radial flows. In: Vafai, K. (Ed.), Handb. Porous Media, second ed Taylor Francis, NY, pp. 485–542.

Clark, C.E., Veil, J.A., 2009. Produced Water Volumes and Management Practices in the United States. Argonne National Laboratory (ANL), Washington, DC.

Cluff, M.A., Hartsock, A., MacRae, J.D., Carter, K., Mouser, P.J., 2014. Temporal changes in microbial ecology and geochemistry in produced water from hydraulically fractured Marcellus Shale gas wells. Environ. Sci. Technol. 48, 6508–6517.

Coday, B.D., Xu, P., Beaudry, E.G., Herron, J., Lampi, K., Hancock, N.T., et al., 2014. The sweet spot of forward osmosis: treatment of produced water, drilling wastewater, and other complex and difficult liquid streams. Desalination 333, 23–35.

Dai, Z., Viswanathan, H., Middleton, R., Pan, F., Ampomah, W., Yang, C., et al., 2016. CO_2 accounting and risk analysis for CO_2 sequestration at enhanced oil recovery sites. Environ. Sci. Technol. 50, 7546–7554.

da Silva, S.S., Chiavone-Filho, O., de Barros Neto, E.L., Foletto, E.L., 2015. Oil removal from produced water by conjugation of flotation and photo-Fenton processes. J. Environ. Manage. 147, 257–263.

de Roever, E.W., Huisman, I.H., 2007. Microscopy as a tool for analysis of membrane failure and fouling. Desalination 207, 35–44.

Dickhout, J.M., Moreno, J., Biesheuvel, P.M., Boels, L., Lammertink, R.G.H., de Vos, W.M., 2017. Produced water treatment by membranes: a review from a colloidal perspective. J. Colloid Interf. Sci. 487, 523–534.

Drewes, J.E., Cath, T.Y., Xu, P., Graydon, J., Veil, J., Snyder, S., 2009. An integrated framework for treatment and management of produced water. RPSEA Project 07122–12.

Du, S., Fung, L.S., Dogru, A.H., 2017. Dual grid method for aquifer acceleration in parallel implicit field-scale reservoir simulation. SPE Reservoir Simulation Conference. Society of Petroleum Engineers, Tulsa, OK.

Ekins, P., Vanner, R., Firebrace, J., 2007. Zero emissions of oil in water from offshore oil and gas installations: economic and environmental implications. J. Clean. Prod. 15, 1302–1315.

El-Gohary, F., Tawfik, A., Mahmoud, U., 2010. Comparative study between chemical coagulation/precipitation (C/P) versus coagulation/dissolved air flotation (C/DAF) for pre-treatment of personal care products (PCPs) wastewater. Desalination 252, 106–112.

Fakhru'l-Razi, A., Pendashteh, A., Abdullah, L.C., Biak, D.R.A., Madaeni, S.S., Abidin, Z.Z., 2009. Review of technologies for oil and gas produced water treatment. J. Hazard. Mater. 170, 530–551.

Fan, W., Liberati, B., Novak, M., Cooper, M., Kruse, N., Young, D., et al., 2016. Radium-226 removal from simulated produced water using natural zeolite and ion-exchange resin. Ind. Eng. Chem. Res. 55, 12502–12505.

Farajzadeh, R., 2004. Produced water re-injection (PWRI), an experimental investigation into internal filtration and external cake build-up. Report TA/PW/04-14, M.Sc. Thesis. Faculty of Civil Engineering and Geoscience, TUDelft.

Fard, A.K., Mckay, G., Chamoun, R., Rhadfi, T., Preud'Homme, H., Atieh, M.A., 2017. Barium removal from synthetic natural and produced water using MXene as two dimensional (2-D) nanosheet adsorbent. Chem. Eng. J. 317, 331−342.

Feng, Y., Gray, K.E., 2016a. A fracture-mechanics-based model for wellbore strengthening applications. J. Nat. Gas Sci. Eng. 29, 392−400.

Feng, Y., Gray, K.E., 2016b. A parametric study for wellbore strengthening. J. Nat. Gas Sci. Eng. 30, 350−363.

Ferrer, I., Thurman, E.M., 2015. Chemical constituents and analytical approaches for hydraulic fracturing waters. Trends Environ. Anal. Chem. 5, 18−25.

Frost, T.K., Johnsen, S., Utvik, T.I., 1998. Environmental Effects of Produced Water Discharges to the Marine Environment. OLF Norway. <http://www.olf.no/static/en/rapporter/producedwater/summary.html>.

Georgie, W.J., Sell, D., Baker, M.J., 2001. Establishing best practicable environmental option practice for produced water management in the gas and oil production facilities. SPE/EPA/DOE Exploration and Production Environmental Conference. Society of Petroleum Engineers, Tulsa, OK.

Ghasemi, M., Naushad, M., Ghasemi, N., Khosravi-Fard, Y., 2014. Adsorption of Pb (II) from aqueous solution using new adsorbents prepared from agricultural waste: adsorption isotherm and kinetic studies. J. Ind. Eng. Chem. 20, 2193−2199.

Gunatilake, U.B., Bandara, J., 2017. Efficient removal of oil from oil contaminated water by superhydrophilic and underwater superoleophobic nano/micro structured TiO_2 nanofibers coated mesh. Chemosphere 171, 134−141.

Hansen, B.R., Davies, S.R., 1994. Review of potential technologies for the removal of dissolved components from produced water. Chem. Eng. Res. Des. 72, 176−188.

Hao, H., Huang, X., Gao, C., Gao, X., 2015. Application of an integrated system of coagulation and electrodialysis for treatment of wastewater produced by fracturing. Desal. Water Treat. 55, 2034−2043.

Harsem, Ø., Eide, A., Heen, K., 2011. Factors influencing future oil and gas prospects in the Arctic. Energ. Pol. 39, 8037−8045.

Hayes, T., Severin, B.F., Engineer, P.S.P., Okemos, M.I., 2012. Barnett and Appalachian Shale water management and reuse technologies. Contract 8122, 05.

He, S., Ning, Y., Chen, T., Liu, H., Wang, H., Qin, G., 2016. Transport properties of natural gas in shale organic and inorganic nanopores using non-equilibrium molecular dynamics simulation. In: International Petroleum Technology Conference.

Henderson, S.B., Grigson, S.J.W., Johnson, P., Roddie, B.D., 1999. Potential impact of production chemicals on the toxicity of produced water discharges from North Sea oil platforms. Mar. Pollut. Bull. 38, 1141−1151.

Henze, M., Harremoes, P., la Cour Jansen, J., Arvin, E., 2001. Wastewater Treatment: Biological and Chemical Processes. Springer Science & Business Media, Berlin, Germany.

Herzig, J.P., Leclerc, D.M., Goff, P.L., 1970. Flow of suspensions through porous media—application to deep filtration. Ind. Eng. Chem. 62, 8−35.

Hoseini, S.M., Salarirad, M.M., Alavi Moghaddam, M.R., 2015. TPH removal from oily wastewater by combined coagulation pretreatment and mechanically induced air flotation. Desal. Water Treat. 53, 300−308.

Houcine, M., 2002. Solution for heavy metals decontamination in produced water/case study in southern Tunisia. SPE International Conference on Health, Safety and Environment in Oil and Gas Exploration and Production. Society of Petroleum Engineers, Tulsa, OK.

Igunnu, E.T., Chen, G.Z., 2012. Produced water treatment technologies. Int. J. Low-Carbon Technol. 9, 157−177.

Ivnitsky, H., Katz, I., Minz, D., Shimoni, E., Chen, Y., Tarchitzky, J., et al., 2005. Characterization of membrane biofouling in nanofiltration processes of wastewater treatment. Desalination 185, 255−268.

Iwasaki, T., Slade, J.J., Stanley, W.E., 1937. Some notes on sand filtration [with discussion]. J. Am. Water Works Assoc. 29, 1591−1602.

Jackson, L.M., Myers, J.E., 2003. Design and construction of pilot wetlands for produced-water treatment. SPE Annual Technical Conference and Exhibition. Society of Petroleum Engineers, Tulsa, OK.

Jackson, R.B., Vengosh, A., Darrah, T.H., Warner, N.R., Down, A., Poreda, R.J., et al., 2013. Increased stray gas abundance in a subset of drinking water wells near Marcellus shale gas extraction. Proc. Natl. Acad. Sci. 110, 11250−11255.

Jacobs, R., Grant, R.O.H., Kwant, J., Marquenie, J.M., Mentzer, E., 1992. The composition of produced water from Shell operated oil and gas production in the North Sea. In: Ray, J.P., Engelhardt, F.R. (Eds.), Produced Water: Technological/Environmental Issues and Solutions. Plenum Press, New York, pp. 13−21.

Janks, J.S., Cadena, F., 1992. Investigations Into the Use of Modified Zeolites for Removing Benzene, Toluene, and Xylene From Saline Produced Water. Plenum Press, New York, pp. 473−487.

Jiménez, S., Micó, M.M., Arnaldos, M., Ferrero, E., Malfeito, J.J., Medina, F., et al., 2017. Integrated processes for produced water polishing: enhanced flotation/sedimentation combined with advanced oxidation processes. Chemosphere 168, 309−317.

Jing, L., Chen, B., Zhang, B., Li, P., 2015. Process simulation and dynamic control for marine oily wastewater treatment using UV irradiation. Water Res. 81, 101−112.

Kalantariasl, A., Schulze, K., Storz, J., Burmester, C., Kuenckeler, S., You, Z., et al., 2015. Injectivity during PWRI and disposal in thick low permeable formations (laboratory and mathematical modelling, field case). SPE European Formation Damage Conference and Exhibition. Society of Petroleum Engineers, Tulsa, OK.

Katz, A., Thompson, A.H., 1985. Fractal sandstone pores: implications for conductivity and pore formation. Phys. Rev. Lett. 54, 1325.

Khatib, Z., Verbeek, P., 2003. Water to value-produced water management for sustainable field development of mature and green fields. J. Pet. Technol. 55, 26−28.

Khoshrou, I., Nasr, M.J., Bakhtari, K., 2017. New opportunities in mass and energy consumption of the multi-stage flash distillation type of brackish water desalination process. Sol. Energ. 153, 115−125.

Klasson, K.T., 2002. Ozone Treatment of Soluble Organics in Produced Water. ORNL Oak Ridge National Laboratory, Oak Ridge, TN.

Kondash, A.J., Albright, E., Vengosh, A., 2017. Quantity of flowback and produced waters from unconventional oil and gas exploration. Sci. Total Environ. 574, 314−321.

Konschnik, K.E., Boling, M.K., 2014. Shale gas development: a smart regulation framework. Environ. Sci. Technol. 48, 8404−8416.

Lake, L.W., 1989. Enhanced Oil Recovery. Society of Petroleum Engineers, Tulsa, OK.

Law, B.E., Curtis, J.B., 2002. Introduction to unconventional petroleum systems. AAPG Bull. 86, 1851−1852.

Lee, C.-M., Frankiewicz, T., 2004. Developing vertical column induced gas flotation for floating platforms using computational fluid dynamics. SPE Annual Technical Conference and Exhibition. Society of Petroleum Engineers, Tulsa, OK.

Liang, Y., 2014. Experimental Study of Convective Dissolution of Carbon Dioxide in Porous Media. The University of Texas at Austin, Austin, TX.

Liang, Y., 2017. Scaling of Solutal Convection in Porous Media. The University of Texas at Austin, Austin, TX.

Liang, Y., Yuan, B., 2017a. Dynamic permeability model for CO_2 injection into dual-porosity system: storage estimation and simulation. In: CMTC-485477-MS. Carbon Management Technology Conference, CMTC.

Liang, Y., Yuan, B., 2017b. A guidebook of carbonate laws in China and Kazakhstan: review, comparison and case studies. In: CMTC-485460-MS. Carbon Management Technology Conference, CMTC. Available from: https://doi.org/10.7122/485460-MS.

Liang, Y., DiCarlo, D.A., Hesse, M.A., 2013. Experimental study of convective dissolution of carbon dioxide in heterogeneous media. In: AGU Fall Meeting Abstracts.

Liang, Y., DiCarlo, D.A., Hesse, M.A., 2015. Experiment and simulation study of hydro-dynamic dispersion and finger dynamics for convective dissolution of carbon dioxide. In: AGU Fall Meeting Abstracts.

Liang, Y., Sheng, J., Hildebrand, J., 2017. Dynamic permeability models in dual-porosity system for unconventional reservoirs: case studies and sensitivity analysis. Presented at the SPE Reservoir Characterisation and Simulation Conference and Exhibition. Society of Petroleum Engineers, Tulsa, OK.

Liu, X., Civan, F., 1996. Formation damage and filter cake buildup in laboratory core tests: modeling and model-assisted analysis. SPE Form. Eval. 11, 26−30.

Lopez, A.M., Williams, M., Paiva, M., Demydov, D., Do, T.D., Fairey, J.L., et al., 2017. Potential of electrodialytic techniques in brackish desalination and recovery of industrial process water for reuse. Desalination 409, 108−114.

Mahmoudi, S., 1997. Legal protection of the Persian Gulf's marine environment. Mar. Pol. 21, 53−62.

McCormack, P., Jones, P., Hetheridge, M.J., Rowland, S.J., 2001. Analysis of oilfield produced waters and production chemicals by electrospray ionisation multi-stage mass spectrometry (ESI-MS n). Water Res. 35, 3567−3578.

McGuire, D., 2016. Apparatus for treating fluids. Google Patents.

Moghaddam, R.N., Rostami, B., Pourafshary, P., Fallahzadeh, Y., 2012. Quantification of density-driven natural convection for dissolution mechanism in CO_2 sequestration. Transp. Porous Media 92, 439−456.

Moraes, J.E.F., Silva, D.N., Quina, F.H., Chiavone-Filho, O., Nascimento, C.A.O., 2004. Utilization of solar energy in the photodegradation of gasoline in water and of oil-field-produced water. Environ. Sci. Technol. 38, 3746−3751.

Morrow, L.R., Martir, W.K., Aghazeynali, H., Wright, D.E., 1999. Process of treating produced water with ozone. Google Patents.

Nadav, N., 1999. Boron removal from seawater reverse osmosis permeate utilizing selective ion exchange resin. Desalination 124, 131−135.

Nair, R.R., Protasova, E., Bilstad, T., Strand, S., 2016. Reuse of produced water by membranes for enhanced oil recovery. SPE Annual Technical Conference and Exhibition. Society of Petroleum Engineers, Tulsa, OK.

Neff, J.M., 2002. Bioaccumulation in Marine Organisms: Effect of Contaminants From Oil Well Produced Water. Elsevier, London, UK.

Nghiem, L.D., Ren, T., Aziz, N., Porter, I., Regmi, G., 2011. Treatment of coal seam gas produced water for beneficial use in Australia: a review of best practices. Desal. Water Treat. 32, 316−323.

Ning, Y., Jiang, Y., Qin, G., 2014. Numerical simulation of natural gas transport in shale formation using generalized lattice Boltzmann method. In: International Petroleum Technology Conference.

Ning, Y., He, S., Chen, T., Jiang, Y., Qin, G., 2015a. Simulation of shale gas transport in 3D complex nanoscale-pore structures using the lattice Boltzmann method. SPE Asia

Pacific Unconventional Resources Conference and Exhibition. Society of Petroleum Engineers, Tulsa, OK.

Ning, Y., Jiang, Y., Liu, H., Qin, G., 2015b. Numerical modeling of slippage and adsorption effects on gas transport in shale formations using the lattice Boltzmann method. J. Nat. Gas Sci. Eng. 26, 345–355.

Ning, Y., He, S., Liu, H., Wang, H., Qin, G., 2016a. A rigorous upscaling procedure to predict macro-scale transport properties of natural gas in shales by coupling molecular dynamics with lattice Boltzmann method. SPE Annual Technical Conference and Exhibition. Society of Petroleum Engineers, Tulsa, OK.

Ning, Y., He, S., Qin, G., Liu, H., Wang, H., 2016b. Upscaling in numerical simulation of shale transport properties by coupling molecular dynamics simulation with lattice Boltzmann method. Unconventional Resources Technology Conference. Society of Exploration Geophysicists, American Association of Petroleum Geologists, Society of Petroleum Engineers, San Antonio, TX, pp. 1706–1717.

Noble, B., Ketilson, S., Aitken, A., Poelzer, G., 2013. Strategic environmental assessment opportunities and risks for Arctic offshore energy planning and development. Mar. Pol. 39, 296–302.

Obe, I., Fashanu, T.A., Idialu, P.O., Akintola, T.O., Abhulimen, K.E., 2017. Produced water re-injection in a non-fresh water aquifer with geochemical reaction, hydrodynamic molecular dispersion and adsorption kinetics controlling: model development and numerical simulation. Appl. Water Sci. 7, 1169–1189.

Onishi, V.C., Carrero-Parreno, A., Reyes-Labarta, J.A., Fraga, E.S., Caballero, J.A., 2017. Desalination of shale gas produced water: a rigorous design approach for zero-liquid discharge evaporation systems. J. Clean. Prod. 140, 1399–1414.

Osborn, S.G., Vengosh, A., Warner, N.R., Jackson, R.B., 2011. Methane contamination of drinking water accompanying gas-well drilling and hydraulic fracturing. Proc. Natl. Acad. Sci. 108, 8172–8176.

Padaki, M., Murali, R.S., Abdullah, M.S., Misdan, N., Moslehyani, A., Kassim, M.A., et al., 2015. Membrane technology enhancement in oil–water separation. A review. Desalination 357, 197–207.

Pang, S., Sharma, M.M., 1995. Evaluating the performance of open-hole, perforated and fractured water injection wells. SPE European Formation Damage Conference. Society of Petroleum Engineers, Tulsa, OK.

Pavasovic, A., 1996. The Mediterranean action plan phase II and the revised Barcelona convention: new prospective for integrated coastal management in the Mediterranean region. Ocean Coast. Manag. 31, 133–182.

Percival, R.V., Schroeder, C.H., Miller, A.S., Leape, J.P., 2015. Environmental Regulation: Law, Science, and Policy. Wolters Kluwer Law & Business, Chicago, IL.

Pichtel, J., 2016. Oil and gas production wastewater: soil contamination and pollution prevention. Appl. Environ. Soil Sci 2016.

Posada, D., Betancourt, P., Fuentes, K., Marrero, S., Liendo, F., Brito, J.L., 2014. Catalytic wet air oxidation of oilfield produced wastewater containing refractory organic pollutants over copper/cerium–manganese oxide. React. Kinet. Mech. Catal. 112, 347–360.

Rawlins, C.H., 2009. Flotation of fine oil droplets in petroleum production circuits. Recent Adv. Miner. Process. Plant Des. 12, 232–246.

Rebello, C.A., Couperthwaite, S.J., Millar, G.J., Dawes, L.A., 2016. Understanding coal seam gas associated water, regulations and strategies for treatment. J. Unconv. Oil Gas Resour. 13, 32–43.

Reed, M., Johnsen, S., 2012. Produced Water 2: Environmental Issues and Mitigation Technologies. Springer Science & Business Media, Berlin, Germany.

Reig, M., Casas, S., Gibert, O., Valderrama, C., Cortina, J.L., 2016. Integration of nano-filtration and bipolar electrodialysis for valorization of seawater desalination brines: production of drinking and waste water treatment chemicals. Desalination 382, 13—20.

Reynolds, R.R., Kiker, R.D., 2003. Produced Water and Associated Issues. Oklahoma Geological Survey, Tulsa, OK.

Roach, R.W., Carr, R.S., Howard, C.L., Cain, B.W., Carrý, R., Station, C., 1993. An Assessment of Produced Water Impacts in the Galveston Bay System. US Fish and Wildlife Service, Clear Lake, TX.

Rowan, E.L., Engle, M.A., Kirby, C.S., Kraemer, T.F., 2011. Radium content of oil-and gas-field produced waters in the Northern Appalachian basin (USA)—Summary and discussion of data. US Geological Survey Scientific Investigations Report 2011-5135 31.

Sacramento, R.N., Yang, Y., You, Z., Waldmann, A., Martins, A.L., Vaz, A.S., et al., 2015. Deep bed and cake filtration of two-size particle suspension in porous media. J. Pet. Sci. Eng. 126, 201—210.

Santander, M., Rodrigues, R.T., Rubio, J., 2011. Modified jet flotation in oil (petroleum) emulsion/water separations. Colloids Surf. Physicochem. Eng. Asp. 375, 237—244.

Saththasivam, J., Loganathan, K., Sarp, S., 2016. An overview of oil—water separation using gas flotation systems. Chemosphere 144, 671—680.

Semiat, R., Sutzkover, I., Hasson, D., 2003. Scaling of RO membranes from silica super-saturated solutions. Desalination 157, 169—191.

Seright, R.S., Lane, R.H., Sydansk, R.D., 2001. A strategy for attacking excess water pro-duction. SPE Permian Basin Oil and Gas Recovery Conference. Society of Petroleum Engineers, Tulsa, OK.

Shannon, M.A., Bohn, P.W., Elimelech, M., Georgiadis, J.G., Mariñas, B.J., Mayes, A.M., 2008. Science and technology for water purification in the coming decades. Nature 452, 301—310.

Sharma, M.M., Yortsos, Y.C., 1987. Transport of particulate suspensions in porous media: model formulation. AIChE J. 33, 1636—1643.

Sharma, M.M., Pang, S., Wennberg, K.E., Morgenthaler, L.N., 2000. Injectivity decline in water-injection wells: an offshore Gulf of Mexico case study. SPE Prod. Facil. 15, 6—13.

Sheng, J., 2013. Enhanced Oil Recovery Field Case Studies. Gulf Professional Publishing, Oxford.

Shepherd, M.C., Shore, F.L., Mertens, S.K., Gibson, J.S., 1992. Characterization of pro-duced waters from natural gas production and storage operations: regulatory analysis of a complex matrix. Prod. Water Technol. Issues Solut. Plenum Publ. Corp, New York, pp. 163—174.

Shutong, P., Sharma, M.M., 1997. A model for predicting injectivity decline in water-injection wells. SPE Form. Eval. 12, 194—201.

Silva, T.L., Morales-Torres, S., Castro-Silva, S., Figueiredo, J.L., Silva, A.M., 2017. An overview on exploration and environmental impact of unconventional gas sources and treatment options for produced water. J. Environ. Manag. 200, 511—529.

Spellman, F.R., 2013. Handbook of Water and Wastewater Treatment Plant Operations. CRC Press, Boca Raton, London, New York, Washington, DC.

Srinivasan, A., Viraraghavan, T., 2008. Removal of oil by walnut shell media. Bioresour. Technol. 99, 8217—8220.

Sutzkover-Gutman, I., Hasson, D., 2010. Feed water pretreatment for desalination plants. Desalination 264, 289—296.

Swisher, M., 2000. Summary of DWS application in Northern Louisiana. In: Proceedings of the Downhole Water Separation Technology Workshop, Baton Rouge, LA.

Tansel, B., Pascual, B., 2011. Removal of emulsified fuel oils from brackish and pond water by dissolved air flotation with and without polyelectrolyte use: pilot-scale investigation for estuarine and near shore applications. Chemosphere 85, 1182−1186.

Tibbetts, P.J.C., Buchanan, I.T., Gawel, L.J., Large, R., 1992. A comprehensive determination of produced water composition. In: Ray, J.P., Engelhardt, F.R. (Eds.), Produced Water: Technological/Environmental Issues and Solutions. Plenum Press, New York, pp. 97−112.

Tromp, D., Wieriks, K., 1994. The OSPAR convention: 25 years of North Sea protection. Mar. Pollut. Bull. 29, 622−626.

Vegueria, S.J., Godoy, J.M., Miekeley, N., 2002. Environmental impact studies of barium and radium discharges by produced waters from the "Bacia de Campos" oil-field offshore platforms, Brazil. J. Environ. Radioact. 62, 29−38.

Veil, J.A., Langhus, B.G., Belieu, S., 1999. Feasibility Evaluation of Downhole Oil/Water Separator (DOWS) Technology. Argonne National Laboratory (ANL), Lemont, IL.

Veil, J.A., Puder, M.G., Elcock, D., Redweik Jr., R.J., 2004. A White Paper Describing Produced Water From Production of Crude Oil, Natural Gas, and Coal Bed Methane. Argonne National Laboratory (ANL), Lemont, IL.

Von Flatern, R., 2003. Troll pilot sheds light on seabed separation. Offshore Eng. 45−46.

Wen, B., 2015. Porous Medium Convection at Large Rayleigh Number: Studies of Coherent Structure, Transport, and Reduced Dynamics. University of New Hampshire, Durham, NH.

Wen, B., Chini, G., 2015. Inclined porous medium convection at large Rayleigh number. In: APS Division of Fluid Dynamics Meeting Abstracts.

Wen, B., Chini, G.P., Dianati, N., Doering, C.R., 2013. Computational approaches to aspect-ratio-dependent upper bounds and heat flux in porous medium convection. Phys. Lett. A 377, 2931−2938.

Wen, B., Corson, L.T., Chini, G.P., 2015. Structure and stability of steady porous medium convection at large Rayleigh number. J. Fluid Mech. 772, 197−224.

Wennberg, K.E., Sharma, M.M., 1997. Determination of the filtration coefficient and the transition time for water injection wells. SPE European Formation Damage Conference. Society of Petroleum Engineers, Tulsa, OK.

Wu, W., Sharma, M.M., 2017. A model for the conductivity and compliance of unpropped and natural fractures. SPE J 22.

Wu, Y., Bryant, S.L., 2014. Optimization of field-scale surface dissolution with thermoelastic constraints. Energ. Procedia 63, 4850−4860.

Wu, Y., Taylor, J.M., Frei-Pearson, A., Bryant, S.L., 2015. Injection Induced Fracturing As a Necessary Evil in Geologic CO_2 Sequestration. 49th US Rock Mechanics/Geomechanics Symposium. American Rock Mechanics Association, San Francisco, CA.

Wu, Y., Bryant, S.L., Lake, L.W., 2017a. Modeling of gas exsolution for CO_2/brine surface dissolution strategy. In: 13th International Conference On Greenhouse Gas Control Technology GHGT-13. 14−18 November 2016, Lausanne, Switzerland. 114, 4106−4118. Available from: https://doi.org/10.1016/j.egypro.2017.03.1551.

Wu, Y., Bryant, S.L., Lake, L.W., 2017b. Pressure management of CO_2 storage by closed-loop surface dissolution. In: 13th International Conference on Greenhouse Gas Control Technology. GHGT-13. 14−18 November 2016, Lausanne, Switzerland. 114, 4811−4821. Available from: https://doi.org/10.1016/j.egypro.2017.03.1620.

Xiao, R., Teletzke, G.F., Lin, M.W., Glotzbach, R.C., Aitkulov, A., Yeganeh, M., et al., 2017. A novel mechanism of alkaline flooding to improve sweep efficiency for viscous oils. SPE Annual Technical Conference and Exhibition. Society of Petroleum Engineers, Tulsa, OK.

Yang, Y., Zhang, X., Wang, Z., 2002. Oilfield produced water treatment with surface-modified fiber ball media filtration. Water Sci. Technol. 46, 165–170.

Yerramilli, R.C., Zitha, P.L.J., Yerramilli, S.S., Bedrikovetsky, P., 2013. A novel water injectivity model and experimental validation using CT scanned core-floods. SPE European Formation Damage Conference & Exhibition. Society of Petroleum Engineers, Tulsa, OK.

Yerramilli, R.C., Zitha, P.L., Yerramilli, S.S., Bedrikovetsky, P., 2014. A novel water-injectivity model and experimental validation with CT-scanned corefloods. SPE J. 20, 1200–1211.

You, Z., Kalantariasl, A., Schulze, K., Storz, J., Burmester, C., Künckeler, S., et al., 2016. Injectivity impairment during produced water disposal into low-permeability Völkersen aquifer (compressibility and reservoir boundary effects). SPE International Conference and Exhibition on Formation Damage Control. Society of Petroleum Engineers, Tulsa, OK.

Young, G.A.B., Wakley, W.D., Taggart, D.L., Andrews, S.L., Worrell, J.R., 1994. Oil–water separation using hydrocyclones: an experimental search for optimum dimensions. J. Pet. Sci. Eng. 11, 37–50.

Yuan, B., Wood, D.A., 2015. Production analysis and performance forecasting for natural gas reservoirs: theory and practice (2011–2015). J. Nat. Gas Sci. Eng. 1433–1438.

Yuan, B., Su, Y., Moghanloo, R.G., Rui, Z., Wang, W., Shang, Y., 2015a. A new analytical multi-linear solution for gas flow toward fractured horizontal wells with different fracture intensity. J. Nat. Gas Sci. Eng. 23, 227–238.

Yuan, B., Wood, D.A., Yu, W., 2015b. Stimulation and hydraulic fracturing technology in natural gas reservoirs: theory and case studies (2012–2015). J. Nat. Gas Sci. Eng. 26, 1414–1421.

Yuan, B., Bedrikovetsky, P., Huang, T.T., Moghanloo, R.G., Dai, C., Venkatraman, A., et al., 2016a. Special issue: formation damage during enhanced gas and liquid recovery. J. Nat. Gas Sci. Eng. 1051–1054.

Yuan, B., Moghanloo, R.G., Shariff, E., 2016b. Integrated investigation of dynamic drainage volume and inflow performance relationship (transient IPR) to optimize multi-stage fractured horizontal wells in tight/shale formations. J. Energ. Resour. Technol. 138, 052901.

Yuan, B., Moghanloo, R.G., Zheng, D., 2016c. Enhanced oil recovery by combined nanofluid and low salinity water flooding in multi-layer heterogeneous reservoirs. SPE Annual Technical Conference and Exhibition. Society of Petroleum Engineers, Tulsa, OK.

Yuan, B., Zheng, D., Moghanloo, R.G., Wang, K., 2017. A novel integrated workflow for evaluation, optimization, and production predication in shale plays. Int. J. Coal Geol. 180, 18–28.

Yuan, T., Ning, Y., Qin, G., 2017. Numerical modeling of mineral dissolution of carbonate rocks during geological CO_2 sequestration processes. SPE Europec Featured at 79th EAGE Conference and Exhibition. Society of Petroleum Engineers, Tulsa, OK.

Zhao, X., Wu, Y., Baiming, Z., Yu, S., Xianbao, Z., Gao, P., 2011. Strategies to conduct steam injection in waterflooded light oil reservoir and a case in Fuyu reservoir, Jilin. SPE Middle East Oil and Gas Show and Conference. Society of Petroleum Engineers, Tulsa, OK.

Zheng, J., Chen, B., Thanyamanta, W., Hawboldt, K., Zhang, B., Liu, B., 2016. Offshore produced water management: a review of current practice and challenges in harsh/Arctic environments. Mar. Pollut. Bull. 104, 7–19.

Zhou, S., Xue, H., Ning, Y., Guo, W., Zhang, Q., 2018. Experimental study of supercritical methane adsorption in Longmaxi shale: insights into the density of adsorbed methane. Fuel 211, 140–148.

Zhu, D., Okuno, R., 2014a. A robust algorithm for isenthalpic flash of narrow-boiling fluids. Fluid Phase Equilibria 379, 26—51.

Zhu, D., Okuno, R., 2014b. A robust algorithm for multiphase isenthalpic flash. SPE Heavy Oil Conference-Canada. Society of Petroleum Engineers, Tulsa, OK.

Zhu, D., Okuno, R., 2015a. Analysis of narrow-boiling behavior for thermal compositional simulation. SPE Reservoir Simulation Symposium. Society of Petroleum Engineers, Tulsa, OK.

Zhu, D., Okuno, R., 2015b. Robust isenthalpic flash for multiphase water/hydrocarbon mixtures. SPE J. 20, 1,350—1,365.

Zhu, D., Okuno, R., 2016. Multiphase isenthalpic flash integrated with stability analysis. Fluid Phase Equilibria 423, 203—219.

Zhu, D., Eghbali, S., Shekhar, C., Okuno, R., 2017. A unified algorithm for phase-stability/split calculation for multiphase PT flash. SPE J.

Zouboulis, A.I., Avranas, A., 2000. Treatment of oil-in-water emulsions by coagulation and dissolved-air flotation. Colloids Surf. Physicochem. Eng. Asp 172, 153—161.

FURTHER READING

Liang, Y. 2017. Scaling of Solutal Convection in Porous Media [phd dissertation], https://www.pge.utexas.edu/images/pdfs/theses17/liangdis17.pdf.

Short-Term Energy Outlook—US Energy Information Administration (EIA) [WWW Document], n.d. <https://www.eia.gov/outlooks/steo/report/global_oil.cfm> (accessed 11.30.17).

CHAPTER FIFTEEN

Integrated Risks Assessment and Management of IOR/EOR Projects: A Formation Damage View

David A. Wood[1] and Bin Yuan[2]
[1]DWA Energy Limited, Lincoln, United Kingdom
[2]Coven Energy Technology Research Institute, Qingdao, Shandong, China

Contents

Formation Damage during Improved Oil Recovery.
DOI: https://doi.org/10.1016/B978-0-12-813782-6.00015-4

© 2018 Elsevier Inc.
All rights reserved.

15.1 INTRODUCTION

Damage of subsurface rock formations has been studied for many years from the perspective of negative impacts on reservoir formations during drilling (e.g., Faergestad, 2016). Such negative impacts related to drilling are mainly the consequences of solids migrating from the wellbore and blocking pores or drilling fluids interacting with reservoir fluids to change their composition and induce some kind of precipitation, or otherwise impede their flow. These can be usefully categorized under four, distinct, high-level mechanisms:

- *Biological*—bacterial contamination; introduction of nutrients that enhance bacterial activity. Increased bacterial activity in the reservoir can lead to souring of the petroleum fluids, erosion of soils, and corrosion of wellbore tubulars.
- *Chemical*—reactions between wellbore and reservoir fluids (e.g., swelling smectite and mixed clays; deflocculation of clays due to changes in pH of the reservoir fluids; dissolution of reservoir minerals causing collapse of pore framework; changes in formation wettability increasing formation water mobility).
- *Mechanical*—indigenous fines migration; introduced external solids; introduced fluids; bit grinding; fines related to perforation damage; fines from degraded proppants introduced during fracture stimulation; solids out after rock failure during production.
- *Thermal*—introduction of higher temperature fluids into the reservoir zones. This can degrade or sour the petroleum fluids and/or enhance mineral dissolution with adverse structural impacts (collapse of pore structure) and/or mineral phase changes. Thermal effects also can lead to changes in mechanical properties due to thermoelasticity effects.

However, these formation damage mechanisms are not mutually exclusive and a drilling and completion operation can induce one or all of them when penetrating specific formations. In improved and enhanced oil recovery (IOR/EOR) operations, a similar set of possible damage mechanisms apply, but the extent of the damage is likely to be more extensive and the nature of specific types of damage more diverse (Fig. 15.1). This is because in IOR/EOR greater quantities of fluids incorporating a wider range of chemical compositions and/or physical properties are injected further into the formation, by choice and design, compared with standard drilling operations. Consequently, the risk

Diverse mechanisms contribute to reservoir formation damage with mainly negative but some positive consequences

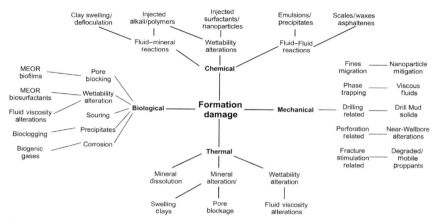

Figure 15.1 Key mechanisms contributing to formation damage in reservoirs classified according to their biological, chemical, mechanical, or thermal origins. All of these mechanisms are likely to be encountered in the wide range of IOR projects conducted by the oil and gas industry.

assessments of formation damage in IOR/EOR operations need to include issues of scale and extent of damage as well as specific damage mechanisms.

Data gathered from laboratory tests on reservoir cores and pilot field tests involving individual wells or small grids of injection and production wells provide valuable direct insights to reservoir behavior. Such data is complemented by three-dimensional (3D) reservoir simulation analysis taking into account reservoir anisotropy. Together these data and analysis help to quantify the formation damage mechanisms, and the likely extent of damage on a reservoir scale associated with specific IOR/EOR methods. However, it is frequently misleading to draw far-reaching conclusions from such laboratory and pilot studies, or to use results of such studies on one reservoir to infer risks of similar outcomes in other reservoirs by means of direct analogy. This is because each reservoir tends to have its own unique mineralogy, porosity/permeability framework, reservoir fluid chemistry, range of pressure—temperature conditions, and stress fields. Indeed, heterogeneity of these characteristics and large variations in depth for individual reservoirs mean that damage behavior observed and quantified in one sector of the reservoir may not apply or apply to different degrees, in other sectors of that reservoir.

Various indicators of formation damage can be used. Some of these can be measured on rock cores and other formation samples in laboratory conditions (e.g., porosity—permeability impairment), others are easier to establish from well-log and well-test data (e.g., skin damage, generally associated with near-wellbore effects in standard drilling operations can be more pervasive in IOR/EOR projects), which provide the ultimate formation damage key performance indicators (KPI) production rate and water cut. It is a comparison of sustainable production rates and water cuts (i.e., from long-term well tests or production test) for the reservoir sector, before and after IOR/EOR operations were initiated, that provide the "acid" test to determine whether a particular IOR/EOR method is both technically and commercially beneficial, despite the potential and actual degrees of formation damage associated with it.

Formation damage mechanisms can have both positive and negative impacts on oil production and ultimate resource recovery from specific formations. Consequently, it is appropriate to conduct analysis that considers both potential downside risks and upside opportunities associated with formation damage induced by specific IOR techniques. This is best done in terms of a comprehensive risk and opportunity framework or workflow (Fig. 15.2) that not only just identifies these formation damage impacts, but assesses them in detail, ranks them, develops and costs possible mitigation (or exploitation) strategies and actions, schedules and implements those actions, monitors the performance of those actions, and adjusts/fine-tunes the response actions. A risk and opportunity register should play a key role in recording, communicating, and updating the assessments, response actions, and their performance.

Clearly, the integrated risk and opportunity assessment and management of IOR projects is complex and requires multiple steps conducted within a systematic framework. This is the case even when it is focused just on the formation damage issues (i.e., ignoring the above ground uncertainties concerning regulatory approvals, land access, availability of materials, access to technologies, cost of materials and equipment, etc., that also need to be part of a full-project-risk—opportunity analysis). The value of a comprehensive and systematic risk and opportunity analysis will depend very much on the detail, rigor, and effort with which the identification, assessment, and ranking steps are conducted, and the level of data, information, and knowledge that is captured, recorded, and regularly updated in the risk/opportunity register.

**Risk and opportunity management framework
focused on formation damage related to IOR techniques**

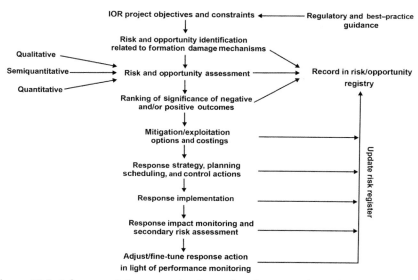

Figure 15.2 A framework for managing formation damage risks and opportunities in IOR projects that focuses the comprehensive use of a risk/opportunity register to aid risk identification and assessment.

In general terms, risk and opportunity assessments benefit by combining several different approaches:

- *Qualitative*—useful for identifying the range of risks and opportunities, and their cause(s) and effect(s).
- *Semiquantitative*—useful for identifying the relative significance (ranking) of specific risks and opportunities, together with the magnitude and likelihood of their impacts on the reservoir
- *Quantitative*—useful for establishing numerical relationships for production rates and resource recovery improvements that should be realistically expected when applying specific methods to a specific reservoir or sector of a reservoir. Initial deterministic methods to quantify risks and opportunities can sometimes be usefully expanded into stochastic (probabilistic) risk assessment models.

A thorough risk and opportunity assessment is likely to include all three types of assessment methods. Influence diagrams usually help in the identification and qualitative assessment of specific events/actions. In that stage of the analysis, it is important to avoid just listing events, and

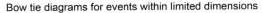

Figure 15.3 Bow tie and butterfly influence diagrams focus attention on causes and effects and thereby aid in the identification and qualitative assessment of risks and opportunities.

influence diagrams help to broaden the focus to causes and impacts (Fig. 15.3). For example, if "fines migration" is considered as the "event," it is clear that multiple causes and multiple impacts (some good and some bad) are linked to that event. Bow tie and butterfly diagrams aid the visualization of potentially complex relationships.

For projects focusing just on the formation damage issues and impacts, bow tie influence diagrams typically suffice, probably with a series of sequential impacts associated with some events. However, for more complex projects such as CO_2 capture and sequestration EOR projects considering all the belowground and aboveground risks, butterfly influence diagrams are likely to be more appropriate.

A natural progression from the information provided by influence diagrams is the qualitative risk/opportunity profiling of the identified events/actions in terms of 2D impact versus likelihood matrices (Fig. 15.4). An initial step with such diagrams is to plot a

Qualitative risk–opportunity assessment using profiling matrices

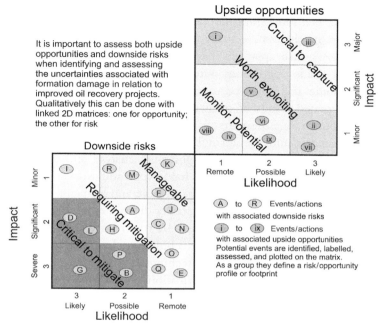

Figure 15.4 2D matrices considering the impact and likelihood of both risk and opportunity events/actions that are useful for constructing profiles for qualitative assessments.

two-dimensional distribution of each event/action identified. This helps in ranking and prioritizing events/actions for further evaluation. A subsequent focus should be to consider how might mitigation strategies move the downside-risk events/actions toward the top-right "manageable" section of the risk matrix, and how might exploitation strategies move the upside-opportunity events/actions toward the top-right "crucial to capture" section of the opportunity matrix. Typically, projects are not easily sanctioned to proceed, from an economic perspective, if risks remain in the "critical to mitigate" region of the risk matrix (Fig. 15.4), after all mitigation options have been considered.

A key application of risk analysis is to use its findings to seek mitigation actions that reduce or eliminate the impacts of particular-formation—damage outcomes (downside-risk) associated with certain IOR/EOR methods when applied to specific reservoirs or reservoir sectors. This is desirable as it is likely to lead to higher productivity and/or resource recovery. However, what also needs to be taken into account are the secondary risks and opportunities that are often associated with the mitigation actions taken, i.e., some mitigation actions may while inhibiting one formation damage mechanism, actually exacerbate another formation damage mechanism. A similar approach is required to consider how best to exploit the potential upside-opportunity outcomes identified, and the secondary risks and opportunities that might be associated with those exploitative actions.

The integrated results of qualitative, semiquantitative, and quantitative risk and opportunity assessment are likely to form the basis of an evolving response strategies and the formulation of mitigation/exploitation plans, schedules, and actions. The risk/opportunity analysis and management is best conducted as part of systematic framework and workflow (Fig. 15.2). However, that framework needs to be flexible and adaptable on a project-by-project basis, where objectives typically differ. For example, the objective of such analysis in one project may be to screen various field locations for deployment of a particular IOR technique, whereas in another project it may be to screen between IOR techniques to apply at a specific field location. In some IOR projects, information from laboratory and well tests, core experiments, reservoir simulations, and pilot field deployment tests of various scales may already be available. On the other hand, in other IOR projects, some or all of such information may be lacking, and/or for whatever reason cannot be obtained before an investment decision is required.

15.2 IDENTIFICATION AND RECOGNITION OF FORMATION DAMAGE RISKS IN IOR PROJECTS

This is the key step in meaningful risk and opportunity analysis of formation damage, and much of this chapter is focused upon identifying the wide range of complex factors responsible for formation damage in IOR and their interactions. Such analysis requires multidisciplinary inputs and the recognition that some facets of formation damage have positive impacts on oil recovery, while the majority have negative, but variable impacts on oil recovery. A flavor of the multifaceted nature of formation damage issues and number of issues to consider for EOR projects can be gained from a 2010 review of EOR projects (Alvarado and Manrique, 2010), and a 2016 special issue on the subject (Yuan, et al., 2016a). This section explores the multifaceted nature of these risks, and in some cases opportunities, in some detail, based upon recently published studies covering a wide range of distinct IOR techniques. While reviewing these cases and recent evaluations, it is worth considering how these might be approached from a risk/opportunity identification and assessment basis, applying the framework outlined in Fig. 15.2 together with the qualitative assessment tools described in Figs. 15.3 and 15.4.

15.2.1 Low-salinity waterflooding risks

Fines migration, particle swelling, and swelling-induced migration are the main formation damage mechanisms that impact sandstone reservoirs subjected to low-salinity waterflooding (LSWF) (Mohan et al., 1993). Song and Kovscek (2016) observed through visualization models that once salt concentration in the formation water was below a critical salinity value, swelling-induced and fines migration could reduce permeability significantly. Clay bridging and fines redeposition in pore throats and small pores in the high-water-cut flow channels promoted the diversion of injected water to higher-oil-cut pathways improving the oil sweep in the reservoir. This mechanism was found to be much more effective in Kaolinite-rich systems leading to IOR, but in montmorillonite-rich systems it did not lead to IOR. So, in LSWF the targeted reduction of permeability, concentrated in the high-water-cut flow pathways can be a benefit rather than a negative outcome for the reservoir, but it depends on the predominant clay mineralogy, which may vary across a reservoir zone. Moreover, this is likely to be only one of the formation damage mechanisms that could be at work when LSW is applied.

There have been several experimental studies (Lager et al., 2008; Austad et al., 2010; Aladasani et al., 2014) and field tests (McGuire et al., 2005; Skrettingland et al., 2010) that demonstrate the benefits of the LSWF method (i.e., increase in oil recovery factors and production rates), but it does not work in all sandstone reservoirs and requires quite specific conditions to work in carbonate reservoirs. The recent experimental studies have established correlations between residual oil saturation, contact angle, interfacial tension (IFT), and capillary pressure as functions of salinity (Salehi et al., 2017; Aladasani and Bai, 2012) improving the accuracy of reservoir simulations involving LSWF. However, the fact that uncertainties exist regarding the main mechanisms at work in specific reservoirs subjected to LSW, means that risk of it being ineffective, and/or the predictions of simulation analysis being inaccurate, remain until pilot testing is performed.

15.2.2 Timing of fines migration

It is well established that fines migration leads to permeability reductions in high- and low-permeability reservoirs (Guo et al., 2015; You et al., 2016), but experiments suggest that it occurs at slow rates, some 500–1000 times slower than injection velocities (Yang et al., 2016a). Fines migration results from mobilization and migration of detached colloidal or suspended fines, and these fines subsequently are concentrated into thin pore throats degrading permeability. The interactions between the fines and the pore walls and rock, which includes rolling, sliding, and short jumps suspended in the pore fluids, tend to slow down the average particle migration speed. The experimental observation of slow-speed fines migration is consistent with the long periods during core flooding before permeability stabilizes.

The velocity of the fluids through the pore space plays a key role with higher flow velocities mobilizing smaller particles. Based on their experimental results, Yang et al. (2016a) developed a function exhibiting high accuracy in predicting the maximum retention time for size-distributed fine particles attached to pore walls. The function establishes the accumulated concentration of particles smaller than the size that will be mobilized by the pore-fluid flux at a specific flow velocity. If the particle size distribution has a high variance, the higher the maximum retention function at high velocities and the lower the maximum retention function at low velocities.

Fines particles migrating in coal bed methane reservoirs have significant mineral content other than coal, particularly clays (Massarotto et al., 2014). Liu et al. (2011) compared produced fines in different phases of gas production from coals in the South Quinshui Basin in China. They found that the size of the fines particle could be as high as 8 mm during dewatering but reduces to <1 mm during the gas production phase, when gas desorption also affects the fines migration. Guo et al. (2015, 2016) conducted experiments to investigate permeability degradation caused by water flow in anthracite and bituminous coal samples using X-ray diffusion (XRD), scanning electron microscope coupled with the thin film energy dispersive X-ray analysis (SEM-EDX), and petrographic methods. Produced fines from the anthracite coal were mainly clay, whereas from the bituminous coal the fines were a mixture of coal and clay.

Water flow tests revealed ~35% reduction in coal permeability at the end of a 33-day flow test. The analytical model developed by Guo et al.

(2016) suggests that fluid flow rate, interaction energy between fines and coal surface, and coal composition are the key factors controlling fines migration in the coals studied. The models also reveal that most fines migration in the coal samples occurred as water flow began, implying that fines migration would be concentrated in the dewatering phase of coal bed methane reservoirs and much permeability damage could occur before significant gas starts to flow. Optimizing wellbore geometry can play a significant role in mitigating fines migration formation damage issues in coal bed methane reservoirs (Azadi et al., 2016).

15.2.3 CO_2 injection—EOR/sequestration formation damage risks

In addition to reservoir risks, there are many aboveground issues that need to be considered in a comprehensive risk assessment of CO_2-EOR projects concerning large-scale and long-term CO_2 transport, storage, and injection. In the United States, the CO_2-EOR industry has established a good safety and CO_2-handling record over its 40 + years of operation (Duncan et al., 2008).

Long-term injection of CO_2 into aquifers can lead to evaporation from the resident formation fluid causing it to precipitate salt in the pore space (Qiao et al., 2016a) and negatively impact effective porosity and permeability in the reservoir, reducing injectivity and oil recovery. Jin et al. (2016) modeled deposition and dissolution in the near-wellbore region under varying CO_2-injection conditions into a saline reservoir and established that only dolomite cement was likely to dissolve with all other minerals precipitating, hence leading to a net increase in deposition. Increasing concentration of CO_2 in the brine and in mineral deposition did effectively secure the CO_2, but halite deposition and re-imbibition of saline brine due to increased capillary pressure have the potential to significantly reduce injectivity around the injection well, while channeling the flow of CO_2 to sweep other areas of the reservoir yet to be impacted by deposition. The process involves a backflow of brine toward the injector wells due to capillary forces which further increases the halite deposition in the vicinity of the injection wells (Pruess, 2009).

Core-flooding experimental studies evaluating and observing mineral dissolution and mobilization during CO_2 injection into the Es1 water-flooded layer of the Pucheng Oilfield, China (Tang et al., 2016), applying scanning electron microscopy (SEM) and XRD analysis, revealed varied mineral dissolution depending upon the prevailing permeability in particular zones of the reservoir. The Pucheng reservoir is a heterogeneous sandstone with a wide range of permeabilities and now with a high prevailing water saturation. In low-permeability zones, mineral migration tended to block the few exiting flow channels, whereas in high-permeability zones, permeability increases as dissolved minerals are easily transported out of these zones.

The net effects of dissolution and deposition within the reservoir cores studied were that calcite from the rock matrix was dissolved and formed from secondary minerals; the contents of both calcite and feldspar were reduced, whereas that of clay minerals increased. Carbonate minerals forming cements were more readily dissolved, compared to the rock matrix and clay minerals, because they are more easily penetrated by the brine$-CO_2$ solutions. These results suggest that mineral migration can occur in the Pucheng reservoir zones with both high and low permeability, but the outcomes for these zones can be quite different in terms of permeability changes and ultimate mineral compositions. Two-phase (gas/water) flow also influences mineral deposition and potential storage capacity in CO_2 storage reservoirs (Moghanloo et al., 2015).

CO_2 gas injection can induce asphaltene deposition in some reservoirs. Kalantari–Dahaghi et al. (2006) described this associated with the miscible CO_2 flood in the Bangestan carbonate reservoir of Kupal field (Iran). When the CO_2 encounters the reservoir oil it alters the asphaltene-to-resin ratio of crude oil which favors precipitation of organic solids, mainly asphaltenes. These solids may flow as suspended particles, or they may deposit resulting in pore plugging and wettability alteration of reservoir matrix. Deposition of asphaltenes in this reservoir was related, at least in part, to adsorption of flocculated asphaltene particles onto active sites such as clay minerals, particularly kaolinite. Oil reservoirs with naturally high CO_2 levels in the associated gas experience similar issues regarding asphaltene deposition and wettability alteration, for example, as observed in some fields in Colombia (Sepulveda et al., 2015).

Li et al. (2015) established a method for screening CO_2-miscible-flood EOR reservoir candidates. They developed a multiattribute decision-making techniques by calculating the combination of weights from an analytical hierarchy process, involving an entropy method, with technique for order preference by similarity to ideal solution (TOPSIS) analysis and gray relational analysis to construct a customized degree of nearness (to the optimal conditions). This weighted gray correlation (GC)-TOPSIS model was used to screen five reservoirs, with known developed outcomes, for potential application of CO_2-miscible flooding. The results correctly identified the ranking of the candidate reservoirs in respect of their suitability for CO_2-miscible flooding.

15.2.4 Formation damage associated with waterflooding and produced water blockage

Many waterflood projects perform suboptimally in terms of reservoir sweep (Aristov et al., 2015) due to a range of issues causing a decline in well injectivity, e.g., poor water quality, lack of subsurface definition of optimal water quality, poorly managed surface facilities, inappropriate producer—injector well-spacing or matrix pattern, short-circuiting of flow between injector and producer wells, out-of-reservoir injection, etc.

For optimal waterflood operations, it is imperative to evaluate and adjust the control metrics for individual injection wells, in addition to the pattern and well-spacing (Wen et al., 2014). Such optimization is likely to lead to IOR, lower costs, and enhanced project net present value (NPV) (Jansen and Durlofsky, 2016). Suspended particles, generally from fines picked up from the formation of the injected water, are one of the factors that cause suboptimal well injectivity as a consequence of the formation damage they induce (Fallah and Sheydai, 2013). The consequence is often a significant increase in injection pressure, potentially exceeding the pressure capabilities of the surface pumps and pipework, and likely to add costs to the operation.

Feng et al. (2016) developed a well-injection-optimization simulation model, taking into account the suspended particles, based upon a covariance matrix adaptation evolution strategy. This model quantified the effects of formation damage induced by suspended particles. They used the model to evaluate two oil field waterflooding cases, i.e., a five-spot model and an egg model, both taking into account the formation damage related to suspended particles in the injection fluids and ignoring it. The results showed that formation damage caused by the suspended particles in the injected water should not be neglected in the optimization of the well-injection strategy for waterflood projects. By taking formation damage into account, lower water injection pressures could be achieved.

Water-blocking formation damage results in much reduced permeability in the near-wellbore zone of a reservoir caused by increased water saturation, associated with the accumulation of produced/formation water (e.g., due to coning, or separation during the production flow) or invasion by water-based wellbore fluids (Bennion et al., 1996). Treatments with antiwater-blocking agents can resolve many water-blocking problems (Liu and Wu, 2013). Fan et al. (2016) conducted repeated forward and reverse core-flooding experiments in order to evaluate the effectiveness of antiwater-blockage agents. These experiments were conducted under initial reservoir conditions, water-blockage conditions, followed by the application of several different antiwater-blockage agents. The performances of the various antiwater-blockage agents tested were expressed in terms of two, easy-to-calculate, evaluation indices: (1) recovered permeability percentage (RP) and (2) ratio of permeability recovery (RR). The RP and RR indices were shown to be correlated with the interfacial properties of the prevailing pore fluids, and variations in these indices related to microscale variations in capillary pressure. This evaluation method establishes an effective and rigorous procedure for screening and developing suitable antiwater-blocking agents.

15.2.5 Formation damage induced during well stimulation and hydraulic fracturing

Significant formation damage can occur during well stimulation operations, much of which is associated with incompatibility between the fracturing fluids and the formation water and/or the reservoir mineralogy. Methods to mitigate this effect include modifying the composition of the injected fluids in various ways.

Waterless fracturing technologies are one option to avoid adverse reactions in the reservoir caused by injected fluids. They not only potentially avoid formation damage, but also reduce the secondary risks of overuse of limited water resources, disposal of flow-back water (Gallegos et al., 2015), and potential for induced seismic activity if that flow-back water is reinjected into subsurface disposal zones (Davies et al., 2013; Ellsworth, 2013). Various waterless fracturing technologies have been tested and evaluated, e.g., oil-based and CO_2 energized oil fracturing (Hlidek et al, 2012), explosive and propellant fracturing (Page and Miskimins, 2009), gelled liquefied petroleum gas and alcohol fracturing (LeBlanc et al., 2011), gas fracturing (Rogala et al., 2013), CO_2 fracturing (Ishida et al., 2013; Middleton et al., 2015), and cryogenic fracturing with liquid nitrogen (McDaniel et al.,1998; Wang, Yao et al., 2016). Various modeling and analysis approaches have been developed for quantifying and optimizing performance of well stimulation and hydraulic fracturing in types of reservoirs (Clarkson, 2013; Yuan et al., 2015, 2017a).

The application of most of these waterless fracturing technologies tends to be hindered by difficulties in either sourcing the specialist fracturing fluids or operating complexity, high cost, environmental issues, and safety concerns. Experimental studies have demonstrated (Wang, Yao et al., 2016) that liquid nitrogen enhances fracture initiation and propagation in concrete samples, and shale and sandstone reservoir rocks, and does not suffer from the problems just listed. The experiments indicate that cryogenic fracturing with liquid nitrogen does not lead to formation damage. Indeed, it can enhance permeability and reduce the breakdown pressures for subsequent stimulation of the zones treated. However, laboratory studies show that in some instances the freezing of formation water can counter the contraction effects of the liquid nitrogen and form near-wellbore barriers to the propagation of liquid–nitrogen–induced fractures (Cha et al., 2014).

Liquid CO_2 fracturing can also avoid formation damage and display rapid post-fracturing reservoir cleanup. However, potential reactions with formation water risks generating CO_2 hydrate at specific low-temperature—high-pressure conditions (Oldenburg, 2007). Sun et al. (2016) developed a dynamically coupled mass and heat transfer model to evaluate the unstable process of CO_2 flowing in reservoir fractures and leaking-off into the surrounding porous rock and its associated throttling effects. The model revealed that injected CO_2 tended to change phase, from a liquid to a supercritical state, as it progressed through the fractures and on into the rock formation evaluated. Simulations indicated that hydrates did form long bullet-like deposits at certain low temperatures and high pressures. When this occurred it significantly reduced the formation permeability and oil production/recovery. However, Sun et al. (2016) established that hydrate formation could be suppressed by additives to the liquid CO_2, in particular by methanol at mole fractions higher than 20% or by NaCl at mole fractions >8% relative to formation water.

Lu et al. (2016) evaluated the Alteration of Bakken reservoir samples subjected to CO_2-based fracturing using autoclave reaction experiments. Samples exposed to the CO_2-saturated brine displayed dissolution of the carbonate minerals, particularly calcite, whereas quartz and feldspars remained intact, and some pyrite framboids underwent slight dissolution. Additionally, slight calcite precipitation was observed in the pores and fractures, but an overall increase in porosity and permeability occurred. On the other hand, samples exposed to supercritical CO_2 only displayed etching of the calcite grain surface and precipitation of salt crystals (halite and anhydrite) with a slight decrease in permeability, but no change to porosity.

Desulfurated seawater flooding has been widely used offshore for decades (McCune, 1982) together with seawater-based drilling fluids, completion fluids, and fracturing fluids (AlMubarak et al., 2016). However, seawater-based acid, for acid stimulations, often reacts with minerals to form precipitates resulting in formation damage (He et al., 2011). During stimulation with hydrofluoric (HF) acid, reactions with seawater tend to produce fluoride silicate precipitates with strong stability making them difficult to migrate (Frayret et al., 2006). Li, Zhang, and Je (2016) compared experimentally the ability of mud acid (almost pure HF) and organophosphonic acid to inhibit such precipitates when mixed with seawater. Organophosphonic acid outperformed mud acid, because it formed stronger bonds with metal ions. Fluoride precipitates (CaF_2 and MgF_2) were inhibited by a reduction in free F-ions and an increase in the seawater dilution and inorganic ions. Moreover, the addition of hydrochloric acid and NH_4Cl was found to inhibit the precipitation of secondary fluorosilicates.

Reinoso et al. (2016) provided lessons learned from more than 100 sandstone acidizing treatments regarding fines migration in the Villeta and Caballos formations in the Putumayo Basin of Colombia. They identified an optimal three-step sequence (pre-flush, primary treatment, and post-flush) with HF acid as the main agent to retard fines migration. For certain reservoirs, particularly high-temperature zones, scaling treatments, biodegradable chelating agents, and aminopolycarboxylic acid (Reyes et al., 2013, 2016), a low-pH fluid is also added to the treatments. Aminopolycarboxylic acid has demonstrated deeper reservoir stimulation due to their lower reaction rates and, in carbonate reservoirs, potentially a more effective wormhole development. These low-pH fluids also chelate reaction products aiding well cleanup and require lower doses of corrosion inhibitors.

Well-developed fracture systems, although ultimately beneficial, can involve loss of circulation during drilling and drill-in fluid penetration of conventional and tight reservoirs, frequently inducing severe formation damage. Remedial actions involving fracture plugging with loss control material (LCM) typically are applied to control lost circulation, but they also can lead to long-term formation damage. Xu et al. (2016) developed a model for fracture propagation pressure accounting for fracture plugging and conducted experiments to establish the appropriate selection of

LCM. They found plugging zone length, width, and permeability to be the major plugging parameters that affect the fracture propagation pressure. LCM combinations of rigid granules, fibers, and elastic particles were found to create synergistic effects that optimized fracture plugging with respect to fracture propagation pressure and minimized formation damage.

15.2.6 Formation damage associated with alkali—surfactant—polymer flooding

Alkali—surfactant—polymer (ASP) methods have been applied with varying degrees of success for several decades ((Saleh et al., 2014; Sharma et al., 2015; Shedid, 2015), but persistent challenges for the methods are the cost of chemicals, their loss within the formation, and the formation damage that they sometimes induce. It is therefore essential to understanding the interactions between the injected ASP chemicals and the reservoir fluids and the formation mineralogy (Zhu et al., 2014; Korrani et al., 2016). Farajzadeh et al. (2013) evaluated the effects of continuous, trapped, and flowing gas on the performance of ASP flooding of an oil reservoir, and concluded that the lower the amount of trapped gas saturation, the better the oil recovery.

Wang, Yu et al. (2016) conducted experiments on high alkaline (NaOH) ASP in rock cores from the heterogeneous, midrange-permeability, high-water-saturation reservoirs of the Daqing oil field (China). The high losses of alkali and surfactant (up to 90%) were positively correlated with the resulting formation damage (i.e., permeability decreased by $>15\%$) and with the degree of heterogeneity of the reservoir zones. ASP chemical losses were greater in more heterogeneous reservoirs, such as those with mudstone interlayer, due to greater absorption and trapping in those low-permeability layers. For alkali and surfactant loss rates of 76.7% and 95.1%, respectively, the permeability reduced by 18%. The core-flooding tests also revealed that injecting ASP in a sequence of slugs enables each ASP component to impact the reservoir more effectively. Such action led to $\sim 4\%$ improvement in oil recovery compared to the same volume injected as a single chemical slug. From a stand-alone polymer flood study of similar medium-permeability type II reservoirs of the Daqing oil field, Zhong et al. (2017) identified various formation damage impacts, but concluded that oil recoveries were improved by 10%—15%.

In another ASP study on the Daqing oil field (Sanan-5 block), Li, Zhang, and Tang (2016) reported that SEM and XRD analysis of rock cores subjected to ASP flooding showed evidence of scaling, adsorption, and mineral dissolution formation damage. Cores drilled along the pathway between an injection well and a production well revealed that permeability initially decreases in the near-injection-well region. However, as distance from the injection well increased so did permeability, but in zones close to the production well permeability again decreased. Porosity changes along the same trajectory showed similar trends to permeability changes but they were less pronounced. Analysis identified that chemical adsorption, fine migration, and scaling were responsible for the formation damage in the vicinity of the injection wells. On the other hand, formation damage near the production wells was a consequence of silicon precipitation, which was caused by the sudden change of reservoir temperature and pressure in that production zone. The permeability increase in intermediate distances between the injection and production wells was caused by cementation dissolution. These findings highlight the complexity of formation damage with such treatments, indicating that different well-spacings and reservoir temperature and pressures could lead to quite different changes to the formations.

Ma et al. (2016) conducted macroscopic and microscopic (i.e., quantitative evaluation of minerals by scanning electron microscopy, Fang et al., 2016) core-flow experiments on low-permeability reservoirs subjected to alkali injection. Their results indicated that pore and throat plugging is the root cause of core alkali sensitivity damage. In detail, big pores were plugged and divided into narrower flow channels. Illite and chlorite are dominant in clay minerals, which are dispersed in alkali environments and then migrate with displacement fluids to plug pores and pore throats. Alkali sensitivity damage occurred at both inlet and outlet sides of the core, suggesting that it would impact both proximal and distal zones from an injection well. The alkali reacted with dolomite, muscovite, and K-feldspar, and generated precipitates, leading to decreases in permeability of $>80\%$ in the highest pH cases.

15.2.7 Geomechanical and reservoir stress-sensitivity impacts on formation damage

Stress sensitivity of a reservoir expresses how sensitive it is to effective pressure in terms of the changes to permeability and pore structure, and consequently fluid flow and storage capacity behavior, that result when pressure changes across a reservoir (Dvorkin et al., 1996). Porosity and permeability tend to display a typical stress–sensitivity behavior (Lei et al., 2015) in which both properties initially decrease quite rapidly with increasing stress, followed by a slower decline with further increases in stress. Porosity tends to be a less stress–sensitivity property than permeability.

Most stress–sensitivity analysis of reservoirs has been performed on the macroscale, with only few studies focusing on the pore scale (An et al., 2016). However, to fully understand the pore structure and framework changes resulting from stress sensitivity, and its implications for formation damage, it is necessary to identify the microstructure deformations to the pore network, which involve stress changes at the pore scale.

Yang et al. (2016b) applied a digital-core technique to determine the pore network in sandstone cores with a range of permeabilities using computed tomography scans. Their model revealed the dynamic effects of confining pressure on the geometric and topological reservoir structure and its absolute permeability. Sandstone cores with a range of permeabilities displayed exponential relationships between rock porosity and permeability and effective stress, with the tight sandstones displaying the strongest stress sensitivity for the same range of effective stress. These findings suggest that stress sensitivity should be taken into account in field-wide reservoir simulations, and when formulating well-spacing strategies that take into account the various levels of formation damage (i.e., range of permeabilities) resulting from the range of stress conditions prevailing throughout a reservoir. This is particularly the case for tight sandstone reservoirs (Ren et al., 2015).

15.2.8 Geochemical reactions leading to salt precipitation and related formation damage

Mixed salt precipitation as mineral deposits within reservoirs and as scale at wellbore perforations and on completion tubulars are major causes of formation damage related to chemical reactions and thermal-dynamic disequilibrium between injected and reservoir fluids and/or reservoir minerals (Safari and Jamialahmadi, 2014). Modeling of chemical reactive flow in porous media typically focuses on two empirical parameters, i.e., a kinetic rate coefficient for the reaction(s) and formation damage coefficient. A kinetic chemical reaction model for single-phase flow involving just two reacting components and irreversible salt precipitation requires mass balance equations for the two reactants and for the precipitant. This progress of the reaction also has to be linked with the formation damage coefficient, which typically is expressed in terms of reduction in formation permeability (Bedrikovetsky et al., 2002).

In order to avoid expensive and time-consuming laboratory concentration measurements for determining kinetic rate coefficients for such reactions, Vaz et al. (2016) developed a low-cost, three-point pressure measurement method with pressures taken at inlet, outlet, and an intermediate point from a core sample. The method is based on an analytical model for quasi-steady-state-commingled flow. The method demonstrated high agreement between the measured and predicted pressure at a fourth point in the core sample. The method was able to determine the formation damage coefficient with greater accuracy than the reaction coefficient, because the reaction coefficients are less sensitive to pressure differences created throughout the core sample. Vaz et al. (2016) applied their method to predict reservoir performance in an example well, with the results indicating a strong sensitivity of the injectivity of that example well to barium and sulfate concentration in the injected seawater and produced water for disposal.

15.2.9 The potential for nanofluids to mitigate certain types of formation damage

Nanofluids possess properties that potentially can solve some of the problems associated with fines migration within reservoirs. In an oil reservoir in its pristine state, fines have established over long periods of time an equilibrium involving the various forces at play within the reservoir fluids and matrix, which typically involves small particles at rest juxtaposed to the pore walls. During production as conditions change that equilibrium is often disrupted. This is particularly the case when a brine of a different composition (higher or lower salinity) and/or temperature is injected back into the formation. The impact of such induced disequilibrium and changes to the electrostatic and mechanical forces, and some chemical reactions, is that some of the fines are released from their formerly static positions at the pore walls and become suspended in the reservoir fluids flowing through the pore spaces, if the flow velocity of those fluids is greater than a critical velocity. The best way to avoid fines-induced formation damage is to prevent disturbing the existing in situ equilibrium conditions, but this is easier said than done. Various acids and polymers have been applied to help remove fines from the near-wellbore areas (Nguyen et al., 2007; Jaramillo et al., 2010).

Huang et al. (2010) evaluated coating proppants with customized nanoparticles for frac-packing operations, designed to hold fines in propped fractures. The nanoparticles used have high surface forces, including van der Waals and electrostatic forces, enabling them to firmly attach themselves to the surface of the proppant during fracturing and pumping treatments. Once in the fracture system the nanoparticles on the proppant capture the fines and limit their ability to flow back toward the wellbore, thereby reducing near-wellbore formation damage. Qiao et al. (2016b) develop a coupled numerical model for multiphase flow, nano-particle transport, fines release, and its dependence on salinity and nano-particle adsorption. Their simulations show that nanoparticle injection can effectively reduce the mobilization of fines and prevent permeability decreases over a range of salinities. In addition, the heterogeneous field scale simulation was performed. In particular, their simulation models demonstrated that the deployment nanoparticles in LSWF could improve oil recovery and injectivity by mitigating formation damage.

Nanoparticles have the potential to alter the wettability of reservoir formations and to reduce the IFT between oil and brine phases (Moradi et al., 2015). Onyekonwu and Ogolo (2010) evaluated the performance of three different polysilicon nanoparticles as agents for wettability alter-ation and EOR. Experimental evidence demonstrating the efficiency of modified silica nanoparticles in enhancing oil recovery was presented by Roustaei et al. (2013). These modified silica nanofluids were shown to work by reducing IFT and shifting wettability toward more oil-wet states.

Yuan et al. (2017b, 2018a, b, c) proposed series of models relating the adsorption of nanoparticles to their geochemical reactions in the reservoir which inhibit fines migration and related sand production at wellbore for both water injection and enhanced oil recovery processes. Chen et al. (2016) used simulation to evaluate the drag reduction and slip effects caused by adsorption of hydrophobic nanoparticles (HNP), on a field-wide scale, to enhance water-phase permeability. They developed a 3D, two-phase reservoir model to determine adsorption and distribution of HNP and its impacts on wettability using core-flood experiments prior to a pilot test in the Jiangsu oilfield (China). Their simulation combined with the subsequent field test confirmed that deploying HNP maintained and in some cases enhanced well injectivity over substantial periods and was an effective formation damage mitigation tool.

15.2.10 Thermal formation damage related to steam injection of heavy oil reservoirs

Three types of formation damage associated with thermal enhanced recovery projects are well documented (Bennion et al., 1992): (1) transformation of kaolinite to swelling smectite; (2) mineral dissolution and re-precipitation, and (3) wettability alteration, which may also be associated with the formation of in situ emulsions. Core-flow studies indicate that permeability may be reduced by up to 95% by the impacts of these mechanisms. Residual oil saturation is typically a function of reservoir temperature and reduces as temperature increases. Steam flooding mobilizes substantially more oil than hot-water flooding, but on the downside tends to move reservoir wettability to a more oil-wet state (increasing permeability to water). For shallow and poorly consolidated reservoirs, in which many heavy oil deposits reside, net confining pressure has a significant effect on fines migration.

Coskuner and Maini (1990) found in core-flow experiments that the critical velocity, at which fines start to be dispersed and permeability starts to decline, decreases with increasing net confining pressure. Low- and high-temperature oil—water relative permeability curves derived from core-flow tests (Romanova et al., 2014) provide a useful technique for identifying a formation's susceptibility to thermal formation damage and permeability impairment, which generally correlates with clay mineral content and type.

Zhou et al. (2016) proposed a method for improving steam-assisted gravity drainage (SAGD) performance in shallow oil sands deposits with a top water zone. This involved using polymer injection and the fishbone well pattern. Without an effective cap rock, excessive heat loss from the top water zone renders conventional SAGD uneconomical. Beneficial formation damage can be induced at the base of the top water zone by injecting high-temperature, viscous polymer to establish a stable high viscosity layer that will prevent steam from leaking through it. Simulation models were used to establish the optimum viscosity of the polymer and injection well design. In order to minimize harmful formation damage, further analysis is required to optimize the steam circulation strategy (Yuan and Mcfarlane, 2009), and the drilling fluid (Warren et al., 2004) to provide optimum wellbore stability, while minimizing accretion of the heavy oil on the drill string and produced solids under extreme temperature conditions.

SAGD is known to induce formation damage of several types, including: (1) fines mobilization; (2) organic deposition (waxes and asphaltenes); (3) inorganic scaling (silicates and/or carbonates). The flow pathways to the producing well can be improved by injecting thermally tuned acids that help to stabilize fines, remediate scale deposits, and degrade heavy oil—water emulsions, and aid in protection against tubular corrosion.

In addition to formation damage, steam injection projects typically involve several challenging aboveground risks that need to be managed and mitigated before projects can be sanctioned (e.g., access to freshwater, ability to processing and recycle produced water for steam generation, disposal of excess produced water, fuel for steam generation, environmental impact, and safety issues).

Chouhdary et al. (2012) discussed the risk management of such issues associated with the giant Wafra heavy oil field (onshore partitioned neutral zone between Kuwait and Saudi Arabia), and the need for exhaustive screening work, and carefully designed pilot projects before a full field project could be planned. These were in addition to the extensive G&G and simulation studies needed to provide detailed reservoir characterization on a field-wide basis with which to assess formation damage issues.

15.2.11 Formation damage associated with microbial activity in the reservoir

Microbial EOR (MEOR) involves bacteria and/or their metabolic by-products being deployed to improve oil mobilization and/or reduce water cut in a reservoir. Metabolic by-products consist of the assortment of compounds produced through microbial metabolic pathways, e.g., biosurfactants. In principle, MEOR is a straightforward concept where increased recovery occurs through inoculation of a reservoir with microorganisms to clog pores and redirect flow, or for mobilization of oil as a result of reduced IFT.

Although awareness of these impacts dates back to the 1930s, and tests have proliferated since the 1970s, results have been inconsistent and sometimes hard to explain, as outlined in reviews of the methods (Lee et al., 1996; Augustinovic et al., 2012). Early focus was on water injection quality control and applications of biocide (Mitchell and Bowyer, 1982), and other beneficial applications in the reservoir of microbe plugging during waterflooding (bioclogging), changes to IFT, wettability (biofilms), and oil viscosity soon followed (Gray et al., 2008). Dietrich et al. (1996) reported five field cases (Argentina, China, United States) where MEOR had increased production.

Soudmand-Asli et al. (2007) conducted experiments in fractured porous media comparing two types of bacteria: *Bacillus subtilis* (a biosurfactant-producing bacterium) and *Leuconostoc mesenteroides* (an exopolymer-producing bacterium). Their results showed that higher oil recovery efficiency was achieved by the biosurfactant-producing bacterium, because it led to greater reductions in oil viscosity and IFT. Armstrong and Wildenschild (2012) conducted experiments on the pore-scale mechanisms of MEOR and concluded that bioclogging and biosurfactants were effective at modifying the pore morphology and flow conditions in a reservoir beyond that predicted by capillary number. Ebigbo et al. (2010) modeled biofilm growth in the presence of carbon dioxide and water flow in the subsurface with a view to simulating potential applications of biofilms in carbon capture and storage reservoirs. Their analysis suggested that biofilms could potentially be used to plug damaged caprock, but the challenge was to make those plugs long-lasting rather than temporary.

15.3 SEMIQUANTITATIVE AND QUANTITATIVE RISK AND OPPORTUNITY ASSESSMENT

In order to help in the ranking and prioritization of risk and opportunity events/actions identified, the qualitative impact versus likelihood matrices can be modified to include a scoring system for each component box. Fig. 15.5 illustrates a semiquantitative 2D risk matrix. A similar modification could also be applied to the qualitative opportunity matrix component of Fig. 15.4 to introduce an arbitrary but discriminatory scoring system.

Fig. 15.5 applies a five-by-five grid in which the ascending numbers (1−5) on the two axes are multiplied to derive a risk assessment number assigned to each box in the grid. The higher the number of the box, the higher the assessed risk associated with that box. The risk actions/events are then placed in the appropriate box (not shown in Fig. 15.5) based upon their respective assessments. Clearly, this numerical scale of risk is

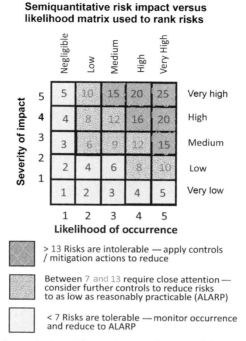

Semiquantitative risk impact versus likelihood matrix used to rank risks

Figure 15.5 A semiquantitative risk matrix provides a useful tool for ranking and prioritizing risk events/actions. The approach can also be applied to opportunities.

arbitrarily constructed and cannot be considered to represent a quantitative analysis. However, such an approach can help to improve rigor and make qualitative risk assessments less subjective. To make the assessments more systematic and related to the magnitude of their impacts (either in terms of associated costs, impacts on production, resource recovery, etc.), each identified risk could be rigorously assessed based upon a risk-magnitude calibration table constructed for each specific IOR project. The assessed costs, volumes, etc. included in that calibration table are derived and defined with scales that are relevant to the specific project, and all risk events/actions should then be assessed for that project using that calibration table.

This semiquantitative assessment does not have to be limited to two dimensions. A third dimension relative to some formation damage processes is "duration," i.e., how long does that damage persist in the reservoir? This dimension is relevant for example in damage instigated to block pores and improve the top seal in shallow bitumen/heavy oil reservoirs for steam injection or reinforcing the top seal to a carbon capture and sequestration reservoir with polymers or biofilms. In such an assessment, a five-by-five-by-five matrix and scoring system could be applied, such as the following:

Dimension 1—I = Impact
    1 = very low
    2 = low
    3 = medium
    4 = high
    5 = very high

Dimension 2—L = Likelihood
    1 = negligible
    2 = low
    3 = medium
    4 = high
    5 = very high

Dimension 3—D = Duration
    1 = very short term
    2 = short term
    3 = medium term
    4 = long term
    5 = semipermanent

(depending upon whether it is a downside-risk or upside-opportunity event/action being assessed, the duration scoring system could be reversed, i.e., 1 would become "semipermanent," 5 would become "very short term," etc.).

Each of these dimensions would require a magnitude calibration table for each project to qualify and clarify what is meant by each qualitative description terms for each level of each dimension. Additional dimensions might also be considered for certain projects (e.g., percentage of the reservoir exposed to the component of formation damage considered).

For the 3D, semiquantitative assessment of a risk or opportunity assessment score (RO), which relates to the overall significance of a specific event/action outcome can be obtained by applying the formula: $RO = I*L*D$. For such a scheme, R scores could range from 1 to 125 and significance levels could be (arbitrarily) assigned to certain ranges of scores. For example, the following (subjective/arbitrary) scoring thresholds could be applied:

High priority where RO scores are ≥ 75

Medium priority where RO scores are $= 27$ to 64

Low priority where RO scores are < 27

It is typically assumed in such assessments that each dimension is independent of the other, and the assessment of each dimension is conducted explicitly. This approach justifies using multiplication rules to combine their impacts. In reality, dimensions L and D may not be totally independent of each other, so this assumption is a simplification of potentially more complex relationships between the dimensions.

While the scoring system derived can be useful in ranking and prioritizing specific events/actions, it no way represents a meaningful quantitative scale of risk or opportunity. For instance, an RO score of 80 for one event/action versus an RO score of 40 for another does indicates that the first event/action is of greater priority than the second, but it should in no way be interpreted to suggest that the first event/action is twice as significant as the second.

Although qualitative and semiquantitative, risk/opportunity assessment methods are easy and transparent to implement, there are a number of disadvantages associated with them:

- They tend to be imprecise and do not always fully differentiate between the spectrum of relevant events/actions identified. The same categories in the 2D and 3D matrices tend to span significant ranges of outcomes.
- They often provide no common basis for comparison of events/actions. Some event outcomes may be financial, some production volume, some resource recovery, etc.
- The holistic/integrated numerical definition of risk/opportunity derived from them is often too simplistic combining several consequences of a single event/action into one number.
- Two events that are assigned the same assessment description or score may have very different reservoir performance, financial, or other outcomes.
- Such assessments cannot be readily integrated with financial/economic/valuation analysis (without further probabilistic input) or deal with complex multiple outcomes of differing severity.

For complex projects with many diverse events/actions to be considered, a quantitative assessment of some or all of those events/actions assessed by qualitative and semiquantitative techniques is almost always desirable and informative.

15.4 A STRUCTURED APPROACH TO RISK/OPPORTUNITY ASSESSMENT

Each identified risk and opportunity should be evaluated on the 2D or 3D (qualitative and/or semiquantitative) matrix supported by project-specific risk/opportunity magnitude calibration tables. These two features form the core of a structured risk/opportunity assessment and ranking process (Fig. 15.6). Based upon the results of such an assessment a decision is then taken as to whether a more detailed quantitative analysis is required for certain high-priority risk/opportunity events/actions. For such events, the likelihood scale is typically transformed into a probabilistic scale and the impact scale is typically transformed into ranges of actual outcomes (e.g., production, resource recovery, costs, etc.). In that form, the quantitative assessments can be initially calculated as deterministic assessments, e.g., a base case, supported by low and high case sensitivities to represent the extreme end of the assessed value ranges.

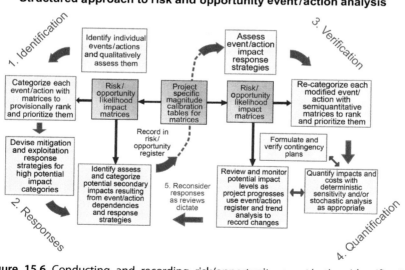

Figure 15.6 Conducting and recording risk/opportunity event/action identification and assessment applying a structured approach involving qualitative, semiquantitative, and quantitative analysis. *Source: Modified after Wood, D.A., Lamberson, G., Mokhatab, S., 2008. Better manage risks of gas-processing projects. Hydrocarbon Process. June, 124–128 (Wood et al., 2008).*

By expressing the assessments of events/actions as probabilistic distributions of potential impact outcomes and costs, the quantitative analysis can be conducted and expressed stochastically.

The results of the assessment for each identified risk/opportunity event/action should be documented in the risk/opportunity register (Fig. 15.7) in a comprehensive and systematic way. The format of that register can be varied to suit the project and organization, but should include dates of assessments, details of impact/likelihood/duration assessments, mitigation or exploitation strategies, secondary risks, dependencies, rankings, costs, contingency plans, etc. Ideally, an individual or department should take

Generic Risk / Opportunity Register of Events / Actions Indentified and Assessed					
Project Name	Events / Actions Could be Arranged in Order of Rank (those with highest assessed impacts first)				
Event / Action Classification Number	#1	#2	#3	...	#N
Event / Action Description					
Department or Contractor Assigned Responsibility					
Manager Responsible for Assessment /Implementation					
Provisional Assessment (Date)					
Likelihood / Probability of Occurrence					
Impact Assessment / Score					
Overall Risk / Opportunity (RO) Score					
Dependencies (if any)					
Costs to Impliment (if any)					
Mitigation / Exploitation Analysis					
Mitigation / Exploitation Options					
Mitigation / Exploitation Costs					
Associated Secondary Risks / Opportunities Identified					
Repeat Assessments (Dates)					
As required (recording details)					
...					
Details of Assessment Used for Investment Decision (Date)					
Likelihood / Probability of Occurrence					
Impact Assessment / Score					
Overall Risk / Opportunity (RO) Score					
Dependencies (if any)					
Costs to Impliment (if any)					
Mitigation / Exploitation Analysis					
Mitigation / Exploitation Options					
Mitigation / Exploitation Costs					
Associated Secondary Risks / Opportunities Identified					
Contingency Plans to Manage Event / Action					
Manager Responsible for Contingency Plan					
Details of Contingency Actions Recommended					
Estimated Cost of Contingency Plan (Date)					
Cost Budget / Costs Incurred					
Proposed Budget to Manage Event / Action (Date)					
Sanctioned Budget to Manage Event / Action (Date)					
Record of Actual Costs Incurred (Date)					
Post-Project Analysis (Learning Outcomes)					
Actual Impacts Incurred / Recorded					
Significance of Impacts on Project Outcome					
Associated Costs Actually Incurred					
Observed Dependencies / Interactions					
Key Lessons Learned					
Ongoing Management Requirements					
Note: Each identified event / action could be assigned its own underlying sheet in a workbook to record details of assessment					

Figure 15.7 Example format of risk/opportunity event/action register for a formation damage IOR project. This should represent a key project control document and support project investment decisions.

responsibility for each event, particularly those assessed as high priority events/actions, and be accountable for the ongoing assessment, implementation, and monitoring of outcomes. This register of information can be beneficial in supporting project investment decisions, and as a learning record by capturing knowledge and understanding of formation damage processes, and specific types of improved recovery projects.

15.5 QUANTITATIVE RISK/OPPORTUNITY ASSESSMENT

There are typically several dimensions to quantitative risk/opportunity analysis of formation damage events/actions in relation to IOR projects. The most fundamental of these are likely to be derived from the available reservoir performance/simulation model. This would be achieved by applying various assumptions related to aspects of formation damage to form a series of sensitivity cases and/or stochastic distributions. The key reservoir performance indicators likely to be quantified in this way are oil production rate over time, reservoir pressure over time, ultimate resource recovery, water cut over time, gas-to-oil ratio over time, etc. The relationships between the cases assessed for these key indicators are likely to be nonlinear (Fig. 15.8) and/or form asymmetrical probability distributions. One advantage of stochastic analysis is that confidence levels can be placed on the likelihood of occurrence of specific values or profiles within the calculated probability distributions.

Other key dimensions to the quantitative risk/opportunity analysis that are typically crucial for a positive project investment decision to be taken are related to aboveground factors and a project's cost—benefit analysis. On the cost side, this requires analysis of capital and operating cost (including any finance charges) requirements to implement/mitigate/exploit key events/actions related to formation damage. On the revenue side, this requires a forecast of oil price (for the specific grade and sale location of the crude oil produced) and production volumes over time. On the fiscal side, deductions for taxes and royalties paid taking account of any deductions for fiscal incentives (e.g., investment credits, tax allowances). On the risk side, assessment will need to extend beyond the reservoir and include a wide range of additional aboveground risks (e.g., permitting and implementation issues that will likely influence project timing; supplier and materials availability that will likely influence costs).

Key outputs from improved oil recovery reservoir simulation models that help to quantify the upside and downside uncertainties

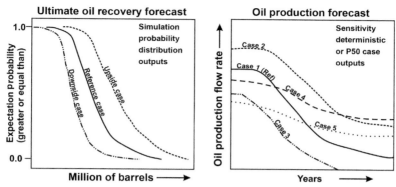

Figure 15.8 Examples of generic reservoir model and simulation outputs for IOR projects in which formation damage events/actions are likely to contribute to both upside and downside cases. Other typical outputs from the reservoir models (not shown) are multiyear profiles for key metrics such as reservoir pressure, water cut, and gas-to-oil ratio.

Armed with this cost, revenue, fiscal, and risk analysis, it is possible to build a quantitative, multiyear cash-flow model for the project, which can be used for discounted cash-flow analysis, applying discount rates appropriate for the cost of capital of the organizations involved. This will make it possible to initially derive unrisked NPVs and internal rates of return (IRRs) for the project in question. Investment decisions are typically based on consideration of the NPV and/or IRR calculated for a project and additionally adjusted for risk and opportunity. That adjustment for risk/opportunity can be made in various ways: (1) taking the mean of a series of deterministic sensitivity cases (base, reference, or most likely case, low-side case, high-side case); (2) adjusting an unrisked deterministic reference case cash-flow profile by a risk adjustment factor (between zero and one) to provide the downside-risk-adjusted value; (3) calculating an expected monetary value by adjusting the cash-flow profile by a chance of success (also 0−1) factor and cost of failure by the chance of failure, which should be assessed together with upside-opportunity case values; and (4) stochastic analysis involving input distributions of production volumes, oil prices, capital costs, and operating costs that cover the spectrum of downside risks and upside opportunities with the reference case situated close to the center of what may be symmetrical or asymmetrical probability distribution of various forms.

Again, a key advantage of stochastic analysis is that confidence levels can be placed on the likelihood of occurrence of specific value within the calculated probability distributions. For example, this could enable the 50 percentile value case (P50) to be selected as the reference case, the 10 percentile value case (P10—10% chance of being equal to or greater than) to be selected as the upside-opportunity case, and 90 percentile case (P90—90% chance of being equal to or greater than) to be selected as the downside-risk case.

Whichever of the above valuation methods are used, it is important to ensure that both downside risks and upside opportunities are given due to consideration in the value and cost analysis used for IOR project decision making. Formation damage events and actions are likely to contribute to both ends of the spectrum.

For some projects, it can be useful to conduct relatively simplistic stochastic cost of implementation analysis, which can be readily performed on a spreadsheet (with or without VBA macro controls). Such analysis is useful because it brackets the expected range of costs associated with identified events (including mitigation and exploitation actions) in IOR projects for which there exists substantial uncertainty while they are being planned for future execution. An example of this type of analysis is included (Figs. 15.9—15.12) here for a hypothetical IOR project involving six identified events/actions that have associated implementation costs,

Improved oil recovery project events/actions with uncertain costs				
Event/ action number	Events/action description	Probability of requirement	Base case cost P50 estimate $(000)	High case cost P95 estimate $(000)
1	Site: additional land	100%	440	985
2	Site: additional facilities/plant	100%	310	835
3	Chemicals/materials	100%	125	510
4	Initial reservoir stimulation event	100%	355	765
5	Second partial repeat stimulation event	60%	260	625
6	Third partial repeat stimulation event	30%	200	500

Figure 15.9 Event/actions associated with an IOR project with associated probabilities of requirement (100% means always required; <100% means only likely to be required in some scenarios). Cost estimates are provided for two confidence levels: P50 (50 percentile; 50% chance of being greater or less than that value) and P95 (95 percentile; 95% chance of being less than or equal to that value). From these two-point-percentile estimates, lognormal cost distributions are extrapolated for the costs of each event.

Simulation of costs associated with 6 IOR project events/actions - which occur, or not, according to defined probabilities

Trial # 1 to 1000 iterations of Simulation	Event #1 Cost Random Number	Event#1 Cost if Event Always Occurs	Event #2 Cost Random Number	Event #2 Cost if Event Always Occurs	Event #3 Cost Random Number	Event #3 Cost if Event Always Occurs	Event #4 Cost Random Number	Event #4 Cost if Event Always Occurs	Event #5 Occurs Random Number	Event #5 Cost Random Number	Event #5 Cost if Event Occurs	Event #6 Occurs Random Number	Event #6 Cost Random Number	Event #6 Cost if Event Occurs	Combined Project Events Costs ($000)
1	0.06	201	0.36	250	0.58	148	0.79	517	0.46	0.82	424	0.16	0.85	360	1900
2	0.49	433	0.38	259	0.98	715	0.38	309	0.36	0.85	451	0.30	0.86	0	2168
3	0.10	232	0.16	171	0.58	150	0.13	208	0.23	0.20	165	0.81	0.72	0	927
4	0.83	702	0.24	204	0.93	431	0.18	231	0.41	0.88	484	0.59	0.32	0	2051
5	0.13	254	0.39	263	0.26	72	0.37	305	0.19	0.24	180	0.13	0.40	173	1246
6	0.21	294	0.53	324	0.05	31	0.47	343	0.55	0.19	164	0.82	0.39	0	1156
7	0.46	419	0.70	428	0.37	95	0.89	627	0.88	0.93	0	0.16	0.30	149	1716
8	0.91	852	0.04	110	0.47	118	0.16	224	0.39	0.52	268	0.69	0.80	0	1572
9	0.84	713	0.07	125	0.13	48	0.18	232	0.38	0.31	200	0.26	0.05	78	1395
10	0.45	416	0.38	257	0.05	30	0.07	178	0.77	0.15	0	0.55	0.27	0	881

Figure 15.10 The results of the first 10 iterations of a 1000-trial simulation to estimate the range of potential cost liability associated with restoration of the processing site based upon the event cost and probability estimates listed in Fig. 15.9. Note that random numbers are shown to only two decimal places for display purposes only.

Statistical analysis of six simulations based upon different numbers of total trials						
Number of iterations (trials):	1000	750	500	250	100	50
Mean US$(000):	1695	1696	1707	1685	1615	1598
% Variation of mean to 1000 trials:	0.00%	0.06%	0.68%	-0.62%	-4.73%	-5.72%
Standard deviation US$(000):	523	506	520	535	443	439
Percentiles						
1%	768	773	805	824	767	818
10%	1114	1128	1127	1123	1107	1098
20%	1274	1288	1285	1262	1250	1260
30%	1393	1404	1408	1377	1331	1296
40%	1502	1515	1522	1512	1467	1424
50%	1624	1640	1636	1619	1598	1557
60%	1732	1744	1741	1700	1669	1684
70%	1896	1896	1899	1849	1793	1840
80%	2076	2050	2077	2050	1963	1963
90%	2362	2350	2372	2361	2243	2179
95%	2661	2597	2689	2566	2373	2357
99%	3260	3132	3116	3150	2786	2582

Figure 15.11 Statistical analysis of the total example IOR project cost distribution calculated by simulation with different numbers of trials (iterations). Note the higher levels of uncertainty in the cases involving fewer trials. There is also a reproducibility issue if too few trials are used.

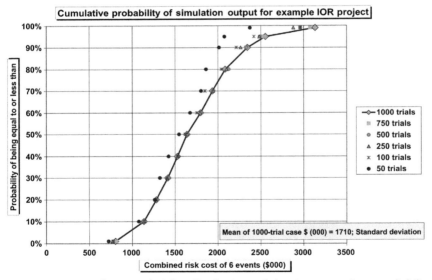

Figure 15.12 Cumulative probability distributions of the site restoration cost liability calculated with the simulation model described for different numbers of simulation trials (iterations).

the magnitude of which are highly uncertain. Indeed, actions #5 and #6 (Fig. 15.9) may never be required and therefore their costs may be considered under contingencies from a budget perspective, but need to be taken into account in a fully costed and risked cash-flow analysis.

There are several cost issues and uncertainties that need to be addressed associated with this project. Until the negotiations with landowners and bids from contractors and suppliers are available, it is difficult to know the exact extent of works (and associated costs) that are required. Also, the requirements (if any) for a second and/or a third reservoir stimulation operation will only become apparent once the initial stimulation operation is completed and reservoir responses and production performance indicators are monitored and assessed. The reason why an assessment of potential costs is required in advance is that the decision makers are able to make an investment decision taking into account estimates for the range of costs that are likely to be incurred and their associated uncertainties.

In a real project, the cost components would likely be broken down further and segregated into capital and operating costs as those categories are dealt with differently for accounting and fiscal purposes. A simulation approach taking into account cost estimates and probabilities of specific rehabilitation work being required provide one useful approach to the problem.

Fig. 15.9 lists cost estimates (in terms of 2 percentiles of probability distributions) for the six-identified event/actions associated with the IOR project. Also specified are the probabilities that each event/action will be required. Events #1 to #4 are deemed to be required with certainty (100% or 1.0 probability), but their costs are highly uncertain. On the other hand, events # 5 and #6 may or may not be required with various degrees of uncertainty, which is governed by the likelihood of certain formation damage events occurring or building up over time. Event #5 (second partial repeat of the reservoir stimulation operation) is assessed to be required in 60% of the outcomes. On the other hand, event #6 (third partial repeat of the reservoir stimulation operation) is assessed to be required in only 40% of the outcomes. Moreover, the cost estimates for event #6, although highly uncertain, are considered to be lower than for event #5, based on improved operating performance and on learning and experience from the previous stimulation operations.

Fig. 15.10 provides details of the results of the first 10 iterations of a simulation that evaluates the total example IOR project's costs based upon the summation of the risked outcomes of each of the six identified events described in Fig. 15.9. The model considers all six events to be independent of each other, which is likely not to be the case for events #5 and #6 in some projects. In a real project, some sensitivity analysis would be required to evaluate the impact of correlations (dependencies) existing between events #5 and #6. It is relatively straightforward to amend the simulation logic to accommodate correlations between events, if necessary. The cost distribution for each event sampled by each trail of the simulation is expressed, in the model results shown (Fig. 15.10) as a lognormal distribution. Other skewed distribution forms could be used (e.g., triangular), but positively skewed lognormal distributions are often useful forms to apply for cost estimates based on several components. These asymmetrical distributions are able to include a wide range of potential cost values, particularly on the high side, with low probabilities of occurrence. They can also be constructed from two-point percentile estimates such as those provided in Fig. 15.9.

Note that two random numbers are involved in the calculation of the cost incurred in each iteration for events #5 and #6. The first random number (0−1) determines whether the event occurs or not. For the assumptions made in Fig. 15.10 event #1 is always going to occur (i.e., probability of 1). For event #6, for example, it will only be deemed to have occurred in a particular iteration of the simulation if the first random number is ≤ 0.3 (i.e., for ∼30% of the iterations). The second random number (0−1) for each event samples the appropriate lognormal cost distribution for the event (a low, second random number samples the low end of the cost distribution; a high, second random number samples the high end of the cost distribution). As events #1 to #4 in this project occur in all iterations, there is no need for the first random number; only the second random number is required to sample the cost distribution. The final column in Fig. 15.10 sums the sampled costs for each event in each iteration to provide a full project cost estimate, taking into account the uncertainties involved on both the downside and upside. In this way 1000 cost estimates of the total-risk-and-opportunity-adjusted costs are regenerated in the 1000-trial simulation.

Note that in the simulation methodology described, an event with very-low probability of occurrence, e.g., say 1% chance of occurring, would be sampled by the simulation in approximately only 1% of the iteration run (i.e., ∼10 iterations in a 1000-trial simulation). In a situation where some events are associated with potentially very-high costs, but at very low probabilities of occurrence (e.g., <1%), such events may not be included as the high priority events considered for the simulation model. This methodology can therefore lead to a low-side estimate of actual cost exposure if only the costs associated with the highest-probability events are modeled (i.e., only the risk/opportunity events/actions prioritized because they are most likely to occur have been deemed worthy of quantitative assessment). This methodology is mathematically correct, but for low probability events (e.g., <1% chance of occurring), even if they are included in the simulation model, few simulation trials will actually sample them as occurring.

The methodology is highly dependent on the accuracy to which low probabilities of occurrence events can be estimated. If there are a lot of high impact/low probability events that could possibly impact the liability, then it is important to test a range of probabilities by running sensitivity cases to gauge the significance of different levels of uncertainty. Dealing with extreme risks (i.e., events of very high potential magnitude if they occur, but very low probability of actually occurring) is often an issue with simulation models based on probability distributions. It is relatively easy for such models to underestimate the overall impacts of extreme events. That is why it is usually appropriate to run some deterministic, sensitivity cases that consider the extreme ends of the cost distributions (i.e., high-side and low-side cases). These complement the simulation results and aid investment decision making.

Fig. 15.11 provides summary statistical analysis for the total example IOR project costs calculated by simulations involving different numbers of iterations. The distribution mean, standard deviation, and a range of percentiles (confidence limits) are listed for six different numbers of trials (iterations), ranging from 1000 to 50 iterations. For this simplistic simulation, cases with 250 or greater trials yield similar results, suggesting that number or more trials lead to a consistent result. For this model, set up on a single Excel workbook, execution times for 1000 trials only require a fraction of a second of computation time. In all cases the standard deviations for the calculated total project cost distribution are high, indicating significant uncertainty. The mean of the simulation is ~US$1.7 million, whereas the P95 value is US$2.7 million for the 1000-trial case.

The data from Fig. 15.11 are illustrated as cumulative probability distributions in Fig. 15.12. Cumulative probability displays are useful for expressing the full range of quantitative uncertainty analysis derived as outputs of simulation models. They enable appropriate levels of confidence to be selected for further analysis or decision-making purposes.

15.6 CONCLUSIONS

Recent and historical studies reveal and confirm that formation damage in IOR and EOR projects leads to a series of complex interactions, with a combination of negative and positive impacts on the production performance and resource recovery of oil reservoirs. The extent and occurrence of these impacts in many cases remains highly uncertain and varies significantly from reservoir to reservoir. The risk and opportunity assessment and management of formation damage phenomena impacting IOR projects has to address this wide range of complex and varied issues. It also needs to integrate the formation-damage-related issues with the broader aboveground, technical, regulatory, and market uncertainties associated with IOR projects, if it is to a useful aid to investment decision making.

Risk and opportunity management of formation damage events and actions associated with IOR projects can be meaningfully and rigorously achieved by applying the multiple steps associated with the structured and systematic framework described. Once the objectives and constraints (below- and aboveground) of an IOR project are fully understood, the proposed risk and opportunity management framework initially focuses on identification, assessment, ranking, and prioritization of formation-damage-related events and activities. This is best conducted using qualitative and semiquantitative matrices and influence diagrams.

The risk and opportunity register is a key part of the proposed framework providing a documentary record of the risk and opportunity events identified, assessed, and prioritized. If regularly updated with information and analysis, the register serves as a useful tool to communicate an integrated review of risk and opportunity management as it progresses.

Devising and implementing strategies to mitigate the priority downside risks and to exploit the priority upside opportunities is essential for optimum outcomes of IOR projects. For those impacts that may or may not occur, costed contingency plans to implement should they occur are viable responses. Quantitative probabilistic analysis of the impacts of the prioritized formation damage events and actions related to an IOR project, in the form of reservoir models/simulations, cost, and cash-flow models/simulations, are generally required to justify and support project investment decisions and their associated expenditure budgets.

REFERENCES

Aladasani, A., Bai, B., 2012. Investigating low-salinity water flooding recovery mechanism(s) in carbonate reservoirs. In: Proceedings of the 2012 SPE EOR Conference at Oil and Gas West Asia. 16—18 April, Muscat, Oman, SPE155560.

Aladasani, A., Bai, B., Wu, Y.-s, Salehi, S., 2014. Studying low-salinity water flooding recovery effects in sandstone reservoirs. J. Petrol. Sci. Eng. 120, 39—51.

AlMubarak, T., AlKhaldi, M., AlGhamdi, A., 2016. Design and application of high temperature seawater based fracturing fluids in Saudi Arabia. In: Paper SPE 26822. Presented at the Offshore Technology Conference Asia. 22—25 March, Kuala Lumpur, Malaysia. Available from: https://doi.org/10.4043/26822-MS

Alvarado, V., Manrique, E., 2010. Enhanced oil recovery: an update review. Energies 3 (9), 1529—1575. Available from: https://doi.org/10.3390/en3091529.

An, S., Yao, J., Yang, Y., Zhang, L., Zhao, J., Gao, Y., 2016. Influence of pore structure parameters on flow characteristics based on a digital rock and the pore network model. J. Nat. Gas Sci. Eng 31, 156—163.

Aristov, S., van den Hoek, P., Pun, E., 2015. Integrated approach to managing formation damage in waterflooding. SPE 174174. SPE European Formation Damage Conference. 3—5 June, Budapest, Hungary.

Armstrong, R.T., Wildenschild, D., 2012. Investigating the pore scale mechanisms of microbial enhanced oil recovery. J. Petrol. Sci. Eng. 94—95, 155—163.

Augustinovic, Z., Birketveit, O., Clements, K., Gopi, S., Ishoey, T., Jackson, G., et al., 2012. Microbes—oilfield enemies or allies. Schlumberger Oilfield Rev. 24 (2), 4—17.

Austad, T., Rezaeidoust, A., Puntervold, T., 2010. Chemical mechanism of low salinity waterflooding in sandstone reservoirs. Paper SPE 129767. Presented at the SPE Improved Oil Recovery Symposium. 24—28 April, Tulsa, OK.

Azadi, M., Aminossadati, S.M., Chen, Z., 2016. Large-scale study of the effect of wellbore geometry on integrated reservoir-wellbore flow. J. Nat. Gas Sci. Eng. 35, 320—330.

Bedrikovetsky, P.G., Lopes Jr., R.P., Rosario, F.F., Bezerra, M.C., Lima, E.A., 2002. Oilfield scaling—Part I: mathematical and laboratory modelling. In: SPE Paper 81127. Presented at the SPE Latin American and Caribbean Petroleum Engineering. 27—30 April, Conference Held in Port-of-Spain, Trinidad.

Bennion, D.B., Thomas, F.B., Sheppard, D.A., 1992. Formation damage due to mineral alteration and wettability changes during hot water and steam injection in clay-bearing sandstone reservoirs. SPE 23783. SPE Formation Damage Control Symposium. 26—27 February, Lafayette, LA.

Bennion, D.B., Thomas, F.B., Bietz, R.F., 1996. Water and hydrocarbon phase trapping in porous media-diagnosis, prevention and treatment. J. Can. Pet. Technol 35 (10), 29—36.

Cha, M., Yin, X., Kneafsey, T., Johanson, B., Alqahtani, N., Miskimins, J., et al., 2014. Cryogenic fracturing for reservoir stimulation—laboratory studies. J. Petrol. Sci. Eng. 124, 436—450.

Chen, H., Di, Q., Ye, F., et al., 2016. Numerical simulation of drag reduction effects by hydrophobic nanoparticles adsorption method in water flooding processes. J. Nat. Gas Sci. Eng 35 (Part A), 1261—1269.

Chouhdary, M.A., Wani, M.R., Al-Mahmeed, A., Al-Rasheedi, H.R., 2012. Challenges and risk management strategy for enhanced oil recovery projects in carbonate reservoirs of a giant field in Middle East. SPE-154631-MS, EOR Conference at Oil and Gas West Asia. 16 — 18 April, Muscat, Oman. Available from: https://doi.org/10.2118/154631-MS.

Clarkson, C.R., 2013. Production data analysis of unconventional gas wells: review of theory and best practices. Int. J. Coal Geol. 109, 101—146.

Coskuner, G., Maini, B., 1990. Effect of net confining pressure on formation damage in unconsolidated heavy oil reservoirs. J. Petrol. Sci. Eng. 4, 105−117.

Davies, R., Foulger, G., Bindley, A., Styles, P., 2013. Induced seismicity and hydraulic fracturing for the recovery of hydrocarbons. Mar. Pet. Geol 45, 171−185.

Dietrich, F.L., Brown, F.G., Zhou, Z.H., Maure, M.A., 1996. Microbial EOR technology advancement: case studies of successful projects. SPE 36746. SPE Annual Technical Conference and Exhibition. 6−9 October, Denver, CO.

Duncan, I.J., Nicot, J.-P., Choi, J.-W., 2008. Risk assessment for future CO_2 sequestration projects based CO_2 enhanced oil recovery in the U.S. Presented at the 9th International Conference on Greenhouse Gas Control Technologies (GHGT-9). November 16−20, Washington, DC. GCCC Digital Publication Series #08-03i.

Dvorkin, J., Nur, A., Chaika, C., 1996. Stress sensitivity of sandstones. Geophysics 61 (2), 444−455.

Ebigbo, A., Helmig, R., Cunningham, A.B., Class, H., Gerlach, R., 2010. Modelling biofilm growth in the presence of carbon dioxide and water flow in the subsurface. Adv. Water Resour. 33, 762−781.

Ellsworth, W.L., 2013. Injection-induced earthquakes. Science 341, 1225942.

Faergestad, I., 2016. Formation damage. Schlumberger Oilfield Rev. 52−53.

Fallah, H., Sheydai, S., 2013. Drilling operation and formation damage. Open J. Fluid Dyn. 3, 38−43.

Fan, H., Lyu, J., Zhao, J., et al., 2016. Evaluation method and treatment effectiveness analysis of anti-water blocking agent. J. Nat. Gas Sci. Eng 33, 1374−1380.

Fang, W., Jiang, H., Li, J., Li, W., Li, J., Zhao, L., et al., 2016. A new experimental methodology to investigate formation damage in clay-bearing reservoirs. J. Pet. Sci. Eng 143, 226−234.

Farajzadeh, R., et al., 2013. Effect of continuous, trapped, and flowing gas on performance of alkaline surfactant polymer (ASP) flooding. Ind. Eng. Chem. Res 52 (38), 13839−13848.

Feng, Q., Chen, H., Wang, X., et al., 2016. Well control optimization considering formation damage caused by suspended particles in injected water. J. Nat. Gas Sci. Eng 35 (Part A), 21−32.

Frayret, J., Castetbon, A., Trouve, G., Potin-Gautier, M., 2006. Solubility of $(NH4)2SiF6$, K_2SiF_6 and Na_2SiF_6 in acidic solutions. Chem. Phys. Lett 427 (4−6), 356−364. Available from: https://doi.org/10.1016/j.cplett.2006.06.044.

Gallegos, T.J., Varela, B.A., Haines, S.S., Engle, M.A., 2015. Hydraulic fracturing water use variability in the United States and potential environmental implications. Water Resour. Res 5839−5845.

Gray, M., Yeung, A., Foght, J., Yarranton, H.W., 2008. Potential microbial enhanced oil recovery processes: a critical analysis. SPE-114676-MS. SPE Annual Technical Conference and Exhibition. 21−24 September, Denver, CO. Available from: https://doi.org/10.2118/114676-MS.

Guo, Z., Hussain, F., Cinar, Y., 2015. Permeability variation associated with fines production from anthracite coal during water injection. Intern. J. Coal Geol 147, 46−57.

Guo, Z., Hussain, F., Cinar, Y., 2016. Physical and analytical modelling of permeability damage in bituminous coal caused by fines migration during water production. J. Nat. Gas Sci. Eng 35 (Part A), 331−346.

He, J., Mohamed, I.M., Nasr-El-Din, H.A., 2011. Mixing hydrochloric acid and seawater for matrix acidizing: is it a good practice? In: Paper SPE 143855. Presented at the SPE European Formation Damage Conference. 7−10 June, Noordwijk, The Netherlands. <https://doi.org/10.2118/143855-MS>

Hlidek, B.T., Meyer, R.K., Yule, K.D., Wittenberg, J., 2012. A case for oil-based fracturing fluids in Canadian Montney Unconventional Gas Development. SPE 159952.

Huang, T., Evans, B.A., Crews, J.B., Belcher, K., 2010. Field case study on formation fines control with nanoparticles in offshore applications. SPE Annual Technical Conference and Exhibition, 19—22 September 2010. SPE, Florence, Italy, 135088 (8 pages).

Ishida, T., Nagaya, Y., Inui, S., Aoyagi, K., Nara, Y., Chen, Y., et al., 2013. AE monitoring of hydraulic fracturing experiments conducted using CO_2 and water. In: ISRM International Symposium—EUROCK 2013. 23—26 October, Wroclaw, Poland.

Jansen, J.D., Durlofsky, L.J., 2016. Use of reduced-order models in well control optimization. Optim. Eng 35, 105—132.

Jaramillo, O.J., Romero, R., Ortega, A., Milne, A., Lastre, M., 2010. Matrix acid systems for formations with high clay content. In: SPE 126719. Presented at the 2010 SPE International Symposium and Exhibition on Formation Damage Control. 10—12 February, Lafayette, LO.

Jin, M., Ribeiro, A., Mackay, E., et al., 2016. Geochemical modelling of formation damage risk during CO_2 injection in saline aquifers. J. Nat. Gas Sci. Eng 35 (Part A), 703—719.

Kalantari-Dahaghi, A.M., Moghadasi, J., Gholami, V., Abdi, R., 2006. Formation damage due to asphaltene precipitation resulting from CO_2 gas injection in Iranian carbonate reservoirs. SPE-99631-MS. SPE Europec/EAGE Annual Conference and Exhibition. 12—15 June, Vienna, Austria.

Korrani, A.K.N., Sepehrnoori, K., Delshad, M., 2016. Significance of geochemistry in alkaline/surfactant/polymer (ASP) flooding. In: SPE SPE-179563-MS Improved Oil Recovery Conference. Society of Petroleum Engineers. 11—13 April, Tulsa, OK. Available from: https://doi.org/10.2118/179563-MS.

Lager, Arnaud, Webb, K.J., Black, C.J.J., Singleton, M., Sorbie, K.S., 2008. Low salinity oil recovery—an experimental investigation. Petrophysics 49 (1), 28—35.

LeBlanc, D., Martel, T., Graves, D., Tudor, E., Lestz, R., 2011. Application of propane (LPG) based hydraulic fracturing in the McCully gas field, New Brunswick, Canada. SPE 144093.

Lee, D, Lowe, D., Grant, P., 1996. Microbiology in the oil patch: A review. Paper 96—109. 10—12 June, The Petroleum Society, Calgary.

Lei, G., Dong, P., Wu, Z., Mo, S., Gai, S., Zhao, C., et al., 2015. A fractal model for the stress-dependent permeability and relative permeability in tight sandstones. J. Can. Petrol. Technol 54 (1), 36—48.

Li, Y., Zhang, Y., Wang, S., Chen, H., Shi, X., Bai, X., 2015. A new screening evaluation method for carbon dioxide miscible flooding candidate reservoirs. J. Petrol. Sci. Technol. 5 (2), 12—28.

Li, X., Zhang, G., Ge, J., et al., 2016. Potential formation damage and mitigation methods using seawater-mixed acid to stimulate sandstone reservoirs. J. Nat. Gas Sci. Eng 35 (Part A), 11—20.

Li, Z., Zhang, W., Tang, Y., et al., 2016. Formation damage during alkaline—surfactant—polymer flooding in Daqing Oilfield, Block Sanan-5, China. J. Nat. Gas Sci. Eng 35 (Part A), 826—835.

Liu, M., Wu, Q., 2013. The advances and prospects on the research of anti-water blocking agent. Guangzhou Chem. Ind 41 (5), 32—33.

Liu, S., He, X., Li, H., 2011. Production mechanism and control measures of coal powder in coalbed methane horizontal well. J. Liaoning Tech. Univ. Nat. Sci 30 (4), 508—512.

Lu, J., Nicot, J.-P., Mickler, P.J., Ribeiro, L.H., Darvari, R., 2016. Alteration of Bakken reservoir rock during CO_2-based fracturing—an autoclave reaction experiment. J. Unconvent. Oil Gas Resour. 14, 72—85.

Ma, K., Jiang, H., Li, J., et al., 2016. Experimental study on the micro alkali sensitivity damage mechanism in low-permeability reservoirs using QEMSCAN. J. Nat. Gas Sci. Eng 36, 1004—1017. Available from: https://doi.org/10.1016/j.jngse.2016.06.056.

Massarotto, P., Iyer, R., Elma, M., Nicholson, T., 2014. An experimental study on characterizing coal bed methane (CBM) fines production and migration of mineral matter in coal beds. Energ. Fuels 28 (2), 766–773.

McCune, C.C., 1982. Seawater injection experience—an overview. J. Pet. Technol. 34 (10), 2265–2270. Available from: https://doi.org/10.2118/9630-PA.

McDaniel, B., Grundmann, S.R., Kendrick, W.D., Wilson, D.R., Jordan, S.W., 1998. Field applications of cryogenic nitrogen as a hydraulic-fracturing fluid. JPT 50 (3), 38–39.

McGuire, P.L.L., Chatham, J.R.R., Paskvan, F.K.K., Sommer, D.M.M., Carini, F.H.H., BP Exploration, 2005. Low salinity oil recovery: an exciting new EOR opportunity for Alaska's North slope. In: SPE Western Regional Meeting, pp. 1–15. Available from: https://doi.org/10.2118/93903-MS.

Middleton, R.S., Carey, J.W., Currier, R.P., Hyman, J.D., Kang, Q., Karra, S., et al., 2015. Shale gas and non-aqueous fracturing fluids: opportunities and challenges for supercritical CO_2. Appl. Energ. 147, 500–509.

Mitchell, R.W., Bowyer, P.M., 1982. Water injection methods. SPE 10028. SPE International Petroleum Exhibition and Technical Symposium. 17–24 March, Beijing.

Moghanloo, R.G., Dadmohammadi, Y., Bin, Y., Salahshoor, S., 2015. Applying fractional flow theory to evaluate CO_2 storage capacity of an aquifer. J. Pet. Sci. Eng 125, 154–161.

Mohan, K., Vaidya, R.N., Reed, M.G., Fogler, H.S., 1993. Water sensitivity of sandstones containing swelling and non-swelling clays. Colloids Surf. A Physicochem. Eng. Aspects 73, 237–254. Available from: https://doi.org/10.1016/0927-7757(93)80019-B.

Moradi, B., Pourafshary, P., Jalali, F., Mohammadi, M., Emadi, M.A., 2015. Experimental study of water-based nanofluid alternating gas injection as a novel enhanced oil-recovery method in oil-wet carbonate reservoirs. J. Nat. Gas. Sci. Eng. 27, 64–73.

Nguyen, P.D., Weaver, J.D., Rickman, R.D., Dusterhoff, R.G., Parker, M.A., 2007. Controlling Formation Fines at Their Sources to Maintain Well Productivity. SPE Production & Operations.

Oldenburg, C.M., 2007. Joule–Thomson cooling due to CO_2 injection into natural gas reservoirs. Energ. Convers. Manag 48 (6), 1808–1815.

Onyekonwu, M.O., Ogolo, N.A., 2010. Investigating the use of nanoparticles in enhancing oil recovery. In: Annual International Conference and Exhibition, Tinapa-Calabar, Nigeria. SPE140744.

Page, J.C., Miskimins, J.L., 2009. A comparison of hydraulic and propellant fracture propagation in a shale gas reservoir. J. Can. Petrol. Technol 48 (5), 26–30.

Pruess, K., 2009. Formation dry-out from CO_2 injection into saline aquifers: 2. Analytical model for salt precipitation. Water Resour. Res 45 (3). Available from: https://doi.org/10.1029/2008wr007102.

Qiao, C., Li, L., Johns, R.T., Xu, J., 2016a. Compositional modeling of dissolution induced injectivity alteration during CO_2 flooding in carbonate reservoirs. SPE J 1 (June), 809–826. Available from: https://doi.org/10.2118/170930-PA.

Qiao, C., Han, J., Huang, T., 2016b. Compositional modeling of nanoparticle-reduced fine-migration. J. Nat. Gas Sci. Eng 35 (Part A), 1–10.

Reinoso, W., Torres, F., Aldana, M., Campo, P., Alvarez, E., Tovar, E., 2016. Removing formation damage from fines migration in the Putumayo Basin in Colombia: challenges, results, lessons learned, and new opportunities after more than 100 sandstone acidizing treatments. SPE-178996-MS. SPE International Conference and Exhibition on Formation Damage Control. 24–26 February, Lafayette, LO.

Ren, X., Li, A., He, B., Wu, S., 2015. Influence of the pore structures on stress sensitivity of tight sandstone reservoir. Adv. Petrol. Explor. Develop. Vol. 10 (No. 2), 13–18. Available from: https://doi.org/10.3968/7663.

Reyes, E.A., Smith, A.L., Beuterbaugh, A., 2013. Properties and applications of an alternative aminopolycarboxylic acid for acidizing of sandstones and carbonates. SPE-165142-MS SPE European Formation Damage Conference & Exhibition. 5–7 June, Noordwijk, The Netherlands.

Reyes, E.A., Rispler, K., Davis, J., Stimatze, R., Ouedraogo, M., Beuterbaugh, A., et al., 2016. Acidizing of high temperature and highly sensitive multilayered carbonate well with aminopolycarboxylic acid low-pH fluid: field implementation and laboratory validation. SPE-175839-MS SPE North Africa Technical Conference and Exhibition. 14–16 September, Cairo, Egypt.

Rogala, A., Krzysiek, J., Bernaciak, M., Hupka, J., 2013. Non-aqueous fracturing technologies for shale gas recovery. Physicochem. Problems Mineral Process 49 (1), 313–322.

Romanova, U., Piwowar, M., Ma, T., 2014. An overview of the low and high temperature water–oil relative permeability for oil sands from different formations in Western Canada. Paper SCA2014-066. International Symposium of the Society of Core Analysts. 8–11 September, Avignon, France.

Roustaei, A., Saffarzadeh, S., Mohammadi, M., 2013. An evaluation of modified silica nanoparticles' efficiency in enhancing oil recovery of light and intermediate oil reservoirs. Egypt. J. Petrol. 22, 427–433.

Safari, H., Jamialahmadi, M., 2014. Thermodynamics, kinetics, and hydrodynamics of mixed salt precipitation in porous media: model development and parameter estimation. J. Transp. Porous Media 101 (3), 477–505.

Saleh, L.D., Wei, M.Z., Bai, B.J., 2014. Data analysis and updated screening criteria for polymer flooding based on oilfield data. SPE Reserv. Eval. Eng 17 (1), 15–25.

Salehi, M.M., Omidvar, P., Naeimi, F., 2017. Salinity of injection water and its impact on oil recovery absolute permeability, residual oil saturation, interfacial tension and capillary pressure. Egypt. J. Petrol. 26, 301–312.

Sepulveda, J.A., Pizon-Torres, C., Galindo, J.M., Charry, C., Chavarro, J.I., 2015. Effect of high CO_2 content on formation damage of oil fields: a field case in a South Western Colombian field. ARPN J. Eng. Appl. Sci. 10 (2), 773–781.

Sharma, H., Dufour, S., Pinnawala-Arachchilage, G.W.P., Weerasooriya, U., Pope, G.A., Mohanty, K., 2015. Alternative alkalis for ASP flooding in anhydrite containing oil reservoirs. Fuel 140, 407–420.

Shedid, S.A., 2015. Experimental investigation of alkaline/surfactant/polymer (ASP) flooding in low permeability heterogeneous carbonate reservoirs. In: SPE 175726, SPE North Africa Technical Conference and Exhibition, Cairo, Egypt.

Skrettingland, K., Holt, T., Tweheyo, M.T. Skjevrak, I., 2010. Snorre low salinity water injection—core flooding experiments and single well field pilot. Paper SPE129877 presented at the 2010 SPE Improved Oil Recovery Symposium. 22–26 April.

Song, W., Kovscek, A., 2016. Direct visualization of pore-scale fines migration and formation damage during low-salinity waterflooding. J. Nat. Gas Sci. Eng 34, 1276–1283.

Soudmand-Alsi, A., Ayatollahi, S., Mohabatkar, H., Zareie, M., Shariatpanahi, F., 2007. The in situ microbial enhanced oil recovery in fractured porous media. J. Petrol. Sci. Eng. 58 (1-2), 161–172.

Sun, X., Wang, Z., Sun, B., et al., 2016. Research on hydrate formation rules in the formations for liquid CO_2 fracturing. J. Nat. Gas Sci. Eng 33, 1390–1401.

Tang, Y., Lv, C., Wang, R., et al., 2016. Mineral dissolution and mobilization during CO_2 injection into the water-flooded layer of the Pucheng Oilfield, China. J. Nat. Gas Sci. Eng. 33, 1364–1373.

Vaz, A., Maffra, D., Carageorgos, T., et al., 2016. Characterisation of formation damage during reactive flows in porous media. J. Nat. Gas Sci. Eng. 34, 1422–1433.

Wang, L., Yao, B., Cha, M., et al., 2016. Waterless fracturing technologies for unconventional reservoirs—opportunities for liquid nitrogen. J. Nat. Gas Sci. Eng 35 (Part A), 160–174.

Wang, Z., Yu, T., Lin, X., et al., 2016. Chemicals loss and the effect on formation damage in reservoirs with ASP flooding enhanced oil recovery. J. Nat. Gas Sci. Eng. 33, 1381–1389.

Warren, B.K., Baltoiu, L.V., Dyck, R.G., 2004. Development and field results of a unique drilling fluid designed for heavy oil sands drilling. SPE/IADC 92462. Drilling Conference. 23–25 February 2005, Amsterdam, The Netherlands.

Wen, T., Thiele, M.R., Ciaurri, D.E., Aziz, K., Ye, Y., 2014. Waterflood management using two-stage optimization with streamline simulation. Comput. Geosci. 18 (3–4), 483–504.

Wood, D.A., Lamberson, G., Mokhatab, S., 2008. Better manage risks of gas-processing projects. Hydrocarbon Process. June, 124–128.

Xu, C., Kang, Y., Chen, F., You, Z., 2016. Fracture plugging optimization for drill-in fluid loss control and formation damage prevention in fractured tight reservoir. J. Nat. Gas Sci. Eng. 35 (2016), 1216–1227.

Yang, Y., Siqueira, F., Vaz, A., et al., 2016a. Slow migration of detached fine particles over rock surface in porous media. J. Nat. Gas Sci. Eng. 34, 1159–1173.

Yang, Y., Zhang, W., Gao, Y., et al., 2016b. Influence of stress sensitivity on microscopic pore structure and fluid flow in porous media. J. Nat. Gas Sci. Eng 36 (Part A), 20–31.

You, Z., Yang, Y., Badalyan, A., Bedrikovetsky, P., Hand, M., 2016. Mathematical modelling of fines migration in geothermal reservoirs. Geothermics 59, 123–133.

Yuan, B., Bedrikovetsky, P., Huang, T., Moghanloo, R.G., Dai, C., Venkatraman, A., et al., 2016a. Special issue: formation damage during enhanced gas and liquid recovery. J. Nat. Gas Sci. Eng. 36, 1051–1054.

Yuan, B., Su, Y., Moghanloo, R.G., 2015. A new analytical multi-linear solution for gas flow toward fractured horizontal well with different fracture intensity. J. Nat. Gas Sci. Eng. 23, 227–238.

Yuan, B., Moghanloo, G.R., Zheng, D., 2017a. A novel integrated production analysis workflow for evaluation, optimization and predication in shale plays. Int. J. Coal Geol. 180, 18–28.

Yuan, B., Rouzbeh, G., Moghanloo, 2017b. Analytical modeling improved well performance by nanofluid pre-flush. Fuel 202, 380–394.

Yuan, B., Moghanloo, R., Wang, W., 2018a. Using nanofluids to control fines migration for oil recovery: nanofluids co-injection or nanofluids pre-flush?—A comprehensive answer. Fuel 215, 474–483.

Yuan, B., 2018b. Nanofluid treatment, an effective approach to improve performance of low salinity water floodingJ. Petrol. Sci. Eng. in press. Available from: https://doi.org/10.1016/j.petrol.2017.11.03213.

Yuan, B., Moghanloo, R.G., 2018c. Nanofluid pre-coating, an effective method to reduce fines migration in radial systems saturated with mobile immiscible fluids. SPE J. SPE-189464-PA. Available from: https://doi.org/10.2118/189464-PA.

Yuan, J-Y., Mcfarlane, R., 2009. Evaluation of Steam Circulation Strategies for SAGD Start-Up. Canadian International Petroleum Conference (CIPC) paper 2009-14. 16–18 June Calgary, Alberta, Canada.

Zhong, H., Zhang, W., Fu, J., Lu, J., Yin, H., 2017. The performance of polymer flooding in heterogeneous type II reservoirs—an experimental and field investigation. Energies 2017 10, 454. Available from: https://doi.org/10.3390/en10040454. 19 pages.

Zhou, X., Zeng, F., Zhang, L., 2016. Improving steam-assisted gravity drainage performance in oil sands with a top water zone using polymer injection and the fishbone well pattern. Fuel 184, 449—465.

Zhu, Y.Y., Cao, F.Y., Bai, Z.W., Wang, Z., Wang, H.Z., Zhao, S.M., 2014. Studies on ASP flooding formulations based on alkylbenzene sulfonate surfactants. In: SPE 171433, SPE Asia Pacific Oil & Gas Conference and Exhibition, Adelaide, Australia.

INDEX

Note: Page numbers followed by "*f*" and "*t*" refer to figures and tables, respectively.